Methods in Molecular Biology™

AF865293

Series Editor
John M. Walker
School of Life Sciences
University of Hertfordshire
Hatfield, Hertfordshire, AL10 9AB, UK

For further volumes:
http://www.springer.com/series/7651

Embryo Culture

Methods and Protocols

Edited by

Gary D. Smith

University of Michigan, Ann Arbor, MI, USA

Jason E. Swain

University of Michigan, Ann Arbor, MI, USA

Thomas B. Pool

Fertility Center of San Antonio, San Antonio, TX, US

Humana Press

Editors
Gary D. Smith
University of Michigan
Ann Arbor, MI, USA

Jason E. Swain
University of Michigan
Ann Arbor, MI, USA

Thomas B. Pool
Fertility Center of San Antonio
San Antonio, TX, USA

ISSN 1064-3745 ISSN 1940-6029 (electronic)
ISBN 978-1-4939-5920-4 ISBN 978-1-61779-971-6 (eBook)
DOI 10.1007/978-1-61779-971-6
Springer New York Heidelberg Dordrecht London

Printed on acid-free paper

Humana Press is a brand of Springer
Springer is a part of Springer Science+Business Media (www.springer.com)

Preface

Assisted reproductive technologies have had a profound impact on biomedical research through transgenic animals, food supply and production, as well as genetic gain of domestic species, and treatment of human infertility. Understanding regulatory pathways of early mammalian embryo development, preimplantation embryo cellular differentiation, and pluripotency of the embryonic inner cell mass has been pivotal to advances in embryo culture, transgenic mouse production, and biomedical advances in basic research, understanding the pathophysiology and progression of diseases, and development of novel treatments and cures. In 2007, Drs. Capecchi, Evans, and Smithies were awarded the Nobel Prize in Physiology or Medicine for their groundbreaking discoveries in relation to utility of embryonic stem cells, DNA recombination, and refinement of gene targeting in mice. Central to transgenic animal technology is mammalian embryo culture.

While assisted reproductive technologies for the treatment of infertility can be controversial and have a relatively brief lifespan, they have also touched many families' lives in very positive ways. The pioneering utility of assisted reproductive technologies by Patrick Steptoe and Robert Edwards at Bourn Hall in Cambridge, England culminated in the birth of Louis Brown on July 25, 1978. This was not a chance happening. Efforts of numerous individuals, including Dr. Edwards, pre-1970s in studying and understanding the basic concepts of oocyte biology, sperm function, processes of fertilization and embryo production, culture, and selection paved the way for the clinical success of treating infertility with assisted reproductive technologies. These scientific and laboratory skills, in conjunction with enhanced understanding and regulation of the reproductive endocrine system and clinical advances in laparoscopy refined by Dr. Steptoe, led to the initial success in treating infertility. Human infertility afflicts approximately 10 % of the world's population. Continued efforts of many basic, translational, and clinical scientists have improved the success of treating infertility with assisted reproductive technologies. Today it is estimated that more than four million children have been born by in vitro fertilization. As is fitting, Dr. Edwards was awarded the 2010 Nobel Prize in Physiology or Medicine for the development of in vitro fertilization therapies. Central to these therapies is mammalian embryo culture.

While significant advances in embryo culture have occurred over the last few decades, there is still room for improvement. This book is targeted to basic science researchers, clinical laboratorians, and fertility healthcare providers. An emphasis is placed on advances, controversies, and opportunities to improve human embryo culture. With this said, important reviews regarding utility, strengths, and weaknesses of model systems in designing and testing culture alterations for improved human embryo culture are essential and discussed. Embryo culture media are dissected, and components discussed, by experts with emphasis on impact on cell function and embryo development. This is followed by debate of benefits and limitations of various embryo culture systems and environmental influence beyond the soluble component of embryo culture. Within most chapters practical laboratory protocols and/or specific procedures are presented. Finally, a discussion of emerging methodologies and impact of embryo culture on epigenetics and offspring health is presented to set the stage for future research and laboratory application involving embryo culture.

As is true in most professional crafts, careers, and science, mentoring and collaboration are keys to passing on knowledge and facilitating discovery. Interaction between scientists and clinicians only hastens advancements in obtaining basic biology information and translating these data to improved clinical care. We are fortunate in the field of embryology and assisted reproductive technologies to have pioneers that truly had vision and recognized the pathway to success was through discovery, training, sharing knowledge, and mentoring the next generation of scientists. Many such mentors are key contributors to this collection of works. We would like to express our appreciation to the distinguished authors who have contributed to this project. Your efforts have made this a great mentoring tool. Finally, we thank Humana Press, Methods in Molecular Biology Series, and Dr. John M. Walker, series editor, for their support and confidence in our abilities to organize, institute, and complete this project.

Ann Arbor, MI — *Gary D. Smith*
Ann Arbor, MI — *Jason E. Swain*
San Antonio, TX — *Thomas B. Pool*

Contents

Contributors

JAY M. BALTZ • *Department of Obstetrics and Gynecology, Ottawa Hospital Research Institute, University of Ottawa, Ottawa, ON, Canada*

BARRY BEHR • *IVF Laboratory, Department of OB/GYN, Division of Reproductive Endocrinology and Infertility, Stanford University, Stanford, CA, USA*

JACQUES COHEN • *Tyho-Galileo Research Laboratories & Reprogenetics, West Orange, NJ, USA*

CATHERINE M.H. COMBELLES • *Biology Department, Middlebury College, McCardell Bicentennial Hall, Middlebury, VT, USA*

MICHELLE M. DENOMME • *Departments of Obstetrics & Gynecology and Biochemistry, Children's Health Research Institute, University of Western Ontario, London, ON, Canada*

KAY ELDER • *Bourn Hall Clinic, Cambridge, UK*

DAVID HAN • *Department of Statistics, University of Texas at San Antonio, San Antonio, TX, USA*

ANDRÉ HAZOUT • *UNILABS France, Paris, France*

APARNA HEGDE • *Department of OB/GYN, Division of Reproductive Endocrinology and Infertility, Stanford University, Stanford, CA, USA*

MARGO L. HENNET • *Biology Department, Middlebury College, McCardell Bicentennial Hall, Middlebury, VT, USA*

FRANCHESCA D. HOUGHTON • *Faculty of Medicine, University of Southampton, Southampton General Hospital, Southampton, UK*

PHOEBE H. LEONARD • *Division of Reproductive Endocrinology and Infertility, Department of Obstetrics and Gynecology, Mayo Clinic, Rochester, MN, USA*

BORUT KOVAČIČ • *Department of Human Reproduction and Gynecologic Endocrinology, University Medical Centre Maribor, Maribor, Slovenia*

REBECCA L. KRISHER • *National Foundation for Fertility, Lone Tree, CO, USA*

MICHELLE LANE • *Department of Obstetrics and Gynaecology, School of Pediatrics and Reproductive Health, University of Adelaide, Adelaide, SA, Australia*

RONIT MACHTINGER • *Obstetrics, Gynecology & Reproductive Biology, Brigham and Women's Hospital, Boston, MA, USA*

MELLISSA R.W. MANN • *Departments of Obstetrics & Gynecology and Biochemistry, Children's Health Research Institute, University of Western Ontario, London, ON, Canada*

MARIUS MEINTJES • *Frisco Institute for Reproductive Medicine, Frisco, TX, USA*

YVES J.R. MÉNÉZO • *UNILABS France, Paris, France*

DEAN E. MORBECK • *Division of Reproductive Endocrinology and Infertility, Department of Obstetrics and Gynecology, Mayo Clinic, Rochester, MN, USA*

THOMAS B. POOL • *Fertility Center of San Antonio, San Antonio, TX, USA*

PATRICK QUINN • *Research and Development, Sage IVF, A CooperSurgical Company, Redmond, OR, USA*

CATHERINE RACOWSKY • *Obstetrics, Gynecology & Reproductive Biology, Brigham and Women's Hospital, Boston, MA, USA*
MICHAEL L. REED • *Center for Reproductive Medicine of New Mexico, Albuquerque, NM, USA*
DON RIEGER • *Research and Development, LifeGlobal, LLC, Guelph, ON, Canada*
ANDRÉ MONTEIRO DA ROCHA • *Department of Cell and Developmental Biology, University of Michigan, Ann Arbor, MI, USA*
GERALD SCHATTEN • *Division of Developmental & Regenerative Medicine, University of Pittsburgh School of Medicine, Pittsburgh, PA, USA*
JOHN SCHOOLFIELD • *Department of Periodontics, University of Texas Health Science Center at San Antonio, San Antonio, TX, USA*
EDOUARD SERVY • *The Servy Institute, Augusta, GA, USA*
CALVIN SIMERLY • *Pittsburgh Development Center of Magee-Women's Research Institute, Departments of Obstetrics, Gynecology & Reproductive Sciences and Cell Biology & Physiology, University of Pittsburgh School of Medicine, Pittsburg, PA, USA*
GARY D. SMITH • *University of Michigan, Ann Arbor, MI, USA*
AMY E. SPARKS • *IVF and Reproductive Testing Laboratories, Department of Obstetrics and Gynecology, Center for Advanced Reproductive Care, University of Iowa Hospitals and Clinics, Iowa City, IA, USA*
JASON E. SWAIN • *University of Michigan, Ann Arbor, MI, USA*
THEODORE THOMAS • *Reproductive Technology Laboratory, Martha Jefferson Hospital, Charlottesville, VA, USA*
ANNA VEIGA • *Servei de Medicina de la Reproducció, Institut Universitari Dexeus, Barcelona, Spain*
BRENNA A. MARKET VELKER • *Departments of Obstetrics & Gynecology and Biochemistry, Children's Health Research Institute, University of Western Ontario, London, ON, Canada*
DEIRDRE ZANDER-FOX • *Department of Obstetrics and Gynaecology, School of Pediatrics and Reproductive Health, University of Adelaide, Adelaide, SA, Australia*

Chapter 1

Historical Background of Gamete and Embryo Culture

Jacques Cohen and Don Rieger

Abstract

The first tissue culture media were developed nearly 150 years ago by Ludwig and Ringer. These were simple salt solutions, which were initially based on the chemical properties of blood serum. The second generation of culture media was developed more than a century later, in the 1970s, aiming to mimic the reproductive environment. In the 1990s, simplex optimization was used to design the third group of media, to some extent ignoring existing formulations and principles. Simultaneous with the development of culture media, it became evident that it was necessary to carefully control the culture conditions, including temperature, pH, osmolarity, and air quality. Equally important was the development of instruments and tools specifically designed for cell tissue culture such as the inverted microscope, the incubator, the Petri dish, sterile plasticware, the laminar flow cabinet, and air filtration equipment.

Key words: Embryo culture, Media, Temperature, pH, Osmolarity, Dishes, Air filtration, Historical aspects

1. Introduction

The first tissue culture media were developed nearly 150 years ago by Ludwig and Ringer. These were simple salt solutions, which were initially based on the chemical properties of blood serum. Knowledge was gained during those early years about the biochemistry of metabolism in mammals and humans, particularly osmolarity, pH and temperature. This knowledge provided the basis for making many changes to the salt solutions that were already in use. The second generation of culture media was developed more than a century later, in the 1970s, aiming to mimic the reproductive environment (the so-called "back to nature" principle). The third group of media was designed to optimize growth in vitro, to some extent ignoring existing formulations and principles. To formulate these media, the performance of each ingredient was evaluated separately using a process referred to as "simplex optimization"—a system

Gary D. Smith et al. (eds.), *Embryo Culture: Methods and Protocols*, Methods in Molecular Biology, vol. 912, DOI 10.1007/978-1-61779-971-6_1, © Springer Science+Business Media, LLC 2012

developed first for the mouse in the laboratory of John Biggers at Harvard University nearly 20 years ago. This work was supported by the United States National Institutes of Health (NIH) in part to define culture media for use in the human while to circumvent the moratorium on human embryo research, which has been in place since 1979. The US National Institute of Child Health and Human Development provided a large program grant bearing the name *National Cooperative Program on Non-Human In Vitro Fertilization and Preimplantation Development*; this was more colloquially known as "The Culture Club."

Simultaneous with the development of culture media for studying fertilization and embryogenesis, scientists and industrialists became more aware of the need to control environmental conditions and started developing or perfecting instruments specifically for cell tissue culture such as the inverted microscope, the incubator, the Petri dish, sterile plasticware, the laminar flow cabinet, and High Efficiency Particulate Air (HEPA) filtration. This chapter reviews historical components of gamete and embryo culture and in three sections: (1) development of culture media, (2) the study of physiological conditions, and (3) the development of common embryology-specific tools.

1.1. Development of Culture Media

1.1.1. Ringer's Saline Solution and Its Derivatives

Sydney Ringer of University College in London, a British clinician and pharmacologist, developed a simple salt solution based on the constituents of blood serum for the study of the beating frog heart in vitro (1, 2). It was composed of sodium chloride (NaCl), potassium chloride (KCl), calcium chloride ($CaCl_2$), and a low concentration of sodium bicarbonate ($NaHCO_3$, 2.7 mM). Ringer's was a modification of another solution used in the 1860s by Carl Ludwig who was trying to develop a perfusion model for organs. The accidental addition of London tap water by Ringer's assistant was found to be crucial to maintain regular heartbeat and resulted in a marked improvement over previous perfusion solutions. The component in tap water that had been missing before was $CaCl_2$. Ringer's solution was modified by Locke and Rosenheim a quarter century later (3), notably by the addition of 11.1 mM glucose, for the study of the perfused rabbit heart. This solution is often referred to as Ringer–Locke's. Maurice Tyrode, an American pharmacologist, added sodium phosphate (NaH_2PO_4) and magnesium chloride ($MgCl_2$) and increased the concentration of $NaHCO_3$ to 11.9 mM for use in the study of rabbit intestine (4). Krebs and Henseleit (5) removed the glucose, but increased the concentration of $NaHCO_3$ to 25 mM, for the study of nitrogen metabolism in rat tissue. This concentration of $NaHCO_3$ in Krebs–Ringers bicarbonate solution (KRB) is what is conventionally used in conjunction with 5 % CO_2 in many tissue culture media in order to effect physiological pH.

1.1.2. Embryo Culture Media Based on Krebs–Ringers Bicarbonate Solution

John Hammond was the first director of the new Animal Reproduction Unit of the Animal Research Station in Cambridge (UK) in 1949 (6). He was the first to observe mouse embryos cleaving in vitro and 8-cell embryos reaching the blastocyst stage. He was presumably also the first to observe the hatching process in vitro. He is considered the father of embryo culture for having set in motion a host of studies undertaken in the 1950s by experimental embryologists interested in obtaining fertilization and extended embryo culture in vitro, allowing continuous observations to be made on developmental stages from just after sperm penetration to immediately preimplantation. The general plan was to model preimplantation development in vitro, although some scientists also suggested continuing to grow the fetus in vitro. Hammond used a salt solution containing sodium chloride, potassium chloride, calcium chloride, magnesium chloride and some glucose for culture. To this he added about 5 % egg white. The embryos were flushed and placed in a serological tube, sealed, and cultured in air for 48 h. Only a few of the 8-cell embryos could develop further under those conditions. Hammond demonstrated that glucose was essential for compaction and development of the blastocyst. He was presumably the first to publish on the use of test tubes for mouse embryo culture.

In 1956, Wesley Whitten from Australia showed that more than 90 % of 8-cell mouse embryos would develop to the blastocyst stage when cultured in KRB supplemented with 5.55 mM glucose, penicillin, streptomycin, and bovine serum albumin (7). McLaren and Biggers (8) were the first to show that such blastocysts would produce live young when transferred to recipient mothers. This was the first step on the road to in vitro fertilization (IVF). Whitten (9) added lactate to his 1956 formulation and was able to culture 2-cell mouse embryos from outbred animals to the blastocyst stage. However, he could not achieve the same with zygotes. This observation led to the discovery of the so-called "2-cell block" in mouse embryo culture. Both Hammond and Whitten used serological or test tubes with volumes of media <1 mL to culture the embryos. Before them, Pincus (10) used Carrel flasks with volumes of 2–10 mL of fluid. Work on fertilization in vitro also proceeded actively during those years. Thibault observed fertilization of rabbit eggs in 1953 (11) and live kits were obtained by Chang at the Worcester Foundation in 1959 (12), but defined embryo culture media were not used for IVF in those days.

In a series of elegant papers, Ralph Brinster showed that phosphoenolpyruvate, pyruvate, lactate, and oxaloacetate, but not glucose, could support the development of the 2-cell mouse embryo to the 8-cell stage, while glucose could support the development of the 8-cell embryo to the blastocyst stage (13). Based on these findings, Brinster modified Whitten's medium by reducing the calcium concentration and adding 0.27 mM pyruvate to produce BMOC2, which would support the development of the mouse 2-cell embryo to the blastocyst stage at high rates (14). Whittingham (15) modified Brinster's medium by decreasing the concentration of lactate and increasing the concentration of pyruvate to produce M16 medium.

Brinster's medium and M16 were significant advances for embryo culture and were widely used by experimental embryologists. Minor modifications were occasionally made with considerable success (16). However, except for outbred or hybrid strains of mice, these media could not overcome the 2-cell block. Blocks to development in vitro were similarly found for hamster (2-cell), cattle (8–16 cell), and pig (4–8 cell). A partial block has been described for human embryos (4–8 cell). All these arresting stages appear to coincide approximately with the major onset of embryonic genome expression (see ref. 17). The 2-cell block to in vitro development of the mouse embryo was finally overcome by Chatot et al. (18) with Chatot–Ziomek–Bavister (CZB) medium, a modified version of BMOC2 with no glucose, 1.0 mM glutamine, and 0.1 mM EDTA. This led to the misconception that glucose is inhibitory to early embryo development.

1.1.3. Embryo Culture Media Based on the Composition of Oviduct and Uterine Fluids

Much of the early development of embryo culture media was based on simple salt solutions. An alternative approach was the "back to nature" concept. In accordance with this principle, the formulation of media was based on components of oviduct and uterine fluids and their naturally occurring concentrations. These media include SOF (synthetic oviductal fluid) based on sheep oviduct fluid (19); B2, based on cattle oviduct and uterine fluids (20); HTF (human tubal fluid), based on human oviduct fluid (21); and MTF (mouse tubal fluid), based on mouse oviduct fluid (22).

The differences in the composition of oviduct and uterine fluids as well as changes in the metabolic activity of embryos during early development led Gardner and Lane (23) to suggest that "in order to optimize mammalian embryo development in culture, sequential media are required, each designed to meet the changing requirements of the developing embryo." Although this seems a logical conclusion to their findings, the need for sequential culture, in particular, and the "back to nature" approach in general are now being seriously challenged not least because the approach is heavily reliant on the analysis of oviductal and uterine components. Such analyses have several shortcomings, including the following.

First, as noted by Summers and Biggers (24), the composition of oviduct and uterine fluids is highly variable and almost certainly subject to changes because of sampling. Second, the measurements only reflect the overall composition of the tract fluids and not the microenvironment to which the embryo is exposed. Third, some biological fluids, such as seminal plasma, are so complex that the constituents can hardly explain its function. Fourth, the physical environment of the embryo in vivo is completely different from the environment in vitro. Except in pathological conditions, there is no significant pool of fluid in the reproductive tract. The embryo is surrounded by a thin layer of fluid and is in close contact with maternal tissue; this allows rapid exchange of nutrients, gases, waste products, and effectors between the embryo and the mother. In contrast, in vitro, the embryo is bathed in a relatively large pool of fluid in which the nutrients are continually decreasing and the waste products are continually increasing. Clearly, the stresses on the embryo in vitro are very different from those in vivo, and culture media must be designed to optimize embryo development under in vitro conditions.

1.1.4. Embryo Culture Media Designed by Simplex Optimization

In a radical departure from the traditional methods for designing embryo culture media, Lawitts and Biggers (25, 26) applied the principles of simplex optimization to determine the optimal concentration of each component. They began the process with a "generating medium," based on M16 and CZB, containing NaCl, KCl, potassium phosphate (KH_2PO_4), magnesium sulfate ($MgSO_4$), lactate, pyruvate, glucose, bovine serum albumin (BSA), ethylenediaminetetraacetic acid (EDTA), and glutamine. Ten other test media were derived from the generating medium, each containing a high concentration of one of the components. These 11 media formed the START simplex. Four cycles of the simplex optimization process showed that high concentrations of NaCl, pyruvate, KH_2PO_4, and glucose were detrimental to mouse embryo development (25). A further 16 cycles of optimization resulted in the formulation of Simplex Optimization Medium (SOM), which was marked by a low NaCl concentration, and was able to overcome the mouse 2-cell block (26). In a subsequent study, it was found that blastocyst development was improved by increasing the concentration of KCl from 0.25 to 2.5 mM and this modified version was called KSOM (27). Of particular interest, glucose was found to have no inhibitory effect on mouse embryo development in KSOM, even at a relatively high concentration (28). Embryo development was further improved by the addition of amino acids to KSOM (KSOM-AA) (29–31). More recently, Biggers et al. (32) and Summers et al. (33) have shown that the development of mouse embryos is improved when the medium includes the glycyl-glutamine dipeptide compared to either glutamine or alanyl-glutamine.

In addition to many studies on mouse embryos, KSOM, with or without amino acids, has been shown to support the development of cattle (34), rabbit (35), rhesus monkey (36), pig (37), rat (38), and human (39) embryos. For more detailed and extensive reviews of the history of the development of embryo culture media, (see refs. (24, 40–42)).

1.2. The Physiology of Embryo Culture

The history of the development of culture media in mammals is closely linked with the discovery of the fundamental physiological aspects of mammalian tissue environment and the maintenance of cellular homeostasis in vivo. Four critical factors must be considered with respect to embryo culture: (a) maintenance of temperature, (b) regulation of pH, (c) regulation of cellular volume and osmolarity, and (d) the effects of environmental pollution and infection. The assessment and manipulation of these factors were essential not only for the evolution of experimental embryology but also for the study of cell tissue culture in general. Usually, it was the latter discipline that led the way, but more recently, and particularly because of the advent of clinical IVF, this relationship has been mostly reversed.

1.2.1. The Maintenance of Temperature

Early thermometers, called thermoscopes, were already known in ancient times, but they lacked a scale. Temperature assessment therefore was mostly subjective and crude until Galileo (1593) developed his thermometer using the expansion and contraction of air in a glass bulb to move water in a cylinder. Galileo was among the rare scientists who show a wide range of interests, curiosities, and talents. As a philosopher he did not shy away from technological development at a time when the concept of technology did not even exist. The Galileo thermometer, replicas of which are still being sold today, was only one of his many technological inventions. It was Santorio Santorio who was the first to put a scale on the thermoscope. He was also the first to use a thermometer to measure body temperature in 1612. Pioneering embryologists assumed that, in vivo, gametes interacted under core body temperature and it therefore seemed reasonable to choose that temperature as the preferred setting for in vitro incubation. Maintenance of temperature was considered very important. However, other evidence suggests that the use of core body temperature for culturing mammalian embryos may be a significant error based on a persistent misinterpretation.

Though mammals are warm-blooded—meaning that they are able to keep their internal body temperature relatively constant—in isolation in vitro, mammalian embryos respond to external temperature changes in a manner that is more similar to that of cold-blooded organisms. The latter use one of three mechanisms for thermoregulation: ectothermy, or regulation via the external environment; poikilothermy, or regulation via variation of body temperature with external temperature; and bradymetabolism, or alteration of metabolic rate in response to changes in external temperature. Warm-blooded organisms, on the other hand, use conduction, convection, and evaporation to disperse heat. When external temperature fluctuates, embryos are not able to fully adjust to the changing conditions because these mechanisms are lacking in embryos. Thus they may be said to respond to the temperature of their surroundings as cold-blooded organisms do through bradymetabolism. A rise in environmental temperature will increase embryonic metabolism and, conversely, a drop in temperature will slow down metabolism.

Relatively speaking, embryos consume more energy than other cell types; however, it is unknown whether nutrient uptake is different in embryos growing in vitro compared to that of their counterparts developing within the reproductive tract. It is known that thermotolerance of embryos growing in vitro differs among different breeds of livestock (43) suggesting some genetic diversity with respect to temperature regulation in mammalian embryos. In the cow, it has been postulated that genetic differences in susceptibility to elevated temperature may be an important determinant of embryonic survival during heat stress. Embryologists working with mouse embryos often perform their manipulations at room temperature. Strict maintenance of body temperature during mouse embryo manipulation such as pipetting and micromanipulation actually reduces developmental potential. These observations, however, are rather recent and studies focused on empirical arrival at optimal temperature are just emerging. This is especially thanks to the work of Henry Leese who has put forth the "quiet embryo hypothesis." This intriguing theory holds that viable mammalian embryos have a "quieter" metabolism than those that arrest (44). Accordingly, it is argued that viable embryos consume less oxygen and nutrients since they do not require significant energy expenditure for cellular repair activities. In contrast, less viable embryos are more likely to have incurred DNA and other cellular damage, requiring repair or the onset of apoptosis. Such activities in turn require increased uptake of nutrients, which is reflected in an increase in metabolism. Interestingly, a parallel can be found in the meristem tips of growing plant roots, which contain so-called initial undifferentiated cells. At the center of the root apical meristem between the cambium and the root cap, is a zone of reduced mitotic activity called the "quiescent center" with clearly reduced metabolic activity (45).

Spermatozoa form and develop at a temperature at least 1–2 °C below core body temperature. Heat stress (defined as an increase in temperature) is known to compromise spermatogenesis. Surprisingly, ovarian follicles are also at least 2 °C cooler than core body temperature (46). Moreover, the fallopian tube has a complex temperature regulation system and is generally cooler than core body temperature. In the rabbit, the differences between core temperature and those of the isthmus and ampulla are −3.1 and −2.0 °C, respectively. Leese et al. (47) have postulated that the reduced temperature in the isthmus might have a role in maintaining sperm in a quiescent state prior to sperm activation and fertilization in the ampulla. It has been suggested that sperm are sensitive to these gradients and that this facilitates their passage from the site of storage to the site of fertilization (48). There are no investigations on the temperature or temperature gradients within the human fallopian tube, but it can be assumed that values and ranges are similar to that in other species. It is therefore likely that human gametes are normally exposed to temperatures that are 1–3 °C cooler than core body temperature. It is also possible that the temperature of the fallopian tube diminishes further during and after ovulation. Does this result in a quieter metabolism in gametes and cleaving embryos? Some data seem to support Leese's arguments and from that it is inferred logically that the core body temperature of 37 °C may act as a stressor when used during embryo culture. It is indeed remarkable how little work has been done to investigate the culture and manipulation of mammalian embryos at lower than core body temperatures. Gametes from most experimental animal species have been kept at a conventional temperature of 37 °C for as long as embryos have been cultured. An exception to this is the cow, the embryos of which are maintained at 38–39 °C. It appears that proponents of the "back to nature" philosophy in experimental embryology may have overlooked this potentially important aspect of embryo culture.

1.2.2. Optimization and Regulation of pH in Blood, Beer, and Embryos

In biology and chemistry, pH is the concentration of hydrogen ions and a measure of the acidity or basicity of an aqueous solution. Pure water is considered neutral with a pH of about 7.0 at 25 °C. When pH is <7 the solution is considered to be acidic, while a pH >7 indicates that a solution is basic or alkaline. The idea of pH was first introduced in 1909 by Danish chemist Søren Sörenson (49), when he was the director of the Carlsberg Laboratory in Copenhagen, Denmark. This fundamental aspect of physiology was discovered in the mid-1800s as an indirect result of the insight of J.C. Jacobsen, the founder of the Carlsberg Brewery, who also established the Carlsberg Laboratory with the aim of advancing the field of biochemistry as related to the science of brewing! Little did he know that his modest initiative would have a profound impact on many areas of biology and medicine.

The modern use of the symbol "pH" was introduced after it was demonstrated that the electromotive force in cells depends on their activity rather than concentration of hydrogen ions. In a solution, pH approximates but is not equal to p(H), the negative logarithm of the concentration of dissolved hydrogen ions (H^+); a low pH indicates an abundance of hydrogen ions, while a high pH indicates a low concentration. This logarithmic negative matches the number of positions after the decimal point. A change of 0.1 is therefore considerable in a range between 7.1 and 7.4, when embryologists seek to define optimal pH for improving clinical outcomes. A base must be dissolved in an aqueous solution in order to achieve elevated pH. This will neutralize many of the hydrogen ions. Blood pH is regulated to stay within the range of

7.35–7.45, making it slightly alkaline. Hydrogen ions tend to interact with other components in solution, which can affect the electrical potential readout using a pH meter—a principle with which embryologists are keenly familiar. As a result, pH can be affected by the ionic strength of a solution and in turn embryo culture media may require different adjustments.

Mammals have several mechanisms at work to prevent pH drift of blood outside its functional range. These mechanisms include the carbonic acid–bicarbonate–carbon dioxide (50) and phosphate buffer systems, but proteins also may have a buffering effect. The blood buffering systems were described in detail by Lawrence Henderson, who in 1908 (51) proposed an equation describing the use of carbonic acid as a buffer solution. Hasselbalch (52) later reexpressed that formula logarithmically, resulting in the Henderson–Hasselbalch equation.

The bicarbonate buffering system is familiar to every embryologist. It is the most effective and dynamic physiological buffering system in mammals. In this system, carbon dioxide (CO_2) combines with water to form carbonic acid (H_2CO_3), which subsequently dissociates to form the hydrogen ion and bicarbonate (HCO_3^-). This process is important in the physiology of animals with blood systems. It manages the many acid and base imbalances that can be produced by both normal and abnormal physiology. It also affects the handling of carbon dioxide, a waste product of cellular respiration. The dynamic process involved in keeping the stasis of this buffer acts according to Le Châtelier's Principle and the well-known formula:

$$CO_2 + H_2O \leftrightarrow H_2CO_3 \leftrightarrow HCO_3^- + H^+.$$

Le Châtelier's Principle is a qualitative law, which states that when a chemical system at equilibrium experiences a change, the equilibrium will compensate by shifting the direction of the process in order to minimize the original variation. Any disturbance will be compensated by a redirection of the chemical equilibrium. For example, if one attempted to acidify blood by releasing many hydrogen ions (acidemia), some of those hydrogen ions will associate with bicarbonate, forming carbonic acid, resulting in a smaller net increase of acidity than would be otherwise expected. This buffering system is a potent regulator of acidity tied with respiratory compensation, in which breathing alterations will modify CO_2 in circulation.

Two excellent reviews describe recent work on the relationship between pH and its effects on embryos developing in vitro (53, 54). Surprisingly, and similar to studies on temperature dynamics in vitro, little attention had been paid to the physiological impact of pH changes in embryology until about 15 years ago. A thorough overview of this often neglected subject is provided elsewhere in this book in Chapter 10.

1.2.3. The Regulation of Cellular Volume and Osmolarity

Rene du Trochet (1776–1847), a disciple of Lazzaro Spallanzani (1729–1799) presented, in 1824, the basic outlines of cellular theory and subsequently discovered osmosis, the passage of water across biological membranes. He designed a device called an osmometer with a membrane separating two compartments, the lower open to a reservoir filled with purified reference water and the upper topped by a vertical glass tube containing the experimental liquid. The maximum level reached by the column indicated the osmotic pressure. This brilliant discovery and invention led to the previously unknown physical principle of osmolarity.

Osmolarity and cell volume homeostasis are extensively discussed in Chapter 5 entitled "Media Composition: Salts and Osmolality." Osmolarity is a measure of the pressure exerted by a solution across a semi-permeable membrane that allows free passage of water and inhibits any movement of solute. Osmolarity is dependent on the number of solute particles but independent of their nature. If two solutions contain the same concentration of particles, they are isoosmotic. If one solution has a greater concentration of particles than another solution, it is hyperosmotic compared to the weaker solution. If a solution has a lesser concentration of particles, then it is hypoosmotic compared to the stronger solution. Although osmolality refers to another unit of measurement, there is a very marginal difference between absolute values and hence the terms osmolarity and osmolality can be interchanged. The normal reference range of osmolarity in human plasma is about 275–299 mOsm (milliosmolar). This measure has guided pioneer clinical embryologists when deciding on the proper osmolarity of culture media for human embryos.

The empirical optimization of culture media in the last 50 years included a lowering of osmolarity and the inclusion of amino acids involved in regulating cell volume. Conditions that inhibit the embryo's ability to control cell volume may block further development. Baltz and Tartia (42) have recently reviewed cell volume regulation and concluded that cell volume homeostasis in vitro is a key factor in normal embryo development. The interest in osmolarity increased in the early 1990s with the development of CZB and SOM. Both media had essentially the same ingredients as the earlier more empirical solutions such as M16. Two components were added to earlier formulations, the amino acid glutamine and the chelator EDTA. Interestingly both CZB and SOM had lower osmolarity compared to the earlier media, which were all balanced at 290 mOsm. SOM was optimized at 250 mOsm and CZB at 275 mOsm. This is, physiologically, a significant decrease of osmolarity, as demonstrated by the fact that spontaneous activation of mature mouse eggs (extrusion of the second polar body with formation of a single female pronucleus) can occur after diluting M16 or similar media with water at a rate of about 10 % resulting in an osmolarity of 260 mOsm (55). The reduced osmolarity was one of the essential characteristics of the media that finally resolved the developmental blocks of the early cleavage stages. The survival of embryos in excessively high osmolarity during culture may have been due to the absence in the culture media of certain amino acids that can protect embryos from increased NaCl (25, 56). The effects of cell shrinkage in embryos exposed to high osmotic stress may have implications for clinical embryology, as the second generation of human IVF culture media has reduced osmolarity by design (reviewed in ref. (42)).

1.2.4. The Effects of Environmental Pollution and Infection

Bacterial contamination in culture and incidences of toxicity of materials and instruments in close contact with gametes and embryos occur periodically in almost all IVF laboratories. This has been typical in research environments where countermeasures, such as activated carbon filtration, are often not in place. The embryo growing in vitro lacks the defense mechanisms that are operative in full organisms. Its exposure to hazardous environmental elements can therefore exact a developmental toll. Embryos have no skin to control air pollution, no lungs to filter microorganisms and volatile organic compounds (VOC), no liver and kidneys to remove toxic compounds. It is indeed left to laboratory designers and embryologists to foresee potential problems and take proactive measures to prevent their occurrence. Antibiotics are used to prevent bacterial infection. Chelators such as EDTA were shown to improve the development of embryos, presumably by removing the any heavy-metal cations and thereby reducing the production of reactive oxygen species (reviewed in ref. (57)). Removal of VOC using central and independent systems such as gas-line filters filled with oxidizing agents or activated carbon has only become routine in some clinical laboratories and such measures are rare in research environments (58). Often, HEPA filtration is thought to remove VOCs as well as particulates, but the effective pore size of HEPA filters is far larger than any VOC molecule. VOC removal as a critical component of laboratory quality control has only been considered for the past 15 years, yet the implementation of major new studies are hampered possibly because of lack of funding. It has been shown that acrolein, one of the most common aldehydes found in smoke and tarmac, can seriously inhibit the development of human and mouse embryos (59). It is also known that under certain circumstances continuous air filtration by activated carbon can improve developmental outcomes (60).

A significant improvement in culture efficiency was achieved in 1956 when Whitten introduced antibiotics to culture media (7). This important step as well as the introduction of sterile techniques during handling of gametes and embryos has resulted in contamination being a relatively rare event and when it does occur in human IVF, it is usually associated with microbial organisms present in the seminal plasma of infected patients. The actual toxicity of antibiotics in embryo culture is not well known although a handful of studies argue in favor of rethinking their use. Magli et al. (61) showed that reducing the concentration of antibiotics during human IVF did not improve cleavage rates; however, development was significantly improved when penicillin and streptomycin were completely removed. In a study on hamster zygotes, it was shown that development was inhibited when penicillin and streptomycin were used simultaneously (62). Recently, it has been shown that removal of penicillin and streptomycin from mouse embryo culture decreased apoptosis and increased chromatin integrity (63). All three teams advocated sterile technique in lieu of antibiotics. However, these reports show only a deleterious effect of penicillin and streptomycin, which cannot be extended to antibiotics in general. Gentamicin is the most commonly used antibiotic in embryo culture media, and to our knowledge, there is no evidence that gentamicin has any effect on embryo development at the usual concentration of 10 μg/mL. In principle, it might be best to culture embryos without antibiotics, but in practice this would be inadvisable because of the possibility of introduction of bacteria from the semen, the follicle flush, and accidental contamination in the lab.

1.3. The Development of Embryology-Specific Tools

1.3.1. Containers

In the late 1940s, Hammond was the first to observe mammalian embryo cleavage and hatching after culturing embryos in small vessels—usually test tubes (6). He used 1–2 mL salt solution supplemented with 5 % egg white in air. The embryos were examined after 2 days. When he started with 2-cell mouse embryos, some developed to the 4-cell stage, but never further. Late 4-cell and 8-cell embryos, however, did develop to the blastocyst stage and some even escaped from the zona pellucida. Test tubes for in vitro fertilization and embryo development were used by some early clinical investigators. Teams in Australia and Europe employed test tubes well into the 1990s, but after switching to Petri dishes and culture under oil at Bourn Hall Clinic in 1982, the practice of culture in test tubes has been largely abandoned.

The Carrel flask has been used infrequently in culture of mammalian ova, perhaps the most famous of its users being Pincus (10). The device is named after its inventor, Alexis Carrel (64). The advantage was that the flask had a relatively small volume of 2–10 mL and could be gassed and sealed. Mintz (65) designed a cumbersome bottle unit, which was placed in a standard laboratory incubator. The bottle was placed horizontally and had a bicarbonate buffered solution on the bottom to maintain humidity. A dish was placed on a small platform for the actual culture.

1.3.2. The Dessicator

One of the more interesting embryo storage units employed by pioneers of embryo culture and IVF was the desiccator. The first laboratory desiccators were made of glass and were sealable units usually containing desiccants used for preserving moisture-sensitive material. A common use for desiccators was to protect chemicals from humidity. Leslie Brown's embryo and other embryos from patients treated by Steptoe and Edwards in Oldham (UK) in the 1970s were held and cultured in dishes or tubes placed in stopcock glass desiccators without desiccants and placed on a removable stainless steel platform several centimeters from the bottom (66). The lid was sealed using minimal low toxicity vacuum grease. After sealing the top, the desiccator was gassed with 5 % CO_2 in air, although reduced oxygen gas mixture was used as well. The stopcocks were opened quickly after closure of the top to release expanded air, the procedure referred to as "burping." Failure to "burp" the desiccators could lead to the lid sliding off or to blow resulting in the loss of the gas mixture and detrimental pH changes. On the other hand, burping too long or too often could reduce the gas content and elevate the pH. Some programs in North America still use this type of culture for IVF (e.g. (67)). The contents of desiccators are exposed to atmospheric moisture whenever the desiccators are opened. Maintaining humidity was of great concern when embryos were cultured without mineral oil. However, with mineral oil, humidity is not readily or rapidly lost.

1.3.3. The Pasteur Pipette

The most common tool used in experimental embryology has probably been the Pasteur pipette or its derivatives. These were originally made of glass tubes stretched to a narrow point, and fitted with a rubber bulb at the wider end. The glass pipettes were invented well before Louis Pasteur, but he popularized an alternative version that is still in use. Basically all pipettes for manipulation, movement and micromanipulation in use today in embryology laboratories are in one way or another derived from the Pasteur pipette. Although the use of pipetting is ubiquitous in clinical embryology, uniform guidelines for pipetting are nonexistent and clinical outcome studies have not been published extensively. A study by Taylor et al. (68) on the effects of pipette diameter during denuding showed a change in polar body position when narrow-bored pipettes were used. Xie et al. (69) showed the effects of pipetting on expression of stress-related markers in the mouse embryo.

1.3.4. The Petri Dish

The Petri dish has been the preferred home for embryos growing in vitro before their ultimate destination—the uterus or the liquid nitrogen tank. The Petri dish is now used in more than 99.9 % of ART procedures. It is estimated that as many as 100 million Petri dishes have been used in IVF laboratories to date. Despite its broad use, even in specialized areas such as cell tissue culture, microbiology and preclinical embryo research, few alterations have been made to the dish since its invention 125 years ago. However, the physical aspects of embryo culture are finally being reconsidered.

One area of interest is to move away from static systems like the Petri dish altogether and replace it with such exciting systems as microfluidics (70), discussed in much more detail elsewhere in this book. Another option, now being explored by several manufacturers, is to improve on the original Petri dish by incorporating new features that address previous design weaknesses. These include providing the means by which embryos can be located in predetermined positions, improving temperature stability, enhancing safety, improving visibility, and growing the embryos in groups while allowing individual tracking.

The Petri dish was developed in the latter part of the nineteenth century because there was a need in vaccine research to grow microorganisms on a solid substrate rather than in a broth. The famous German scientist, physician and Nobel Prize recipient (in 1905) Robert Koch, known as the master of bacteriology, was the first to suggest replacing the commonly used liquid phase. The problem was that Dr. Koch's assistant had difficulty using glass flasks for this purpose. Koch's assistant was Julius Richard Petri (1852–1921). He decided one day in 1887 to cut off the flask and use only the bottom for containing the solid media. A larger glass cover was loosely placed over the flask bottom and the first glass Petri dish was "born." Chronologically Petri was not the first inventor. At least three individuals would claim to have invented the dish within about 5 years. The inventors developed the dish independently from one another, but only Mr. Petri had the advantage of working in the most famous experimental microbiology laboratory of his time. The reputation of his employer elevated his name and put it in every biology laboratory in the world!

The dishes were first manufactured in glass. In the mid 1960s, injection molding technology sufficiently evolved so that Petri dishes could be manufactured from clear sterilized polystyrene plastic. However, the plastic dish was merely a clone of the first glass Petri dish. It was large, thick, and bulky. It weighed almost twice as much as the conventional plastic dish in use today. Objectives on older inverted microscopes were sometimes unable to accurately focus through these thick layers of plastic. The dish's main advantages included its cost—relatively inexpensive—and its durability as compared to the easily breakable glass dish.

In the mid-1970s, this cost advantage was lost as the price of petroleum-based products, including plastic Petri dishes, skyrocketed. Customer pressure forced Petri dish manufacturers to cut back on the excess size and weight to reduce cost. By the mid-1980s, further developments in the processing of raw materials and a better understanding of the injection molding process allowed most manufacturers to reduce the weight of their Petri dishes to between 15 and 17 g. These new thin plastic Petri dishes have remained largely unchanged although clinical embryology-specific dishes are now being manufactured allowing better visualization and faster setup times (71, 72).

1.3.5. Recent Changes to the IVF Dish

One of the most important steps towards contemporary embryo culture was developed by the maverick scientist Ralph Brinster. In 1963, he (13) successfully cultured mouse zygotes to blastocysts. He decided to do away with "open" culture and protect small amounts of culture medium using a transparent viscous fluid overlaying the media. He used paraffin oil for this purpose. This system moved away from Koch's solid substrate approach for which the Petri dish was designed. The advantage of the oil–medium bilayer was enormous. Oil prevented most microbial infections, and allowed fertilization and embryo growth to take place in less stringent conditions. The gametes and embryos could be observed for long periods without concerns about evaporation or easy contamination. The method also allowed the study of minute quantities of metabolites released or absorbed by the cells. It later facilitated the introduction of micromanipulation methods such as ICSI. The high heat capacity of oil also helped to maintain temperature when the dishes were moved around for observation or manipulation.

It is estimated that more than 90 % of clinical IVF laboratories now use some variation of the method originally proposed by Brinster (13). The current changes to the method address some of its key issues: oil toxicity and batch to batch variation. The issue of batch to batch variation persists but can be addressed by prewashing the oil with water or media to remove toxic compounds. Also pregassing with 5 % or 6 % CO_2 will be advantageous as equilibration with CO_2 is slower with the use of oil; otherwise, dishes must be set up well in advance of intended use.

1.3.6. The Inverted Microscope

The progress and development of experimental and clinical embryology would have been considerably slower, if not for the invention of the inverted microscope in 1852 by J.L. Smith of Tulane University, Louisiana (73). An inverted microscope is a type of compound microscope with its light source and condenser in a reversed position on top of the stage and pointing down. The objectives and turret are below the stage pointing up. This configuration allows for the study of sedimented tissue and living cells on the bottom of containers rather than on glass slides. Inverted microscopes have also been used in micromanipulation applications where space above the specimen is required for manipulator mechanisms and the microtools they hold. The development of culture conditions of gametes and embryos would not have been possible without these specialized microscopes, now a standard feature in embryology laboratories often equipped with sophisticated optics, cameras, and lasers.

1.4. Conclusions

It appears, at least from a historical perspective, that all aspects of culture systems and culture media can be further refined. Although media, tools, and environmental conditions have improved markedly during the past 50 years, this may have occurred perhaps more as a result of standardization and quality control than by design. There is agreement among basic and clinical scientists that further improvements in pH regulation and optimization of osmolarity and temperature are critically needed, and this should be aimed at reducing stress and possibly reducing metabolic rates in embryos. These changes can only be facilitated by targeted research using different animal and clinical models. We would welcome a Culture Club 2.0!

Acknowledgment

We thank Dr. Mina Alikani whose edits, comments, and discussions are reflected throughout this manuscript.

References

1. Ringer S (1882) Concerning the influence exerted by each of the constituents of the blood on the contraction of the ventricle. J Physiol Lond 3:380–393
2. Ringer S (1883) A further contribution regarding the influence exerted of the different constituents of the blood on the contraction of the heart. J Physiol (Lond) 4:29–42
3. Locke FS, Rosenheim O (1907) Contributions to the physiology of the isolated heart. The consumption of dextrose by mammalian cardiac muscle. J Physiol 36:205–220
4. Tyrode MV (1910) The mode of action of some purgative salts. Arch Int Pharmacodyn Ther 20:205–223
5. Krebs A, Henseleit K (1932) Untersuchungen über die Harnstoffbildung im tierköper. Z Phys Chem 210:33–66
6. Hammond J Jr (1949) Recovery and culture of tubal mouse ova. Nature 163:28
7. Whitten WK (1956) Culture of tubal ova. Nature 177:96
8. McLaren A, Biggers JD (1958) Successful development and birth of mice cultivated in vitro as early as early embryos. Nature 182: 877–878
9. Whitten WK (1957) Culture of tubal ova. Nature 179:1081–1082
10. Pincus G, Enzmann EV (1934) Can mammalian eggs undergo normal development in vitro. Proc Natl Acad Sci USA 20:121–122
11. Thibault C, Dauzier L, Wintenberger S (1954) Etude cytologique de la fecondation in vitro de l'oeuf de la lapine. C R Seances Soc Biol Fil 148:789–790
12. Chang MC (1959) Fertilization of rabbit ova in vitro. Nature 184(Suppl 7):466–467
13. Brinster RL (1963) A method for in vitro cultivation of mouse ova from two-cell to blastocyst. Exp Cell Res 32:205–208
14. Brinster RL (1968) In vitro culture of mammalian embryos. J Anim Sci 27(Suppl 1):1–14
15. Whittingham DG (1971) Culture of mouse ova. J Reprod Fertil Suppl 14:7–21
16. Hoppe PC, Pitts S (1973) Fertilization in vitro and development of mouse ova. Biol Reprod 8:420–426
17. Rieger D (1992) Relationships between energy metabolism and development of early mammalian embryos. Theriogenology 37:75–93
18. Chatot CL, Ziomek CA, Bavister BD, Lewis JL, Torres I (1989) An improved culture medium supports development of random-bred 1-cell mouse embryos in vitro. J Reprod Fertil 86:679–688
19. Tervit HR, Whittingham DG, Rowson LEA (1972) Successful culture in vitro of sheep and cattle ova. J Reprod Fertil 30:493–497
20. Ménézo Y (1976) Milieu synthétique pour la survie et la maturation des gamètes et pour la culture de l'oeuf fécondé. Comptes Rendus Acad Sci Paris 282 Serie D:1967–1970
21. Quinn P, Warnes GM, Kerin JF, Kirby C (1984) Culture factors in relation to the success of human in vitro fertilization and embryo transfer. Fertil Steril 41:202–209
22. Gardner DK, Leese HJ (1990) Concentrations of nutrients in mouse oviduct fluid and their effects on embryo development and metabolism in vitro. J Reprod Fertil 88:361–368

23. Gardner DK, Lane M (2002) Development of viable mammalian embryos in vitro: evolution of sequential media. In: Cibelli JB, Lanza RP, Campbell KHS, West MD (eds) Principles of cloning. Academic, Amsterdam, pp 187–213
24. Summers MC, Biggers JD (2003) Chemically defined media and the culture of mammalian preimplantation embryos: historical perspective and current issues. Hum Reprod Update 9:557–582
25. Lawitts JA, Biggers JD (1991) Optimization of mouse embryo culture media using simplex methods. J Reprod Fertil 91:543–556
26. Lawitts JA, Biggers JD (1992) Joint effects of sodium chloride, glutamine, and glucose in mouse preimplantation embryo culture media. Mol Reprod Dev 31:189–194
27. Erbach GT, Lawitts JA, Papaioannou VE, Biggers JD (1994) Differential growth of the mouse preimplantation embryo in chemically defined media. Biol Reprod 50:1027–1033
28. Summers MC, Bhatnagar PR, Lawitts JA, Biggers JD (1995) Fertilization in vitro of mouse ova from inbred and outbred strains: complete preimplantation embryo development in glucose-supplemented KSOM. Biol Reprod 53:431–437
29. Ho Y, Wigglesworth K, Eppig JJ, Schultz RM (1995) Preimplantation development of mouse embryos in KSOM: augmentation by amino acids and analysis of gene expression. Mol Reprod Dev 41:232–238
30. Summers MC, McGinnis LK, Lawitts JA, Raffin M, Biggers JD (2000) IVF of mouse ova in a simplex optimized medium supplemented with amino acids. Hum Reprod 15:1791–1801
31. Biggers JD, McGinnis LK, Raffin M (2000) Amino acids and preimplantation development of the mouse in protein-free potassium simplex optimized medium. Biol Reprod 63:281–293
32. Biggers JD, McGinnis LK, Lawitts JA (2004) Enhanced effect of glycyl-l-glutamine on mouse preimplantation embryos in vitro. Reprod Biomed Online 9:59–69
33. Summers MC, McGinnis LK, Lawitts JA, Biggers JD (2005) Mouse embryo development following IVF in media containing either l-glutamine or glycyl-l-glutamine. Hum Reprod 20:1364–1371
34. Liu Z, Foote RH (1995) Development of bovine embryos in KSOM with added superoxide dismutase and taurine and with five and twenty percent O_2. Biol Reprod 53:786–790
35. Liu Z, Foote RH, Simkin ME (1996) Effect of amino acids and alpha-amanitin on the development of rabbit embryos in modified protein-free KSOM with HEPES. Mol Reprod Dev 45:157–162
36. Weston AM, Wolf DP (1996) Differential preimplantation development of rhesus monkey embryos in serum-supplemented media. Mol Reprod Dev 44:88–92
37. Machaty Z, Day BN, Prather RS (1998) Development of early porcine embryos in vitro and in vivo. Biol Reprod 59:451–455
38. Zhou Y, Galat V, Garton R, Taborn G, Niwa K, Iannaccone P (2003) Two-phase chemically defined culture system for preimplantation rat embryos. Genesis 36:129–133
39. Biggers JD, Racowsky C (2002) The development of fertilized human ova to the blastocyst stage in KSOM(AA) medium: is a two-step protocol necessary? Reprod Biomed Online 5:133–140
40. Biggers JD (1998) Reflections on the culture of the preimplantation embryo. Int J Dev Biol 42:879–884
41. Hammer RE (1998) Egg culture: the foundation. Int J Dev Biol 42:833–839
42. Baltz JM, Tartia AP (2010) Cell volume regulation in oocytes and early embryos: connecting physiology to successful culture media. Hum Reprod Update 16:166–176
43. Hernandez-Ceron J, Chase CC Jr, Hansen PJ (2004) Differences in heat tolerance between preimplantation embryos from Brahman, Romosinuano, and Angus breeds. J Dairy Sci 87:53–58
44. Leese HJ (2002) Quiet please, do not disturb: a hypothesis of embryo metabolism and viability. Bioessays 24:845–849
45. Clowes FAL (1956) Localization of nucleic acid synthesis in root meristems. J Exp Bot 7:307–312
46. Grinsted J, Kjer JJ, Blendstrup K, Pedersen JF (1985) Is low temperature of the follicular fluid prior to ovulation necessary for normal oocyte development? Fertil Steril 43:34–39
47. Leese HJ, Baumann CG, Brison DR, McEvoy TG, Sturmey RG (2008) Metabolism of the viable mammalian embryo: quietness revisited. Mol Hum Reprod 14:667–672
48. Bahat A, Tur-Kaspa I, Gakamsky A, Giojalas LC, Breitbart H, Eisenbach M (2003) Thermotaxis of mammalian sperm cells: a potential navigation mechanism in the female genital tract. Nat Med 9:149–150
49. Sörenson SPL (1909) Enzymstudien. II. Über die Messung und die Bedeutung der Wasserstoffionkonzentration bei enzymatischen prozessen. Biochem Z 21:131–200
50. Delory GE, King EJ (1945) A sodium carbonate-bicarbonate buffer for alkaline phosphatases. Biochem J 39:245

51. Henderson LJ (1908) Concerning the relationship between the strength of acids and their capacity to preserve neutrality. Am J Physiol 21:173–179
52. Hasselbalch KA (1917) Die Berechnung der Wasserstoffzahl des Blutes aus der freien und gebundenen Kohlensäure desselben, und die Sauerstoffbindung des Blutes als Funktion der Wasserstoffzahl. Biochemische Zeitschrift 78:112–144 (From Wikepedia)
53. Pool TB (2004) Optimizing pH in clinical embryology. Clin Embryologist 7:1–17
54. Swain JE (2010) Optimizing the culture environment in the IVF laboratory: impact of pH and buffer capacity on gamete and embryo quality. Reprod Biomed Online 21:6–16
55. Kaufman MH, Surani MA (1974) The effect of osmolarity on mouse parthenogenesis. J Embryol Exp Morphol 31:513–526
56. Van Winkle LJ, Haghighat N, Campione AL (1990) Glycine protects preimplantation mouse conceptuses from a detrimental effect on development of the inorganic ions in oviductal fluid. J Exp Zool 253:215–219
57. Olson SE, Seidel GE Jr (2000) Reduced oxygen tension and EDTA improve bovine zygote development in a chemically defined medium. J Anim Sci 78:152–157
58. Cohen J, Gilligan A, Esposito W, Schimmel T, Dale B (1997) Ambient air and its potential effects on conception in vitro. Hum Reprod 12:1742–1749
59. Hall J, Gilligan A, Schimmel T, Cecchi M, Cohen J (1998) The origin, effects and control of air pollution in laboratories used for human embryo culture. Hum Reprod 13(Suppl 4):146–155
60. Merton JS, Vermeulen ZL, Otter T, Mullaart E, de Ruigh L, Hasler JF (2007) Carbon-activated gas filtration during in vitro culture increased pregnancy rate following transfer of in vitro-produced bovine embryos. Theriogenology 67:1233–1238
61. Magli MC, Gianaroli L, Fiorentino A, Ferraretti AP, Fortini D, Panzella S (1996) Improved cleavage rate of human embryos cultured in antibiotic-free medium. Hum Reprod 11:1520–1524
62. Zhou H, McKiernan SH, Ji W, Bavister BD (2000) Effect of antibiotics on development in vitro of hamster pronucleate ova. Theriogenology 54:999–1006
63. Liu J, Tang S, Xu W, Wang Y, Yin B, Zhang Y (2011) Detrimental effects of antibiotics on mouse embryos in chromatin integrity, apoptosis and expression of zygotically activated genes. Zygote 19:137–145
64. Carrel A (1912) On the permanent life of tissues outside of the organism. J Exp Med 15:516–528
65. Mintz B (1964) Synthetic processes and early development in the mammalian egg. J Exp Zool 157:85–100
66. Edwards RG, Steptoe PC, Purdy JM (1980) Establishing full-term human pregnancies using cleaving embryos grown in vitro. Br J Obstet Gynaecol 87:737–756
67. Reed ML, Hamic A, Thompson DJ, Caperton CL (2009) Continuous uninterrupted single medium culture without medium renewal versus sequential media culture: a sibling embryo study. Fertil Steril 92:1783–1786
68. Taylor TH, Chang CC, Elliott T, Colturato LF, Kort HI, Nagy ZP (2008) Effect of denuding on polar body position in in-vitro matured oocytes. Reprod Biomed Online 17:515–519
69. Xie Y, Wang F, Puscheck EE, Rappolee DA (2007) Pipetting causes shear stress and elevation of phosphorylated stress-activated protein kinase/jun kinase in preimplantation embryos. Mol Reprod Dev 74:1287–1294
70. Suh RS, Phadke N, Ohl DA, Takayama S, Smith GD (2003) Rethinking gamete/embryo isolation and culture with microfluidics. Hum Reprod Update 9:451–461
71. Rieger D, Schimmel T, Cohen J, Cecchi M (2007) Comparison of GPS and standard dishes for embryo culture: set-up and observation times, and embryo development. In: 14th world congress on in vitro fertilization & the 3rd world congress on in vitro maturation, Montreal, QC, Canada, pp 141–142 (Abstr. P1202)
72. Ebner T, Shebl O, Moser M, Mayer RB, Arzt W, Tews G (2010) Group culture of human zygotes is superior to individual culture in terms of blastulation, implantation and life birth. Reprod Biomed Online 21:762–768
73. Smith J (1852) The inverted microscope—a new form of microscope. Am J Sci Arts 14:233–241

Chapter 2

Utility of Animal Models for Human Embryo Culture Development: Rodents

Gary D. Smith

Abstract

Advancements in clinical human embryo culture over the last 30–40 years have been supported by research conducted with embryos from rodent and domestic species. The mouse has been the primary rodent species that has contributed to improved embryo culture outcomes. Numerous parameters applied in the beginning of experiments, during progress of experiments, and as end-point measures provide varying degrees of rigor and interpretive strength and/or complexity. A nonexhaustive discussion of these parameters is presented with important emphasis on experimental design to obtain the greatest power of intraexperimental interpretation of inferior, equivalent, or improved culture conditions in the mouse model. Additionally, data are presented demonstrating the inherent flaw of overinterpretation of interexperimental outcome comparisons and caution of expectations of data translation from the mouse to the human embryo culture scenario. Finally, a "materials, methods, and notes" discussion enumerates important steps in use of mouse embryos as a bioassay tool, independent of whether they are being used in an experiment focused on quality control or improving culture conditions.

Key words: Mouse, Embryo culture, Bioassay

1. Introduction

In the mid- to late 1950s three independent research groups made landmark reports of the ability to fertilize rodent ova (1), culture rodent embryos (1–3), and produce normal offspring from cultured embryos (1, 2). These works were performed at a time when chemically defined media were replacing their predecessors, biological media, which contained biological fluids of unknown chemical composition. Chemically defined media for embryo culture are solutions of water containing compounds of known composition and concentration that can recapitulate, at least partially, the natural microenvironment encountered by an embryo as it develops from a zygote to a blastocyst. The use of chemically defined embryo

Gary D. Smith et al. (eds.), *Embryo Culture: Methods and Protocols*, Methods in Molecular Biology, vol. 912,
DOI 10.1007/978-1-61779-971-6_2, © Springer Science+Business Media, LLC 2012

culture media provided immeasurable value in the 1950–1970s toward understanding the developmental requirements of preimplantation mammalian embryos. Although beyond the scope of this review, an eloquent discussion of this chronology has been previously provided (4).

Early work in mouse embryo development in vitro had to contend with the "2-cell block" (3). While investigators could culture mouse 2-cell embryos to the blastocyst stage, the same was not true for culture of 1-cell zygotes. This suggested that the oviduct provided a specialized and unique microenvironment required to support complete developmental competence of zygotes. In support of this hypothesis, experiments using oviductal explants could overcome the "2-cell block" of mouse zygote to blastocyst development (5). Further investigation into the "2-cell block" demonstrated that the event was mouse strain specific (6), and that addition of glutamine and/or EDTA to media would overcome the event (7, 8). While advances in embryo culture media composition overcame the "2-cell block," these studies in the 1970s and 1980s emphasized the point that mammalian preimplantation embryo culture media were unbalanced and not optimized. This lead to the milestone work performed in Dr. John Biggers' laboratory, performed as part of the National Cooperative Program on Non-Human In Vitro Fertilization and Preimplantation Development, sponsored by the National Institute of Child Health and Human Development in the 1980–1990s. With the application of sequential simplex optimization, Dr. Biggers and his colleagues have produced an optimized medium for mouse preimplantation embryo culture, originally called SOM, then KSOM. A similar version of this medium has been used for human embryo culture (9). Whether this is an optimal medium for human embryo culture is difficult to know or predict. Numerous other human embryo culture medium exist that support efficient mouse embryo development.

Whether one is utilizing the mouse preimplantation embryo development model system to optimize media for human embryo development or as an indicator of clinical laboratory quality control (previously reviewed (10)), it is important to recognize the combined impact of mouse strain, inbreeding or outcrossing, and culture media (11, 12). Most human media manufacturer's and human in vitro fertilization and embryo culture laboratories use the mouse embryo assay (MEA) as a tool for quality control, toxicology testing, and to identify shortcomings of culture media and components, instruments, and laboratory procedures (13–15). The outcome of mouse in vitro fertilization and the MEA can be significantly different when one uses embryos derived from CF1-hybrids, CBA/B6-F1 hybrids, C57BL6-F1/F2 hybrids, or B6C3/D2F1-hybrids (11, 12, 16, 17). This is an important consideration when one attempts to collectively draw conclusions regarding best embryo culture practices from multiple independent reports in the literature.

Another initial parameter that must be considered in using the MEA for developing or testing culture media, quality control, and toxicology testing is the initial stage of development when exposure and/or culture begins. Quinn and coworkers (15) suggested that mouse 1-cell zygotes were a more sensitive testing system for quality control in comparison to 2-cell embryos. In addition, these authors also suggested assessment of rate of fully expanded blastocyst development (discussed further below). This premise of the zygote being more stringent compared to the 2-cell embryo has been demonstrated when comparing sensitivity to suboptimal changes in media pH and osmolality (18) as well as influence of percentage of oxygen in the gas mixture (19). Furthermore, it has been reported that zona-free embryos were more responsive to perturbations compared to zona-intact embryos (20). These reports, as well as others, demonstrate the importance of starting material and susceptibility of the MEA to test parameters.

As mentioned above the selection of end-point measures of embryo development also can significantly affect interpretations. Rates of development are calculated as developmental stage progression, from point A to point B, during a specific time. Thus beyond comparative treatments, embryo origin in relation to inbred, outbred, and strain, beginning developmental stage, period or time of culture, the category of blastocyst development assessed is important. Such categories included total blastocyst development and/or percentage of blastocysts determined to be early, full, expanded, hatching, or fully hatched. In addition, one can assess total number of cells per blastocyst (21, 22) or partitioned distribution of cells to the inner cell mass and trophectoderm with differential staining (23). It has also been suggested that blastocyst development may not suffice for determining added benefit of one culture platform over another; but instead their ability to implant, and develop into normal fetuses after transfer into recipients (22, 24). Finally, other end-point measures that provide detail on culture system testing included gene expression (25), metabolism (26, 27), and imprinted gene expression (28, 29). Independent of which end-point measure is selected, one should be extremely cautious about conducting interexperimental comparisons. If one looks at the literature there are very few reports where all parameters mentioned previously are kept constant with one single variable altered. For this reason, it is imperative that research on culture media components, systems, or platforms include intraexperimental "gold-standard" comparisons and proper controls (22).

2. Materials

The mouse embryo bioassay, in the context of this review, is used to evaluate culture media and system modifications, to detect toxins or suboptimal quality in contact materials, and it serves as a quality control measure to assess culture conditions used in human assisted reproductive technology laboratories. The following are materials needed, methods performed, and noted important points of clarity for performing this bioassay. As is often the case, this is one of many ways of performing this bioassay (30).

1. Chemicals/media
   (a) Pregnant mare serum gonadotropin (PMSG).
   (b) Human chorionic gonadotropin (hCG).
   (c) Media that support stable pH at atmospheric conditions (collection media, HEPES-buffered media, MOPS-buffered media). This will usually contain a protein source or other surfactant.
   (d) Media that support stable pH in incubator conditions (culture media, bicarbonate-buffered media). These will usually contain a protein source or other surfactant. One may want to overlay this media with mammalian embryo-tested oil.
   (e) Protein source (bovine serum albumin, human serum albumin, or other biomolecules that provide surfactant properties).
2. Supplies
   (a) 35 mm disposable Petri dishes, tissue culture grade.
   (b) Syringes and needles for injection of mice (usually 1 cm^3 tuberculin syringe with 25 gauge needle).
   (c) Dissecting instruments (coarse scissors, coarse forceps, iris scissors, and fine watchmaker forceps).
   (d) 30 Gauge × 1 in. hypodermic needle and 1 cm^3 syringe for flushing oviducts.
   (e) Fine pulled pipettes for handling mouse embryos and pipetting device.
3. Equipment
   (a) Dissecting microscope with heated stage (~37 °C).
   (b) Slide warmer and/or warming plate (~37 °C).
   (c) CO_2 incubator (~37 °C, 5 % CO_2 in air or whichever percent CO_2 that combined with the buffering system used yields media pH of 7.25–7.35, replacement of atmospheric air with 5–7 % O_2, with remainder N_2, may also be used).

4. Animals
   (a) F_1 of C_3B_6: Males and females have been suggested in original MEA (30). This strain of mouse was originally proposed due to its consistent response to exogenous gonadotropins, ease of handling and overall semipredictable and consistent embryo development in vitro. As mentioned earlier, different strains and/or crosses may be more desirable for individual study goals.

3. Methods

1. All mice should be housed under a 12 h light–12 h dark cycle and receive standard laboratory chow and water ad libitum. Female mice may be group-housed with up to five females per cage. Male mice should be individually housed (see Note 1).
2. For gonadotropin stock solution preparations reconstitute lyophilized hormone with saline or distilled water to a concentration of 100 IU/ml (see Note 2).
3. Female mice may be randomly selected for injection without attention to the stage of the estrous cycle. Females should be injected with 5–10 IU of PMSG, intraperitoneally between 3 and 6 p.m. Forty-eight hours post-PMSG injection, females should be injected with 5–10 IU of hCG, again intraperitoneally (see Note 3).
4. When breeding always add female mice to the male cage, not the reverse. Add one female, or two, into the cage with the male of proven fertility (see Note 4).
5. The day before zygotes/embryos are collected prepare *collection media* and *culture media*. Culture media dishes should be prepared at least 10 h prior to use and allowed to preequilibrate in a CO_2 incubator.
6. For 1-cell zygotes and 2-cell embryos, females should be sacrificed by cervical dislocation 24 or 36 h post-hCG, respectively. Collection of 1-cell zygotes and 2-cell embryos has been previously described Hogan and coworkers (31).
7. There can be a significant incidence of intermouse variation in fertilization rates and/or developmental competence of collected zygotes and/or embryos. For this reason all the zygotes/embryos from a single female should be pooled and equally distributed among treatment comparisons (see Note 5).
8. Mouse embryos should be observed with a dissecting microscope, on a heated stage at 24 h intervals after placing into culture (see Notes 6 and 7).

9. Considering the suggested mouse strain used above, 2-cell embryos should progress to hatching/hatched blastocyst at a rate of ≥75 %. If <75 % progress to this stage of blastocyst development in this time period the culture system should be considered suboptimal. One-cell zygotes take longer to reach the hatching/hatched blastocyst stage with above referenced expectations taking approximately 96 h.

4. Notes

1. All mice should be acclimated to their housing environments at least 1 week before gonadotropin stimulation.
2. Freeze in small aliquots that are useful for single thaw and inject scenarios. Do not refreeze gonadotropin aliquots; discard what is not used at the time of warming. Store gonadotropin aliquots in a −80 °C freezer, in liquid nitrogen vapor, or liquid nitrogen. Do not store gonadotropin aliquots in a −20 freezer or a frost-free freezer. Finally, do not filter sterilize reconstituted gonadotropins.
3. Immediately after injection with hCG place females into the male's cage.
4. Use only sexually mature animals (6–8 weeks of age); avoid using female and males that are over 3 and 8 months of age, respectively.
5. To reduce variability, frozen–thawed mouse embryos can be obtained for use in an MEA.
6. Care should be taken to reduce the time of embryo/culture media exposure to atmospheric conditions during observation periods. A general rule is to not have embryos out of the incubator for >2 min. Shifts in media pH can stress embryos and compromise their development. For this reason, it has been suggested that embryo development only be assessed at the end of the experiment—for example 96 h of culture of 1-cell zygotes or 72 h of culture of 2-cell embryos. This option may be useful, but consistency within and between experiments and/or comparisons is necessary.
7. Assessment of early time points or more often than 24 h intervals may be useful if examining rates of embryo development as an added measure of system efficacy.

References

1. Chang M (1959) Fertilization of rabbit ova in vitro. Nature 193:446–467
2. McLaren A, Biggers J (1958) Successful development and birth of mice cultivated in vitro as early embryos. Nature 182:877–878
3. Whitten W (1956) Culture of tubal mouse ova. Nature 177:96
4. Biggers J (1998) Reflections on the culture of the preimplantation embryo. Int J Dev Biol 42:879–884
5. Biggers J, Gwatkin B, Brinster R (1962) The development of mouse embryos in organ cultures of fallopian tubes on a chemically defined medium. Nature 194:747–749
6. Whitten W, Biggers J (1968) Complete development in vitro of the pre-implantation stages of the mouse in a simple chemically defined medium. J Reprod Fertil 17:399–401
7. Chatot C, Ziomek C, Bavister B, Lewis J, Torres I (1989) An improved culture medium supports development of random-bred 1-cell mouse embryos in vitro. J Reprod Fertil 86:679–688
8. Abramczuk J, Solter D, Koprowski H (1977) The beneficial effect of EDTA on development of mouse one-cell embryos in chemically defined medium. Dev Biol 61:378–383
9. Biggers J, Racowsky C (2002) The development of fertilized human ova to the blastocyst stage in KSOM(AA) medium: is a two-step protocol necessary? Reprod Biomed Online 5:133–140
10. Leibo S (1990) Ova and embryos for quality control in reproductive biology. Theriogenology 33:67–76
11. Dandekar P, Glass R (1987) Development of mouse embryos in vitro is affected by strain and culture medium. Gamete Res 17:279–285
12. Punt-van der Zalm J, Hendriks J, Westphal J, Kremer J, Teerenstra S, Wetzels A (2009) Toxicity testing of human assisted reproduction devices using the mouse embryo assay. Reprod Biomed Online 18:529–535
13. Wood C, Trounson A (1982) In vitro fertilization and embryo transfer. In: Bonnar J (ed) Recent advances in obstetrics and gynaecology. Chruchill Livingstone, Edinburgh, pp 259–282
14. Ackerman S, Swanson R, Stokes G, Veeck L (1984) Culture of mouse preimplantation embryos as a quality control assay for human in vitro fertilization. Gamete Res 9:145–152
15. Quinn P, Warnes G, Kerin J, Kirby C (1984) Culture factors in relation to the success of human in vitro fertilization and embryo transfer. Fertil Steril 41:202–209
16. Rawlins R, Bassuray R, Binor Z, Rana N, Henig I, Radwanska E, Dmowski W (1985) Biological assay of culture media for human in vitro fertilization (IVF): differential effects of mouse strain and media. In: 41st Annual Meeting Americal Fertility Society, Chicago
17. Summers M, Bhatnagar P, Lawitts J, Biggers J (1995) Fertilizaton in vitro of mouse ova from inbred and outbred strains: complete preimplantation embryo development in blucose-supplemented KSOM. Biol Reprod 53:431–437
18. Davidson A, Vermesh M, Lobo R, Paulson R (1988) Mouse embryo culture as quality control for human in vitro fertilization: the one-cell versus the two-cell model. Fertil Steril 49:516–521
19. Pabon J Jr, Findley W, Gibbons W (1989) The toxic effect of short exposures to the atmospheric oxygen concentration on early mouse embryonic development. Fertil Steril 51:896–900
20. Fleming T, Pratt H, Braude P (1987) The use of mouse preimplantation embryos for quality control of culture reagents in human in vitro fertilization programs: a cautionary note. Fertil Steril 47:858–860
21. Ebert K, Hammer R, Papaioannou V (1985) A simple method for counting nuclei in the preimplantation mouse embryo. Experientia 41:1207–1209
22. Heo Y, Cabrera L, Bormann C, Shah C, Takayama S, Smith G (2010) Dynamic microfunnel culture enhances mouse embryo development and pregnancy rates. Hum Reprod 25:613–622
23. Du F, Sung L, Tian X, Yang X (2002) Differential cytoplast requirement for embryonic and somatic cell nuclear transfer in cattle. Mol Reprod Dev 63:183–191
24. Arny M, Nachtigal L, Quagliarello J (1987) The effect of preimplantation culture conditions on murine embryo implantation and fetal development. Fertil Steril 48:861–865
25. Lane M, Gardner D (2003) Ammonium induces aberrant blastocyst differentiation, metabolism, pH regulation, gene expression and subsequently alters fetal development in the mouse. Biol Reprod 69:1109–1117
26. Biggers J, Whittingahm D, Donahue R (1967) The pattern of energy metabolism in the mouse oocyte and zygote. Proc Natl Acad Sci USA 58:560–567
27. Hardy K, Hooper M, Handyside A, Rutherford A, Winston R, Leese H (1989) Non-invasive

measurement of glucose and pyruvate uptake by individual human oocytes and preimplantation embryos. Hum Reprod 4:188–191

28. Doherty A, Mann M, Tremblay K, Bartolomei M, Schultz R (2000) Differential effects of culture on imprinted H19 expression in the preimplantation mouse embryo1. Biol Reprod 62:1526–1535
29. Fernandez-Gonzalez R, Moreira P, Bilbao A, Jimenez A, Perez-Crespo M, Ramirez M, De Fonseca F, Pintado B, Gutierrez-Adan A (2004) Long-term effect of *in vitro* culture of mouse embryos with serum on mRNA expression of imprinting genes, development, and behavior. Proc Natl Acad Sci USA 101:5880–5885
30. Gerrity M (1988) Mouse embryo culture bioassay. In: Wolf D, Bavister B, Gerrity M, Kopf GS (eds) In vitro fertilization and embryo transfer. A manual of basic techniques. Plenum, New York, NY, pp 57–76
31. Hogan B, Beddington R, Costantini F, Lacy E (1994) Manipulating the mouse embryo: a laboratory manual, 2nd edn. Cold Spring Harbor Laboratory Press, Woodbury, NY

Chapter 3

Utility of Animal Models for Human Embryo Culture Development: Domestic Species

Rebecca L. Krisher

Abstract

Although the mouse is widely used as an animal model for the in vitro culture of human preimplantation embryos, arguments can be made for the validity of many domestic species as more relevant, applicable models of early human development. Here, we describe the advantages and disadvantages of domestic species as animal models to develop in vitro culture methods and conditions that successfully support human embryo development. In vitro embryo culture procedures are described and compared between species for the bovine, ovine, caprine, and porcine species.

Key words: Embryo, In vitro culture, Domestic animal species, Animal model

1. Introduction

Due to the ethical limitations of human embryo experimentation, it is necessary to use animal models to study early embryo development and develop assisted reproductive technologies for both basic knowledge and clinical application. There is no disagreement that animal models have considerably advanced our understanding of basic embryology and early developmental events. In fact, development of hamster IVF medium directly led to the first human IVF (1). However, there are limitations to animal models, in that it is difficult to predictably model the complex function of a biological process, such as embryo development, particularly when it may be affected and confounded by the external environment so dramatically. Despite these complications, a variety of arguments can be made for the applicability of various domestic animals as models of human embryo culture. No clear consensus has been reached as to which is most appropriate, as each animal model has both similarities and differences when compared with human reproductive function. In general, closer phylogenetic relationships result in

Gary D. Smith et al. (eds.), *Embryo Culture: Methods and Protocols*, Methods in Molecular Biology, vol. 912,
DOI 10.1007/978-1-61779-971-6_3, © Springer Science+Business Media, LLC 2012

more validity and applicability of results obtained in an animal model to human medicine, and phylogeny has typically been used when selecting model species (2). For embryo culture, physiological parameters such as the length of follicular recruitment, the number of follicles ovulated, the timing of oocyte maturation and fertilization events, the timing of the maternal to zygotic transition, the length of time required for the embryo to form a blastocyst, hatch and interact with the maternal tissues, pregnancy recognition, and the type of implantation are all important to consider when choosing an appropriate animal model of early embryonic development in the human.

1.1. Appropriateness of Domestic Species as a Model for Human Embryos

Domestic animal models offer some clear advantages for modeling human embryo development. In many cases, we know a great deal about the reproductive physiology of these mammals due to their agricultural applications. In addition, because they are used as foodstuffs, reproductive organs are available in large quantity at a low price. Because postmortem material can often be utilized, large numbers of embryos can be produced in vitro at relatively little cost from material collected at abattoirs. If in vivo produced embryos are required, both surgical and nonsurgical techniques are well developed for embryo collection in these species, again due to their agricultural value. Many land grant universities maintain herds or flocks of domestic species, typically dairy and beef cattle and pigs, as well as horses, sheep and to a lesser extent goats, for both agricultural and biomedical research. Production operations are often willing to participate in research projects. Overall, access to large numbers of domestic animals is quite good, an important facet of a good model species. Most domestic species are easily handled and manipulated using commonly available equipment, another by-product of their long domestication by humans. Finally, the life span of most domestic species is closer to that of humans than rodent models, which may be important particularly when attempting to model the effects of age on gamete and embryo viability.

However, the critical piece of information to assess is, can domestic animal embryos reliably inform us about conditions for human embryo culture? Do domestic animal embryos sufficiently represent the human embryos we encounter in a clinical IVF setting? Can we dependably translate findings from domestic animal models in the research laboratory to human clinical IVF application? Even findings that have a large impact in domestic animal embryo culture require validation in randomized clinical trials before being widely applied to human embryos. This is in part due to the potential impact of any in vitro manipulation of the preimplantation embryo on the health of future offspring. Importantly, studies using animal models must be properly designed and conducted, with impartial randomization, appropriate sample size and correct statistical analysis, and interpreted without bias to translate efficiently into clinical practice (2, 3). We must also evaluate whether there exists innate disparities between embryos derived from any animal model and the human embryos encountered in an infertility clinic. For example, most of the domestic animals we work with are not infertile, as we define infertility in humans. This disparity is not limited to domestic animal models. To be applicable to human embryology as it is currently practiced, embryos should ideally be produced in vitro from in vivo matured oocytes, which have been obtained from females that have been exposed to large amounts of gonadotropins to superstimulate follicle growth. Primarily due to availability, most domestic animal embryos are produced from oocytes gathered from abattoir derived ovaries. These oocytes, collected from follicles in various stages of folliculogenesis, may be intrinsically different from those collected from ovulatory size follicles particularly following hyperstimulation, and likely produce an embryo of varying quality as well. However, even given these limitations, it is important to remember that embryos produced in this manner are capable of producing healthy offspring (4). Animals should be at mid to late reproductive age, but agricultural production is optimal when young animals are harvested. One advantage of a domestic animal model system is that, compared to the inbred and hybrid mice commonly used in embryo research, most domestic animals (with the possible exception of the dairy cow) are comparatively outbred and thus heterogeneous, and as such more accurately reflect the human population. But with this similarity comes a great deal of variability, which may make experimental results more difficult to interpret correctly. In contrast, the high genetic uniformity of the inbred/hybrid mouse provides researchers with considerably lower experimental variability, although the results obtained may not necessarily be as applicable to the heterogeneous human population. The use of domestic animal species as a model for humans has also suffered from the lag in genome sequencing, compared to mouse and human. This has been a disadvantage when studying gene expression in embryos using microarray and quantitative PCR technologies. However, this gap is now rapidly closing. In addition, although mice are clearly easy to manipulated genetically, nuclear transfer and stem cell technology has improved in domestic species making genetic modification, including transgenics, knockouts, and knock-ins, quite possible (5, 6). One primary drawback of domestic animal models is that embryo culture outcome for humans is first and foremost live birth of a healthy offspring. Unfortunately, large scale use of this outcome as a study parameter in domestic species is not widespread, primarily due to cost considerations. Finally, to better apply data obtained using domestic animal models to human IVF, it would be helpful to clinicians for researchers to publish critical components of their studies in a format similar to that of clinical trials (3).

There are clear physiological advantages and disadvantages that can be argued for and against each domestic animal species as a model for human embryo culture. For example, cows and horses may be an excellent model as they are similarly a monoovulatory species, whereas sheep and goats ovulate 1–3 follicles per cycle and pigs up to 20 follicles per cycle. However, the timing of the maternal to zygotic transition (MZT) is the same in the pig, horse and human (4 cell stage), but one cell cycle later in the cow, goat and sheep (7). In contrast, MZT in mice is quite rapid, occurring at the 2 cell stage. Thus, the domestic species may be more relevant to the human situation in this regard (8). No domestic species is identical to the human in the time the embryo takes to optimally form a blastocyst, although the pig embryo may be the closest. In the human, blastocyst formation should occur on day 5 of culture (120 h), while in the pig it requires 6 days and in the ruminants almost 7 days (150 h). Mouse preimplantation embryo development is considerably faster, as blastulation can be observed as early as 70 h, and embryos are typically cultured for only 4 days (7). An excellent argument for the horse as a model for human reproductive aging has been made, as horses have similar anatomy, long follicular phase length and are monoovulatory, as well as having a long life span (9). In addition, because horses may be considered a companion and not a production animal, one is much more likely to find old horses than old cows with which to work. However, there are no longer any abattoirs for horses in the United States, necessitating the use of ultrasound guided ovum pick up to recover oocytes in this country, standard in vitro fertilization is very difficult, and the kinetics of preimplantation development are unique (over 1,000 cells by day 7).

Metabolism of the embryo may be quite critical when determining an animal model for human embryo culture, as this will dictate the effectiveness of any medium developed. Embryos from different species have slightly different metabolic needs. We know more details in this regard about most of the domestic species than we do about human embryos, due primarily to the invasive nature of many of the current assays required to determine embryo metabolism. It seems fairly consistent that embryos from most mammalian species studied to date have low glucose metabolism via glycolysis and the pentose phosphate pathway during the early cleavage stages, relying more on oxidative phosphorylation of pyruvate via the TCA cycle for energy (ATP) production (10–19). An exception to this general rule is the pig embryo, which uses significant amounts of glucose even at the early cleavage stages (20, 21). As the embryo approaches compaction and blastulation, the contribution of glycolysis, and the uptake of glucose from the culture environment, increases (19, 22). The time of the switch from pyruvate to glucose does vary between species, and may reflect the degree of blastocyst expansion and/or the length of the preimplantation period. In these regards, the human is intermediate to that of the mouse and the domestic species (19). In respect to endogenous fatty acid content, the domestic species contain significantly more fatty acids in the oocyte and embryo cytoplasm than do mice; morphologically, human embryos more closely resemble the clear cytoplasm of a mouse embryo. The relative abundance of specific fatty acids should also be considered; the three most abundant fatty acids in human, cattle, pig, and sheep oocytes are palmitic, oleic, and stearic (23, 24). What relationship endogenous fatty acid content has to metabolism and energy production is still unclear, although there is accumulating evidence suggesting that humans, mice and domestic species require fatty acids as an energy substrate during preimplantation development (25–27).

Ultimately, it may be that the physiological events of preimplantation embryo development in humans are best studied using a variety of animal models, tailored to the specific question or mechanism being investigated. Be it media composition, gene transcription, protein function, redox regulation, blastocyst expansion or a multitude of other physiological parameters, domestic animal models may provide a more relevant animal model than the mouse and a more cost effective animal model than the nonhuman primate for human embryo research and clinical application. Although the argument has been made that while excellent for basic scientific research, animal models are not terribly useful as human biomedical models (2), by carefully choosing the most appropriate model for any given question and observing strict experimental guidelines we may hope to attain the best applicability possible. As always, care must be taken to avoid overgeneralization between model species. It may be that we need to alter our expectations given our current level of knowledge regarding embryo development in multiple species, and accept that we may never achieve exact extrapolation from an animal model to human, but that we can provide good evidence for target mechanisms or techniques that will ultimately be investigated in clinical trials in the human IVF setting.

2. Materials

2.1. Culture Media

All concentrations given (mM) are those present in the final working solutions. Water used for all media is 18.2 mΩ with TOC <50 ppb. Reagents are weighed on an analytical balance. All disposable, sterile plasticware is from BD Falcon, unless otherwise specified. Final filtration of all stock solutions, as well as preparation of working solutions, is performed using sterile technique under a biological safety cabinet or laminar flow hood to maintain sterility. All glassware used in preparation of media solutions are dedicated for embryo use, and are heat sterilized in an oven at 120 °C for a minimum of 4 h. After use, glassware is immediately rinsed, then washed seven times with reverse osmosis (RO) water and seven times with ultrapure water, then dried completely and covered with aluminum foil before subsequent resterilization.

2.1.1. Synthetic Oviductal Fluid (SOF)-4-(2-Hydroxyethyl)-1-Piperazineethanesulfonic Acid (HEPES) Bench Medium

Combine 100 mL SOF base stock, 75 mL glucose stock, 84 mL HEPES stock, 10 mL nonessential amino acids without glutamine (NEAA, 100×, MEM; MP Biomedicals), 20 mL essential amino acids without glutamine (EAA, 50×, MEM; MP Biomedicals), 0.336 g $NaHCO_3$ (4.0 mM), 0.036 g Na-pyruvate (0.33 mM), 0.2703 g L-Lactate (3.0 mM; MP Biomedicals, LLC), 0.2172 g alanyl-glutamine (Ala-Gln, 1.0 mM), 1 mL gentamicin (50 μg/mL), 1 g fraction V BSA (0.1 %, w/v). Fraction V BSA may be substituted with 0.1 g polyvinyl alcohol (PVA) to make SOF-HEPES-PVA, a completely defined medium. QS to 1,000 mL with ultrapure water, adjust pH to 7.35 and filter (0.22 μm). Store at 4 °C and use within 1 month (see Note 1).

Stock solutions required for this medium are:

(a) *SOF base stock*. 58.270 g NaCl (99.70 mM), 5.340 g KCl (7.16 mM), 1.62 g KH_2PO_4 (1.19 mM), 0.996 g $MgCl_2{\cdot}6H_2O$ (0.49 mM), 1.891 g $CaCl_2$ (1.71 mM), 2.71 g L-lactate (3 mM). Prepare in 1,000 mL ultrapure water and filter in 0.22 μm 1 L Stericup GV (Millipore) (see Note 2). Store at 4 °C and use within 1 month.

(b) *Glucose stock*. 3.6 g Glucose (1.5 mM). Prepare in 1,000 mL ultrapure water and filter in 1 L Stericup GV. Store at 4 °C and use within 1 month.

(c) *HEPES stock*. 5.958 g HEPES (21 mM), 0.19 g Phenol red (0.19 mM). Prepare in 1,000 mL ultrapure water and filter (0.22 μm). Store at 4 °C and use within 3 months.

2.1.2. Modified Synthetic Oviductal Fluid (SOF) Culture Medium (28, 29)

Modified SOF culture medium is prepared in two versions, step 1 and step 2, for culture of early cleavage stage and blastocysts, respectively. Step 1: to 3,150 μL ultrapure water, add 500 μL SOF base, 500 μL bicarb, 500 μL glucose, 50 μL each pyruvate, alanyl-glutamine, taurine, EDTA, NEAA, EAA, and 0.04 g crystallized BSA. Step 2: to 2,650 μL ultrapure water, add 500 μL SOF base, 500 μL bicarb, 1,000 μL glucose, 50 μL pyruvate, 50 μL alanyl-glutamine, 50 μL NEAA, 100 μL EAA, 50 μL vitamins, and 0.04 g crystallized BSA. Prepare these solutions (total volume 5 mL) and filter (0.22 μm).

Stock solutions required for this medium are the following:

1. *SOF base stock*. See above.
2. *Supplement stocks*. Each stock should be made individually with 10 mL ultrapure water. No need to filter. Store at 4 °C.
   (a) *Bicarbonate*. 0.21 g $NaHCO_3$ (25.07 mM), 1 week expiration.
   (b) *Glucose*. 0.027 g glucose (1.5 mM), 1 month expiration.
   (c) *Pyruvate*. 0.0360 g Na-pyruvate (0.33 mM), 1 week expiration.
   (d) *Alanyl-glutamine*. 0.2172 g aln-gln (1 mM), 1 week expiration.
   (e) *Taurine*. 0.0125 g taurine (0.1 mM), 1 month expiration.
   (f) *EDTA*. 0.029 g EDTA (0.1 mM), 1 month expiration.

2.1.3. NCSU23 Culture Medium (30)

To 10 mL NCSU23 stock solution, add 0.04 g crystallized BSA and filter (0.22 μm) (see Note 2).

Stock solutions required for this medium are:

(a) *NCSU23 stock*. 0.635 g NaCl (108.73 mM), 0.0356 g KCl (4.78 mM), 0.0162 g KH_2PO_4 (1.19 mM), 0.0189 g $CaCl_2$ (1.7 mM), 0.0294 g $MaSO_4 \cdot 7H_2O$ (1.19 mM), 0.2106 g $NaHCO_3$ (25.07 mM), 0.0217 g Ala-Gln (1 mM), 0.1 g glucose (5.55 mM), 0.0876 g taurine (7 mM), 0.001 g phenol red. Prepare in 100 mL ultrapure water and filter (0.22 μm). Store at 4 °C and use within 1 month.

3. Methods

3.1. In Vitro Culture of Bovine, Caprine, Ovine, and Porcine Preimplantation Embryos

1. One cell zygotes should be placed into culture, either immediately upon retrieval if in vivo fertilized zygotes are collected, or at a designated time post insemination if in vitro fertilization is used to produce the embryos (see Notes 3–5).
2. Wash zygotes three times in culture medium prior to placing them in the final culture drop.
3. Incubator conditions for embryo culture are 38.7 °C in 6 % CO_2 and 5 % O_2 in a humidified environment, except for porcine embryos which are cultured at 6 % CO_2 and 10 % O_2. Embryos are typically cultured 10 per 50 μL drop under mineral oil (see Notes 6–8). In contrast, human and mouse embryos are typically cultured in 20 μL drops at 37 °C in 6 % CO_2 and 5 % O_2.
4. Two stage culture media (cleavage and blastocyst stages) are commonly used for bovine, ovine, and caprine (see Notes 9

and 10). If using a two step culture medium, move the cleaved embryos into the second step after 72 h of culture (bovine, caprine, ovine), washing the embryos three times in second stage medium prior to placing them in the final culture drops. Assess and record embryonic cleavage at this time, including the number of cleaved embryos moved and the number of uncleaved embryos not moved to step 2.

5. After 168 h of culture (72 hours in stage 1 and 96 h in stage 2 medium for bovine, ovine, and caprine), score embryos for blastocyst development. A record of the stage reached by each embryo should be recorded. Pig embryos, typically cultured in a one stage medium, are assessed after 144 h of culture. Embryos reaching the compact morula or blastocyst stages should be given a quality grade. Domestic animal embryos are typically graded 1–4 using the International Embryo Transfer Society's grading system (see Note 11).
6. After assessment, embryos can be used for a variety of end points, including but not limited to transfer to a recipient female to establish a pregnancy, or more commonly for research purposes, staining for inner cell mass and trophectoderm cell number to indirectly assess quality, or snap freezing to assess gene expression.

4. Notes

1. When QS'ing and filtering SOF-HEPES with BSA, bubbles can be a problem. Let the BSA fully dissolve slowly with gentle stirring. Pour the solution slowly down the side of the volumetric flask.
2. All media is filtered with Millipore PVDF membranes, which are low protein binding. Each product has been tested for embryo toxicity. We do not discard the first 2 mL of medium filtered through the syringe filters when we make our working media, although this is common practice in some laboratories to ensure that any toxins are washed from the filter.
3. The time (hours post insemination, hpi) when IVF embryos are placed into culture varies by species: bovine, 18 hpi; porcine, 6 hpi; caprine, 18–24 hpi; ovine, 24 hpi.
4. All work should be carried out using either a heated microscope stage or nearby warming plate.
5. If IVF embryos are used for bovine, ovine, or caprine culture, it is necessary to denude the zygotes of any remaining cumulus cells prior to placing them into culture medium. In porcine, oocytes are usually denuded prior to fertilization. This is typically

done by vortexing the zygotes in 100 μL HEPES buffered medium with 1 μL hyaluronidase (final concentration 0.01 %, w/v, 80–160 U/mL; or 0.1 mg/mL) for approximately 2.5 min. The type of maturation medium used may affect the length of time necessary to remove all of the cumulus cells; maturation medium with serum typically requires longer vortexing times to achieve complete denuding, as the cumulus cells are very expanded and sticky. If any cumulus cells remain on the zygotes when they are placed into culture, they may form a monolayer culture. If monolayers are allowed to form, it is not possible to conduct controlled culture experiments. Cumulus cells may also be removed using hyaluronidase solution and a small bore pulled glass pipette or commercially available "stripper tip," such that the inner diameter of pipette is slightly larger than the zona pellucida. The zygote is rapidly pulled into and pushed out of the pipette to strip the cumulus cells off of the zona pellucida. However, for large numbers of embryos, this may necessitate the embryos being on the bench top for too long a period of time.

6. Embryo culture medium can be either prepared in house or purchased. Examples of commonly used embryo culture medium for pigs are North Carolina State University 23 (NCSU23) and Porcine Zygote Medium (PZM); for cattle, sheep, and goats, the G series media, Synthetic Oviductal Fluid (SOF), CR1aa, or KSOMaa are commonly used. Many of these media are available commercially from companies catering to both bovine and human laboratories.
7. The oil used to cover embryo culture drops is a critical component and must be part of the laboratory QC system. Sterile embryo tested mineral or paraffin oil is commercially available. Oil should be stored at 2–8 °C, and used with sterile technique. Oil may be washed to ensure that no toxins are present, by mixing oil with sterile D-PBS and allowing the two phases to separate. Working aliquots of washed oil are then kept with a loose cap in the incubator to decrease drop equilibration time.
8. Dishes should be prepared and equilibrated in the incubator at least 4 h prior to use. Culture dishes should be kept outside of the incubator for a minimal amount of time when loading embryos.
9. Move embryos in as small a volume as possible from wash to wash and into culture drops.
10. Some protocols call for addition of 10 % fetal bovine serum to the second stage culture medium to increase blastocyst development.
11. Embryo stages and grades commonly used to assess development of domestic livestock species embryos are defined in the IETS Manual, http://www.iets.org/pubs_educational.asp.

References

1. Bavister BD (2002) How animal embryo research led to the first documented human IVF. RBM Online 4(1):24–29
2. Wall RJ, Shani M (2008) Are animal models as good as we think? Theriogenology 69:2–9
3. van der Worp HB, Howells DW, Sena ES, Porritt MJ, Rewell S, O'Collins V, CMacleod MR (2010) Can animal models of disease reliably inform human studies? PLoS Med 7:e1000245
4. Lonergan P, Fair T (2008) In vitro-produced bovine embryos: dealing with the warts. Theriogenology 69:17–22
5. Keefer CL (2008) Lessons learned from nuclear transfer (cloning). Theriogenology 69:48–54
6. Prather RS, Shen M, Dai Y (2008) Genetically modified pigs for medicine and agriculture. Biotech Genet Eng Rev 25:245–266
7. Menezo YJR, Herubel F (2002) Mouse and bovine models for human IVF. RBM Online 4:170–175
8. Betteridge KJ, Rieger D (1993) Embryo transfer and related techniques in domestic animals, and their implications for human medicine. Hum Reprod 8:147–167
9. Carnevale EM (2008) The mare model for follicular maturation and reproductive aging in the woman. Theriogenology 69:23–30
10. Biggers JD, Whittingham DG, Donahue RP (1967) The pattern of energy metabolism in the mouse oocyte and zygote. PNAS 58:560–567
11. Leese HJ, Barton AM (1984) Pyruvate and glucose uptake by mouse ova and preimplantation embryos. J Reprod Fertil 72:9–13
12. Rieger D, Loskutoff NM, Betteridge KJ (1992) Developmentally related changes in the uptake and metabolism of glucose, glutamine and pyruvate by cattle embryos produced in vitro. Reprod Fertil Dev 4:547–557
13. Rieger D, Loskutoff NM, Betteridge KJ (1992) Developmentally related changes in the metabolism of glucose and glutamine by cattle embryos produced and co-cultured in vitro. J Reprod Fertil 95:585–595
14. Kim JH, Funahashi H, Niwa K, Okuda K (1993) Glucose requirement at different developmental stages of in vitro fertilized bovine embryos cultured in semi-defined medium. Theriogenology 39:875–886
15. Gardner DK, Lane M, Batt P (1993) Uptake and metabolism of pyruvate and glucose by individual sheep preattachment embryos developed in vivo. Mol Reprod Dev 36:313–319
16. Hardy K, Hooper MA, Handyside AH, Rutherford AJ, Winston RM, Leese HJ (1989) Non-invasive measurement of glucose and pyruvate uptake by individual human oocytes and preimplantation embryos. Hum Reprod 4:188–191
17. Conaghan J, Handyside AH, Winston RM, Leese HJ (1993) Effects of pyruvate and glucose on the development of human preimplantation embryos in vitro. J Reprod Fertil 99:87–95
18. Bavister BD (1995) Culture of preimplantation embryos: facts and artifacts. Hum Reprod Update 1:91–148
19. Leese HJ, Conaghan J, Martin KL, Hardy K (1993) Early human embryo metabolism. Bioessays 15:259–264
20. Swain JE, Bormann CL, Clark SG, Walters EM, Wheeler MB, Krisher RL (2002) Use of energy substrates by various stage preimplantation pig embryos produced in vivo and in vitro. Reproduction 123:253–260
21. Gandhi AP, Lane M, Gardner DK, Krisher RL (2001) Substrate utilization in porcine embryos cultured in NCSU23 and G1.2/G2.2 sequential culture media. Mol Reprod Dev 58:269–275
22. Harvey AJ, Kind KL, Thompson JG (2002) REDOX regulation of early embryo development. Reproduction 123:479–486
23. Matorras R, Ruiz JI, Mendoza R, Ruiz N, Sanjurjo P, Rodriguez-Escudero FJ (1998) Fatty acid composition of fertilization-failed human oocytes. Hum Reprod 13:2227–2230
24. McEvoy TG, Coull GD, Broadbent PJ, Hutchinson JS, Speake BK (2000) Fatty acid composition of lipids in immature cattle, pig and sheep oocytes with intact zona pellucida. J Reprod Fertil 118:163–170
25. Sturmey RG, Reis A, Leese HJ, McEvoy TG (2009) Role of fatty acids in energy provision during oocyte maturation and early embryo development. Reprod Dom Anim 44(3):50–58
26. Dunning KR, Cashman K, Russell DL, Thompson JG, Norman RJ, Robker RL (2010) Beta-oxidation is essential for mouse oocyte developmental competence and early embryo development. Biol Reprod 83:909–918
27. Haggarty P, Wood M, Ferguson E, Hoad G, Srikantharajah A, Milne E, Hamilton M, Bhattacharya S (2006) Fatty acid metabolism in human preimplantation embryos. Hum Reprod 21:766–773

28. Gandhi AP, Lane M, Gardner DK, Krisher RL (2000) A single medium supports development of bovine embryos throughout maturation, fertilization and culture. Hum Reprod 15: 395–401
29. Tervit HR, Whittingham DG, Rowson LEA (1972) Successful culture in vitro of sheep and cattle ova. J Reprod Fertil 30:493–497
30. Petters RM, Wells KD (1993) Culture of pig embryos. J Reprod Fertil 48:61–73

Chapter 4

Utility of Animal Models for Human Embryo Culture: Nonhuman Primates

Calvin Simerly and Gerald Schatten

Abstract

Nonhuman primates are the closest relatives to humans and therefore our most evolutionary close cousins. While marvelous insights are gleaned from studying rodents and other systems, it is impossible to envision how those mechanistic findings can be responsibly translated to the clinic without the appropriate use of nonhuman primates. Thankfully, noninvasive technologies now permit nonhuman primate studies without endangering the model itself. Work with primates is predicted to continue to lead the fields of reproductive and regenerative medicine for the rest of the twenty-first century.

Key words: Rhesus, baboon, nonhuman primates, Fertilization, Development, Stem cells, Regeneration

1. Introduction

As we continue to celebrate the 2010 Nobel Prize in Physiology or Medicine awarded to Professor Robert G. Edwards for the development of in vitro fertilization in humans, we also recognize that further optimization of clinical IVF can be achieved by responsible investigations with primate gametes—human and nonhuman alike. The implications of these investigations are enormous, ranging from single embryo transfer, reduced use of ovarian stimulations as well as tremendous insights in both regeneration medicine and epigenetics for a host of disorders and diseases. This chapter provides crucial background for the researcher interested in entering this important, dynamic field.

Gary D. Smith et al. (eds.), *Embryo Culture: Methods and Protocols*, Methods in Molecular Biology, vol. 912,
DOI 10.1007/978-1-61779-971-6_4, © Springer Science+Business Media, LLC 2012

2. Materials

Protocols for rhesus (*Macaca mulatta*) and baboon (*Papio anubis*) stimulation, oocyte retrieval, and methods for in vitro fertilization is presented here, representing nonhuman primate (NHP) research for more than 40 years (1–26).

2.1. Nonhuman Primates

Nonhuman primates should be disease-free, of appropriate reproductive age (5–17 years old), and purchased from accredited, licensed nonhuman primate dealers or directly from nonhuman primate research facilities such as National Primate Research Centers, pharmaceutical companies, or universities (see Note 1).

2.2. Hormones

Acquire the following for stimulating rhesus and baboon females:

1. Gonadotropin-releasing hormone (GnRH) agonist such as Leuprolide Acetate (Sicor Pharmaceuticals, Irvine, CA).
2. Recombinant human follicle stimulating hormone (r-FSH; Gonal-F; Serono, Randolph, MA).
3. Recombinant human luteinizing hormone (r-hLH; Luveris; Serono).
4. Recombinant human chorionic gonadotropin (r-hCG; Ovidrel; Serono), or equivalents (reviewed by (27)).
5. Reconstitute hormones according to manufacturer's instructions in provided diluents and store at 4 °C until use. Prepare hormones fresh on the day of injections to maintain maximum activity (see Note 2).

2.3. Surgical Equipment

1. Ketamine hydrochloride anesthesia.
2. Hair clippers.
3. Cuffed endotracheal tube.
4. Anesthesia machine.
5. Medical quality isoflurane gas.
6. Betadine; sterile surgical drapes.
7. Medical grade 5 % CO_2 gas.
8. Laparoscopic equipment, including monitor, light cable and source, camera head, insufflations unit and tubing.
9. Aspiration pump with appropriate tubing.
10. Double-hole stopper for connecting aspiration pump to sterile 15 cm^3 collection tubes.
11. 37 °C heating block that holds these tubes.
12. Disposable cannulas.

13. Sterile standard surgical instruments including trocars, toothed (traumatic) grasper forceps, curved scissors, standard forceps, scalpel with blades.
14. Standard suture pack.
15. Water jacketed heating blankets.
16. Sterilize all equipment by autoclave where permitted. Components that cannot be autoclaved can be sterilized by 100 % ethylene oxide gas sterilization in bags 72 h in advance of surgery; ultrasound machine with appropriate transducer probe to visualize ovaries.

2.4. Sperm Equipment

1. A restraint chair (Primate Products, Immokalee, Florida).
2. Electroejaculator unit (Model S-9; Grass Instruments, Quincy, MA) with aluminum electrode (rhesus only). Alternative is a rectal probe electroejaculator unit with probe (2-cm diameter) available from several commercial medical companies (for rhesus or baboon males).
3. Ketamine hydrochloride anesthesia.
4. Sterile 50 cm^3 glass beakers.
5. PureCeption™ Sperm gradient kit (Sage In-Vitro Fertilization, Inc., Trumbull, CT).
6. Hemocytometer for counting spermatozoa.
7. Upright microscope with Hoffman Modulation Contrast optics and 4×, 10×, and 20× objectives (see Note 3).

2.5. Oocyte Collection, Staging, and Sorting Equipment

1. Dissecting scope with variable zoom to 63×.
2. A Stripper pipette with 125 μm tips for removing adhering cumulus cells (Mid-Atlantic Diagnostics, Mt Laurel, NJ).
3. 100 μl Hamilton syringe with pulled Unopette™ tips for transferring oocytes.
4. 35-mm tissue culture dishes.
5. 60-mm tissue culture dishes.
6. Mineral oil.
7. 37 °C transport incubator (CryoLogic BioTherm INC-RB1; AgTech, Inc., Manhattan, KS).
8. EM Con™ filter system (Immuno Systems, Inc., Spring Valley, WI).

2.6. Culture Medium

1. Prepare Tyrode's-lactate (TL) stock media (114 mM NaCl, 2.0 mM $CaCl_2{\cdot}2H_2O$, 3.2 mM KCl, 0.5 mM $MgCl_2{\cdot}6H_2O$, 0.4 mM $NaH_2PO_4{\cdot}H_2O$, 11.0 mM lactic acid (sodium salt; 60 %, w/w, syrup), 25 mM $NaHCO_3$, 0.2 mM sodium

pyruvate, 5 mM glucose, 50 μg/ml gentamicin, 1 mg/ml phenol red pH indicator), and TL-HEPES (114 mM NaCl, 2.0 mM $CaCl_2 \cdot 2H_2O$, 3.2 mM KCl, 0.5 mM $MgCl_2 \cdot 6H_2O$, 0.4 mM $NaH_2PO_4 \cdot H_2O$, 11.0 mM lactic acid (sodium salt; 60 %, w/w, syrup), 2 mM $NaHCO_3$, 10 mM 4-(2-hydroxyethyl)-1-piperazine ethane sulfonic acid (HEPES), 0.2 mM sodium pyruvate, 5 mM glucose, 50 μg/ml gentamicin, 1 mg/ml phenol red pH indicator) (5, 8) (see Note 4).

2.7. Other Chemicals

1. Hyaluronidase.
2. Heparin.
3. Caffeine.
4. N6,2′-O-dibutyryladenosine 3′:5′ cyclic monophosphate (dbcAMP).
5. Bovine serum albumin, Fraction V, fatty-acid free.

2.8. In Vitro Fertilization and Intracytoplasmic Sperm Injection (ICSI) Equipment

1. Inverted microscope equipped with Hoffman Modulation Contrast optics (or equivalent); 4×, 10× and 20× HMC objectives (or equivalent).
2. Temperature controlled thermo plate stage insert (Tokai-HT, Shizuoka-Ken, Japan).
3. Dual micromanipulators for holding and microinjection needles.
4. Microinjectors for suction control and a sperm injection pipette; tubing.
5. Micropipettes (Humagen, Charlottesville, VA) for both holding (O.D. 100 μm; I.D. 20 μm) and injection of a single spermatozoon (O.D. 6–7 μm and I.D. 4–5 μm).
6. Fluorinert FC-70; 100 mm sterile dish lid.
7. 10 % Polyvinylpyrrolidone (PVP).
8. TALP-HEPES medium.
9. Mineral oil.
10. Needle flame composed of a blunted 18-guage needle hooked up to a gas supply for creating a small flame (17, 28).

3. Methods

3.1. Exogenous Hormonal Stimulation of Nonhuman Primates

An example of rhesus ovarian stimulation with exogenous human recombinant gonadotropins is shown in Table 1 (29). Beginning at signs of menses (Day 1), females are gently immobilized in a squeeze cage and given a shot at 0700 in the morning with a subcutaneous (sub-Q) injection of a gonadotropin-releasing hormone (GnRH) antagonist (0.5 mg/kg body weight) to block any

Table 1
Ovarian stimulation protocol for rhesus and baboon

Days post-mense	GnRH agonist given	Gonadotropin administration	Procedure
1	250 μg-once daily; Sub Q[a]	60 IU FSH (30 IU twice daily)	25-Guage needle IM[b]
2	250 μg-once daily; Sub Q	60 IU FSH (30 IU twice daily)	25-Guage needle IM
3	250 μg-once daily; Sub Q	60 IU FSH (30 IU twice daily)	25-Guage needle IM
4	250 μg-once daily; Sub Q	60 IU FSH (30 IU twice daily)	25-Guage needle IM
5	250 μg-once daily; Sub Q	60 IU FSH (30 IU twice daily)	25-Guage needle IM
6	250 μg-once daily; Sub Q	60 IU FSH (30 IU twice daily)	25-Guage needle IM
7	250 μg-once daily; Sub Q	60 IU FSH; 60 IU LH (twice daily)[c]	25-Guage needle IM ultrasound for follicular development
8	250 μg-once daily; Sub Q	60 IU FSH; 60 IU LH (twice daily)[c]	25-Guage needle IM
9	250 μg-once daily; Sub Q	60 IU FSH; 60 IU LH (twice daily)[c]	25-Guage needle IM
10	250 μg-once daily; Sub Q	1,000–5,000 IU r-hCG (single dose in morning)	25-Guage needle IM (surgery: 27–36 h post-r-hCG injection)

[a]Subcutaneous injection
[b]IM = intramuscular injection
[c]FSH and LH mixed and given as a single injection

premature endogenous LH surge in cycling females (30). The sub-Q GnRH antagonist is given each morning throughout the 10-day stimulation cycle. At the same time on Day 1, an intramuscular (thigh) dose of 30 IU recombinant human FSH (r-hFSH) is administered for follicular recruitment. About 12 h later on Day 1, a second dose of 30 IU r-hFSH is given intramuscularly. This regime is repeated for five additional days keeping to the 12-h interval between doses of r-hFSH as consistently as possible. Beginning on Day 7, a combined intramuscular shot of 30 IU each of r-hFSH and recombinant human luteinizing hormone (r-hLH) is administered twice daily for an additional 3 days. Ultrasonography may be performed on Day 7 of the stimulation to confirm several follicles 3–4 mm in diameter on each ovary, indicating adequate follicular response. If the presence of follicles is not confirmed, further stimulation is halted. On Day 10 of the successful stimulation cycle, a single intramuscular injection of 1,000 IU of recombinant human chorionic gonadotropin (r-hCG) is administered for oocyte maturation and ovulation induction. Ovarian aspiration is typically performed 27–30 h post-r-hCG administration. We typically stimulate two rhesus females per day if available (see Note 5).

3.2. Oocyte Aspiration. 24 h Prior to Ovarian Aspiration

Prepare complete TALP culture medium (5, 8). To 50-ml TL stock, add 0.5 mM sodium pyruvate (from a 100 mM stock) and 200 mg BSA Fraction V. Let the BSA dissolve into the medium at 37 °C for about 5 min without agitation. With the complete TALP, prepare 4–6 insemination and culture dishes. First, use a permanent marker pen to draw intersecting lines on the bottom of the dish, dividing the plate into four equal sections (11). Pipette 90 μl of the sterile complete TALP medium into each section plus one center drop (five droplets total) and quickly overlay with warm mineral oil. Label each plate with medium type and date before placing in 5 % CO_2–5 % O_2–90 % air (5–5–90) incubator for pH equilibration for a minimum of 16 h before use. For postfertilization culture plates, prepare 4-well sterile plates and add 500 μl of complete TALP to each well overlaid with 300 μl of warm mineral oil. Label as before and place in incubator. Store any remaining TALP in the incubator with a loose fitting cap to permit pH equilibration. TALP insemination and culture plates should not be kept outside of the incubator except for very short periods (<5 min) to avoid detrimental shifts in the pH of the medium (see Note 6).

3.2.1. Prior to Ovarian Aspiration

Prior to the start of surgery, prepare 100 ml of complete TALP-HEPES medium by adding 400 mg BSA and 0.5 mM sodium pyruvate to the TL-HEPES stock solution. After the BSA is dissolved, filter sterilize medium through a 0.22 μm filter tower system. The TALP-HEPES can be incubated in a 37 °C water bath for a maximum of 6 h. Just before the start of surgery, take 20 ml of complete TALP-HEPES and add 5 IU/ml heparin (TALP-HEPES-HEP) to prevent blood clotting in aspiration collection tubes. Distribute 1 ml of the TALP-HEPES-HEP into ten 15-cm^3 sterile conical tubes and place into the portable transfer incubator to keep at 37 °C. The remaining 10-ml of TALP-HEPES-HEP is used to wash the tubing of the follicle aspiration pump prior to and during surgery. About 10 min prior to the anticipated arrival of aspirated oocytes into the lab, prepare 50-ml of warm TALP-HEPES with 2.0 mg/ml hyaluronidase enzyme (TALP-HEPES-Hyal) to assist in the removal of cumulus cells from aspirated oocytes. Filter the TALP-HEPES-Hyal solution with a 0.22 μm syringe filter, and store in a 37 °C water bath.

3.2.2. Follicular Aspiration (4, 8, 14, 23, 25, 31)

Stimulated females are fasted for 12 h prior to surgery. Anesthetize the female with an intramuscular dose of ketamine (10 mg/kg). The female is then quickly prepped for surgery by clipping the hair from the ventral abdomen, and intubated with a cuffed endotracheal tube. Next, the female is fitted with a 22-gauge angiocath in the radial or saphenous vein. Maintenance anesthesia is provided by connecting to isoflurane gas vaporized in 100 % oxygen and delivered via the cuffed endotracheal tube. The female is transferred to the operation table in the Trendelenburg position and the abdomen skin sterilized by washing with betadine. The female is then sterile draped. The abdomen is insufflated with medical 5 % carbon dioxide (CO_2) gas to 15 mmHg pressure via a Verres needle inserted approximately 1–3 cm cranial to the umbilicus. After the Verres needle is removed, a 5 or 10 mm laparoscopic viewing telescope is inserted at the same site, and a 5 mm accessory cannula is placed in the right ventral paralumbar region. A short pair of grasping forceps is inserted to mobilize the visualized ovary. The 22-gauge short beveled aspiration needle (8) is attached with tubing to the aspiration pump and dual hole stopper for connecting to the 15-cm^3 tubes with 1 ml TALP-HEPES-HEP. The needle and tubing is first flushed with sterile TALP-HEPES supplemented with 5 IU/ml heparin prior to insertion transabdominally through the cannula and insertion into the gravid ovarian follicles. Multiple individual follicles are aspirated with continuous vacuum at approximately 40–60 mmHg pressure into blood collection tubes containing 1 ml of TALP-HEPES/5 IU/ml heparin medium maintained at 37 °C in the heating block. Collection tubes are transported to the dedicated primate lab in a portable 37 °C incubator for oocyte recovery and evaluation after both ovaries are aspirated. To complete the surgery, the internal abdomen is lavaged with warmed saline to remove any blood clots. All incisions are closed with synthetic absorbable sutures. The laparoscopic instruments are removed and the small punctures in the body wall are closed using 3-0 vicryl in a mattress pattern (see Note 7).

3.3. Collection of Sperm

Rhesus or baboon males of proven fertility are utilized for collection of semen samples (8, 12, 13, 25, 32–36). For rhesus males appropriately chair trained (see Note 3), ejaculates can routinely be produced by penile band electroejaculation (8, 11, 13, 36). Stimulation is performed using a Grass S-5 stimulator connected to a disposable electrode made of ordinary aluminum foil folded into a strip approximately 1 × 6.0 cm and place around the base of the penis. Activation pulses of 20 Hz for 20 ms are applied to a maximum of three times with 1–2 min rest between pulses (13). The produced ejaculate is collected in a sterile 50 cm^3 glass beaker and immediately transported to the laboratory for semen analysis. For baboons, semen samples are obtained by rectal probe ejaculation (20, 25, 33, 37). After ketamine anesthesia is applied to sedate the male, electroejaculation was initiated with a clinical unit connected to a 2-cm probe containing two 35 mm longitudinal electrodes. A built-in temperature probes monitors heat to prevent injury during the electrical current application. The probe was lubricated with KY Jelly®, inserted ventrally within the anus, and the prostate/accessory glands stimulated with rhythmic pulse stimulations applied at a constant voltage up to 7.5 V. The stimulation cycle was repeated up to five cycles with 1 min rest increments between stimulations. Semen samples collected in the 50 cm^3 glass beakers are then transported to the lab for analysis.

3.4. Analysis of Sperm Samples

After liquefaction of the coagulated ejaculates, the liquid semen is removed using a 1,000 μl sterile micropipette and transferred to a 15 ml conical tube of an 80:40 (v/v) discontinuous gradient sperm washing medium. The sperm suspension is then spun for 5 min at 250 × *g* in a clinical centrifuge at room temperature to separate the live versus moribund spermatozoa. After centrifugation, the top layers of the gradient is removed with a 5 ml pipette and discarded. About 10 ml of warm complete TALP-HEPES is added to the sperm pellet and the tube gently inverted five times to resuspend the contents. The sperm solution is centrifuged at 200 × *g* at room temperature for 5 min, the supernatant removed with a 10 ml sterile serological pipette, and the final sperm pellet suspended in 1 ml warm TALP-HEPES. A sperm count is prepared using the hemocytometer to estimate the sample concentration. Approximately 10 μl of the sperm is added to 190 μl of distilled H_2O to immobilize the actively swimming sperm and the suspension mixed by gentle swirling. About 10 μl of this suspension is then applied to both sides of the hemocytometer with overlaid coverslip. After 1 min, the first grid is located on an upright microscope using 20× Hoffman Modulation contrast (HMC) optics and all intact sperm counted in five squares of the smallest grid pattern. This is repeated for the opposite grid pattern on the other side of the hemocytometer. The two counts are averaged by adding the total numbers and dividing by two. This is the sperm count in millions per milliliter. Morphology and motility are also determined by taking 10 μl of the swimming sperm in the TALP-HEPES, adding to a clean glass slide, and quickly covering with a 22 mm^2 cover glass. Immediately, an estimate of vigorous forward motility is made with a 10× HMC objective. A minimum of 5–8 fields are observed with >50 sperm observed per field to estimate forward linear motility. If sample is <50 % motile, it is typically rejected. Morphology is also determined by recording ten fields on the slide and counting 10–20 sperm per field at 20× HMC. Notes are made regarding the number of bent tails, curled tails, agglutinated sperm, presence of cytoplasmic droplets, detached heads and tails, abnormal swollen or mis-shaped heads, or tail defects (short tails, thick mid-piece, splayed tails). If defects >30 % are observed, the sample is rejected. The washed and diluted sperm is kept in TALP-HEPES for no more than 4 h at room temperature (see Note 8).

3.5. Analysis of Oocyte Aspirates

Upon delivery in the lab, the aspiration tubes are diluted with 5 ml of warm TALP-HEPES containing 2 mg/ml hyaluronidase enzyme (TALP-HEPES-Hyal). While incubating for 3–4 min, assemble the EM Con™ filter system by attaching the tubing onto the lower portion of the filter and applying the pinch clamp tightly. Wash the filter by adding 5 ml of TALP-HEPES-Hyal solution, swirling for 2–3 min and then draining out the fluid into a 1-l sterile wide-mouth flask by releasing the pinch clamp. After draining, reapply the pinch clamp. Next, combine all of the diluted aspirates from a single female into the top of the filter, gently swirl, and then release the pinch clamp to drain out the medium. Quickly pour in 10–15 ml of warm TALP-HEPES-Hyal into the top of the filter, gently swirl, and then release the pinch clamp. Repeat this step once more. After the last wash, add about 5–10 ml of warm TALP-HEPES, gently swirl, and quickly pour the contents into three 60-mm sterile Petri dishes which have a drawn grid pattern on the bottom with a permanent marker pen. Wash the filter once more and transfer to another grid dish. Quickly screen the supernatant using the dissecting scope with transmitted light to locate oocytes and pick them up using a Hamilton syringe fitted with a sterile pulled Unopette tip slightly larger than the oocyte diameter. Transfer the oocytes in a 35-mm dish of fresh TALP-HEPES kept warm on a slide warmer at 37 °C. After collection of all oocytes, any remaining adhering cumulus cells can quickly be stripped using the Stripper pipette fitted with a 125-μm sterile tip by pulling 2–4 oocytes into the tip at one time in a rapid in-out pattern to dislodge tightly adhering cumulus cells. Once stripped, transfer the oocytes to another 35-mm dish using the Hamilton syringe with tip and examine oocytes under the dissecting scope. Sort the oocyte clutch into four categories under the dissecting scope: (1) immature oocytes that have a visible germinal vesicle (GV), (2) maturing oocytes that have neither visible GV or first polar body, (3) fully mature oocytes with a clearly visible extruded first polar body, and (4) atretic oocytes with large vacuoles, dark and/or highly concentrated centralized cytoplasm, or abnormal shapes or completely lysed. Place all maturing oocytes into TALP culture medium at 37 °C in a 5%CO_2–5%O_2–90%N_2 incubator (ten per droplet) in a TALP culture plate that was prepared the previous day. Monitor their development over a few hours for elicitation of the first polar body (briefly examine them every 30–45 min to determine if the first polar body has been elicited). Transfer all fully mature oocytes with first polar bodies to a separate gas-equilibrated TALP plate and inseminate within <4 h (see Note 9).

3.6. Insemination In Vitro

First, prepare hyperactivation solution for rhesus spermatozoa by adding 2 mg of caffeine and 3 mg of dbcAMP to 1 ml of gas-equilibrated TALP medium (14, 38–40). Mix well by gentle swirling, filter with a syringe filter, and store at 37 °C in the 5–5–90 incubator for not longer than 30 min. Adjust the sperm suspension to a concentration of 20×10^6 sperm/ml in TALP in the 15-ml conical tube. To do this, divide the averaged sperm count (millions per milliliter) as determined by the hemocytometer by 20 to determine the final milliliters of TALP medium to add. For example, if the sperm count is 270 million/ml, dividing by 20 yields the number 13.5 ml. Since our original volume is 1 ml, just add 12.5 ml of TALP to produce a final count of 20×10^6 sperm/ml. Incubate the adjusted sperm for 1 h at 37 °C. Using an insemination plate with 90 μl of gas-equilibrated TALP, add 10 μl of the hyperactivating solution of caffeine and dbcAMP directly to the droplet. Swirl *gently* to mix but avoid dislodging the droplets of medium from the plate bottom. Transfer in about 3–10 mature metaphase-II arrested oocytes into the medium and immediately add 1 μl of the 20×10^6 sperm suspension, again with gentle swirling. Place dish at 37 °C in the 5–5–90 incubator and culture for 12–15 h. After incubation, investigate fertilization success. Using the Stripper pipette with the 125-μm sterile tip, gently strip off excess sperm from all oocytes and transfer to a fresh gas-equilibrated TALP plate. Determine the number of zygotes with two pronuclei and two polar bodies, the number of oocytes without evidence of any pronuclei (failed ferts), and the number of zygotes with only one pronucleus (parthenogenotes) or more than two pronuclei (polyspermy). The rate of normal insemination success is the total number of zygotes with two pronuclei + two polar bodies divided by the total number inseminated (see Note 10).

3.7. ICSI

ICSI is performed on an inverted microscope platform with long working distances to aid in culture dish and microneedle placement (17–19, 25, 28, 41–46). Optics for performing ICSI should be Hoffman modulation contrast (HMC) with 4×, 10×, and 20× optics (or equivalent). First, assemble hydraulic manipulators onto an inverted microscope with appropriate brackets and attach a stainless steel needle holder to each side. Now, fill the Hamilton syringes (one for holding the oocyte; the other for sperm delivery) half-way with either water or mineral oil and expel a tiny amount of fluid from the end (avoid bubbles in syringe or barrel). Then, assemble the loaded Hamilton syringes into a syringe holder and connect the plunger to the screw actuator. Next, connect tubing to needle holder and fill both holder and tubing with oil or water, also avoiding bubbles at either end. Do this for both sides. Finally, connect the lines to each Hamilton syringe assembled in the syringe holder. These units are now ready for loading needles.

Remove a holding pipette and with the needle flame bend the needle in a Z-pattern so that it can be maneuvered around the condenser of the microscope. Do the same with the 50° beveled, short, sharp point microinjection needle. To ensure the sperm microinjection pipette bevel is in the correct position during the bending of the pipette, first turn the mark on the top of the microinjection pipette (denotes bevel side) 90° away from you, then apply the first bend. Rotate the micropipette back towards you 90° and make second Z-bend. Backfill both holding and microinjecting pipettes with Fluorinert FC-70 to help with the fluid control of the micropipettes and then assemble into each needle holder. Place needles into TALP-HEPES media, align needles in the same focal plane in the microscope and then turn the syringes to expel any bubbles at the tip of the micropipettes. Test moving medium up and down in both the holding and microinjecting pipettes to ensure smooth delivery in and out of the needles with easy control. Fluid movement should be immediate with a turn of the syringe actuator. Keep the media: Fluorinert meniscus visible at all times.

Sperm injections are carried out at 37 °C. First, prepare an injection plate consisting of two adjacent 100 μl drops of TALP-HEPES placed in the lid of 100 mm tissue culture dish followed by a third drop of 10 % PVP in warm TALP-HEPES. Cover all in light mineral oil (17). The first drop of medium is for microinjecting the oocytes, the second medium drop is for washing off the outside of the sperm injection micropipette tip prior to introducing it into the oocyte drop. The PVP is for a 1:10 dilution of the sperm solution (1×10^6 sperm/ml) to slow the forward speed of spermatozoa, thus aiding in the capture of a sperm with the micropipette. The protocol of sperm injection is as follows: first, move the holding pipette out of the field of view and bring the sperm microinjecting tip into the PVP drop. Pull a small amount of the PVP solution (no sperm) into the pipette tip (keep the meniscus in view). Now, capture a single motile sperm of good morphology by aspirating tail-first from the sperm-PVP drop into the microinjection needle. Move the tip to the center drop and rinse off the outside of the sperm microinjection pipette. Also, return the holding pipette to the field of view. Now move both micropipettes to a site just below the unfertilized oocytes in the first medium drop and use the holding pipette to gently immobilize one oocyte with the first polar body at either the 12 or 6 o'clock position. This positions the second meiotic spindle away from the sperm microinjection needle to avoid damaging this vital structure. Now, gently insert the microinjection needle through the zona pellucida and into the cytoplasm with a continuous, steady forward motion. Once positioned near the center of the cytoplasm, the oolemma is then breached by gentle cytoplasmic aspiration, accomplished by pulling back on the syringe actuator a small amount of cytoplasm is viewed in the microinjection tip. Then, gently expel the cytoplasm and captured spermatozoon into the oocyte. It is important to visually confirm that the single spermatozoan is introduced into the cytoplasm and does not "float" out of the oolemma breach. This is accomplished using the 20× HMC objective to verify the presence of a single sperm within the cytoplasm. Place the injected oocyte at the top of the drop and away from other unfertilized oocytes and repeat until all oocytes are microinjected with a single sperm (see Note 11).

ICSI insemination is scored between 12 and 15 h post-sperm injection by confirming extrusion of the second polar body and by the presence of two pronuclei in the cytoplasm. Zygotes are cultured in fresh gas-equilibrated TALP medium until the 2-cell stage. After completion of the first cleavage division (24–28 h post-injection), 2-cell embryos can be cultured in advanced media with or without feeder cells for development to late preimplantation stages (8, 21, 25, 26) (see Figs. 1, 2 and 3).

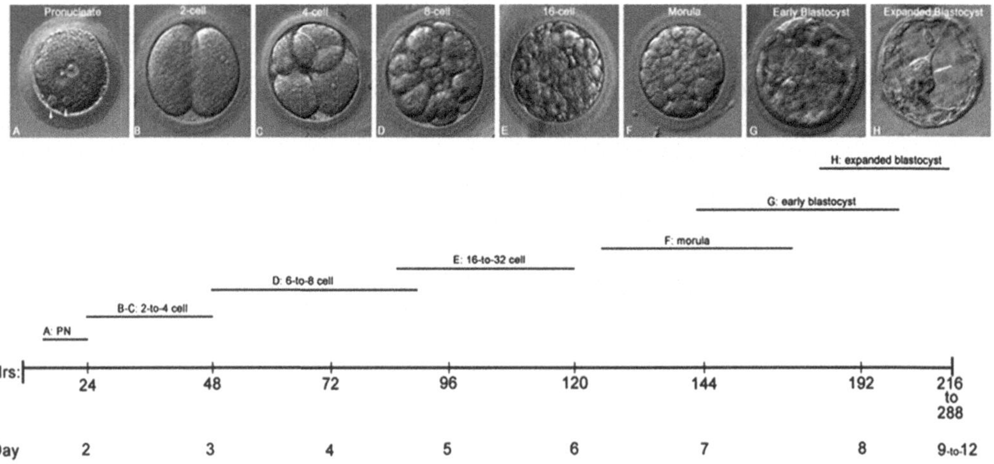

Fig. 1. Baboon embryonic development and timeline after intracytoplasmic sperm injection (ICSI). ICSI-derived baboon zygotes developed in vitro from the 1-cell zygote to the expanded blastocyst stage, the point where embryos can be used for immunosurgical isolation of the inner cell mass and attempted derivation of baboon embryonic stem cell lines. (**a**) Fertilized 2-pronuclear stage with two polar bodies (*arrowheads*); (**b**) 2-cell; (**c**) 4-cell; (**d**) 8-cell; (**e**) 16-cell; (**f**) morula; (**g**) early blastocyst with a small blastocoel; (**h**) fully expanded blastocyst with prominent inner cell mass (*arrow*). *Below*: Timeline in hours and days for various in vitro developmental stages in the baboon. *Bar* in (**a**): 20 μm (reprinted with permission of Sage Publications from Simerly et al. (25)).

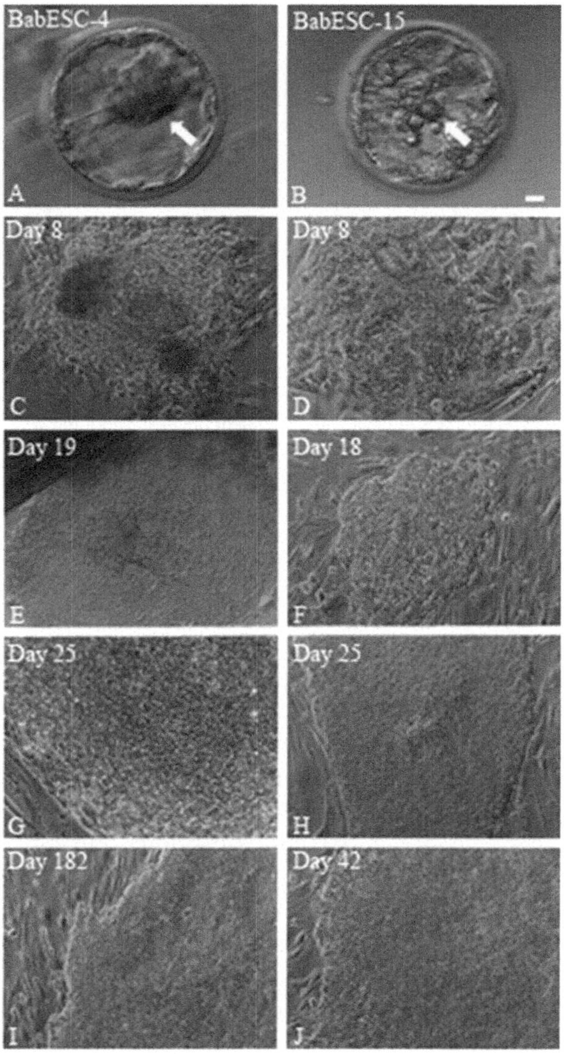

Fig. 2. Derivation of baboon embryonic stem cell lines, BabESC-4 and BabESC-15. (**a** and **b**) Baboon blastocysts produced from ICSI zygotes cultured in vitro for 9 days (*arrows*: prominent inner cell masses). (**c** and **d**) Day 8 post-ICM plating, showing rapid expansion of both nascent cell lines in culture. (**e** and **f**) Colony growth observed after the first mechanical passage ~Days 18–19 post-ICM plating. (**g** and **h**) Both baboon cell lines have colonies demonstrating definitive morphological characteristics of embryonic stem cells, including large nuclear-to-cytoplasmic ratios, tightly packed cells with prominent nucleoli, and distinct colony borders. (**i** and **j**) BabESC colonies at 182 and 42 days post-plating, respectively. *Left column*: BabESC-4. *Right column*: BabESC-15. *Bar*: 20 μm; (**c** and **j**) 100× inverted phase contrast optics (reprinted with permission of Elsevier Limited from Simerly et al. (2009) Establishment and characterization of baboon embryonic stem cell lines: an old world primate model for regeneration and transplantation research. Stem Cell J 2:178–187).

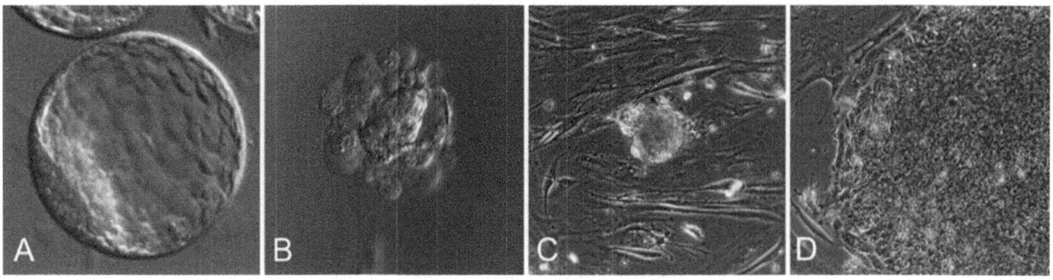

Fig. 3. Rhesus embryo to embryonic stem cells. (**a**) Nonhuman primate blastocysts should be fully expanded with a large distinct inner cell mass (ICM) prior to use for embryonic stem cell derivation. (**b**) After the complement is added the trophectoderm, cells are lysed, and the blastocyst will collapse. Lysed trophectodermal cells are only loosely associated with the ICM. (**c**) Isolated inner cell mass plated onto mouse embryonic feeder cells. (**d**) Early passage nhpES cells for passaging should have very tightly packed cells with prominent nucleoli (reprinted with permission of John Wiley & Sons, Inc. from Navara et al. (2007) Derivation and characterization of nonhuman primate embryonic stem cells. Curr Protoc Stem Cell Biol, Chapter 1: Unit 1A.1).

4. Notes

1. Cycling rhesus and baboon females should be >5 years old with accurate daily recordings of menses cycles taken for a minimum of 2 months to ensure females are cycling normally (rhesus: 28 days (47); baboon: 33 ± 0.5 days (7, 48)). Male rhesus and baboons should also be >5 years old and of known fertility (offspring). Health checks by a qualified veterinarian are advised prior to any hormonal stimulation and include general physical examinations, complete blood counts (CBC), chemistry panel, and palpation of reproductive structures. Animals expressing diseases such as Cercopithecine herpes virus or Simian Retrovirus Type D (SRV) should be avoided where possible (49). Rhesus and baboons are typically housed indoors with controlled temperature (72 °C), diurnal cycle (12:12, light–dark), and ad libitum clean water in appropriate size caging. For baboons, group housing in harems of 12–15 females with a vasectomized male to maintain social order is possible (25). The decision for single versus group house caging is generally outlined by the Institutional Oversight Committees with responsibility for animal welfare. Rhesus, with appropriate social enrichment, can be single or pair caged but the more socially inclined baboons should have minimum times separated from their troop (5 months) if single caged. Male baboons for sperm samples are housed in single caging on a short-term basis (typically about 2 weeks) before rotating with other males to avoid stress that can affect the quality of semen samples. For all single caged females, signs of stress include fecal smearing, hair plucking, and aggression, among others. If extra enrichment with toys, mirrors, or foraging

devices does not alleivate stress, the female should be returned to group housing. With single caging, it is worthwhile to keep room caging arrangements constant (animals will always have the same neighbors), and all animals should be allowed to see other animals at any given time to reduce stresses that may interfer with normal cyclity.

2. The use of human recombinant hormones generally limits the number of times a single female can be stimulated for oocyte production. Typically, a maximum of 3–6 stimulations per nonhuman primate female is performed before circulating antibodies block further useful oocyte recruitment. If recombinant hormones are not employed, a single stimulation may only be achieved (27, 50, 51).

3. For rhesus sperm collection by penile band electroejaculation, it is necessary for males to be fully trained by the pole-and-collar method in order for the male to accept sitting in a restraint chair (Primate Products). This training method is a stepwise process beginning with attachment of a lose fitting collar around the neck for attaching a pole. Adaptation to collaring can take a few weeks. Next, males must be trained to accept a pole attached to the collar and learn "walking" to the restraint chair for immobilization. This process can be lengthy and not all males will be candidates for chair restraint. Nevertheless, it is fairly well accepted that a semen sample collected by penile band electroejaculation is superior to semen samples collected by rectal probe (13, 34). Baboon males cannot be chair trained owing to the dangers to animal staff of working with such large animals, and thus should only be stimulated to produce a semen sample by rectal probe electroejaculation method after light anesthesia.

4. Where possible, purchase chemicals, salt solutions, and sugars that have been certified to support in vitro development of mouse embryos. BSA for fertilization should be Fraction V fatty acid free. The sodium pyruvate is best diluted from a 100 mM stock solution to avoid errors in weighing small amounts of powder. Check pH of TL-HEPES and adjust to between 7.2 and 7.4 with fresh 1 N HCl or sodium hydroxide as required. The pH of TL stock medium will equilibrate in the 5 % CO_2 incubator and should not be adjusted. For both media, check the osmolarity (acceptable range: 285–295 mOsm). Osmolarity slightly higher can be adjusted according to the methods of Boatman (11). Finally, filter the TL and TL-HEPES stocks through 0.22 μm filter system and store at 4 °C. The sterile stocks are good for 2 weeks.

5. Perineal sex skin color alterations during normal cycles are a noninvasive method for staging baboon females (1, 6, 7, 16, 25, 52). There are also many variations to the stimulation

regime that work equally well in both rhesus and baboons (8, 12, 20, 21, 25–27, 53). An alternative stimulation regime is to give a single dose of 60 IU r-hFSH and r-hLH along with the antide once daily in the morning (25). Regardless of the stimulation regime followed, baboons are especially variable in their response to human recombinant hormones with significant numbers of immature oocytes collected at aspiration (25, 26). Baboons also require higher doses of hCG (2,500–5,000 IU) than reported for rhesus (25). For both rhesus and baboons, second and third stimulations with the recombinant hormones nearly always produce fewer overall oocyte numbers and fewer mature oocytes at collection, most likely because of an immune response to the human drugs (27). Even naïve females of good health with normal cycles and good stable environments can be poor responders to exogenous human recombinant hormones, with failure rates as high as 33 % (27). It may be advisable to monitor daily serum estradiol levels and follicular numbers/diameter by ultrasonography. However, limits of staff time as well as the permitted number of times a single animal can be anesthetized (based on local Institutional Oversight Committee regulations) can restrict such measurements. Note that multiple interventions can also create stress (i.e., capture into squeeze cages, sedations, movements to new environments, etc.) that impacts response to stimulation.

6. Mineral oil varies according to supplier, with some having stabilizers added that are toxic to gametes and impede oocyte insemination and development. The best types are sold by human IVF commercial entities that have pretested the mineral oil for compatibility with gametes. Do not store mineral oil in a hood under UV illumination as this can degrade the mineral oil leading to oocyte or zygote lysis. Keep mineral oil in a cool dry environment away from strong light sources. It may also be advisable to test new lots of mineral oil for supporting mouse embryo development to the blastocyst stage before using in nonhuman primate experiments.
7. Note that the time of oocyte aspiration is longer in baboons (32–36 h phCG) than reported for rhesus (27–30 h post-hCG). A minimum of four individuals are typically needed during a laparoscopic procedure on a nonhuman primate—an anesthesiologist, two individuals for laparoscopy, and an instrument assistant to help with unwrapping surgical packs and connecting equipment. The anesthesiologist should monitor isofluorane anesthesia as well as record oxygen saturation, respiratory rate, end-tidal carbon dioxide levels, cardiac rate and rhythm, and body temperature throughout surgery. Circulating water heating blankets are used to ensure that body temperature is maintained throughout surgical procedures.

The head veterinarian and assistant will split aspiration duties, with one operating the laparoscope with camera and grasping forceps while the other concentrates on ovarian needle aspiration. Duties for the fourth member of the team include unwrapping sterile packs, connecting tubing to aspiration pump, connecting the laparoscopic camera to the monitor, and holding/exchanging tubes during follicle aspiration. All females are monitored closely post-surgery and returned to their home cage only when vital statistics have returned to acceptable levels. A variety of analgesics (flunixin meglumine, buprenorphine, butorphanol, and oxymorphone) can be employed preemptively and postoperatively to relieve postsurgical pain and distress. Antibiotics are also utilized preemptively, intraoperatively, and postoperatively when necessary. All administered drugs and analgesics should be logged and recorded for each female.

8. If practical, it is best to have 2–3 males on standby for sperm collections as not every ejaculate produced will be usable in the lab. This is especially true using rectal probe ejaculations which typically have smaller sample sizes and sometimes can have inadvertent urine expressed during the ejaculation event, rendering a moribund semen sample. It is also best if the sperm samples are collected first thing in the morning prior to the start of oocyte aspiration so that washing, counts, morphology and motility estimates can all be completed prior to the arrival of the oocytes. This way, oocytes are not aging while sperm are being collected and analyzed.

9. Be extra careful not to let all of the media drain out of the top of the EM Con™ filter by retaining 2–3 ml of the fluid or else the oocytes may lyse. Since the ovarian aspiration contents are typically bloody, filtering should drastically improve the clarity of fluid while pooling the contents, aiding in the identification of oocytes in your dishes. Samples can quickly be screened (<15 min with training), minimizing time at room temperature and atmospheric oxygen. Maturation of rhesus and baboon GV-oocytes harvested from unstimulated ovaries is not successful while in vitro GV maturation after ovarian stimulation is still suboptimal material for fertilization and long-term culture (54). Maturing oocytes that do not elicit the first polar body in the first few hours post-in vitro culture typically have lower in vitro fertilization and embryonic development rates compared to fully matured in vivo oocytes (55).

10. IVF in the baboon is inconsistent, perhaps owing to issues with sperm hyperactivation (20, 40, 56). Currently, ICSI is the preferred method for insemination in the baboon with consistently high rates of insemination leading to offspring (25). It is advantageous to separate the normal fertilized zygotes from the rest

of the clutch before the onset of first mitosis if developments to advance preimplantation stages or embryo transfer are to be performed. Activated zygotes exhibiting 1, 3 or more pronuclei can divide from one-to-two, making it difficult to determine which embryos are chromosomally compromised. To segregate zygotes, first transfer to a 100-μl droplet of TALP-HEPES medium overlaid with mineral oil and examine on an inverted microscope at 20×. Remove unfertilized (failed ferts) or zygotes with 1, 3, or more pronuclei visible in the cytoplasm as well as any atretics. The normal 2-pronucleate oocytes can then be transferred to a fresh TALP plate gas equilibrated in the 5–5–90 incubator for normal development.

11. Typically, 5–8 oocytes undergo ICSI at one time. This avoids exposing the oocytes to air and temperature fluctuations too long. Sperm motility can vary from different samples. Sometimes, very active swimming sperm in the PVP solution can first be immobilized with the injection needle, accomplished by raking the pipette across the sperm axoneme midpiece in a rapid motion prior to picking up the spermatozoa. It is imperative that only one sperm be injected per oocyte to avoid polyspermy. Thus, it is important to visualize the number of sperm picked up in the microinjection pipette as well as observing sperm delivery into the cytoplasm. The PVP solution should be made fresh on the day of ICSI and discarded at the end of the day. Prepare the injection needles before the oocytes and sperm arrive in the lab on the day of aspiration to avoid delays in insemination. Since the position of the second meiotic spindle is not always accurately predicted based on the location of the second polar body (28), some researchers prefer to use a Pol-Scope to dynamically image the meiotic spindle in the live oocyte. This scope uses circular polarization optics and the anisotropic properties of the microtubules to view the bright meiotic spindle against the darker cytoplasm (57–59).

Acknowledgments

We are grateful to the many colleagues over the years who have contributed to our work in nonhuman primates including Drs. Cassandra Bauer, Buddy Capuano, Carlos Castro, Tanja Dominko, John Fanton, Pat Frost, Laura Hewitson, and Chris Navara. We also thank Jamie Champagne, Kevin Grund, Ethan Jacoby, Christy Martinovich, Dave McFarland, Janet Turpin, and Diana Takahashi for excellent technical support with NHP investigative studies. We thank Ms. Angela Palermo Lauff for administrative assistance. NIH sponsorship is gratefully acknowledged.

References

1. Hendrickx AG, Kraemer DC (1968) Preimplantation stages of baboon embryos (Papio sp.). Anat Rec 162:111–120
2. Hendrickx AG (1971) Embryology of the baboon. The University of Chicago Press, Chicago, IL
3. Hendricks RD, Sawyer M (1975) Embryology of the macaque monkey. In: Bourne GH (ed) The rhesus monkey. Academic, New York, NY, pp 141–169
4. Dierschke DJ, Clark JR (1976) Laparoscopy in *Macaca mulatta*: specialized equipment employed and initial observations. J Med Primatol 5:100–110
5. Bavister BD, Yanagimachi R (1977) The effects of sperm extracts and energy sources on the motility and acrosome reaction of hamster spermatozoa *in vitro*. Biol Reprod 16:228–237
6. Pope CE, Pope VZ, Beck LR (1982) Development of baboon preimplantation embryos to post-implantation stages *in vitro*. Biol Reprod 27:915–923
7. Shaikh A, Shaikh S, Celaya C, Goldzieher J (1982) Ovulation pattern in successive cycles in the baboon. Primates 23:592–959
8. Bavister BD, Boatman DE, Leibfried L, Loose M, Vernon MW (1983) Fertilization and cleavage of rhesus monkey oocytes *in vitro*. Biol Reprod 28:983
9. Bavister BD, Boatman DE, Collins K, Dierschke DJ, Eisele SG (1984) Birth of rhesus monkey infant after *in vitro* fertilization and nonsurgical embryo transfer. Proc Natl Acad Sci USA 81: 2218–2222
10. Clayton O, Keuhl TJ (1984) The first successful *in vitro* fertilization and embryo transfer in a nonhuman primate. Theriogenology 21:228 [abstract]
11. Boatman DE (1987) In vitro growth of nonhuman primates pre- and peri-implantation embryos. Plenum, New York, NY
12. Wolf DP, Vandevoort CA, Meyer-Haas GR, Zelinski-Wooten MB, Hess DL, Baughman WL, Stouffer RL (1989) *In vitro* fertilization and embryo transfer in the rhesus monkey. Biol Reprod 41:335–346
13. Lanzendorf SE, Gliessman PM, Archibong AE, Alexander M, Wolf DP (1990) Collection and quality of rhesus monkey semen. Mol Reprod Dev 25:61–66
14. VandeVoort CA, Tarantal AF (1991) The macaque model for *in vitro* fertilization: superovulation techniques and ultrasound-guided follicular aspiration. J Med Primatol 20:110–116
15. Zelinski-Wooten MB, Hutchison JS, Hess DL, Wolf DP, Stouffer RL (1995) Follicle stimulating hormone alone supports follicle growth and oocyte development in gonadotrophin-releasing hormone antagonist-treated monkeys. Hum Reprod 10:1658–1666
16. Stevens VC (1997) Some reproductive studies in the baboon. Hum Reprod Update 3:533–540
17. Hewitson L, Takahashi D, Dominko T, Simerly C, Schatten G (1998) Fertilization and embryo development to blastocysts after intracytoplasmic sperm injection in the rhesus monkey. Hum Reprod 13:3449–3455
18. Hewitson L, Schatten G (2002) The use of primates as models for assisted reproduction. Reprod Biomed Online 5:50–55
19. Hewitson L, Martinovich C, Simerly C, Takahashi D, Schatten G (2002) Rhesus offspring produced by intracytoplasmic injection of testicular sperm and elongated spermatids. Fertil Steril 77:794–801
20. D'Hooghe TM, Spiessens C, Chai DC, Mwethera PG, Makokha AO, Mwenda JM (2004) Ovarian stimulation, egg aspiration, *in vitro* fertilization and embryo transfer in the baboon (*Papio anubis*): a pilot project at the Institute of Primate Research, Nairobi, Kenya. Gynecol Obstet Invest 57:23–26
21. Wolf DP (2004) Assisted reproductive technologies in rhesus macaques. Reprod Biol Endocrinol 2:37
22. Honore EK, Tardif SD (2009) Reproductive biology of baboons. In: VandeBerg JF (ed) The baboon in biomedical research. Springer, New York, NY
23. Nyachieo A, Spiessens C, Chai DC, Mwenda JM, D'Hooghe TM (2009) Menstrual cycle synchronization, ovarian stimulation, and *in vitro* fertilization in olive baboons (*Papio anubis*): a prospective randomized study. Fertil Steril 91:602–610
24. Nyachieo A, Spiessens C, Mwenda JM, Debrock S, D'Hooghe TM (2009) Improving ovarian stimulation protocols for IVF in baboons: lessons from humans and rhesus monkeys. Anim Reprod Sci 110:187–206
25. Simerly CR, Castro CA, Jacoby E, Grund K, Turpin J, McFarland D, Champagne J, Jimenez JB Jr, Frost P, Bauer C, Hewitson L, Schatten G (2010) Assisted Reproductive Technologies (ART) with baboons generate live offspring: a nonhuman primate model for ART and reproductive sciences. Reprod Sci 17:917–930
26. Chang TC, Eddy CA, Ying Y, Liu YG, Holden AE, Brzyski RG, Schenken RS (2011) Ovarian stimulation, *in vitro* fertilization, and effects of culture conditions on baboon preimplantation

embryo development. Fertil Steril 95: 1217–1223
27. Stouffer RL, Zelinski-Wooten MB (2004) Overriding follicle selection in controlled ovarian stimulation protocols: quality vs quantity. Reprod Biol Endocrinol 2:32
28. Hewitson L, Dominko T, Takahashi D, Martinovich C, Ramalho-Santos J, Sutovsky P, Fanton J, Jacob D, Monteith D, Neuringer M, Battaglia D, Simerly C, Schatten G (1999) Unique checkpoints during the first cell cycle of fertilization after intracytoplasmic sperm injection in rhesus monkeys. Nat Med 5:431–433
29. VandeVoort CA, Baughman WL, Stouffer RL (1989) Comparison of different regimens of human gonadotropins for superovulation of rhesus monkeys: ovulatory response and subsequent luteal functions. J In Vitro Fertil Embryo Transfer 6:85–91
30. Davenport AT, Lees CJ, Green HL, Grant KA (2003) Long-acting depot formulation of luprolide acetate as a method of hypothalamic down regulation for controlled ovarian hyperstimulation and oocyte production in *Macaca fascicularis*. Biol Reprod 68:2261–2266
31. Tarantal AF (1990) Interventional ultrasound in pregnant macaques: embryonic/fetal applications. J Med Primatol 19:47–58
32. Mastroianni L Jr, Manson WA Jr (1963) Collection of monkey semen by electroejaculation. Proc Soc Exp Biol Med 112:1025–1027
33. Amboka JN, Mwethera PG (2003) Characterization of semen from olive baboons. J Med Primatol 32:325–329
34. VandeVoort CA (2004) High quality sperm for nonhuman primate ART: production and assessment. Reprod Biol Endocrinol 2:33
35. Sanchez-Partida LG, Simerly CR, Ramalho-Santos J (2008) Freeze-dried primate sperm retains early reproductive potential after intracytoplasmic sperm injection. Fertil Steril 89:742–745
36. Wolf DP (2009) Artificial insemination and the assisted reproductive technologies in non-human primates. Theriogenology 71:123–129
37. Cseh S, Chan PJ, Corselli J, Bailey LL (2000) Electroejaculated baboon (*Papio anubis*) sperm requires a higher dosage of pentoxifylline to enhance motility. J Assist Reprod Genet 17:449–453
38. Boatman DE, Bavister BD (1984) Stimulation of rhesus monkey sperm capacitation by cyclic nucleotide mediators. J Reprod Fertil 71:357–366
39. VandeVoort CA, Tollner TL, Overstreet JW (1994) Separate effects of caffeine and dbcAMP on macaque sperm motility and interaction with the zona pellucida. Mol Reprod Dev 37:299–304
40. Nyachieo A, Spiessens C, Chai DC, Kiulia NM, Mwenda JM, D'Hooghe TM (2010) Separate and combined effects of caffeine and dbcAMP on olive baboon (*Papio anubis*) sperm. J Med Primatol 39:137–142
41. Hewitson LC, Simerly CR, Tengowski MW, Sutovsky P, Navara CS, Haavisto AJ, Schatten G (1996) Microtubule and chromatin configurations during rhesus intracytoplasmic sperm injection: successes and failures. Biol Reprod 55:271–280
42. Nusser KD, Mitalipov S, Widmann A, Gerami-Naini B, Yeoman RR, Wolf DP (2001) Developmental competence of oocytes after ICSI in the rhesus monkey. Hum Reprod 16:130–137
43. Ng SC, Martelli P, Liow SL, Herbert S, Oh SH (2002) Intracytoplasmic injection of frozen-thawed epididymal spermatozoa in a nonhuman primate model, the cynomolgus monkey (*Macaca fascicularis*). Theriogenology 58:1385–1397
44. Hewitson L, Simerly CR, Schatten G (2003) ICSI, male pronuclear remodeling and cell cycle checkpoints. Adv Exp Med Biol 518:199–210
45. Wolf DP, Thormahlen S, Ramsey C, Yeoman RR, Fanton J, Mitalipov S (2004) Use of assisted reproductive technologies in the propagation of rhesus macaque offspring. Biol Reprod 71:486–493
46. Yeoman RR, Mitalipov S, Gerami-Naini B, Nusser KD, Wolf DP (2005) Low temperature storage of rhesus monkey spermatozoa and fertility evaluation by intracytoplasmic injection. Theriogenology 63:2356–2371
47. Slayden OD, Brenner RM (2006) A critical period of progesterone withdrawal precedes menstruation in macaques. Reprod Biol Endocrinol 4(Suppl 1):S6
48. VandeBerg JF, Williams-Blangero S, Tardif SD (2008) The baboon in biomedical research series: developments in primatology: progress and prospects. Springer, New York, NY
49. Schroder M, Fisk SK, Lerche NW (2000) Eradication of simian retrovirus type D from a colony of cynomolgus, rhesus, and stump-tailed macaques by using serial testing and removal. Contemp Top Lab Anim Sci 39:16–23
50. Bavister BD, Dees C, Schultz RD (1986) Refractoriness of rhesus monkeys to repeated ovarian stimulation by exogenous gonadotropins is caused by nonprecipitating antibodies. Am J Reprod Immunol Microbiol 11:11–16

51. Bavister BD (2004) ARTs in action in nonhuman primates: symposium summary—advances and remaining issues. Reprod Biol Endocrinol 2:43
52. Zuckerman S, Parkes A (1932) The social life of monkeys and apes. In: Proc Zool Soc, Kegan Paul, Trench, Trubner and Company, London, pp XII, 357
53. Nyachieo A, Spiessens C, Chai DC, Debrock S, Mwenda JM, d'Hooghe TM (2011) Randomized comparison of different ovarian stimulation regimens for assisted reproductive technology in baboons (*Papio anubis*). Fertil Steril 95:1354–1359
54. de Prada JK, Hill DL, Chaffin CL, VandeVoort CA (2009) Nuclear maturation and structural components of nonhuman primate cumulus-oocyte complexes during *in vivo* and *in vitro* maturation. Fertil Steril 91:2043–2050
55. Dupont C, Bavister BD, Armant DR, Brenner CA (2009) Rhesus macaque embryos derived from MI oocytes maturing after retrieval display high rates of chromosomal anomalies. Hum Reprod 24:929–935
56. Fourie FR, Snyman E, van der Merwe JV, Grace A (1987) Primate in vitro fertilization research: preliminary results on the folliculogenic effects of three different ovulatory induction agents on the chacma baboon, *Papio ursinus*. Comp Biochem Physiol A Comp Physiol 87:889–893
57. Keefe D, Liu L, Wang W, Silva C (2003) Imaging meiotic spindles by polarization light microscopy: principles and applications to IVF. Reprod Biomed Online 7:24–29
58. Montag M, van der Ven H (2008) Symposium: innovative techniques in human embryo viability assessment. Oocyte assessment and embryo viability prediction: birefringence imaging. Reprod Biomed Online 17:454–460
59. Shen Y, Betzendahl I, Tinneberg HR, Eichenlaub-Ritter U (2008) Enhanced polarizing microscopy as a new tool in aneuploidy research in oocytes. Mutat Res 651:131–140

Chapter 5

Media Composition: Salts and Osmolality

Jay M. Baltz

Abstract

The main components of embryo culture media are salts, which dissociate into their component inorganic ions in aqueous solution. All embryo culture media contain the same six inorganic ions: Na^+, K^+, Cl^-, Ca^{2+}, Mg^{2+}, and SO_4^{2-}, while most also contain PO_4^{2-}. The salts that are used to formulate embryo culture media can be traced back to classic saline solutions, particularly Krebs–Ringer Bicarbonate (KRB), that were developed for somatic cells in the first half of the twentieth century. The salt and inorganic ion concentrations in the first successful defined mouse embryo culture medium, Whitten's medium, were identical to those in KRB. These remained largely unchanged in embryo culture media for decades, with similar levels found in the standard mouse embryo culture medium, M16, formulated in the 1970s. Human embryos were initially cultured in undefined somatic cell media such as Earle's and Ham's F-10 with serum added. This changed in the mid-1980s, however, with the development of Quinn's HTF, a defined medium specifically formulated for human embryo culture, in which the inorganic ion concentrations are similar to those in M16 and Whitten's. While these media were useful both for experimental work and clinically, embryos suffered developmental blocks in all of them, with mouse embryos blocking at the 2-cell stage and human embryos at the 4- to 8-cell stage. Starting in the late 1980s, however, mouse embryo culture media were first developed that alleviated these developmental blocks. These media, CZB and KSOM, had much lower osmolalities than previous media, mainly due to lower inorganic ion concentrations. Indeed, lowering total inorganic ion concentration and osmolality proved key to understanding how media that supported complete preimplantation development in vitro can be formulated. A subsequent improvement was the addition of amino acids to culture media for both mouse and human embryos. At least in part, their beneficial effect during the cleavage stages of development is due to the presence in early preimplantation embryos of mechanisms for cell volume regulation that depend on the accumulation of amino acids as "organic osmolytes" to provide intracellular osmotic support. These amino acids, principally glycine, replace a portion of the intracellular inorganic ions that would otherwise be needed to maintain cell size, preventing the intracellular ionic strength from rising to deleterious levels and blocking development. Thus, the optimum salts levels, osmolality, and amino acid contents of culture media are not independent, but interact strongly because of their roles in cell volume regulation. In the absence of compounds that preimplantation embryos can use as organic osmolytes, embryos will develop only at lower osmolalities and salt concentrations in the medium. However, when organic osmolytes such as some amino acids are present, embryos will develop in culture at higher osmolarities that are similar to those they experience in tubal fluid in vivo.

Key words: Amino acids, Culture, Embryo, Ions, Media, Osmolality, Osmolarity, Osmolyte, Preimplantation, Salts

Gary D. Smith et al. (eds.), *Embryo Culture: Methods and Protocols*, Methods in Molecular Biology, vol. 912,
DOI 10.1007/978-1-61779-971-6_5,

1. Introduction

1.1. Salts in Human Embryo Culture Media

The overwhelming majority of the solutes present in embryo culture media consists of inorganic ions from salts. All commercial human embryo culture media share the same six inorganic ions: Na^+, K^+, Cl^-, Ca^{2+}, Mg^{2+}, and SO_4^{2-}, which are the same ones that were included in the first defined mouse embryo culture media formulated in the 1950s (1, 2). Most media also contain PO_4^{2-}, although several that were created specifically for eggs and cleavage stage embryos are phosphate-free. Of these six or seven inorganic ion species, Na^+ and Cl^- are present at by far the largest concentrations in every medium. In addition, much lower concentrations of other inorganic ions are likely present during embryo culture, mainly introduced when the media are supplemented with protein sources such as serum or serum albumin, since serum albumin has a high binding capacity for ions such as Cu^{2+} and Zn^{2+} (3). Serum albumin and similar medium constituents, and EDTA, also exert a beneficial effect by chelating these and other ions that can be toxic when present above trace levels.

The sources of the inorganic ions in human embryo culture media are largely the salts used in their formulation (Table 1). The major sources of Na^+ are NaCl and $NaHCO_3$, with smaller amounts contributed by sodium pyruvate and the sodium salts of lactate, phosphate, citrate, or acetate where present. Relatively negligible amounts would also be introduced with other compounds that are often complexed with sodium (e.g., EDTA). The Cl^- present in culture media is almost entirely derived from NaCl, with minor amounts contributed by KCl and, in media where they are present, $CaCl_2$, $MgCl_2$, or choline chloride. The final monovalent ion, K^+ is derived mainly from the KCl that is used in formulating all media, with a small additional amount from potassium phosphate in many. The divalent ion Ca^{2+} is added as $CaCl_2$ or calcium lactate, with minor amounts sometimes introduced in the form of calcium pantothenate. Mg^{2+} and SO_4^{2-} are always added together, as $MgSO_4$, although in at least one set of human embryo culture media, the Mg^{2+} content is further increased with $MgCl_2$. When present, PO_4^{2-} is derived from either the sodium or potassium salt.

It is important to appreciate, however, that the source of each of the inorganic ions in culture media is not physiologically important (assuming that all of the starting material is equally pure and suitable for embryo culture). This is because ionic compounds—salts—are nearly completely dissociated into their component ions in aqueous solutions (4). Therefore, once the medium is made, all ions of a given species are essentially interchangeable, and it is only the total concentration of each type of ion that is important for the physicochemical properties of the medium and for healthy embryo development.

Table 1
Salts used in formulating human and mouse embryo culture media

Company[a]	Medium	Salt: Sodium chloride	Sodium bicarbonate	Sodium pyruvate	Sodium lactate	Sodium phosphate	Sodium citrate	Sodium acetate	Potassium chloride	Potassium phosphate	Calcium chloride	Calcium lactate	Calcium pantothenate	Magnesium chloride	Magnesium sulfate	Choline chloride
Cook®IVF[b]	Sydney IVF Cleavage	x	x	x					x	x		x	x	x	x	
	Sydney IVF Blastocyst	x	x	x					x	x		x	x	x	x	
Cooper Surgical (SAGE®)[c]	Quinn's Advantage® Cleavage	x	x	x			x		x			x			x	
	Quinn's Advantage® Blastocyst	x	x	x					x	x		x	x		x	x
GyneMed[d]	GM501	x	x	x	x				x	x	x				x	
FertiPro N.V.[e]	FertiCult™ IVF	x	x	x	x	x			x		x				x	
	FertiCult™ G3	x	x	x	x	x			x		x				x	
Irvine Scientific[f]	SSM™	x	x	x	x				x	x	x				x	
	P1®	x	x	x	x		x		x		x				x	
	ECM®	x	x	x	x		x		x		x				x	
	Multiblast®	x	x	x	x		x		x	x	x				x	

(continued)

Table 1 (continued)

Company[a]	Medium	Salt: Sodium chloride	Sodium bicarbonate	Sodium pyruvate	Sodium lactate	Sodium phosphate	Sodium citrate	Sodium acetate	Potassium chloride	Potassium phosphate	Calcium chloride	Calcium lactate	Calcium pantothenate	Magnesium chloride	Magnesium sulfate	Choline chloride
LifeGlobal® (IVFonline)[g]	global®	x	x	x	x				x	x	x				x	
	HTF	x	x	x	x				x	x	x				x	
	Blastocyst	x	x	x	x				x	x	x				x	
Origio (MediCult)[h]	Universal IVF	x	x	x		x			x		x	x			x	
	ISM1™	x	x	x		x	x	x	x	x		x			x	
	ISM2™	x	x	x		x	x	x	x	x	x	x	x		x	x
	Embryo Assist™	x	x	x					x			x	x		x	x
	BlastAssist®	x	x	x					x				x		x	x
VitroLife[i]	G-1™ (G5 series™)	x	x	x	x	x	x		x		x		x		x	
	G-2™ (G5 series™)	x	x	x	x	x	x		x		x				x	
n/a (mouse)[j]	Whitten's	x	x	[k]	[k]				x	x	x	[k]			x	
	M16	x	x	x	x				x	x	x				x	
	CZB	x	x	x	x				x	x	x				x	
	KSOM	x	x	x	x				x	x	x				x	

[a]Company Web sites as specified for each medium were accessed on January 11–12, 2011 and reconfirmed March 31–April 4, 2011. No information about medium salt composition was found for the InVitroCare® IVC™ series of media on the company Web site (http://www.invitrocare.com), except that they are phosphate-free, and information could not be obtained from the company, so these media are not included in the table

[b]Cook® IVF: http://www.cookmedical.com/wh/familyListingAction.do?family=Assisted+Reproduction&subFamily=Embryo+Culture; Sydney IVF Cleavage: http://www.cookmedical.com/wh/content/mmedia/WH-BWE-EMC-EN-200812-13.pdf; Sydney IVF blastocyst: http://www.cookmedical.com/wh/content/mmedia/WH-BWE-EMC-EN-200812-17.pdf

[c]CooperSurgical SAGE®: http://www.coopersurgical.com/OURPRODUCTS/ASSISTREPROD/Pages/default.aspx; Quinn's Advantage® Cleavage: http://www.coopersurgical.com/Documents/ART%201026-27%20DFU.pdf; Quinn's Advantage® Blastocyst: http://www.coopersurgical.com/Documents/ART1029DFU.pdf

[d]GyneMed: http://www.gynemed.de/Home.14+M52087573ab0.0.html; GM501 Basic: http://www.gynemed.de/GM501-Basic.98+M52087573ab0.0.html and http://www.gynemed.de/uploads/media/MediaLine-1_05.pdf

[e]Fertipro: http://www.fertipro.com/index.php?page=home&sub=general; Ferticult™ IVF: http://www.fertipro.com/index.php?page=media&sub=fecu and http://www.fertipro.com/msds/FC.pdf; Ferticult™ G3: http://www.fertipro.com/index.php?page=media&sub=g3 and http://www.fertipro.com/msds/G3.pdf

[f]IrvineScientific: http://www.irvinesci.com/sub.cfm?sec=3&loc=0; SSM™ (Single Step Medium™): http://www.irvinesci.com/ecom/app_ecom_irvinesci.cfm?sec=3&loc=11&goto=detail&cn=90161; P1®: http://www.irvinesci.com/ecom/app_ecom_irvinesci.cfm?sec=3&loc=11&goto=detail&cn=99242; ECM® (Early Cleavage Medium®): http://www.irvinesci.com/ecom/app_ecom_irvinesci.cfm?sec=3&loc=11&goto=detail&cn=90138; Multiblast®: http://www.irvinesci.com/ecom/app_ecom_irvinesci.cfm?sec=3&loc=11&goto=detail&cn=90139

[g]LifeGlobal® (IVFonline): http://www.lifeglobal.com/default2.asp; global®: http://www.lifeglobal.com/asp/Products/ProductDetail.asp?ID=LGGG; Blastocyst: http://www.lifeglobal.com/asp/Products/ProductDetail.asp?ID=LGGB; HTF: http://www.lifeglobal.com/asp/Products/ProductDetail.asp?ID=GMHT

[h]Origio (formerly MediCult): http://www.origio.com/; Universal IVF: http://www.origio.com/products/medicult%20media/universal%20ivf%20medium.aspx; ISM1™: http://www.origio.com/products/medicult%20media/ism1%E2%84%A2.aspx; ISM2™: http://www.origio.com/products/medicult%20media/ism2%E2%84%A2.aspx; EmbryoAssist™: http://www.origio.com/products/medicult%20media/embryoassist%E2%84%A2.aspx; BlastAssist®: http://www.origio.com/products/medicult%20media/blastassist%C2%AE.aspx

[i]VitroLife: http://www.vitrolife.com/fertility/; G-1™ (G5 series™): http://www.vitrolife.com/fertility/index.cfm?page=94EAC99D-078A-D29F-AA0D81665A337290; G-2™ (G5 series™): http://www.vitrolife.com/fertility/index.cfm?page=94EB2598-E1C6-234C-9D5B1558257B8000

[j]For mouse culture media, salt compositions were determined from the references cited in the text and listed in Table 2

[k]Later versions of Whitten's medium also had calcium lactate or sodium lactate added, and sodium pyruvate added (see text)

1.2. History of Embryo Culture Media: Why These Salts?

In vitro embryo culture dates back to 1912, beginning with Brachet's culture of rabbit blastocysts in vitro (5). The early experimental embryo culture work that followed, including the initial culture of primate (rhesus monkey) cleavage stage embryos in 1933 by Lewis and Hartmann (6), all used complex biological fluids such as serum or blood plasma (7). The undefined nature of such biological fluids was not only highly problematic for experimental investigations, but preimplantation embryos of most mammals, including the important laboratory rodents, would not develop in plasma or serum (reviewed by Pincus (8) and Alexandre (7)). The first successful completely defined embryo culture medium was developed by Whitten for mouse embryos in 1956 (2). Whitten's medium was based on Krebs–Ringer Bicarbonate (KRB), which had been initially formulated in 1932 by Krebs and Henseleit (9). KRB is a simple bicarbonate/CO_2-buffered saline solution based on the earlier Warburg's solution that similarly employed approximately physiological concentrations of bicarbonate and CO_2. KRB differed from Warburg's solution in that the inorganic ion concentrations were adjusted so that they more closely approximated those of mammalian serum, based on averaged measurements of inorganic ions in the blood serum of several mammals that became available in the contemporary literature of the 1920s and early 1930s (9). These earliest media, therefore, attempted to replicate the in vivo environment of cells, anticipating the much later efforts to formulate embryo culture media by mimicking the concentrations of metabolites in oviductal and uterine fluids.

Mouse embryos will not develop in KRB at any stage, but Whitten's medium (Tables 1 and 2), which differed from KRB only by the addition of glucose and BSA, supported development from the 8-cell through blastocyst stages (2). These blastocysts were viable, since normal young were produced following embryo transfer (10). Soon after, Whitten extended the range of in vitro development that was possible in a simple defined medium by adding lactate (as either calcium lactate or sodium lactate) to his initial medium, which then permitted development from the 2-cell stage onwards, but not from the fertilized egg to 2-cell stage (11). Nearly a decade later, Biggers, Whittingham, and Donohue found that adding pyruvate produced a medium that supported development from the fertilized egg to 2-cell stage (12, 13). The only changes in these media were in the addition of energy substrates (glucose, lactate and pyruvate), since the salts used in their formulation remained identical in type and concentration to those in KRB.

Through the 1960s and into the early 1970s, a series of media were developed, including Whittingham's M16 medium (14, 15) that became the standard for mouse embryo culture for several decades (reviewed in (1, 14, 16)). M16 is essentially identical in its

Table 2
Comparison of concentrations of components of culture media

Component	Concentration (mM)							
	KRB	Whitten's[a]	M16	CZB	SOM	KSOM	HTF	G-1[b]
NaCl	118	118	95	82	85	95	102	85
KCl	4.7	4.7	4.8	4.8	0.25	2.5	4.7	5.5
KH_2PO_4	1.2	1.2	1.2	1.2	0.35	0.35	0.37	
NaH_2PO_4								0.5
$NaHCO_3$	25	25	25	25	25	25	25	25
$CaCl_2$	2.5	2.5	1.7	1.7	1.7	1.7	2.0	1.8
$MgSO_4$	1.2	1.2	1.2	1.2	0.2	0.2	0.2	1.0
Na lactate[c]			23	31	10	10	21	10.5
Na pyruvate			0.33	0.27	0.2	0.2	0.33	0.32
EDTA (tetra Na)				0.11	0.01	0.01		0.01
Glucose		5.6	5.6		0.2	0.2	2.8	0.5
Glutamine				1.0	1.0	1.0		1.0
Total Na^+	143	143	143	139	120	130	148	121
Total Cl^-	128	128	103	90	89	101	111	94
Total K^+	5.9	5.9	6.0	6.0	0.60	2.8	5.1	5.5
Total dissolved particles[d]	308	313	312	298	248	273	316	262
References	(2, 9)	(2)	(14, 15)	(24)	(26)	(28, 29)	(18)	(37)

[a]Later versions of Whitten's medium also had calcium lactate or sodium lactate added, and sodium pyruvate added (see text)
[b]Original formulation of G-1. G-1 also contains taurine and Eagle's nonessential amino acids, all at 0.1 mM (contributing ~0.5 mM to total dissolved particles). The current commercial G-1™ of the G5 Series™ has additional components, and the concentrations of the components may differ from those shown here
[c]dl-Lactate is used, except for G-1, where L-lactate is specified
[d]total of all dissolved species if ionic compounds were completely dissociated (excluding H^+)

inorganic ion concentrations to KRB, except that Na lactate (23 mM) was substituted for isomolar NaCl leading to a reduction in total Cl^- by that amount, and Ca^{2+} has been modestly reduced (Tables 1 and 2). Considerable work was also done to develop defined culture media for hamster embryos, particularly by Bavister's group, yielding media that were similar in salt composition to KRB and M16, although they omitted sulfate (17). Interestingly, early work had indicated that, of the inorganic ions in KRB, only the omission of sulfate had no effect on mouse embryo development (14). The invention of these media not only greatly facilitated experimental embryology, but also were key to the development of the first generation of media specifically formulated for human embryo culture such as HTF (18), which has similar salt and inorganic ion compositions (see below).

These media, however, did not support complete development of fertilized eggs all the way through the blastocyst stage. Development of fertilized mouse eggs was supported to the 2-cell stage, and in vivo-produced 2-cell embryos were able to develop into viable blastocysts, but most embryos would not develop from the fertilized egg to blastocyst. Instead, they became arrested at the 2-cell stage (19). This artifact of culture was known as the "2-cell block," and affected embryos from most genotypes of female mice, except embryos from some F_1 hybrids (1). Other mammalian species suffered similar developmental blocks at characteristic points in their development, such as the 2- to 4-cell block in hamster (20), 2-cell block in rat (21), 8-cell block in bovine (22), and 4- to 8-cell block in human (23) embryos.

These culture-induced developmental blocks were not overcome until the late 1980s and early 1990s. During that period, two new media for mouse embryo culture were developed as a result of a large research program organized and funded by the US National Institute of Child Health and Human Development, NIH, that was informally known as the "Culture Club" (officially, the National Cooperative Program on Non-human In Vitro Fertilization and Preimplantation Development). The first of these media was CZB, developed by *C*hatot, *Z*iomek, *B*avister, and colleagues (24). CZB is very similar to M16, but with a few key differences (Tables 1 and 2). Glutamine was introduced, based on previous work that showed it was beneficial for hamster embryo development, EDTA was added, the lactate/pyruvate ratio was increased, and NaCl was reduced. In addition, CZB had to be used as a two-step medium to overcome the developmental block in random-bred mouse embryos, with glucose absent for the first 2 days of culture and then introduced at the 4-cell to 8-cell stage, since development was blocked if glucose was present before this point.

The second mouse medium that arose from the Culture Club was KSOM, developed by Biggers's group. KSOM (Tables 1 and 2) was the end product of an extensive series of media that were produced using Simplex optimization, a computer-assisted algorithm developed for optimizing yields of chemical reactions (25). The penultimate product of the optimization was SOM (*S*implex *O*ptimized *M*edium), which supported development of random-bred embryos through the 2-cell block in a single medium in which glucose was present throughout development (26). Subsequent work showed that increasing the K^+ concentration in the medium normalized the intracellular K^+/Na^+ ratio, which was abnormally low in SOM (27). This led to increased K^+ in the final formulation, KSOM (K^+ supplemented SOM), that is now in common use for mouse embryos culture (28, 29). Like CZB, KSOM contains glutamine and EDTA in addition to the constituents that were already present in Whitten's media. However, the concentrations of most of the salts were altered during the optimization process, and deviate from the levels in KRB, Whitten's, or M16 (Table 2).

Another important change from the older media like M16 and those that preceded it was the addition of amino acids. None of the standard mouse embryo culture media before CZB and KSOM contain amino acids, while CZB and KSOM each included glutamine. Work by Bavister's group during the 1990s established that a subset of amino acids was beneficial to hamster embryo development in vitro, while some other amino acids were detrimental (17), and Gardner and Lane showed around the same time that the set of Eagle's "nonessential" amino acids was beneficial for early mouse embryo development (30). A modification of KSOM, denoted KSOMaa, was produced that contained all 20 common α-amino acids, which improved the health and developmental potential of the blastocysts that developed (31, 32) and mouse media containing only Eagle's nonessential amino acids during the cleavage stages also supported preimplantation development (30, 33). As discussed below, the beneficial effect of the addition of a subset of amino acids is not independent from the changes made in the salt compositions of media like CZB and KSOM from earlier media. Instead, the effects of the inorganic ion levels on embryo development depend on the amino acids present, and vice versa.

1.3. History of Human Embryo Culture Media

The media used for the first decade or so of clinical human embryo culture were media that had been formulated for mammalian somatic cells, and included simple media such as Earles' medium plus pyruvate and complex media like Ham's F-10, all with serum added, so that they were not defined media (34). In the mid-1980s, however, media were developed specifically for human embryo culture based on earlier analyses of human fallopian tubal fluid. Quinn's HTF (*H*uman *T*ubal *F*luid) medium (18) was made using the same salts as the mouse embryo culture media current at the time (M16) and the earlier Whitten's, but the amounts added to formulate the medium differed for most of them (Table 2). However, when the total amounts of each ionic species are compared, the net differences are not very profound, with the exceptions of much lower Mg^{2+} and PO_4^{2-} (Table 2). The other main difference was lowered glucose, present in HTF at about half the level of M16 and Whitten's.

While the concentrations of metabolites were adjusted to approximate levels measured in vivo and differed from previous media, the inorganic ion concentrations of HTF do not more closely match those of tubal fluid than the other culture media. Various measurements have been reported of the ion contents of human tubal fluid (reviewed by Borland et al. (35)). These give values for Na^+ of ~130–150 mM, for Cl^- of ~110–130 mM, for K^+

of ~6–20 mM, and for Ca^{2+} of ~1–3 mM. The ion concentrations in HTF lie within these ranges, but so do almost all values for the other culture media stretching back to Whitten's (Table 2). Thus, inorganic ion concentrations do not clearly distinguish any culture medium as being closer to the physiological norm than another, particularly given the variations between values reported in different studies, and between patients (35, 36).

Following the success of HTF, a wide array of different human embryo culture media have been developed. Most current media were formulated or substantially refined following the major advances in laboratory rodent culture media that eliminated in vitro developmental blocks. Because, for human embryo culture media, the precise compositions are considered proprietary information, it is unfortunately not possible to make direct comparisons between the inorganic ion concentrations in most of the diverse media, nor to compare them to those in culture media developed for mouse embryos or to those in human tubal fluid. However, the history of the development of several widely used media does imply their likely compositions. At least two single-step human embryo culture media are closely related to KSOMaa. LifeGlobal's Global® medium and GyneMed's GM501 are modifications of KSOMaa, and therefore likely have very similar inorganic ion compositions to the mouse medium. Vitrolife's G-1™ series of media are modifications of the original G-1, whose composition had been published (37). The amount of NaCl (85 mM) used to make the original G-1 (Table 2) may have been based on that of SOM, since SOM also contained 85 mM NaCl (26) instead of the 95 mM in KSOM, giving G-1 total Na^+ and Cl^- concentrations very close to those of SOM (c.f. 120 mM and 89 mM for Na^+ and Cl^-, respectively, in SOM). The total K^+ concentration in the original G-1 is higher than in SOM or even KSOM, due to KCl added at a somewhat greater level, but is close to that in CZB. Overall, then, the inorganic ion concentrations in G-1 would appear to be similar to those of the current media developed for mouse embryos. Thus, at least some of the current generation of media used for human embryo culture would appear to have made use of the experimental findings that led to the development of media for mouse embryos to eliminate the 2-cell block to in vitro development, and based their salt compositions on these media.

1.4. Culture Medium Osmolality

Inorganic ions serve a very large variety of functions in cells and are vital for a vast number of cell physiological processes. One major function is in establishing the characteristic intracellular environment, in which K^+ concentrations are high while Na^+ and Cl^- concentrations are kept low, mediated by the ubiquitous Na^+, K^+-ATPase (38). The ion gradients thus established are required for the maintenance of cell membrane potential and also serve as the energy source for the large array of transporters that are coupled to transmembrane ion gradients, including those that transport metabolites and amino acids, and others that perform homeostatic functions such as intracellular pH regulation (39). Another fundamental contribution of inorganic ions in physiological solutions is to the total osmolality.

Because the combined concentration of inorganic ions is so high compared to other solutes, they are by far the major determinant of osmolality both in the cell's external environment (either in vivo or the in vitro medium) and in the cytoplasm. Osmolality is a measure of the total osmotic pressure in a solution, with units of osmoles/kg. For an ideal 1 M solution containing a single solute that is completely dissolved in water, the osmolality would be 1 osmole/kg. Another way of expressing the osmotic contribution in a solution is osmolarity, measured in osmoles/L (osmolar; OsM). For concentrated solutions, these measures diverge, since the density of the solution increases significantly above 1 kg/L. However, for the types of dilute solutions used in embryo culture or in the embryo's in vivo environment, osmolality and osmolarity are essentially interchangeable, with osmolarity often used because of the convenience of the units (usually expressed as mOsM) and their correspondence with the commonly used expression of concentration as molarity.

Although in the ideal example given above, osmolality (or osmolarity) and molality (or normality) are equal, this is not generally the case. First, the osmotic pressure exerted by any solute is dependent on the total number of solute particles, so that in the case of salts, which are nearly completely dissociated in aqueous solution, the osmotic contribution will be a multiple of the normal concentration of salt added. As an example, 0.1 N NaCl in water would have an osmolality of nearly 0.2 osmoles/kg, since each NaCl dissociates into two osmotically active, independent particles in aqueous solution. Second, there is often a significant deviation from the ideal osmotic behavior of any given solute due to interactions with the solvent and/or possible incomplete dissociation. For example, the 0.1 N NaCl solution used as an example would have an actual osmolality of 0.186 osmoles/kg or ~186 mOsM (4) rather than the ideal 0.200 mosmoles/kg (~200 mOsM). This is accounted for by empirically defining an osmotic coefficient (φ) that is specific for each solute and that varies with concentration, whose value in this example is 0.93 (4). Thus, the osmolality of a given solution is calculated from $\varphi \times n \times$ concentration, where n is the number of particles into which the solute dissociates (e.g., 2 for NaCl). This effect is what accounts for the measured osmolalities of culture media being somewhat lower than the sum of all dissolved particles. As an example, the total number of dissolved particles in KSOM medium is 273 mM (Table 2). However, its osmolality is approximately

256 osmoles/kg (Table 3). Thus, the composite φ for its components is 256/273 = 0.94, essentially that of NaCl alone. In practice, and because there will also be interactions between solutes in a mixture, culture medium osmolality is always determined by measurement using a laboratory osmometer.

A major change in the mouse embryo culture media that were developed to overcome developmental blocks in culture over previous media was that their osmolality was substantially decreased compared to previous media in which the block occurred. While M16 has an osmolality of ~290 mosmoles/kg (15), that of KSOM is ~255 (28), that of CZB is ~275 (40), and that of SOM is only ~230 (26). Current human embryo culture media have a wide range of osmolalities (Table 3). The lower acceptable limit of the commercial human embryo culture medium of lowest osmolality among those for which data are available, G-2 with HSA, is 248 mosmoles/kg, while the upper acceptable limit of the highest, P1, is 298 (Table 3). Thus, human embryos are successfully cultured to yield healthy offspring in media that span a range of 50 mosmoles/kg. This wide range of osmolalities that support human embryo development is similar to the range of osmolalities over which mouse embryos can develop to blastocysts (reviewed by Baltz (41)).

Table 3
Osmolality of embryo culture media

Company	Medium	Stages[a]	Osmolality range (mean) mosmoles/kg
Cook®IVF[b]	Sydney IVF Cleavage	Cleavage	285–295 (290)
	Sydney IVF Blastocyst	8c to Bl	285–295 (290)
Cooper Surgical[c] (SAGE®)	Quinn's Advantage® Cleavage	Cleavage	257–273 (265)
	Quinn's Advantage® Blastocyst	8c to Bl	257–273 (265)
Fertipro[b]	Ferticult™ IVF	Cleavage	275–285 (280)
	Ferticult™ G3	8c to Bl	270–290 (280)
GyneMed[b]	GM501	All	270–290 (280)
InVitroCare[d]	IVC-ONE™, IVC-TWO™	Cleavage	270–290 (280)
	IVC-THREE™	8c to Bl	270–290 (280)
Irvine Scientific[c]	SSM™ (plain, SSS, DSS)	All	260–270 (265)
	P1® (plain, SSS, DSS)	Cleavage	282–298 (290)
	ECM® (plain, SSS, DSS)	Cleavage	282–295 (288)
	Multiblast® (plain, SSS, DSS)	8c to Bl	281–291 (286)
	HTF (plain, SSS)	Cleavage	272–288 (280)

(continued)

Table 3 (continued)

Company	Medium	Stages[a]	Osmolality range (mean) mosmoles/kg
LifeGlobal® (IVFonline)[b]	global®	All	260–270 (265)
	HTF	Cleavage	280–292 (286)
	Blastocyst	8c to Bl	260–270 (265)
Origio (MediCult)[f]	Universal IVF	Cleavage	277–293 (285)
	ISM1™	Cleavage	272–288 (280)
	ISM2™	8c to Bl	272–288 (280)
	EmbryoAssist™	Cleavage	272–288 (280)
	BlastAssist®	8c to Bl	272–288 (280)
VitroLife[g]	G-1™ (G5 series™)	Cleavage	256–266 (261)
	G-1™ with HSA (G5 series™)[h]	Cleavage	249–259 (254)
	G-2™ (G5 series™)	8c to Bl	255–265 (260)
	G-2™ with HSA (G5 series™)[h]	8c to Bl	248–258 (253)
Mouse media[i]	M16		(290)
	CZB		(275)
	SOM		(229)
	KSOM		(256)

[a]"Cleavage" indicates embryos after fertilization to the time of compaction. "8c" denotes 8-cell embryos usually around the time of compaction, and "Bl" denotes blastocyst. Media listed as "cleavage" and "8c to Bl" are paired as sequential media. Media listed as "all" stages are single media that can be used for culture from fertilized egg to blastocyst

[b]Osmolarity ranges for media from these companies were obtained from the Web sites listed in Table 1 for each individual medium

[c]Patrick Quinn, PhD, HCLD, CooperSurgical, Inc., personal communication, January 13, 2011

[d]Information on osmolality could not be located on the InVitroCare (IVC[V] series of media) Web site (http://www.invitrocare.com) and could not be obtained directly from the company. A brochure for InVitroCare products is available online at http://www.genycell.com/images/productos/protocolos/__94.pdf (accessed April 4, 2011) that gives the osmolalities of IVC™ media as 270–290 mosmoles/kg, which is reported here

[e]Laura Mena, Associate Product Manager, and Wayne Caswell, Field Scientist, Irvine Scientific, personal communication, January 26, 2011

[f]Rasmus Kiil-Nielsen, International Product Manager, ORIGIO a/s, personal communication, January 15, 2011

[g]Susie Oliver, T.S., VitroLife, Inc., personal communication, January 13, 2011

[h]The lower osmolalities for media with HSA added likely are due to the dilution of the media with 5 % (v/v) HSA solution™ added to the media as directed (http://www.vitrolife.com/documents/HSA%20solution%2020493.04.pdf, accessed March 31, 2011)

[i]Osmolalities for mouse embryo culture media were taken from references also given in the text (15, 26, 28, 40)

1.5. Effect of Osmolality on Embryo Development

Although mammalian embryos, including those of mouse and human, will develop normally over a wide range of osmolalities, there are nonetheless very strong effects of osmolality on embryo development. At least in part, the apparent broad tolerance of embryos for diverse osmolalities reflects the strong interaction between osmolality and other components of a medium, particularly some amino acids, which results in different optimum osmolalities in different media (discussed below). As could be inferred from the observation that media which permitted preimplantation mouse embryo development in vitro beyond the 2-cell developmental block had lower osmolalities than previous media with similar components in which such blocks occurred, increasing osmolality above a threshold level prevents embryo development in culture.

Soon after the first media were formulated that eliminated the 2-cell block to mouse embryo development in culture, Lawitts and Biggers (26, 27) showed that increasing the NaCl concentration in SOM from 85 to 125 mM, which can be calculated to raise the osmolality by about 75 mosmoles/kg from 230 to about 305 mosmoles/kg, almost completely blocked development beyond the 2-cell stage. That this was due to increased osmolality and not increased NaCl per se was demonstrated soon afterwards by Dawson and Baltz (42) who showed that increasing osmolality had the same effect whether the osmolyte used was NaCl or the inert trisaccharide, raffinose. Subsequently, it was shown by Hadi et al. (43) that the deleterious effect of increased osmolality was not restricted to embryos from those genotypes of mice that had suffered the 2-cell block—so-called "blocking strain" embryos. Instead, "nonblocking strain" embryos, such as those from some F1 hybrids, simply tolerated a somewhat higher osmolality before becoming blocked. Thus, increased osmolality above a threshold always has a detrimental effect, but the thresholds are different for different genotypes of embryos within species, and likely between species. An investigation of the effect of varied osmolality on human embryo development has not been reported, so the corresponding osmolality thresholds in human embryo culture media are not known.

Why is lowered osmolality beneficial? There are at least two possibilities. One was that the newer culture media replicated the in vivo environment. That would imply that the fluid in oviducts has a low osmolality somewhere in the range of 230–275 mosmoles/kg like SOM, KSOM, and CZB, and that lowering medium osmolality to this level provided embryos with their normal osmolality. This is unlikely, however, since determinations of mouse oviductal fluid osmolality indicated that it was in the 300–310 mosmoles/kg range (44, 45). Similarly, porcine oviductal fluid is reportedly ~320 mosmoles/kg (46) while that of rat is ~290 mosmoles/kg (47). Fallopian tube fluid osmolality in humans does not appear to have been measured, however (nor does it appear to have been determined in any primate). However, where it has been measured, the values of oviductal fluid osmolality are similar to those of the blood of the same species, about 300–320 mosmoles/kg for blood in female mice, 280–290 in rats, and 300–330 in pigs (48–50). If the similarity with blood plasma also applies to humans, fallopian tube fluid should be approximately 290 mosmoles/kg (51), so that culture media with osmolalities substantially below this level lie outside the likely in vivo range.

The second possibility is that the low osmolality needed for optimal embryo development in culture was an in vitro artifact that is due to a lack of some component present in vivo that was critical to allow development at normal osmolality. In this scenario, the artificially lowered osmolality of culture media would be due to the need to compensate for the original absence of this component. This indeed appears to be the case, with the components missing from classical embryo culture media being several amino acids. These amino acids participate in cell volume regulation in early preimplantation embryos.

1.6. Cell Volume Regulation: The Roles of Inorganic Ions and Amino Acids

Animal cells regulate their volumes osmotically, by adjusting the cytoplasmic concentrations of osmolytes to control intracellular osmotic pressure (52). The initial response of cells to decreased cell volume is to activate transporters that mediate the accumulation of inorganic ions (53). In mammalian cells, the main rapid defense mechanism against an unwanted cell volume decrease is the activation of Na^+/H^+ exchangers of the NHE gene family (54), which mediates the uptake of Na^+ into the cell. This also raises intracellular pH (pH_i) due to the export of H^+ in exchange for imported Na^+, which in turn causes the secondary activation of HCO_3^-/Cl^- exchangers of the AE gene family, which mediate the uptake of Cl^- (54, 55). The overall result is the coupled import of Na^+ and Cl^-, which increases intracellular osmotic pressure and causes the cell to swell. During such NaCl uptake, there is little net change in pH_i since each H^+ removed from the cell in exchange for Na^+ is coupled with efflux of a HCO_3^- removed from the cell in exchange for Cl^-, so that each H^+ is neutralized by a HCO_3^-. Upon cell shrinkage, this mechanism is immediately activated and uptake of NaCl occurs quickly, restoring cell volume within minutes. Embryos would appear to also employ this acute cell volume regulatory mechanism, since an experimentally induced cell decrease in mouse 2-cell embryos immediately induces activation of Na^+/H^+ exchange, while inhibiting Na^+/H^+ exchanger activation impairs volume regulation (C. Zhou and JM Baltz, unpublished). Therefore, one of the key functions of inorganic ions in the extracellular environment of embryos, including in culture media, is to provide the Na^+ and Cl^- that are accumulated during acute cell volume regulation.

While uptake of inorganic ions restores cell volume rapidly, this is not always an optimal long-term solution. In cells where osmolality is high in the external environment (e.g., in the kidney), the level of intracellular inorganic ions needed to balance external osmotic pressure can be highly detrimental over an extended time. In such cells, a large portion of the inorganic ions in the cytoplasm are replaced by benign, uncharged "organic osmolytes," which prevents intracellular ionic strength from remaining at damaging levels (56, 57). Organic osmolytes are a set of otherwise unrelated compounds including diverse amino acids, amino acid derivatives, and sugars, that are compatible with biochemical functions such as enzyme activity and macromolecular assembly even at very high concentrations. By using organic osmolytes instead of inorganic ions to provide a portion of cytoplasmic osmolality, intracellular ionic strength can be kept at an optimal level while allowing the cell to independently control intracellular osmolality. In general, accumulation of organic osmolytes takes place much more slowly than the acute cell volume regulation by inorganic ion transport (58), and thus, in cells which use both, acute cell volume regulation is accomplished by inorganic ion transport, while longer term homeostasis requires organic osmolyte uptake. While there are many different compounds that are used as organic osmolytes by cells of animals, plants, unicellular eukaryotes, fungi and bacteria, each type of cell tends to use only a small subset of compounds to provide osmotic support.

Biggers and Van Winkle first proposed that preimplantation embryos require organic osmolytes for normal development (26, 59). One of the possible organic osmolytes initially identified was the glutamine that had been added to the culture media such as CZB and KSOM that first allowed mouse embryos to develop beyond the 2-cell block. Glutamine was shown to have a very significant interaction with NaCl concentration in SOM medium, so that only at high NaCl was it beneficial (26), consistent with a role as an organic osmolyte. Glycine also was shown to similarly protect embryos against increased NaCl in the medium (59). Two other compounds, proline and betaine (*N*,*N*,*N*-trimethylglycine), were subsequently shown to also have robust protective effects against increased osmolality in culture (27, 42). Finally, the β-amino acid β-alanine also had similar protective properties, with a marginal effect by the β-amino acid taurine (60). Subsequent investigations, mainly in mouse, have shown that these amino acids and derivatives can function as organic osmolytes in embryos.

It has been established that the principal organic osmolyte used by embryos is glycine. A high affinity glycine transporter, GLYT1, is expressed and active in mouse embryos until the 4-cell stage, and mediates the osmoregulated accumulation of glycine in embryos (61). The intracellular soluble glycine level in in vivo-developed mouse embryos is extremely high through the 2-cell stage, in the range of 20–30 mM (and therefore contributing ~20–30 mosmoles/kg to the total osmotic pressure), indicating that glycine is also accumulated as an organic osmolyte in vivo (62). Glycine transport via GLYT1 is also present in human embryos (63), implying a similar role there. GLYT1 also transports glutamine in mouse embryos, although with much lower affinity than glycine (64), likely accounting for the ability of glutamine to protect early preimplantation embryos against increased osmolality when glycine is not present (e.g., in KSOM and CZB). The GLYT1-dependent cell volume-regulatory mechanism is apparently unique to early embryos, since GLYT1-mediated uptake of glycine has not been found in mammalian somatic cells, which instead use one of the four identified mammalian somatic cell organic osmolyte transporters (65, 66). Betaine and proline, however, are not taken up via GLYT1 in embryos. Instead, a transporter in the same gene family, SIT1, is responsible for their uptake as organic osmolytes (67, 68). SIT1 is active only in 1-cell and 2-cell stage embryos, and likely acts to support the main role of GLYT1, since betaine is accumulated only to lower levels of about 6–7 mM (68). Like for GLYT1, SIT1-mediated organic osmolyte uptake is unique to early embryos. The function of these organic osmolyte transporters in early preimplantation embryos and their developmental regulation has recently been reviewed in detail by Baltz and Tartia (16).

The wide range of culture media that support development of preimplantation embryos in part reflects the interactions between osmolality, largely determined by the total concentration of salts in the medium, and amino acids that can function as organic osmolytes. In media with high salt contents and high osmolality, organic osmolytes such as glycine must be present to offset the detrimental effects of the intracellular inorganic ions that would otherwise be required to maintain cell volume. However, it is also possible for embryos to develop in media without organic osmolytes if the total osmolality is kept below the threshold at which deleterious effects would occur. Thus, there is not one optimal osmolality for in vitro embryo development. Rather, there is a set of media that contain organic osmolytes used by early preimplantation embryos, that support development at higher osmolality, and a separate set of media devoid of organic osmolytes that support development only at lower osmolality.

1.7. Conclusion

The salts that are used to formulate current embryo culture media, including for human embryos, are those that were originally used in early somatic cell and tissue culture media. Changes to the concentrations of these salts were a key part of the advance that led to our ability to support preimplantation embryo development from fertilized eggs to blastocysts in vitro. In general, increased osmolality is detrimental to early preimplantation embryos, likely because they accumulate intracellular inorganic ions to a level that is deleterious, in order to maintain their cell volume. This has been established to be a major cause of the developmental blocks that occurred in classic embryo culture media, including for human embryos, at osmolalities similar to those of their natural in vivo environment. Embryos normally use organic osmolytes, the major one of which is glycine, to replace a portion of inorganic ions. When organic osmolytes are provided, preimplantation embryos can develop at osmolalities in vitro that are the same or higher than those in vivo.

Thus, compounds that function as organic osmolytes are a key component of most current culture media.

Acknowledgments

The following people kindly provided helpful information on the commercial human embryo culture media described in this chapter: Wayne Caswell (Irvine Scientific), Björn Comhaire (FertiPro NV), Rasmus Kiil-Nielsen (ORIGIO), Aidan McMahon (Cook Medical, Ireland), Laura Mena (Irvine Scientific), Susie Oliver (VitroLife), Thomas Pool (Fertility Centre of San Antonio), Francis Tekpetey (London Health Science Centre, Ontario), Patrick Quinn (CooperSurgical), and Jason Swain (University of Michigan).

References

1. Biggers JD (1998) Reflections on the culture of the preimplantation embryo. Int J Dev Biol 42:879–884
2. Whitten WK (1956) Culture of tubal mouse ova. Nature 177:96
3. Giroux EL, Henkin RI (1973) Macromolecular ligands of exchangeable copper, zinc and cadmium in human serum. Bioinorg Chem 2:125–133
4. Burton, RF (1975) Appendix A: osmolality and osmotic coefficients. In: Ringer solutions and physiological salines. Wright Scientechnica, Bristol
5. Brachet A (1912) Développement in vitro de blastodermes et de jeunes embryons de Mammifères. C R Acad Sci Paris 155:1191–1193
6. Lewis WH, Hartmann CG (1933) Early cleavage stages of the egg of the monkey (Macacus rhesus). Contrib Embryol (Carnegie Inst Wash) 24:187–201
7. Alexandre H (2001) A history of mammalian embryological research. Int J Dev Biol 45:457–467
8. Pincus G (1936) The eggs of mammals. MacMillan, New York, NY
9. Krebs HA, Henseleit K (1932) Untersuchen über die Harnstoffbildung im Tierkörper. Z Phys Chem 210:33–66
10. McLaren A, Biggers JD (1958) Successful development and birth of mice cultivated in vitro as early as early embryos. Nature 182:877–878
11. Whitten WK (1957) Culture of tubal ova. Nature 179:1081–1082
12. Biggers JD, Whittingham DG, Donahue RP (1967) The pattern of energy metabolism in the mouse oocyte and zygote. Proc Natl Acad Sci USA 58:560–567
13. Whittingham DG, Biggers JD (1967) Fallopian tube and early cleavage in the mouse. Nature 213:942–943
14. Whittingham DG (1971) Culture of mouse ova. J Reprod Fertil Suppl 14:7–21
15. Hogan B et al (1994) Manipulating the mouse embryo: a laboratory manual. Cold Spring Harbor Press, Plainview, NY
16. Baltz JM, Tartia AP (2010) Cell volume regulation in oocytes and early embryos: connecting physiology to successful culture media. Hum Reprod Update 16:166–176
17. McKiernan SH, Clayton MK, Bavister BD (1995) Analysis of stimulatory and inhibitory amino acids for development of hamster one-cell embryos in vitro. Mol Reprod Dev 42:188–199
18. Quinn P, Kerin JF, Warnes GM (1985) Improved pregnancy rate in human in vitro fertilization with the use of a medium based on the composition of human tubal fluid. Fertil Steril 44:493–498
19. Goddard MJ, Pratt HPM (1983) Control of events during early cleavage of the mouse embryo: an analysis of the '2-cell block'. Development 73:111–133
20. Schini SA, Bavister BD (1988) Two-cell block to development of cultured hamster embryos is caused by phosphate and glucose. Biol Reprod 39:1183–1192
21. Kishi J et al (1991) Block to development in cultured rat 1-cell embryos is overcome using medium HECM-1. Hum Reprod 6: 1445–1448
22. Camous S et al (1984) Cleavage beyond the block stage and survival after transfer of early

bovine embryos cultured with trophoblastic vesicles. J Reprod Fertil 72:479–485
23. Bolton VN et al (1989) Development of spare human preimplantation embryos in vitro: an analysis of the correlations among gross morphology, cleavage rates, and development to the blastocyst. J In Vitro Fertil Embryo Transfer 6:30–35
24. Chatot CL et al (1989) An improved culture medium supports development of random-bred 1-cell mouse embryos in vitro. J Reprod Fertil 86:679–688
25. Lawitts JA, Biggers JD (1991) Optimization of mouse embryo culture media using simplex methods. J Reprod Fertil 91:543–556
26. Lawitts JA, Biggers JD (1992) Joint effects of sodium chloride, glutamine, and glucose in mouse preimplantation embryo culture media. Mol Reprod Dev 31:189–194
27. Biggers JD, Lawitts JA, Lechene CP (1993) The protective action of betaine on the deleterious effects of NaCl on preimplantation mouse embryos in vitro. Mol Reprod Dev 34:380–390
28. Lawitts JA, Biggers JD (1993) Culture of preimplantation embryos. Methods Enzymol 225:153–164
29. Erbach GT et al (1994) Differential growth of the mouse preimplantation embryo in chemically defined media (published erratum appears in Biol Reprod 1994 Aug;51(2):345). Biol Reprod 50:1027–1033
30. Gardner DK, Lane M (1996) Alleviation of the '2-cell block' and development to the blastocyst of CF1 mouse embryos: role of amino acids, EDTA and physical parameters. Hum Reprod 11:2703–2712
31. Ho Y et al (1995) Preimplantation development of mouse embryos in KSOM: augmentation by amino acids and analysis of gene expression. Mol Reprod Dev 41:232–238
32. Biggers JD, McGinnis LK, Raffin M (2000) Amino acids and preimplantation development of the mouse in protein- free potassium simplex optimized medium. Biol Reprod 63:281–293
33. Lane M, Gardner DK (1997) Differential regulation of mouse embryo development and viability by amino acids. J Reprod Fertil 109:153–164
34. Edwards RG (1981) Test-tube babies, 1981. Nature 293:253–256
35. Borland RM et al (1980) Elemental composition of fluid in the human Fallopian tube. J Reprod Fertil 58:479–482
36. Lippes J et al (1972) The collection and analysis of human fallopian tubal fluid. Contraception 5:85–103
37. Gardner DK, Lane M (1999) Embryo culture systems. In: Trounson AO, Gardner DK (eds) Handbook of in vitro fertilization. CRC, Boca Raton, FL
38. Kaplan JH (2002) Biochemistry of Na, K-ATPase. Annu Rev Biochem 71:511–535
39. Hediger MA et al (2004) The ABCs of solute carriers: physiological, pathological and therapeutic implications of human membrane transport proteins: introduction. Pflugers Arch 447: 465–468
40. Devreker F, Hardy K (1997) Effects of glutamine and taurine on preimplantation development and cleavage of mouse embryos in vitro. Biol Reprod 57:921–928
41. Baltz JM (2001) Osmoregulation and cell volume regulation in the preimplantation embryo. In: Schatten GP (ed) Current topics in developmental biology. Academic, San Diego, CA
42. Dawson KM, Baltz JM (1997) Organic osmolytes and embryos: substrates of the Gly and beta transport systems protect mouse zygotes against the effects of raised osmolarity. Biol Reprod 56:1550–1558
43. Hadi T et al (2005) Similar effects of osmolarity, glucose, and phosphate on cleavage past the 2-cell stage in mouse embryos from outbred and F1 hybrid females. Biol Reprod 72:179–187
44. Collins JL, Baltz JM (1999) Estimates of mouse oviductal fluid tonicity based on osmotic responses of embryos. Biol Reprod 60:1188–1193
45. Fiorenza MT et al (2004) Early transcriptional activation of the hsp70.1 gene by osmotic stress in one-cell embryos of the mouse. Biol Reprod 70:1606–1613
46. Li R et al (2007) Concentration and composition of free amino acids and osmolalities of porcine oviductal and uterine fluid and their effects on development of porcine IVF embryos. Mol Reprod Dev 74:1228–1235
47. Waring DW (1976) Rate of formation and osmolality of oviductal fluid in the cycling rat. Biol Reprod 15:297–302
48. Williams N, Kraft N, Shortman K (1972) The separation of different cell classes from lymphoid organs. VI. The effect of osmolarity of gradient media on the density distribution of cells. Immunology 22:885–899
49. Waymouth C (1970) Osmolality of mammalian blood and of media for culture of mammalian cells. In Vitro 6:109–127
50. Knudsen JF et al (1979) Follicular fluid electrolytes and osmolality in cyclic pigs. J Reprod Fertil 57:419–422
51. Anon (1980) Normal reference laboratory values. N Engl J Med 302:37–48
52. Hallows KR, Knauf PA (1994) Principles of cell volume regulation. In: Strange K (ed) Cellular

and molecular physiology of cell volume regulation. CRC, Boca Raton, FL
53. Hoffmann EK, Lambert IH, Pedersen SF (2009) Physiology of cell volume regulation in vertebrates. Physiol Rev 89:193–277
54. Alexander RT, Grinstein S (2006) Na+/H+ exchangers and the regulation of volume. Acta Physiol (Oxf) 187:159–167
55. Jiang L, Chernova MN, Alper SL (1997) Secondary regulatory volume increase conferred on Xenopus oocytes by expression of AE2 anion exchanger. Am J Physiol 272: C191–C202
56. Yancey PH et al (1982) Living with water stress: evolution of osmolyte systems. Science 217:1214–1222
57. Lang F, Busch GL, Volkl H (1998) The diversity of volume regulatory mechanisms. Cell Physiol Biochem 8:1–45
58. Yancey PH (1994) Compatible and counteracting solutes. In: Strange K (ed) Cellular and molecular physiology of cell volume regulation. CRC, Boca Raton, FL
59. Van Winkle LJ, Haghighat N, Campione AL (1990) Glycine protects preimplantation mouse conceptuses from a detrimental effect on development of the inorganic ions in oviductal fluid. J Exp Zool 253:215–219
60. Hammer MA, Baltz JM (2003) Beta-alanine but not taurine can function as an organic osmolyte in preimplantation mouse embryos cultured from fertilized eggs. Mol Reprod Dev 66:153–161
61. Steeves CL et al (2003) The glycine neurotransmitter transporter GLYT1 is an organic osmolyte transporter regulating cell volume in cleavage-stage embryos. Proc Natl Acad Sci USA 100:13982–13987
62. Tartia AP et al (2009) Cell volume regulation is initiated in mouse oocytes after ovulation. Development 136:2247–2254
63. Hammer MA et al (2000) Glycine transport by single human and mouse embryos. Hum Reprod 15:419–426
64. Lewis AM, Kaye PL (1992) Characterization of glutamine uptake in mouse two-cell embryos and blastocysts. J Reprod Fertil 95:221–229
65. Kwon HM, Handler JS (1995) Cell volume regulated transporters of compatible osmolytes. Curr Opin Cell Biol 7:465–471
66. Pastor-Anglada M et al (1996) Long-term osmotic regulation of amino acid transport systems in mammalian cells. Amino Acids 11:135–151
67. Anas MKI et al (2007) The organic osmolytes betaine and proline are transported by a shared system in early preimplantation mouse embryos. J Cell Physiol 210:266–277
68. Anas MK et al (2008) SIT1 is a betaine/proline transporter that is activated in mouse eggs after fertilization and functions until the 2-cell stage. Development 135:4123–4130

Chapter 6

Media Composition: Energy Sources and Metabolism

Deirdre Zander-Fox and Michelle Lane

Abstract

The preparation of defined culture media for embryo development has progressed from simple chemically defined media based on Krebs–Ringer bicarbonate, supplemented with glucose, bovine plasma albumin, antibiotics and utilizing a CO_2-bicarbonate buffering system to more complete systems based around studies on the physiology and metabolism of the mammalian embryo. Although the concentration of substrates used in media can vary, there are many components that are quintessentially important for embryo development such as energy sources, that play a vital role in regulation of metabolism and hence viability. Here we describe the role of energy substrates within culture media and outline assays which can be utilized to measure embryo metabolism as a mechanism for determining embryo physiology and viability.

Key words: Glucose, Pyruvate, Lactate, Amino acid, Developmental competence

1. Introduction

In vivo, the preimplantation embryo develops in a complex and constantly changing environment containing a variety of nutrients, including pyruvate, lactate and glucose, amino acids, ions, growth factors and macromolecules, as the embryo transverses from the oviduct into the uterus in preparation for implantation (1, 2). Concomitantly, with this change in environment the embryo undergoes major changes in physiology. Preimplantation embryo development is frequently discussed in two phases: precompaction and postcompaction. These stages roughly correlate with the time at which the embryo is in the oviduct and when it makes the transition into the uterus, respectively (2). The precompaction stage embryo is undifferentiated and divides repeatedly however displays no net growth, and genetic control relies largely on maternal mRNA before the embryonic genome is fully activated, which in the human, occurs at the 4- to 8-cell stage (2, 3). At these early

Gary D. Smith et al. (eds.), *Embryo Culture: Methods and Protocols*, Methods in Molecular Biology, vol. 912,
DOI 10.1007/978-1-61779-971-6_6, © Springer Science+Business Media, LLC 2012

Table 1
The nutrient concentration in human oviduct and uterine fluid

	Pyruvate (mM)	Lactate (mM)	Glucose (mM)	Amino acid
Oviduct (midcycle)	0.32	10.5	0.5	Primarily nonessential (alanine, aspartate, glutamate, glycine, serine, taurine)
Uterus	0.10	5.87	3.15	Mix of nonessential and essential

Data obtained from: (5, 7)

stages, metabolic activity in terms of oxygen consumption, DNA/RNA synthesis and protein synthesis is low (4). In comparison, the later stages of development are regulated by the embryonic genome, and metabolism increases dramatically (2). The oocyte and early embryo have the metabolic quotient similar to that of brain tissue (QO_2 of 4 µg/mg dry weight/h). In comparison, the blastocyst has a metabolic quotient similar to tumor cells and has increased six-fold compared to the early stages of development (4). This increase in energy consumption is due to the increase in metabolically costly processes such as compaction and blastocoel cavity generation.

Interestingly, it appears that these changes in metabolism mirror changes in composition of the reproductive tract especially changes between the oviduct and the uterus, where there are gradients in carbohydrates, pH, oxygen, amino acid content as well as differences in many growth factors and proteins (Table 1) (5–9).

1.1. Human Embryo Metabolism

Metabolic activity and regulation is important in all cells for the generation of ATP, production of intermediates for biosynthesis as well as providing substrates for processes such as acetylation and methylation. The metabolic state of any cell is an important determinate of cell health and viability. The mammalian embryo is no different, as the ability to regulate its metabolism is directly related to its ability to implant and form a viable pregnancy (10). Interestingly the mammalian embryo can adapt to its environment however this plasticity can come at the expense of viability. Loss of control of the balance between metabolic pathways, which compete for intermediates, can result in the embryo exhausting substrate supplies, thereby reducing the capacity for production of ATP. This is best highlighted in the precompaction stage embryo where uncontrolled high rates of metabolism and use of nonpreferred substrates, such as precocious glucose metabolism, are associated with a loss of viability, while a quieter more controlled and balanced metabolism is associated with maintenance of viability (11).

Therefore, it is important to think about energy metabolism by the embryo as a balancing act between different pathways,

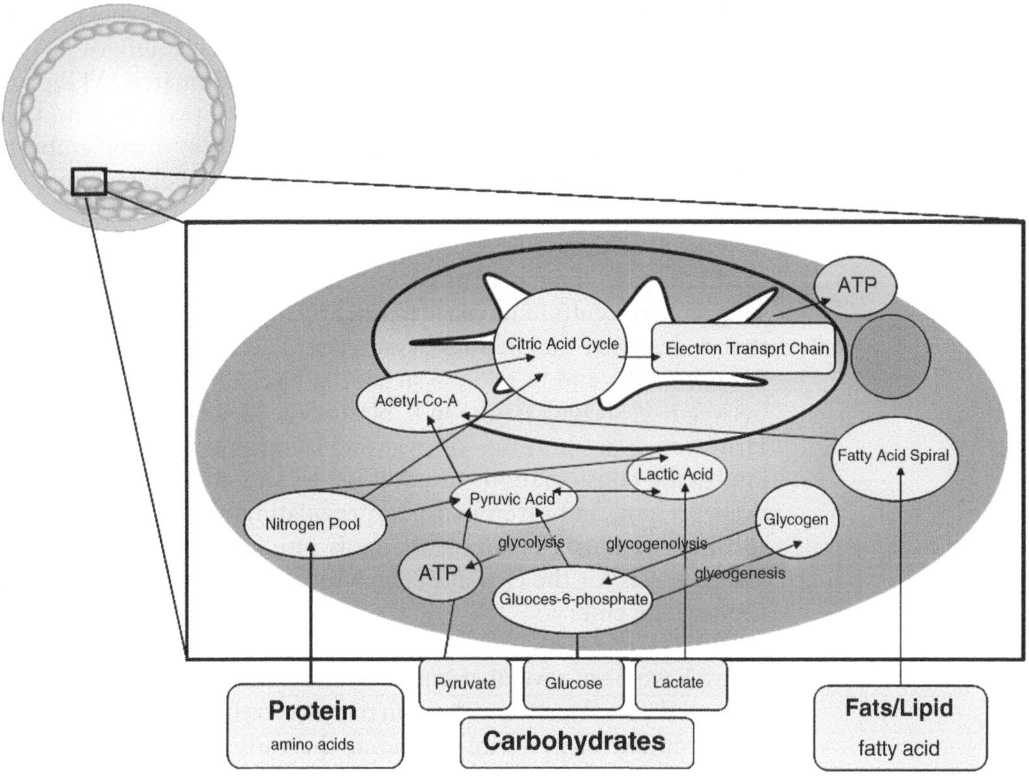

Fig. 1. A diagram depicting the metabolic flow and interaction of the three crucial energy substrate groups for the developing embryo.

where substrates cannot be thought of in isolation, as any shift in metabolism by one pathway can have an effect downstream on another (Fig. 1).

For the purposes of this review we have divided metabolism of the human embryo into four substrates/pathways:

1. Pyruvate and lactate.
2. Glucose.
3. Amino acids.
4. Fatty acids/β-oxidation.

1.2. Pyruvate and Lactate

1.2.1. Basic Discovery

Studies from the 1960s and 1970s were the first to establish the importance of pyruvate as an energy source for the precompaction stage embryo (12). Pyruvate is oxidized directly by mitochondria via oxidative phosphorylation to yield ATP, and the early embryo can only utilize this mechanism to derive energy. Lactate can be an important energy source supporting development from the 2-cell stage. Pyruvate and lactate metabolism in animal models has been shown to act in balance, with optimal development in vitro occurring when both pyruvate and lactate were present in the media compared to either energy source alone (13). However, excess concentrations of lactate in the medium can alter the rate of pyruvate oxidation and therefore the relative availability of NAD^+ and NADH (redox potential) in the cell and ultimately ATP production (14). Therefore, the balance or ratio of pyruvate and lactate in the medium can alter metabolic regulation of the embryo affecting ATP output and therefore developmental competence.

1.2.2. Human Embryo Culture Media

As a result of these studies, pyruvate is a key component of culture media for the human embryo and its concentration in cleavage stage embryo culture media is frequently higher than that in media that is designed for the blastocyst stage. Lactate also is present in culture media and in many systems the levels of lactate also varies between cleavage stage and blastocyst stage culture media. However, it is the ratio of pyruvate:lactate in media that is an important consideration, in addition to the absolute concentrations present, and given its role in regulating redox potential and substrate fate it is common that this ratio is increased in media formulations for the cleavage stage embryo compared to the blastocyst stage embryo.

1.2.3. Viability Assessment

There have been several attempts in the human to correlate pyruvate uptake and lactate production to embryo viability. Interestingly, measurement of pyruvate uptake at the early cleavage stages has determined an inverse relationship between pyruvate uptake by 2- to 8-cell embryos and subsequent pregnancy (15). It was also discovered that fertilized embryos which arrested at the cleavage stage had significantly lower uptake of pyruvate and decreased lactate production at all stages measured than those that developed to the blastocyst (16, 17). However, in all of these studies to date substrate uptake only was assessed and therefore it maybe that, similar to animal models, the metabolic fate of the pyruvate is an important consideration in addition to the raw uptake, as high levels of uptake may indicate inappropriate or inefficient use of the substrate by the embryo.

In contrast, at the postcompaction stage when there is an increasing demand for ATP and biosynthetic activity, embryos with higher levels of pyruvate uptake were associated with increased rates of blastocyst development (18). However, once again the actual metabolic fate of the substrate was not assessed.

1.3. Glucose

1.3.1. Basic Discovery

Glucose is used by almost all mammalian cells as the primary energy source for the production of ATP. Glucose is taken up by cells and converted to pyruvate by a series of reactions in the cytosol by the Embden-Meyerhof pathway. This results in the production of two ATP molecules. Pyruvate is then converted to acetyl-CoA and enters the TCA cycle and, in the presence of oxygen, is oxidized to produce CO_2 and H_2O within the mitochondria. Complete oxidation

Table 2
Effect of culture in mMTF medium on glucose uptake and lactate production by mouse blastocysts (27)

	Glucose uptake (pmol/e/h)	Lactate production (pmol/e/h)	Glycolytic activity (%)
Control	4.89 ± 0.29	2.71 ± 0.34	27.4 ± 2.5
6 h stress	4.55 ± 0.56	8.11 ± 0.59*	93.9 ± 8.5*

$n = 20$ blastocysts, data expressed as mean ± SEM
*Like pairs are significantly different $p < 0.05$

of glucose produces 38 ATP molecules, 2 in glycolysis and 36 in the TCA cycle.

Studies on animal embryos have demonstrated that the zygote and 2-cell mouse embryo cannot use glucose as an energy source (19–21). In contrast the 8-cell mouse embryo will develop in the presence of glucose alone and is the preferred substrate at the blastocyst stage (22–24). It has also been suggested that exposure to glucose is essential for blastocyst development (25, 26). Further, it has been shown in animal embryos that it is the pathway that glucose follows that is essential to viability. Exposure of mouse blastocysts to a medium lacking key regulators (mMTF medium) for 6 h did not alter glucose uptake of the blastocysts however the fate of the glucose was significantly altered with reduced oxidation by the TCA cycle and electron transport chain. Instead metabolism was redirected to cytoplasmic glycolytic conversion to lactate severely compromising ATP production although glucose uptake was not affected (Table 2) (27).

Glucose uptake in the human embryo mirrors what has been reported in other animal models with uptake initially low however rising at the stage of embryonic genome activation (4- to 8-cell stage) and increasing exponentially throughout blastocyst development (17).

1.3.2. Human Embryo Culture Media

Given the data demonstrating that exposure to glucose is essential to blastocyst development, even for as short a time period as a few minutes (25, 26), glucose is a key ingredient in all media for the development of the human blastocyst. The need for glucose in the media for the precompaction stage embryos is less convincing however given its role in the pentose-phosphate pathway and for other pathways of biosynthesis, most media systems have glucose present throughout development, albeit at a lower concentration in media designed specifically for the cleavage stage embryo (16).

1.3.3. Viability Assessment

Studies in animal models from several species have correlated both glucose uptake and more recently glucose metabolic fate with embryo viability. In initial studies in the bovine, it was determined that blastocysts with >5 μg/h uptake developed better than those with uptake below this value (28). In the mouse model it was determined that nonviable embryos had significantly lower rates of glucose uptake than those embryos that went on to form live young (29) and that embryos with balanced glucose metabolism were able to be prospectively selected to improve implantation rates (10).

Similarly, the glucose uptake of the human embryo has been correlated to quality. Arrested embryos are unable to undergo the switch to glucose-based metabolism on day 4 compared to embryos that subsequently reached the blastocyst stage (17, 18). Polyspermic and parthenogenetic embryos also have significantly lower glucose uptake on day 5.5 compared to high quality embryos (16). Further, on both day 5 and day 6, high quality blastocyst (those with well defined inner cell mass and trophectoderm) had significantly higher glucose uptake than those with a lower morphological appearance (18). Glucose uptake following blastocyst transfer has also been shown to be strongly correlated to viability as blastocysts that formed a viable pregnancy had elevated glucose uptake compared to those that failed to implant (30).

1.4. Amino Acids

1.4.1. Basic Discovery

Oviduct and uterine fluid contain significant levels of amino acids and the oocyte and embryo also contain specific transporters for amino acid uptake (8, 31, 32). Amino acids serve a variety of functions in the embryo; not only as energy substrates but also as osmolytes, antioxidants, chelators, regulators of pH as well as their importance for the synthesis of proteins and nucleotides (33–40).

Amino acid transport and metabolism play a vital role in embryo development with many studies on animal embryos including rodents (41–46), domestic animals (47–49) as well as the human (50, 51) demonstrating that their addition to culture media significantly improves development in vitro as well as viability. These studies have also demonstrated that there is a different requirement for amino acids as development proceeds from the cleavage stage to the postcompaction stage.

1.4.2. Human Embryo Culture Media

As a result of these studies on animals and humans which demonstrate such drastic improvements in embryo developmental competence when amino acids are included, all culture media for human embryos contain amino acids. The composition of the amino acids varies between systems, with the most common situation being a biphasic system, with media for cleavage stage embryos containing different amino acids (commonly alanine, aspartate, asparagine, glutamate, a stable derivative of glutamine, proline, serine, and taurine: bearing some similarity to Eagle's nonessential amino acids) (52) while those designed for the postcompaction embryo commonly contain a more complete array of amino acids more similar to medium designed for somatic cells (commonly Eagle's 20 amino acids) (52).

1.4.3. Viability Assessment

This understanding that amino acids are so critical for development in vitro has lead to a series of studies from Leese and colleagues to determine if amino acid metabolism and turnover by embryos could be used as a measure of embryo viability. Using HPLC-based technology to assess the spent culture media, they have demonstrated that cleavage stage embryos which have the potential to develop to the blastocyst stage exhibit a different amino acid profile than those which undergo developmental arrest (53). It has also been further demonstrated that amino acid turnover is associated with the ability of an embryo to implant and give rise to a clinical pregnancy and live birth (53, 54). However, although these early studies were promising there is currently little knowledge as to how an individual culture system may reflect the individual amino acids that are altered.

1.5. Fatty Acids

1.5.1. Basic Discovery

While much is known about the oocyte and embryos ability to metabolize exogenous nutrients such as carbohydrates and amino acids, the role of metabolism of endogenous stores of metabolites such as glycogen and lipid has been largely uninvestigated.

Previous studies have demonstrated that the oocyte and embryo contains endogenous lipid stores and the levels vary between species. This is thought to be linked to the amount of time between fertilization and embryo attachment with lower levels in the rodent and human compared to ruminants as attachment in ruminant species can occur weeks after fertilization (55). The role for lipid metabolism in embryo development is postulated to be multifaceted as lipids play important roles in hormone production, ion channel modulation, DNA polymerase inhibition, gene expression, and function as a nutrient source as well as being crucial in membrane lipid formation (56).

Ruminant studies have demonstrated that oocyte and embryo endogenous lipid content can be influenced by the composition of the culture medium (57, 58) with cow oocytes having increased triglyceride levels after culture in media containing fetal calf serum (59). There is also direct evidence of β-oxidation in cow and pig oocytes (59–61). Also studies on the human have demonstrated that both linoleic and palmitic acid can also be oxidized by the human embryo (56, 62).

However, there are limited studies showing that addition of different unsaturated free fatty acids or their derivatives improve development to the blastocyst in vitro, and to date there are no studies demonstrating an increase in pregnancy outcomes.

1.5.2. Human Embryo Culture Media

To date no culture media that is available commercially for the development of the human embryo contains fatty acids as a component. However, this may be somewhat misleading in that all media contain serum albumin as a protein source which has been reported to contain up to 2–3 mol of fatty acids per mole of protein (63).

1.5.3. Viability Assessment

There is increasing interest in role of fatty acid metabolism in oocytes and embryos, however the research in this area is in its infancy with further needed before fatty acids can be directly used as a marker for embryo selection in human IVF.

2. Materials

The most common methods for determining embryo viability in the embryology laboratory are the use of morphological scoring systems. Although this method is valuable, the interpretation of embryo morphology can be imprecise and highly varied not only between clinics but also between individual scientists. Therefore there is a need to develop more rigorous methods of embryo assessment that are both quantitative and noninvasive.

As mentioned above, metabolism is directly linked to embryo homeostasis and may be a valuable indicator of implantation potential. Microfluorescence techniques, developed in the 1980s, enabled the measurement of nutrient uptake from single cells and have been adapted for the embryo. The use of these microfluorescence assays provides a platform for the assessment of spent culture media using enzymatic analysis. Although many of these assays are performed using submicroliter droplets (10–20 nl), it is possible to up-scale these assays to assess nutrient uptake in spent culture media after 1–2 days culture of human embryos.

It is possible to use these assays to measure the levels of any substrate, carbohydrate, amino acid or fatty acid that can be linked to a linear change in a fluorescent tag.

The following methods explain how a microfluorometric technique can be used to analyze the uptake of a substrate from spent culture media and how this can be used to assist in embryo selection.

2.1. Preparation of Reagents

For the purposes of this chapter we have included pyruvate and glucose as examples for assessment however, this principle can be adapted for many other substrates.

2.1.1. Pyruvate: Assessed on Day 2 or 3 of Culture

Fluorometric assays are based on either the consumption or production of reduced pyridine nucleotides NADH or NADPH in coupled enzymatic reactions as these nucleotides fluoresce when excited and their oxidized forms, NAD^+ and $NADP^+$, do not.

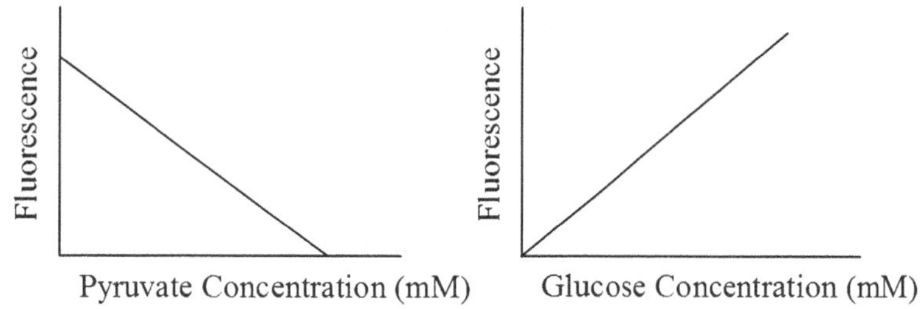

Fig. 2. Standard curve for pyruvate and glucose. Levels of pyruvate can be assessed by a decrease in fluorescence and levels of glucose can be assessed by a linear increase in fluorescence.

The change in fluorescence is linked to this reaction by the substrate to be measured.

Lactate dehydrogenase

Pyruvate + NADH + H^+ → → → Lactate + NAD^+

Therefore the amount of pyruvate in the media will be directly correlated to the amount of fluorescence i.e. increasing pyruvate levels will result in reducing fluorescence (Fig. 2).

EPPS buffer

1. 2.52 g EPPS.
2. 200 ml MilliQ water.
3. pH to 8.0 with 1 M NaOH.
4. Filter and store at 4 °C.

5 mM NADH

1. 17.7 mg NADH.
2. 5 ml MilliQ water.

Working solution

1. 14 ml of EPPS buffer.
2. 0.3 ml 5 mM NADH.
3. 220 IU lactate dehydrogenase (EC 1.1.1.27).
4. Pipette into eppendorf tubes in 500 μl aliquots and freeze at −20 °C.

Standard curve

Prepare a 1 mM pyruvate solution (add 0.0110 g pyruvate to 100 ml Milli-Q water). Filter and store at 4 °C.

Serially dilute this 1 mM solution to give 0.5, 0.25, 0.125 and 0.0625 mM as well as zero water controls. These standards should be prepared weekly, stored at 4 °C and can be used to generate a standard curve.

2.1.2. Glucose: Assessed on Day 4/5/6 of Culture

The glucose assay works on the same principal as the pyruvate assay however the amount of glucose is related to a linear increase in fluorescence.

Hexokinase

Glucose + ATP → → Glucose-6-phosphate + ATP

Glucose-6-phosphate dehydrogenase

Glucose-6-phosphate + $NADP^+$ ↔ 6-phosphogluconate + NADPH + H^+

Therefore as the amount of glucose in the media increases there will be an increase in fluorescence (Fig. 2).

Epps buffer

As above.

5 mM dtt

1. 7.72 mg dtt.
2. 10 ml MilliQ water.

37 nM $MgSO_4$

1. 91.2 mg $MgSO_4$.
2. 10 ml MilliQ water.

10 mM ATP

1. 30.3 mg ATP.
2. 5 ml MilliQ water.

10 mM NADP

1. 39.4 mg NADP.
2. 5 ml MilliQ water.

Working solution

1. 15 ml EPPS buffer.
2. 2 ml 5 mM dtt.
3. 2 ml 37 nM $MgSO_4$.
4. 1 ml 10 mM ATP.
5. 3 ml 10 mM NADP.

6. 340 IU hexokinase (EC 2.7.1.1)/170 IU glucose-6-phosphate dehydrogenase (EC 1.1.1.49).
7. Pipette into eppendorf tubes in 500 μl aliquots and freeze at −20 °C.

Standard curve

As above except prepare a 1 mM glucose solution (0.018 g glucose in 100 ml MilliQ water).

3. Methods

In order to link nutrient uptake from spent culture media to embryo viability embryos must be cultured individually. It is advisable to maintain a consistent embryo:volume ratio that is recommended by the manufacturer to ensure that embryos have sufficient substrate for development. This is especially important if the composition (mM) of the medium is not disclosed by the manufacturer as you could inadvertently perturb embryo development by exhausting an essential substrate.

3.1. Preparation of Assay

1. Remove pyruvate/glucose cocktail from −20 °C and allow thawing on ice. Use a fresh aliquot of cocktail each day.
2. Ensure that the fluorescent microscope is switched on (and that it has been calibrated for a uniform signal across the objective) and ready as well as the imaging program.
3. Label duplicate (*a* and *b*) eppendorf tubes with 0, 0.125, 0.25, 0.5, and 1 for the standard curve.
4. Initially run a standard curve. Aliquot 9 μl of working reagents into the bottom of the labeled eppendorfs being careful to ensure the entire cocktail is sitting in the bottom of the eppendorf and no small drops are on the side of the tube.
5. Obtain the pyruvate/glucose standards (0, 0.125, 0.25, 0.5, and 1 mM) from the fridge (standards should be prepared weekly).
6. Ensure the standards are mixed well and add 1 μl of each standard to the 9 μl cocktail in the corresponding labeled eppendorf and start timer (make sure that the standard is added directly to the cocktail, and is not sitting on the side of the tube and ensure the standard is mixed well throughout the cocktail).
7. Remove 2× glass slide (non polylysine as the polylysine coating can be uneven and result in uneven fluorescence) from the box and clean to remove any dust, etc.

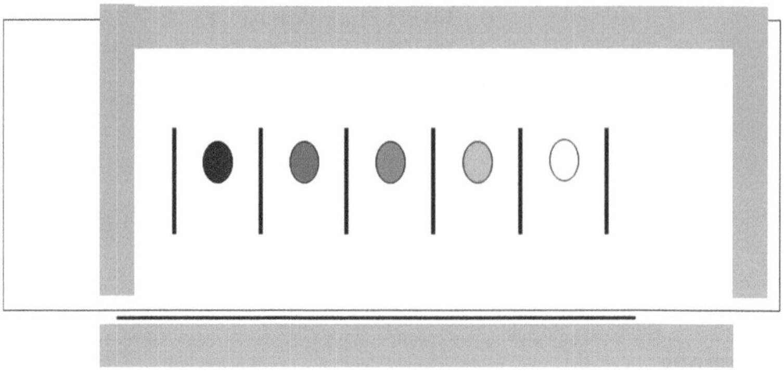

Fig. 3. Schematic depicting the suggested layout of the assay slide.

8. Place a thin layer of aluminum tape around the edge of each of the slides to create a barrier so that the oil does not run off. Draw black lines on the slide to indicate which drop is which standard and using a glass pipette and syringe, place a thin strip of heavy mineral oil (Sigma) on the slide (Fig. 3).
9. After 8 min, pipette the pyruvate/glucose and cocktail mixture up and down to ensure it is mixed well and place a 1 μl drop of each standard/cocktail mix (*a*) onto the slide under the oil, one drop between each black line. Repeat with the duplicate standard/cocktail mix (*b*) on the second slide. Ensure that the drops are adhered to the bottom of the slide and are a uniform shape. The size of the drop is important as a larger, flatter drop will fluoresce differently to a smaller, higher drop. If the drops look different in size, you will need to pipette another drop onto the slide.
10. Image the drops after 10 min using a microscope with a mercury lamp with excitation wavelengths in the UV range (340 nm).

    Images can be captured by either a microscope with photometer attachment (to capture fluorescence and convert to a numerical value based on the standard curve data) or with a CCD camera. It is important that in both cases there is an identical time for the excitation (usually 200–500 ms).

    For systems utilizing a camera with software such as analysis LS Research software (Olympus), the mean levels of fluorescence of each drop can be quantified utilizing image analysis software such as Image J (NIH: http://imagej.nih.gov/ij/).
11. After the standard curve has been analyzed, it can be accepted if the $R^2 > 0.98$ and the standard error of duplicates is <5 %. If the standard curve is not accepted, it must be run again before any patient samples are run.

3.2. Pyruvate/Glucose Assay Patient Samples

1. Dilute patients spent media into a labeled eppendorfs if necessary. The dilution will depend on which substrate and on the composition of the media that is used. This will likely vary between manufacturers (when removing the aliquot of media using a 1–10 μl pipette, carefully wipe the pipette tip on the edge of the dish to remove oil before pipetting media into the water to dilute) (see Note 1).
2. Also remove the same volume of media from the control drop and dilute in the same manner. This will give you a control level of pyruvate/glucose which will be used to calculate what pyruvate/glucose was taken up by the embryo (see Note 2).
3. Run the assay as instructed above adding 1 μl of media (neat or diluted as above) to 9 μl of cocktail, incubating and assaying on the glass slide.
4. Run the patients samples through the assay, one patient at a time, with a maximum of five samples at a time, using the above procedure for the standard curve (see Note 3).
5. Patient sample triplicate can be accepted if the standard error of duplicates is <5 %.
6. If the standard error of duplicates is >5 %, repeat.
7. The amount of pyruvate/glucose uptake from the spent culture media can then be calculated using the standard curve.
8. Calculate the uptake by the embryo by subtracting the value of the embryo spent medium from the control medium.
9. Embryos can then be ranked according to pyruvate/glucose uptake.

4. Notes

1. For accuracy, care must be taken in setting up of patient dishes.
   To accurately determine the uptake of energy sources such as pyruvate, glucose or amino acid from spent culture media, a degree of care must be taken when setting up culture dishes as even small variations in drop size between embryos will result in significant changes to nutrient levels within the drop after culture. Therefore, all culture dishes should be prepared utilizing an accurate pipette (such as a Gilson multi-pipette) and should immediately be covered with oil to prevent evaporation and alterations in osmolarity and concentrations of nutrients. Each dish should also contain a wash drop and two extra drops per dish as a nonembryo control.

2. Care must be used when putting embryos into the drops. Each embryo should be washed well in the same medium that the culture drops are made from to avoid carryover of components from a different medium. Each embryo should be added with minimal new media being transferred (<2 % of the drop size). Make sure that the control drops are not used for washing or culture. When you wish to perform the assay, move each embryo individually to a new culture dish. This must be done with great care as you do not want to transfer fresh media from the new drops into the spent drops that you wish to assay. Pick up the embryo from the drop you are planning on assaying ensuring that the majority of the drop volume is undisturbed. Move the embryo to a fresh dish, washing through the wash drop first. Pick up your next embryo and repeat the above step until all embryos are moved into your fresh dish. The spent culture media dish is now ready to be assayed.
3. These experiments can be utilized to detect alterations in uptake over short periods of time (h) or after longer culture periods (1–2 days). On the day of embryo transfer, embryos should be scored based on morphology. If multiple embryos exhibit the same level of morphology, the spent culture media can be assessed and used to determine the most advanced embryo from the cohort of morphologically identical embryos. At this stage the embryos should be moved to new culture media so the dish can be taken away for analysis. During the move due care must be take to minimize the amount of culture media removed from the drop or transferred in via the handling tip or the mouth pipette. Once the spent media has been analyzed and uptake has been determined a minimum and maximum uptake should also be calculated based on −10 % and +10 % of the actual uptake (this will take into account the assays limits). Overlapping embryos should be ranked equivalent.

References

1. Leese HJ (1988) The formation and function of oviduct fluid. J Reprod Fertil 82(2): 843–856
2. Leese HJ (1995) Metabolic control during preimplantation mammalian development. Hum Reprod Update 1(1):63–72
3. Brinster RL (1967) Protein content of the mouse embryo during the first five days of development. J Reprod Fertil 13(3):413–420
4. Leese HJ (1991) Metabolism of the preimplantation mammalian embryo. Oxf Rev Reprod Biol 13:35–72
5. Gardner DK et al (1996) Environment of the preimplantation human embryo in vivo: metabolite analysis of oviduct and uterine fluids and metabolism of cumulus cells. Fertil Steril 65(2):349–353
6. Fischer B, Bavister BD (1993) Oxygen tension in the oviduct and uterus of rhesus monkeys, hamsters and rabbits. J Reprod Fertil 99(2):673–679
7. Casslen BG (1987) Free amino acids in human uterine fluid. Possible role of high taurine concentration. J Reprod Med 32(3):181–184
8. Miller JG, Schultz GA (1987) Amino acid content of preimplantation rabbit embryos and fluids of the reproductive tract. Biol Reprod 36(1):125–129
9. Maas DH, Storey BT, Mastroianni L Jr (1977) Hydrogen ion and carbon dioxide content of

the oviductal fluid of the rhesus monkey (*Macaca mulatta*). Fertil Steril 28(9): 981–985

10. Lane M, Gardner DK (1996) Selection of viable mouse blastocysts prior to transfer using a metabolic criterion. Hum Reprod 11(9): 1975–1978
11. Lane M, Gardner DK (2005) Understanding cellular disruptions during early embryo development that perturb viability and fetal development. Reprod Fertil Dev 17(3):371–378
12. Biggers JD, Stern S (1973) Metabolism of the preimplantation mammalian embryo. Adv Reprod Physiol 6:1–59
13. Brinster RL (1965) Studies on the development of mouse embryos in vitro. IV. Interaction of energy sources. J Reprod Fertil 10(2):227–240
14. Wales RG, Whittingham (1973) The metabolism of specifically labelled lactate and pyruvate by two-cell mouse embryos. J Reprod Fertil 33(2):207–222
15. Conaghan J et al (1993) Selection criteria for human embryo transfer: a comparison of pyruvate uptake and morphology. J Assist Reprod Genet 10(1):21–30
16. Hardy K et al (1989) Non-invasive measurement of glucose and pyruvate uptake by individual human oocytes and preimplantation embryos. Hum Reprod 4(2):188–191
17. Gott AL et al (1990) Non-invasive measurement of pyruvate and glucose uptake and lactate production by single human preimplantation embryos. Hum Reprod 5(1):104–108
18. Gardner DK et al (2001) Noninvasive assessment of human embryo nutrient consumption as a measure of developmental potential. Fertil Steril 76(6):1175–1180
19. Biggers JD, Whittingham DG, Donahue RP (1967) The pattern of energy metabolism in the mouse oocyte and zygote. Proc Natl Acad Sci USA 58(2):560–567
20. Whitten WK (1957) Culture of tubal ova. Nature 179(4569):1081–1082
21. Brinster RL (1965) Studies on the development of mouse embryos in vitro. II. The effect of energy source. J Exp Zool 158:59–68
22. Whitten WK (1956) Culture of tubal mouse ova. Nature 177(4498):96
23. Brinster RL, Thomson JL (1966) Development of eight-cell mouse embryos in vitro. Exp Cell Res 42(2):308–315
24. Daniel JC Jr, Krishnan RS (1967) Amino acid requirements for growth of the rabbit blastocyst in vitro. J Cell Physiol 70(2):155–160
25. Martin KL, Leese HJ (1995) Role of glucose in mouse preimplantation embryo development. Mol Reprod Dev 40(4):436–443
26. Pantaleon M, Scott J, Kaye PL (2008) Nutrient sensing by the early mouse embryo: hexosamine biosynthesis and glucose signaling during preimplantation development. Biol Reprod 78(4):595–600
27. Lane M, Gardner DK (1998) Amino acids and vitamins prevent culture-induced metabolic perturbations and associated loss of viability of mouse blastocysts. Hum Reprod 13(4): 991–997
28. Renard JP, Philippon A, Menezo Y (1980) In-vitro uptake of glucose by bovine blastocysts. J Reprod Fertil 58(1):161–164
29. Gardner DK, Leese HJ (1987) Assessment of embryo viability prior to transfer by the noninvasive measurement of glucose uptake. J Exp Zool 242(1):103–105
30. Van den Bergh M et al (2001) Glycolytic activity: a possible tool for human blastocyst selection. Reprod Biomed Online 3(Suppl 1):8
31. Menezo Y, Laviolette P (1972) Amino constituents of tubal secretions in the rabbit. Zymogram—proteins—free amino acids. Ann Biol Anim Biochim Biophys 12(3): 383–396
32. Gardner DK, Leese HJ (1990) Concentrations of nutrients in mouse oviduct fluid and their effects on embryo development and metabolism in vitro. J Reprod Fertil 88(1):361–368
33. Epstein CJ, Smith SA (1973) Amino acid uptake and protein synthesis in preimplantation mouse embryos. Dev Biol 33(1):171–184
34. Lane M, Gardner DK (1997) Differential regulation of mouse embryo development and viability by amino acids. J Reprod Fertil 109(1):153–164
35. Van Winkle LJ, Haghighat N, Campione AL (1990) Glycine protects preimplantation mouse conceptuses from a detrimental effect on development of the inorganic ions in oviductal fluid. J Exp Zool 253(2):215–219
36. Dumoulin JC et al (1997) Taurine acts as an osmolyte in human and mouse oocytes and embryos. Biol Reprod 56(3):739–744
37. Nasr-Esfahani MH, Winston NJ, Johnson MH (1992) Effects of glucose, glutamine, ethylenediaminetetraacetic acid and oxygen tension on the concentration of reactive oxygen species and on development of the mouse preimplantation embryo in vitro. J Reprod Fertil 96(1):219–231
38. Edwards LJ, Williams DA, Gardner DK (1998) Intracellular pH of the mouse preimplantation embryo: amino acids act as buffers of intracellular pH. Hum Reprod 13(12):3441–3448
39. Dawson KM, Collins JL, Baltz JM (1998) Osmolarity-dependent glycine accumulation indicates a role for glycine as an organic osmolyte

in early preimplantation mouse embryos. Biol Reprod 59(2):225–232

40. Wu F, Cholewa B, Mattson DL (2000) Characterization of l-arginine transporters in rat renal inner medullary collecting duct. Am J Physiol Regul Integr Comp Physiol 278(6):R1506–R1512
41. Dumoulin JC et al (1992) Temporal effects of taurine on mouse preimplantation development in vitro. Hum Reprod 7(3):403–407
42. Gardner DK, Lane M (1993) Amino acids and ammonium regulate mouse embryo development in culture. Biol Reprod 48(2):377–385
43. Gardner DK, Lane M (1996) Alleviation of the '2-cell block' and development to the blastocyst of CF1 mouse embryos: role of amino acids, EDTA and physical parameters. Hum Reprod 11(12):2703–2712
44. Kane MT, Carney EW, Bavister BD (1986) Vitamins and amino acids stimulate hamster blastocysts to hatch in vitro. J Exp Zool 239(3):429–432
45. Bavister BD, Arlotto T (1990) Influence of single amino acids on the development of hamster one-cell embryos in vitro. Mol Reprod Dev 25(1):45–51
46. Carney EW, Bavister BD (1987) Stimulatory and inhibitory effects of amino acids on the development of hamster eight-cell embryos in vitro. J In Vitro Fertil Embryo Transfer 4(3):162–167
47. Gardner DK et al (1994) Enhanced rates of cleavage and development for sheep zygotes cultured to the blastocyst stage in vitro in the absence of serum and somatic cells: amino acids, vitamins, and culturing embryos in groups stimulate development. Biol Reprod 50(2):390–400
48. Takahashi Y, First NL (1992) In vitro development of bovine one-cell embryos: influence of glucose, lactate, pyruvate, amino acids and vitamins. Theriogenology 37(5):963–978
49. Steeves TE, Gardner DK (1999) Temporal and differential effects of amino acids on bovine embryo development in culture. Biol Reprod 61(3):731–740
50. Devreker F, Winston RM, Hardy K (1998) Glutamine improves human preimplantation development in vitro. Fertil Steril 69(2):293–299
51. Devreker F et al (1999) Effects of taurine on human embryo development in vitro. Hum Reprod 14(9):2350–2356
52. Eagle H (1959) Amino acid metabolism in mammalian cell cultures. Science 130(3373): 432–437
53. Houghton FD, Leese HJ (2004) Metabolism and developmental competence of the preimplantation embryo. Eur J Obstet Gynecol Reprod Biol 115(Suppl):S92–S96
54. Brison DR et al (2004) Identification of viable embryos in IVF by non-invasive measurement of amino acid turnover. Hum Reprod 19(10):2319–2324
55. Sturmey RG et al (2009) Role of fatty acids in energy provision during oocyte maturation and early embryo development. Reprod Domest Anim 44(Suppl 3):50–58
56. Haggarty P et al (2006) Fatty acid metabolism in human preimplantation embryos. Hum Reprod 21(3):766–773
57. Abe H et al (2002) Accumulation of cytoplasmic lipid droplets in bovine embryos and cryotolerance of embryos developed in different culture systems using serum-free or serum-containing media. Mol Reprod Dev 61(1): 57–66
58. Reis A et al (2003) Consequences of exposure to serum, with or without vitamin E supplementation, in terms of the fatty acid content and viability of bovine blastocysts produced in vitro. Reprod Fertil Dev 15(5):275–284
59. Ferguson EM, Leese HJ (1999) Triglyceride content of bovine oocytes and early embryos. J Reprod Fertil 116(2):373–378
60. Kim JY et al (2001) Lipid and fatty acid analysis of fresh and frozen-thawed immature and in vitro matured bovine oocytes. Reproduction 122(1):131–138
61. Sturmey RG, Leese HJ (2003) Energy metabolism in pig oocytes and early embryos. Reproduction 126(2):197–204
62. Matorras R et al (1998) Fatty acid composition of fertilization-failed human oocytes. Hum Reprod 13(8):2227–2230
63. Chen RF (1967) Removal of fatty acids from serum albumin by charcoal treatment. J Biol Chem 242(2):173–181

Chapter 7

Media Composition: Amino Acids and Cellular Homeostasis

Franchesca D. Houghton

Abstract

Amino acids are beneficial for the developing preimplantation embryo and therefore form an important component of culture media. This chapter will critically review the importance of amino acids for preimplantation embryos and the impact of this research for the development of sequential culture media used in many assisted conception units. The advantages of culturing embryos in a full complement of amino acids, at close to physiological concentrations will be considered. Moreover, the noninvasive measurement of amino acid turnover by individual embryos, a method which holds great promise to assess developmental competency prior to transfer, will also be discussed. Thus, this chapter highlights the fundamental role of amino acids for the metabolic and homeostatic regulation of the preimplantation embryo.

Key words: Amino acids, Homeostasis, Preimplantation embryo development, Embryo culture medium, IVF, Assisted Reproduction, Embryo selection

1. Introduction

Amino acids are an important component of human embryo culture media. They form the building blocks of proteins, may be used as a potential energy source and have numerous other functions such as chelating heavy metals, as osmolytes, in pH regulation, to name a few. It is therefore not surprising that with so many intrinsic roles, amino acids are beneficial for preimplantation embryo development and thus routinely added to culture medium. This chapter will critically review the current literature regarding

Gary D. Smith et al. (eds.), *Embryo Culture: Methods and Protocols*, Methods in Molecular Biology, vol. 912, DOI 10.1007/978-1-61779-971-6_7, © Springer Science+Business Media, LLC 2012

the inclusion and role of amino acids in embryo development. Due to limitations in the ability to perform large-scale studies to test varying concentrations and combinations of culture medium components on human embryos, much of our knowledge has been obtained using animal models. Thus, although perhaps not ideal, many of the nutrients included in human embryo culture medium are present at concentrations based on results obtained largely from animal studies. However, where possible, data pertinent to human embryos will be highlighted and compared to that of animal models.

2. Amino Acid Classification

Amino acids can be categorized as being either essential, or nonessential (Table 1). Essential amino acids are unable to be synthesized de novo and must be obtained from the diet, whereas nonessential amino acids are not required in the diet as they can be synthesized in vivo. Although widely used, this classical classification should be used cautiously since in humans some nonessential amino acids have been found to be essential under certain situations and disease states. These amino acids are called conditionally essential and must be obtained from the diet (1, 2).

Table 1
Classification of essential, nonessential, and conditionally essential (C) amino acids

Essential	Nonessential
Histidine	Alanine
Isoleucine	Arginine (C)
Leucine	Asparagine
Lysine	Aspartic acid
Methionine	Cysteine (C)
Phenylalanine	Glutamic acid
Threonine	Glutamine (C)
Tryptophan	Glycine (C)
Valine	Proline (C)
	Serine (C)
	Tyrosine (C)

3. Amino Acids: Medium Composition and Effect on Embryo Development

One of the earliest reports to describe the importance of amino acids for embryo development found that a fixed nitrogen source, either from BSA or amino acids, was required for 2-cell mouse embryos to develop to the blastocyst stage (3). Subsequently, it was found that ten amino acids (arginine, histidine, leucine, lysine, methionine, phenylalanine, serine, threonine, tryptophan, and valine) were essential for rabbit blastocyst development in vitro (4).

During the early 1990s a series of seminal publications proved to be significant for the future development of human embryo culture medium. Initial studies investigated the effect of adding either essential, nonessential, or a complete mixture of amino acids to murine embryo culture medium. Somewhat surprisingly, it was found that embryo development in terms of blastocyst formation, cell number, and hatching rate were significantly increased by the presence of only nonessential amino acids compared to embryos cultured in the presence of all 20 amino acids (5). Moreover, mouse zygotes cultured for 48 h (to the 6- to 8-cell stage) in nonessential amino acids prior to transfer supported the greatest rate of implantation and fetal development. However, when zygotes were cultured to the blastocyst stage for 93 h prior to transfer, the greatest rate of implantation was observed when all 20 amino acids were present in the medium (6). This suggests that embryos have a differential requirement for amino acids as development progresses. Initially, it may appear that cleavage stage embryos favor the inclusion of nonessential amino acids whereas all amino acids are preferred for blastocyst formation. However, this data should be considered carefully since:

1. The segregation of amino acids into essential or nonessential is largely academic since in vivo the embryo would be exposed to a full complement of amino acids present in the secretions of the reproductive tract.
2. The concentration of amino acids used in the above studies was that found in Eagle's minimal essential medium (MEM) and hence not physiological. It would be interesting to revisit these studies in light of more recent data describing the concentrations of amino acids found in the murine oviduct and uterus (7). This would allow the results to be interpreted at physiological concentrations of a complete mixture of amino acids.

Despite several reports describing improved embryo development with the inclusion of amino acids in the culture medium, an overarching concern remains regarding the accumulation of toxic ammonium. Ammonium is a spontaneous breakdown product of amino acids and can be measured in culture drops incubated at 37 °C even in the absence of an embryo (5). Interestingly, when mouse zygotes were cultured to the blastocyst stage in either the presence of nonessential amino acids, or in a full complement of amino acids, the level of ammonium produced was similar (5) and upon transfer both groups gave rise to fetuses with exencephaly (6). However, when the medium used to culture the zygotes was replaced after 48 h, there was a significant increase in the number of implantation sites, fetal mass, and the development of exencephaly was eliminated (6). This suggests that ammonium accumulation is a significant concern with the inclusion of amino acids in embryo culture medium. However, these data should be interpreted with caution since the deleterious effect observed with continued embryo culture may simply reflect the high concentration of amino acids used in these studies resulting in increased levels of spontaneous breakdown and significant ammonium accumulation.

In 1997, a landmark study compared the effect of culturing mouse preimplantation embryos in different combinations of amino acids. It was found that the presence of nonessential amino acids up to the 8-cell stage, followed by the inclusion of all 20 amino acids until the blastocyst stage produced the highest rates of mouse preimplantation development in vitro, and importantly also increased embryo viability after transfer (8). This series of experiments proved highly significant and was the advent of the 2-step, or sequential media system currently used to culture human preimplantation embryos in many assisted reproduction units.

Whether a single medium is sufficient to support development to the blastocyst stage, or whether a sequential system should be used, remains controversial. This has been confounded by data which showed that mouse embryos cultured in a single, potassium simplex optimization medium (KSOM (9)) in the presence of both essential and nonessential amino acids displayed an improved rate of blastocyst formation, hatching, and cell number compared to those cultured in the absence of amino acids (10). However, what was perhaps even more interesting was that at the molecular level, blastocysts cultured in the presence of all 20 amino acids expressed eight out of the nine genes investigated at similar levels to in vivo developed embryos (10). This finding was reassuring and suggests that the presence of a full complement of amino acids ensure that embryos develop at a rate which is quantitatively similar to that which occurs in vivo.

The discrepancies over which amino acids should be included in embryo culture medium and when, are complex. It is likely to be dependent on the type of medium used, the presence and concentration of amino acids as well as other energy substrates, the ability of embryos to adapt to a new environment, and the inherent compensatory mechanisms of the embryo. Support for this theory was presented when the concentration of amino acids in human follicular fluid was measured and found to be approximately half of that found in Eagle's MEM (11). When mouse zygotes were cultured in modified human tubal fluid (HTF) medium containing amino acids at the concentration found in follicular fluid, the rate of blastocyst formation increased compared to those cultured in HTF alone. In comparison, embryo development was inhibited when embryos were cultured in HTF containing amino acids at the concentrations found in either Ham's F-10 or MEM. Interestingly, significantly less ammonium was produced when embryos were cultured in HTF containing amino acids at levels found in follicular fluid compared to those found in MEM or Ham's F-10 (11). This highlights the importance of considering the concentration of amino acids present for optimal embryo development; with too many, or too few, being inhibitory to blastocyst formation.

Currently, KSOM and KSOM supplemented with amino acids present at half the concentration found in Eagle's MEM and designated as $KSOM^{AA}$ (12, 13) are perhaps the most widely used media to culture mouse preimplantation embryos. In terms of development to zona-enclosed blastocysts, there was little difference between these two media. However, culture in $KSOM^{AA}$ significantly increased the rate of blastocyst hatching and increased the number of cells in both the trophectoderm and inner cell mass without affecting apoptosis or cell death (12, 13). Moreover, the global gene expression pattern of embryos cultured to the blastocyst stage in $KSOM^{AA}$ was found to be similar to in vivo developed embryos (14).

4. Amino Acid Requirement of Human Preimplantation Embryos

In comparison to the mouse and other animal models, relatively less is known about the amino acid requirements of human embryos. Initial studies investigated the effect of adding single amino acids such as glutamine or taurine to the medium used to culture human preimplantation embryos. Individually, both amino acids were found to increase blastocyst formation but there was no further augmentation of development when added in combination (15, 16). However, of note in these experiments was the absence of glucose which should be taken into consideration, as it is likely that, along with pyruvate, glutamine was being used predominantly as an energy source. When human embryos were cultured in a sequential medium containing nonessential amino acids for the cleavage stages followed by the inclusion of all 20 amino acids, blastocyst cell number was significantly increased compared to those cultured in the presence of glutamine only (17). This suggests that, like the mouse, human blastocyst development is promoted by amino acids.

In 1997 the concentration of amino acids present in the Fallopian tube was reported (18) and for the first time allowed human embryos to be cultured in a physiological mixture of amino acids. It was found that embryos had an increasing requirement for amino acids as development progressed; leucine was the only amino acid depleted from day 2 to day 3 of development, whereas leucine, serine, and arginine were required over the compact 8-cell to morula stage and leucine, serine, arginine, methionine, and valine were used from the morula to blastocyst stage (19). This was particularly interesting since, as discussed above, most of the sequential media used to culture human embryos include only nonessential amino acids until day 3 of development before including both essential and nonessential amino acids until the blastocyst stage (20). However, if a complete mixture of amino acids is provided, only leucine, an essential amino acid, is taken up by the embryo, suggesting that these media do not provide the preferred nutritional requirements for the embryo.

As well as being beneficial for development, amino acid utilization provided the first noninvasive method to determine the developmental competency of human embryos (19). Total amino acid depletion was significantly less for those embryos which developed to the blastocyst stage compared to those which arrested prior to blastocyst formation. Specifically, embryos developing from day 2 to day 3 displayed a significant difference in the utilization of asparagine, glutamine, arginine, and alanine than those which arrested prior to blastocyst formation (19). This was despite there being no significant difference in the morphology of the embryos in these two groups. Counter-intuitively, it was found that developmentally competent embryos had a reduced requirement for amino acids compared to those which arrested leading to the quiet embryo hypothesis (19, 21).

Similar results were also obtained when frozen-thawed human embryos were cultured in the same near physiological mixture of amino acids (22). Again, despite there being no difference in morphology, it was possible to retrospectively differentiate between developmentally competent embryos and those which arrested prior to blastocyst formation. A further remarkable finding of this study was the ability of amino acid profiling to differentiate between developmentally viable and nonviable embryos of the highest quality, i.e., Grade 1 embryos (22).

To determine whether metabolic and morphological criteria are associated with trophectoderm differentiation in the human blastocyst, Eckert et al. (23) performed amino acid profiling of individual embryos followed by an assessment of junctional protein assembly using immunocytochemistry and confocal microscopy. An association was found between amino acid turnover and the assembly of junctional proteins in the trophectoderm; an affect that was independent of morphology (23). This intriguing study was the first to demonstrate the ability of amino acid turnover to predict trophectoderm integrity.

A further remarkable finding of amino acid profiling was the ability of asparagine, glycine, and leucine utilization to correlate with live birth rate after transfer of ICSI-derived embryos (24). Moreover, these correlations were found to be independent of other known predictors of pregnancy outcome such as embryo grade and cell number, maternal age and basal FSH levels, a measure of ovarian reserve. This data has significant implications for the selection of the best embryo for transfer in assisted reproduction and highlights the importance of amino acids to embryo development. The sector should therefore carefully consider the requirement for current sequential media as opposed to using a single medium to satisfy the amino acid requirements of the embryo as:

1. The initial premise of sequential media was based on studies in the mouse.
2. Embryos will never be exposed to only a subset of amino acids, be it essential or nonessential amino acids, but instead will experience a physiological mixture of all amino acids.
3. Data using a physiological mixture of amino acids shows that leucine is consumed by developmentally competent human embryos and is one of the three amino acids which correlated with live birth after transfer. However, the presence of only nonessential amino acids during the early stages of development in sequential media suggests the provision of a suboptimal nutritional environment.
4. The risks of embryo loss associated with moving embryos are negated with the use of a single medium.

Ultimately, further research is required to determine whether a single or sequential medium is preferential for culturing human preimplantation embryos and will require manufacturers to release the complete composition of each embryo culture medium.

5. Roles of Amino Acids in Embryo Development

Amino acids have several potential roles in the preimplantation embryo. Despite there being no increase in the protein content of preimplantation embryos as development progresses (25), protein synthesis is likely to be one of the predominant uses of amino acids as protein is continually turned over. Another major role for amino acids is as a potential source of energy for the developing embryo. For example, glycine taken up by embryos may be incorporated into macromolecules, or converted into serine and alanine for energy metabolism (26). Amino acids also have several other functions; glycine has been shown to act as an intracellular osmolyte in both mouse (27, 28) and human early cleavage stage embryos (29) and thus may protect against the high levels of inorganic ions found in the reproductive tract (30). Taurine, when included as the sole amino acid, was found to significantly improve mouse 2-cell development to the blastocyst stage (31, 32). Moreover, taurine and proline both have a role in regulating osmotic imbalances and hence cell volume in embryos (33, 34), while glutamine can protect against high NaCl concentrations present in some mouse embryo culture medium (35). Amino acids, particularly nonessential amino acids, are also able to buffer intracellular pH to prevent acidification, at least until the 4-cell stage (36). A further important role for amino acids is in cell signaling. For example, arginine is as a precursor of nitric oxide which is an important signalling molecule obligatory for preimplantation development (37). Other exogenous amino acids, particularly leucine also regulate the protrusive activity of the trophectoderm and its ability to spread through a mammalian target of rapamycin (mTOR)-dependent pathway and hence may act to control implantation (38, 39). Thus, amino acids are involved in several physiological processes which ensure maintenance of cellular homeostasis.

6. Concluding Remarks

This chapter has concentrated on the importance of amino acids for embryo development. However, it is important to remember that rather than discussing each constituent in isolation, a complete culture medium represents a complex milieu where the presence and concentration of each component may affect the nutrition of the developing embryo. As discussed above, the major concern with the inclusion of amino acids in embryo culture medium is the production of toxic ammonium. Thus, to prevent the spontaneous breakdown of amino acids, medium should be stored at 4 °C. Moreover, care should be taken to limit the amount of time medium is exposed to 37 °C prior to addition of the embryo. If media are handled appropriately, amino acids confer immense metabolic and homeostatic benefits to the developing embryo.

Acknowledgement

I thank the UK Medical Research Council (G0701153) for financial support.

References

1. Reeds PJ (2000) Dispensable and indispensable amino acids for humans. J Nutr 130: 1835S–1840S
2. Fűrst P, Stehle P (2004) What are the essential elements for the determination of amino acid requirements in humans? J Nutr 134(Suppl): 1558S–1565S
3. Brinster RL (1965) Studies on the development of mouse embryos *in vitro*. III. The effect of fixed-nitrogen source. J Exp Zool 158: 69–77
4. Daniel JC, Krishnan RS (1967) Amino acid requirements for growth of the rabbit blastocyst *in vitro*. J Cell Physiol 70:155–160
5. Gardner DK, Lane M (1993) Amino acids and ammonium regulate mouse embryo development in culture. Biol Reprod 48:377–385
6. Lane M, Gardner DK (1994) Increase in postimplantation development of cultured mouse embryos by amino acids and induction of fetal retardation and exencephaly by ammonium ions. J Reprod Fertil 102:305–312
7. Harris SE, Gopichandran N, Picton HM, Leese HJ, Orsi NM (2005) Nutrient concentrations in murine follicular fluid and the female reproductive tract. Theriogenology 61:992–1006
8. Lane M, Gardner DK (1997) Differential regulation of mouse embryo development and viability by amino acids. J Reprod Fertil 109:153–164
9. Lawitts JA, Biggers JD (1993) Culture of preimplantation embryos. Methods Enzymol 225:153–164
10. Ho Y, Wigglesworth K, Eppig JJ, Schultz RM (1995) Preimplantation development of mouse embryos in KSOM: augmentation by amino acids and analysis of gene expression. Mol Reprod Dev 14:232–238
11. Nakazawa T, Ohashi K, Yamada M, Shinoda S, Saji F, Murata Y, Araki H (1997) Effect of different concentrations of amino acids in human serum and follicular fluid on the development of one-cell mouse embryos *in vitro*. J Reprod Fertil 111:327–332
12. Biggers JD, McGinnis LK, Raffin M (2000) Amino acids and preimplantation development of the mouse in protein-free potassium simplex optimized medium. Biol Reprod 63:281–293
13. Summers MC, McGinnis LK, Lawitts JA, Raffin M, Biggers JD (2000) IVF of mouse ova in a simplex optimized medium supplemented with amino acids. Hum Reprod 15:1791–1801
14. Rinaudo P, Schultz RM (2004) Effects of embryo culture on global pattern of gene expression in preimplantation mouse embryos. Reproduction 128:301–311
15. Devreker F, Winston RM, Hardy K (1998) Glutamine improves human preimplantation development *in vitro*. Fertil Steril 69:293–299
16. Devreker F, Van den Bergh M, Biramane J, Winston RL, Englert Y, Hardy K (1999) Effects of taurine on human embryo development *in vitro*. Hum Reprod 14:2350–2356
17. Devreker F, Hardy K, Van den Bergh M, Vannin AS, Emiliani S, Englert Y (2001) Amino acids promote human blastocyst development *in vitro*. Hum Reprod 16:749–756
18. Tay JI, Rutherford AJ, Killick SR, Maguiness SD, Partridge RJ, Leese HJ (1997) Human tubal fluid: production, nutrient composition and response to adrenergic agents. Hum Reprod 12:2451–2456
19. Houghton FD, Hawkhead JA, Humpherson PG, Hogg JE, Balen AH, Rutherford AJ, Leese HJ (2002) Non-invasive amino acid turnover predicts human embryo developmental capacity. Hum Reprod 17:999–1005
20. Gardner DK, Lane M (1998) Culture of viable human blastocysts in defined sequential serum-free media. Hum Reprod 13(Suppl 3): 148–159
21. Leese HJ (2002) Quiet please, do not disturb: a hypothesis of embryo metabolism and viability. Bioessays 24:845–849
22. Stokes PJ, Hawkhead JA, Fawthrop RK, Picton HM, Sharma V, Leese HJ, Houghton FD (2007) Metabolism of human embryos following cryopreservation: implications for the safety and selection of embryos for transfer in clinical IVF. Hum Reprod 22:839–835
23. Eckert JJ, Houghton FD, Hawkhead JA, Balen AH, Leese HJ, Picton HM, Cameron IT, Fleming TP (2007) Human embryos developing *in vitro* are susceptible to impaired epithelial junction biogenesis correlating with abnormal metabolic activity. Hum Reprod 22: 2214–2224
24. Brison DR, Houghton FD, Falconer D, Roberts SA, Hawkhead J, Humpherson PG, Lieberman BA, Leese HJ (2004) Identification of viable human embryos in IVF by non-invasive measurement of amino acid turnover. Hum Reprod 19:2319–2324
25. Sellens MH, Stein S, Sherman MI (1981) Protein and free amino acid content in preimplantation mouse embryos and in blastocysts under various culture conditions. J Reprod Fertil 61:307–315
26. Hobbs JH, Kaye PL (1985) Glycine transport in mouse eggs and preimplantation embryos. J Reprod Fertil 74:77–86

27. Dawson KM, Collins JL, Baltz JM (1998) Osmolarity-dependent glycine accumulation indicates a role for glycine as an organic osmolyte in early preimplantation mouse embryos. Biol Reprod 59:225–232
28. Steeves CL, Baltz JM (2005) Regulation of intracellular glycine as an organic osmolyte in early preimplantation mouse embryos. J Cell Physiol 204:273–279
29. Hammer MA, Kolajova M, Léveillé M, Claman P, Baltz JM (2000) Glycine transport by single human and mouse embryos. Hum Reprod 15:419–426
30. Van Winkle LJ, Haghighat N, Campione AL (1990) Glycine protects preimplantation mouse conceptuses from a detrimental effect on development of the inorganic ions in oviductal fluid. J Exp Zool 253:215–219
31. Dumoulin JC, Evers JL, Bras M, Pieters MH, Geraedts JP (1992) Positive effect of taurine on preimplantation development of mouse embryos *in vitro*. J Reprod Fertil 94:373–380
32. Dumoulin JC, Evers JL, Bakker JA, Bras M, Pieters MH, Geraedts JP (1992) Temporal effects of taurine on mouse preimplantation development *in vitro*. Hum Reprod 7: 403–407
33. Dumoulin JC, van Wissen LC, Menheere PP, Michiels AH, Geraedts JP, Evers JL (1997) Taurine acts as an osmolyte in human and mouse oocytes and embryos. Biol Reprod 56: 739–744
34. Anas MH, Hammer MA, Lever M, Stanton JA, Baltz JM (2007) The organic osmolytes betaine and proline are transported by a shared system in early preimplantation mouse embryos. J Cell Physiol 210:266–277
35. Lawitts JA, Biggers JD (1992) Joint effects of sodium chloride, glutamine and glucose in mouse preimplantation embryo culture media. Mol Reprod Dev 31:89–194
36. Edwards LJ, Williams DA, Gardner DK (1998) Intracellular pH of the mouse preimplantation embryo: amino acids act as buffers of intracellular pH. Hum Reprod 13: 3441–3448
37. Manser RC, Leese HJ, Houghton FD (2004) Effect of inhibiting nitric oxide production on mouse preimplantation embryo development and metabolism. Biol Reprod 71:528–533
38. Martin PM, Sutherland AE (2001) Exogenous amino acids regulate trophectoderm differentiation in the mouse blastocyst through an mTOR-dependent pathway. Dev Biol 240:183–193
39. Martin PM, Sutherland AE, Van Winkle LJ (2003) Amino acid transport regulates blastocyst implantation. Biol Reprod 69:1101–1108

Chapter 8

Media Composition: Macromolecules and Embryo Growth

Marius Meintjes

Abstract

Most embryo culture media are still supplemented with proteins rather than with nonprotein macromolecules or recombinant protein products. HSA is probably the most common supplement followed by globulin-enriched preparations. Serum supplementation and Co-Culture of embryos belong to the past. Defined nonprotein or recombinant protein supplements are becoming a viable alternative during gamete and embryo manipulation procedures. Biological protein supplements are still preferred for any extended period of embryo culture. Understanding the goals and purpose of supplemented macromolecules in embryo culture media during each step of the laboratory IVF process should assist us in choosing the safest and most consistent macromolecule for each step, but also selecting a product that has the capability of delivering the best clinical outcome. Each batch of biological protein supplement is unique, even if supplied by the same manufacturer. Each lot of protein supplement typically contains many lot-specific, potentially harmful, and unintended hormone and protein contaminants. Macromolecular embryo culture medium supplements should be identified as one of the highest risk factors in an IVF laboratory that may contribute towards clinical compromise. All efforts should be made to use a proven batch of supplement for as long as the expiration date will allow. The beneficial effect of more complex protein supplements is evident after the activation of the embryonic genome and probably due to the presence of growth factors. Lower live-birth rates due to suboptimum protein supplementation may be a direct result of the preferential loss of female embryos. When deciding on a culture system, thought should be given specifically to the interaction between the culture medium and the macromolecular supplement. Ready-to-use pre-supplemented culture media may be advisable over a more complex product if a comprehensive macromolecular quality management program is not feasible. However, the question remains as to whether the increasing simplification of embryo culture media supplements is ready for large-scale clinical use.

Key words: Macromolecule, Protein, Human, Embryo, Blastocyst, Quality control

1. Introduction

Current IVF laboratory practice is to supplement all commercial IVF embryo culture media with some protein. Actually, most if not all mainstream commercial media manufacturers will provide

Gary D. Smith et al. (eds.), *Embryo Culture: Methods and Protocols*, Methods in Molecular Biology, vol. 912, DOI 10.1007/978-1-61779-971-6_8,

"complete" or "ready to use" protein-supplemented products. It may be prudent to consider the history and origins of culture media protein supplementation, the reasons for supplementing our culture media with protein and to reconsider the sources and types of not only proteins in our culture media, but also the possible use of other macromolecules to further supplement or replace proteins as the sole source of macromolecules in solution.

We tend to think of the birth of the first human IVF baby on July 25, 1978 as the start of human IVF (1). However, human IVF and in vitro embryo culture were already performed by Menkin in 1944 (2, 3). Human blastocysts were grown in vitro as early as 1971 (4). During these early years, in vitro produced embryos were grown in culture media designed for somatic cell culture and typically supplemented with high levels of various sources of serum. Serum in cell culture media was historically required for successful cell proliferation and cell-dish surface attachment. Embryo culture was historically attempted with whole serum or plasma as the sole culture medium. The combination of inadequate, simple salt solutions and sera with, sometimes, unbeknown adverse effects on embryos (5), was partly responsible for not achieving the first human IVF live birth before 1978. However, some years later a pregnancy was reported using only pure human serum as the sole culture medium (6).

During the 1980s and 1990s, media supplementation with ~10 % bovine serum albumin (BSA), human fetal cord serum, fertile donor serum or later, maternal serum became the norm (7, 8). Some of the challenges faced during this time were the immense lack of consistency in the source of supplements (9, 10), occult embryo toxicity of certain batches of protein or serum (8), and the risk of disease transmission (11–13). The 1990s were also characterized as the era of Co-Culture. With the culture media of the day, it was not uncommon to demonstrate superior results with Co-Culture. Embryos were Co-Culture with a variety of cells such as fetal bovine uterine cells (14), ampullary cells (15), bovine oviductal epithelial cells (16), Vero cells (17), and oviductal cells (18). To be able to maintain these cell cultures in a healthy state, medium supplementation with serum was essential. Another dilemma encountered with embryo-somatic cell Co-Culture was that somatic cells and embryos had very different metabolic requirements and; therefore, required different media compositions for optimized development. Media formulations at the time attempted to cater for both somatic cells and embryos at the same time, never being able to fully satisfy the requirements for the cells nor the embryos.

Realizing the very specific and unique requirements for human embryos, stage-specific sequential culture media were developed. Some of these include Universal IVF/M3 (19), G-1/G-2 (20), and P-1/Blastocyst Medium (21). Even though appearing to be an emerging novel approach to in vitro embryo culture, Bavister's Medium/Ham's F-10 sequential culture was used to culture blastocysts much earlier (4). With this advent of enhanced embryo-specific culture media, it now also became feasible to move away from serum supplementation and Co-Culture. Several studies reported that supplementation of IVF culture media with human serum albumin (HSA) or synthetic serum substitute (SSS) now resulted in the same (22) or mostly better (23, 24) clinical outcomes when compared with serum. It must be noted that SSS has since been renamed as serum substitute supplement. Today, many commercial IVF media contain HSA as the sole protein supplement, claiming that improvements in the formulation of modern embryo culture media obviate the need for more complex protein supplements.

With increasing regulation, the quest for consistency, an aversion for any possibility of disease transmission, routine clinical application of more invasive procedures such as ICSI and embryo biopsy, and the emergence of metabolomic/proteonomic technologies to select embryos, there are continuous efforts to even further simplify and define the macromolecular supplementation of human embryo culture media (10, 25). The use of recombinant human albumin (26, 27) or hyaluronate (28) as the sole macromolecular supplement has been advocated by some. In fact, it has been demonstrated that reasonable pregnancy rates can be obtained without any protein in the embryo culture medium (29). Another approach is to supplement only the transfer medium destined for the host uterus with recombinant protein (30) or alternate macromolecules such as hyaluronic acid (31), polyvinyl pyrrolidone (PVP) or polyvinyl alcohol (PVA) (25).

Even though the benefits of using simple or more defined macromolecular supplements for in vitro embryo culture media is clear, it may be possible that oversimplification can compromise clinical outcomes. Tanikawa et al., demonstrated that the addition of α and β globulins to the culture medium of mouse embryos in a dose-dependent manner, significantly increased mouse blastocyst development and in vitro blastocyst hatching (32). There was no similar growth-promoting response when only the concentration of albumin was increased. It was recently demonstrated that culture media supplemented with more complex SSS yielded significantly higher implantation- and live-birth rates in the human compared with media supplemented with HSA alone (33).

Preimplantation embryos are naturally exposed to and designed to respond to various growth factors, paracrine and autocrine substances. In vivo human embryos will be exposed to many different growth factors expressed by the embryos themselves, follicles, the oviduct, and the endometrium. An indication of the presence of appropriate growth factors in the culture media is the ability of embryos to hatch in vitro (24). Studies in both the human and animals have demonstrated that culture media devoid of certain growth factors (34) or supplemented with inappropriate growth factors (35) or other substances such as lipoproteins (9) will compromise embryo development and; may even induce a variety of short-term and long-term developmental defects. Inclusion of growth factors to embryo culture media can substantially improve in vitro embryo development and subsequent clinical outcomes (36).

Knowing then that firstly, more defined macromolecular culture medium supplements tend to outperform traditional serum and secondly, understanding that embryos require the timely and appropriate growth factors at physiological concentrations, careful consideration of our choice of in vitro embryo culture medium supplements is necessary. The question remains if increasingly simplified supplements are better for current clinical use or are we just starting the quest to define the specific temporal growth factor requirements for in vitro preimplantation embryo development. With the exact growth factor requirements yet indeterminate for in vitro embryo culture, oversimplification for the sake of consistency alone must be questioned.

1.1. Purpose and Goals of Macromolecular Medium Supplementation

In general, embryo development in vitro is compromised in the absence of any protein supplementation. When considering the in vitro purpose of protein or other macromolecular supplements, it is helpful to distinguish between in vitro fertilization (insemination), gamete/embryo manipulation, and in vitro culture.

During in vitro insemination, oocytes are still surrounded, encapsulated, and protected in a microenvironment created by thousands of cumulus cells. The purpose of macromolecular supplements in insemination medium at this time may be limited to lubrication, allowing sperm cells to move freely, enabling them to reach the target oocytes. Furthermore, macromolecular supplementation will further reduce the insemination concentration required for successful in vitro fertilization. Similarly, the main goal of macromolecular supplementation of handling medium (e.g., for micromanipulation, embryo transfer) is lubrication, provided that the medium is otherwise appropriately formulated and buffered to include required amino acids and other nutrients as appropriate (37). Supplementing with proteins specifically may protect and coat stripped oocytes or cumulus cell-free embryos when manipulated in oversimplified salt solutions (38). When used in insemination medium or embryo transfer medium, it appears that proteins can successfully be replaced by other macromolecules such as hyaluronic acid (31) or PVP (25). Since any proteins in culture media will exert an osmoregulatory effect, it is wise to never move embryos or oocytes from higher concentrations of protein to lower concentrations. This principle is often violated when thawing embryos or oocytes. Final-step thaw solutions frequently contain more protein than post-thaw culture medium. This may be one reason that hyaluronan-enriched transfer medium improved clinical frozen-thawed outcomes (39). It appears that when deciding on protein or macromolecular supplementation in insemination medium, simpler is better (24) and that the beneficial effects seen from more complex protein supplementation is only observed after fertilization (33, 40). Serum or SSS is not recommended for in vitro insemination as more cases of unexplained no-fertilization or very low fertilization can be expected for reasons not always easy to explain.

Proteins are thought to stabilize the oocyte and embryonic cell membranes in an in vitro environment. This may be evident by improved fertilization and embryo quality after preincubation of oocytes before ICSI or IVF (41, 42). Furthermore, in vitro culture and recovery of frozen-thawed blastocysts for up to 20 h before transfer increased the blastocyst implantation rate three-fold (43). One mechanism of stabilizing and protecting cell membranes may be through the effective inhibition of lipid peroxidation by binding hydroperoxy fatty acids. Because of the large surface area and abundant binding sites of proteins, they readily act as a trap for toxic substances such as heavy metals, antibodies, lipoproteins, and other non-defined toxins. Proteins may mask the toxic effects of a culture environment and; therefore, quality control bioassays can be sensitized by performing these tests with reduced or no protein supplementation (44, 45). Furthermore, proteins (specifically HSA) may serve as an essential and direct nutrient/nitrogen source for blastocyst-stage embryos (46). This function may not be replaced by other macromolecular supplements such as hyaluronic acid, PVA or PVP.

Proteins are complex macromolecules, binding many hormones and growth factors for later release into the in vitro culture environment. Platelet activating factor (PAF) is a great example of the interaction between the embryo, albumin, and a growth factor. PAF is produced by the embryo, and at the same time, the embryo has high-affinity PAF receptors. However, it appears that PAF directly released into the environment is subject to prompt metabolism by PAF-acetyl-hydrolase (PAFAH) (47). Released PAF is uniquely protected from this hydrolysis by the protective action of albumin in the environment. The secondary and tertiary structure of albumin is determined by the presence of 17 interchain disulfide bonds that link almost all of albumin's cysteine residues. Interestingly, albumin that was exposed to embryos or to medium with embryos, express more reactive thiol residues. Albumin is subject to embryo-induced conformational changes involving cysteine–cysteine disulfide bonds, allowing for the loading of embryo-derived PAF into a very special solvent-protected site between amino acids 240 and 386 of the albumin molecule (48). It appears likely that the special nature of embryo-derived PAF-albumin binding ensures that PAF is not immediately inactivated by PAFAH, but can accumulate in vitro to biologically relevant concentrations. The amount of PAF released by the 2-cell mouse embryo is directly dependent upon the extracellular albumin concentration, with the amount of PAF released increasing as the albumin concentration increases (49). Knowing that growth factors are essential for viable embryo development (36) and observing the unique interaction between the embryo, albumin, and PAF to facilitate the embryotrophic effects of PAF, one must consider the uniqueness of protein as an essential macromolecular supplement of next-generation embryo culture media.

1.2. Dangers and Complications of Protein as Media Supplements

From a review of the literature, it appears that HSA supplementation of embryo culture media leads to improved and more consistent clinical outcomes when compared with traditional serum (23). However, there are some clear indications that the addition of globulins (32) and/or growth factors to embryo culture media improves the outcomes over that of supplementation with HSA alone (33, 50). Indications are that proteins have very special functions and it may not be feasible to totally replace proteins in human embryo culture media with nonprotein macromolecules without compromising clinical outcomes. Conversely, even though we appreciate the need and some of the unique functions of protein supplements in human embryo culture media, proteins in the media may pose significant risks not only to consistent clinical outcomes, but also to the embryo recipient.

Various sources of serum, serum extracts such as Plasmatein®, Plasmanate®, and SSS® or blood-derived HSA, all invariably carry the risk of transmitting infectious substances to the embryo or patient. Supplementation of embryo culture medium with hepatitis-B contaminated donor serum did result in the infection of 79 women (11). Even though of grave concern, there are yet no documented cases of media-mediated HIV or prion contamination of human embryos. However, in the mid-1990s, there was a recall of some batches of HSA by Baxter International, used for IVF culture media supplementation. The death of one of the blood donors used for the manufacture of those batches of HSA was thought to be from possible Creutzfeldt–Jakob Disease (13). The risk of prion or viral contamination of embryos through protein supplements is exacerbated by the development and the routine application of invasive micromanipulation procedures that breach the zona pellucida such as blastomere biopsy and ICSI. Disclaimers by the commercial human IVF media manufacturers ensure that the liability from using protein supplements that may contain animal or human-derived proteins is shared by the IVF laboratory. The risk of infectious contamination by commercial HSA products are minimized by heat treatment at 60 °C for 10 h or can also be cold ethanol extracted (51).

One has to question the possible adverse effects of these relative harsh treatments on the tertiary structure and potential functionality of the various proteins in in vitro embryo culture as well as the possible compromise of otherwise beneficial hormones, growth factors, and beneficial "contaminants" of protein suspensions. If we consider that contaminating growth factors may be partially responsible for improved outcomes when supplementing with more complex proteins such as SSS® (33, 34, 52), these sterilization techniques may have a particularly adverse effect on the growth-promoting characteristics of some of these batches of protein supplements. To preserve the conformation and solubility of albumin after sterilization, all HSA solutions approved by the FDA are supplemented with sodium caprylate and sodium acetyltryptophanate (53) and in addition to these stabilizers; many products also contain the preservative malelic acid (25). With a volume–volume supplementation of as much as 20 % HSA or SSS® to human IVF culture and cryopreservation media, one has to question the wholesomeness of these unintended added chemicals as well as the solvent medium which may be simple 0.9 % physiological NaCl.

The protein supplement of IVF culture medium can be a most likely cause of batch-to-batch variation and a frequent reason for compromised IVF outcomes. This may be due to the inherent variability of the protein donor pool, the processing technique, degree of purification, the addition of various stabilizers, and preservatives to the protein solution or even occult endotoxin contamination (54).

In a recent in-house evaluation of SSS®, we found physiological levels of some common steroid and protein hormones to include, progesterone, FSH, TSH, DHEA-S, hCG, and insulin as well as some growth factors such as VEGF and IGF-II. The levels of these substances were not consistent between different batches of SSS® or HSA (Table 1). If true for these hormones, it follows logically that all other protein supplement components may be subject to the same batch-to-batch variation. The notion that commercial protein supplements or even pre-supplemented ready-to-use media is a pure extract of the targeted protein is not accurate. An evaluation of a batch of IVF-ready HSA and SSS® by SDS-polyacrylamide gel electrophoresis, clearly demonstrates that commercial IVF HSA supplements are not composed of pure HSA and even less so when supplementing with a more complex product such as SSS® (Fig. 1). For both HSA and SSS®, a predominant band can be observed between the 50 and 75 kDa markers that is presumed to be albumin. In the HSA sample, further bands can be observed as low as 25 kDa and as high as 250 kDa which represent proteins other than HSA. In the SSS® sample, one can clearly observe a ladder of protein bands down to as low as 12 kDa. Both HSA and SSS® show abundant high molecular weight proteins between 100 and 250 kDa. These electrophoresis results demonstrate that overall, SSS® appears to have many more lower molecular weight proteins compared with HSA, particularly in the range 10–37 kDa. When these lower molecular weight proteins were further identified using nanoflow LC-MS/MS, a very diverse group of cytokines, growth factors, and carrier proteins were identified. For the HSA and SSS® samples

Table 1
Comparisons of selected protein hormones, steroid hormones, and growth factors measured in different batches of commercial human serum albumin (HSA) and serum substitute supplement (SSS®)

	HSA		SSS		
Hormone/protein	Company 1	Company 2	Lot 1	Lot 2	Lot 3
FSH (mIU/ml)	8.2	6.5	4.0	3.9	4.0
LH (mIU/ml)	–	–	–	–	–
hCG (mIU/ml)	51.4	28.0	18.0	10.9	10.7
Progesterone (ng/ml)	–	–	0.31	0.4	0.32
Estradiol (pg/ml)	–	–	–	–	–
DHEA-S (ug/dl)	240.6	68.1	63.5	216.7	224.4
TSH (uIU/ml)	0.057	0.110	0.144	0.138	0.143
PRL (ng/ml)	–	–	–	–	–
IGF-1 (ng/ml)	–	–	–	–	–
IGF-2 (ng/ml)	–	–	328	281	284
VEGF (pg/ml)	–	17	30	27	23
Interleukin-6 (pg/ml)	–	–	–	–	–
Insulin (uIU/ml)	–	–	2.6	4.1	4.5

tested, SSS® showed an overall cytokine signal strength of 25 % and HSA a signal strength of approximately 1 % when compared with that observed in control plasma.

The levels of these unintended protein contaminants in commercial IVF protein supplements varies greatly with the processing batch and the method of product purification. Common methods used to purify albumin preparations include ion-exchange precipitation or a Cohn fractionation process (25). Each of these approaches will render a final product with a different mix of proteins. As proposed previously, using recombinant protein supplements instead of biological proteins will drastically reduce the inclusion of contaminants and consequent batch-to-batch variation (28, 55), preventing the unintended presence of a specific detrimental contaminant or group of contaminants. However, the beneficial effect of some batches of protein may very well be just because of some of these unintended contaminants such as insulin, citrate (56), or growth factors (33, 34).

Protein supplements are notorious as a source of endotoxin contamination of embryo culture media. Even an otherwise

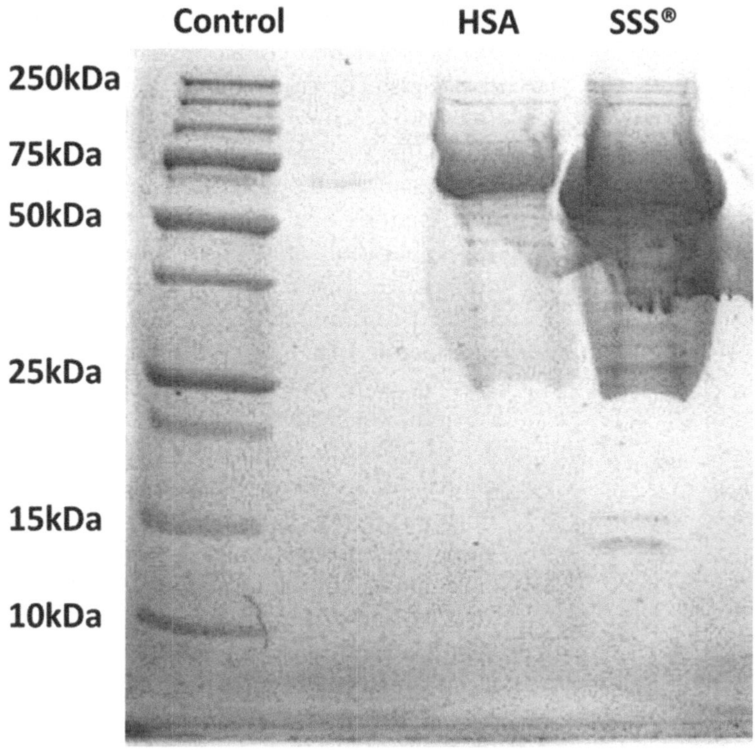

Fig. 1. 10–20 % SDS-PAGE of HSA and SSS®. Molecular weight markers (Mw) in the range 2–250 kDa is displayed for reference. Courtesy Vitrolife Inc., Inglewood Colorado.

embryotrophic batch of protein supplement may contain some endotoxins as a left-over from past gram-negative bacterial contamination. Lipopolysaccharide endotoxin molecules can be as small as 50 kDa, are extremely heat stabile and; therefore, are not removed from culture media by standard sterilization and micropore filtration processes. Higher levels (≥0.1 EU/ml) of endotoxin contamination of IVF culture media may be readily identified by observed in vitro effects on the embryo such as blastomere fragmentation and delayed development (54, 57); however, more subtle endotoxin contamination (as low as 0.02 EU/ml) may not affect observable in vitro embryo quality at all, but may only be manifested as an unexplained reduction of the live birth rate (58).

1.3. The Importance of Quality Assessment of Macromolecular IVF Media Supplements

In our quest to identify the most appropriate macromolecular supplement for today's embryo culture media, we must balance the ability of a macromolecular supplement to result in the best clinical outcomes with that of its safety, consistent performance, availability, and cost. Once a desired macromolecular supplement has been identified, it does not mean that supplementation with that specific batch will culminate in the expected in vitro embryo development. Rigorous in-house quality control is essential to confirm that a specific batch of a chosen macromolecular supplement will deliver the desired clinical outcomes. Batch-specific macromolecular quality assessment should confirm an optimum mix of components from the specific donor pool, the absence of embryotoxicity due to pollutants such as stabilizers and preservatives, acceptable levels of endotoxins; and proper handling during transportation.

As a first line of macromolecular supplement quality control, the application of a sensitive contact bioassay is compulsory. An appropriate contact bioassay or multiple bioassays should be performed to confirm both the expected embryotrophic effect of a macromolecular IVF culture medium supplement, as well as the absence of potential embryotoxic effects. The absence of embryotrophic characteristics and/or the presence of embryotoxic properties may be due to the innate nature of the original macromolecular source or; importantly, may be secondarily due to mishandling during transportation or storage. For this reason, each shipment (even of the same batch) should be tested to confirm safety for clinical use. Testing for embryo toxicity can be accomplished using at least one sensitive contact bioassay such as an extended-time sperm survival assay or a sensitive mouse embryo bioassay (44, 45). Confirming the absence of embryo toxicity of a supplement by the lack of damage to sperm or mouse embryos does not necessarily guarantee the desired embryotrophic effect. The sensitivity and predictive value of a mouse embryo bioassay can be optimized by starting with one-cell embryos instead of two-cell embryos, identifying sensitive strains of mice, using a simple culture medium and/ or by adding additional endpoints besides blastocyst formation rate such as the rate of embryo development, total cell numbers, in vitro hatching, and differentiation between ICM- and trophoblast cells. Since we know that most of the protein supplements are contaminated by beneficial growth factors such as IGF-II, VEGF, and insulin (Table 1), endpoints measured such as in vitro hatching may be an indirect indicator of the presence of embryo-friendly substances in a particular batch of supplement.

With a volume–volume supplementation of as much as 20 % HSA or SSS® solution to human IVF culture and/or cryopreservation media, one has to appreciate that more is added to the solution than just the macromolecules and that the supplement may significantly alter the composition and other characteristics of the final culture medium preparation. Besides adding other protein contaminants as discussed earlier, the electrolyte and pH balance of the final product may be affected. As an example, G-media (Vitrolife, Gotheburg, Sweden) were originally formulated to be supplemented with 0.5 mg/ml HSA or 5 % (v/v) of a 10 % HSA solution. However, when G-media version 2 are supplemented with 10 % SSS® in addition, the equilibrated pH is lowered with about 0.1 pH units, sodium and chlorine concentrations are decreased, potassium and calcium concentrations are

Table 2
Expected pH values and range (mean ± 2SD) of electrolyte concentrations in G-media, version 2 supplemented with HSA or SSS®

Parameter (mean ± 2SD)	Serum	G-1.2 + HSA	G-1.2 + SSS	G-2.2 + HSA	G-2.2 + SSS
pH (U)	7.35–7.45	7.35–7.45	7.25–7.35	7.35–7.45	7.25–7.35
Na^+ (mmol/l)	135–145	125–133	122–130	125–133	119–127
K^+ (mmol/l)	3.5–5.2	4.7–5.1	4.9–5.5	4.7–5.1	5.1–5.5
Cl^- (mmol/l)	97–108	100–106	95–103	104–110	99–107
Ca^{2+} (mg/dl)	8.7–10.2	3.4–4.2	6.7–7.5	3.4–4.2	6.8–7.6
Glucose (mg/dl)	65–99	<25	<25	45–55	47–55

significantly increased and even glucose concentrations may be slightly elevated (Table 2). A sound macromolecular quality assessment program should identify and monitor a target pH and ensure that the osmolarity and electrolyte concentrations are still within target limits.

Even though the formulation of a supplement and previous experience of a supplement may suggest that it should be good for clinical use, an otherwise sound macromolecular supplement may fail to support adequate embryo development due to the presence of batch-specific embryo toxic substances. As an example, the presence of lipoproteins in macromolecular supplements has been shown to adversely affect in vitro human and mouse embryo development (9). Batch-specific contamination of hormones such as hCG and/or progesterone is troublesome (Table 1), as the potential beneficial, adverse effects or interactions of these hormones with different brands of culture media, other growth factors or the in vitro embryos are largely unknown and unpredictable. The consistent and abundant presence of hCG in the donor pool (HSA and globulin-enriched supplements) is surprising and desirability during extended in vitro embryo culture must be questioned when taking into consideration its pronounced long-acting in vivo LH-activity.

A common and often overlooked toxin present in macromolecular supplements is bacterial endotoxin. Endotoxins may be found floating free in the embryo culture medium, or more commonly, may be attached to the proteins in embryo culture medium supplements. The adverse effects of endotoxins on human embryos in vitro and subsequent clinical outcomes have been well described (54, 57, 58). Interestingly, we have used Percoll gradients with endotoxin levels in excess of 100 EU/ml for many years to process IVF sperm samples with no documented impairment on the sperm or subsequent in vitro embryo development.

This is provided; however, that the residual Percoll was thoroughly removed from the processed sperm suspension prior to in vitro insemination. It is possible that the negative effects of bacterial endotoxins are oocyte/embryo specific with the embryos of all species not necessarily equally affected. Identifying a laboratory-specific tolerance limit and independently monitoring the endotoxin concentration of batches of macromolecular supplements are essential to guarantee the consistent in vitro performance of macromolecular supplements. Based on the reported effects of bacterial endotoxins on in vitro human embryos, an upper limit of no more than 0.1 EU/ml in the stock solution of the macromolecular supplement appears appropriate. Several challenges are presented should one choose to accept the endotoxin levels as reported on the certificate of analysis: (1) the endotoxin level reported may not be for the stock solution, but rather the expected value in the working solution if used according to the manufacturer's recommendations. (2) The standard for acceptable endotoxin levels in protein supplements may not be the same for the manufacturer as for the consuming IVF laboratory. The manufacturer's tolerance limit for product release may be based on levels recommended by regulatory agencies such as the FDA on the basis of good manufacturing practices, rather than on what is acceptable for safe human embryo culture. According to the certificate of analysis, some commercial protein supplements such as SSS® may be released for in vitro human embryo culture as long as the endotoxin levels are <12 EU/ml, even though detrimental effects on human embryos have been reported with levels as low as 0.02 EU/ml (58). (3) The most common method for endotoxin testing is the Limulus Amebocyte Lysate (LAL) gell-clott method which relies on the observation of clotting or precipitation in serial dilutions of the test solution. Interpretation is somewhat subjective and experience plays an important role in reading the results. Outsourcing endotoxin testing to specialized testing facilities may be more reliable than in-house testing or testing done by the manufacturing company. (4) The error of the gell-clott method of endotoxin testing is ±one two-fold dilution. This implies that the real endotoxin value of the supplement may be the same as the one reported on the certificate of analysis, half of the value reported or even twice the value reported. Retesting will at least give a second value for consideration. (5) Manufacturing the macromolecular supplement and testing for endotoxins by the same company may represent a conflict of interest. Independent endotoxin testing, verifying the quality presented by the manufacturer, should be considered as an integral part of an IVF quality management program. Confirmatory endotoxin testing is mostly deferred by IVF laboratories because of no regulatory requirement to do so and the perceived logistical and financial resources required to do so.

It is becoming more popular to purchase pre-supplemented, ready-to-use human embryo culture medium products. IVF laboratories appreciate the resultant ease of use of ready products and not having to deal with the additional quality testing of a separate macromolecular supplement or the accompanied higher risk of introducing culture variation when acquiring their own stock solutions of macromolecular supplements. We must consider that when quality testing the ready-to-use media, we no longer have access to the original stock solution. It may be possible that signs of a sub-optimum macromolecular supplement (such as the presence of endotoxins) will be diluted at least 20 times and; therefore, may not be detectable by the end user. Quality management of macromolecular supplement should be one of the highest priorities in the general quality management program for any IVF program as the macromolecular supplement is a most common culprit of performance variation in embryo culture media as no two protein supplements are identical in composition or performance. The shelf life of most macromolecular supplements by far outlasts the shelf life of currently available embryo culture media. When routinely using pre-supplemented embryo culture media, the shelf life of the protein supplement is limited by the shelf life of the embryo culture medium. Furthermore, one can potentially have a different batch of protein supplement with each lot of supplied embryo culture medium. In addition, in vitro insemination, micromanipulation, and the different stages of embryo culture may potentially all be performed using different batches of macromolecular supplements. This implies that the embryos of a single patient will be exposed to these different batches of macromolecular supplements, introducing unnecessary variation into the culture system. It may be prudent to accept the additional headache of screening and testing a macromolecular supplement and then to purchase and safely store enough of that specific batch of approved supplement to last through its expiration date. This selected supplement should then span multiple batches of embryo culture media as well as provide consistency in supplementation across medias used for different purposes such as initial and extended culture.

Rigorous quality management and testing of macromolecular embryo culture media supplements are essential to ensure not only quality clinical outcomes, but to achieve those desired outcomes consistently. If not willing to invest the time, personnel, and financial resources into a rigorous quality management program, a laboratory may be better off to use pre-supplemented media where the manufacturing company performed at least some level of testing on the supplement. Even though there are indications that more complex macromolecular supplements may outperform standard HSA under certain circumstances, simpler supplements, in general, may yield more consistent results if the IVF laboratory is not committed to rigorous testing and batch optimization strategies as described.

A proposed algorithm for macromolecule quality management is as follows:

1. Start screening candidate samples at least 6 months in advance.
2. Samples with certificates of analysis with endotoxin levels ≥0.1 EU/ml are immediately discarded.
3. Outsource samples for independent endotoxin testing to confirm the value reported on the certificate of analysis. Example company is: Associates of Cape Cod (www.acciusa.com).
4. Perform a 5-day sperm survival assay.
5. Perform a one-cell mouse embryo assay, using the IVF laboratory's standard human embryo culture medium and macromolecule supplementation protocol. Make sure to measure in vitro hatching as an endpoint to gain some indication of growth factor activity.
6. After passing the bioassays, test for FSH, DHEA-S, hCG, and insulin as rough markers for hormone contamination. Over time, these values can be correlated to outcomes and meaningful ranges established for lot screening.
7. Perform a standard electrolyte panel of final supplemented media to include Na^+, Ca^{2+}, Cl^-, K^+, and glucose. Though many media companies do not list concentrations of these components in their media, over time these values can be calculated by the individual laboratory to establish meaningful reference ranges for lot screening.
8. Confirm final equilibrated and supplemented working pH according to the laboratory's standards.
9. Consider introducing the new supplement for clinical use while still having some of the previous batch available for clinical use and comparison. A few transitional sibling embryo studies can be performed during this introductory period to confirm acceptable clinical performance on real patients.
10. Calculate the anticipated volume of the supplement needed to be able to supplement all appropriate culture media throughout the shelf life of this batch of supplement. Order enough of this supplement to provide a stable source of supplement for the duration of the shelf life.
11. All newly shipped and received product should be resubmitted to the sperm survival and mouse embryo assay to confirm sound shipment.
12. Ensure appropriate and safe storage of the bulk of the product, according to the manufacturer's recommendations. Be sure not to unnecessarily freeze any protein supplement, unless recommended by the manufacturer. Usually stored at 2–8 °C.

13. Consider a one-time aliquoting exercise of the supplement into smaller, safe, and tested containers to reduce the risk of bacterial and/or endotoxin contamination from repeated entry into bulk containers.
14. Make sure to receive new samples for testing no later than 6 months from the expiration date as new batches are limited and release dates far apart. Furthermore, at least 75 % of samples are expected to fail the testing.

1.4. Clinical Considerations

In a recent randomized study of 528 patients, an 8.6 % higher live birth rate was observed when supplementing commercial culture media with HSA + SSS® compared with supplementation with HSA alone (33). Interestingly, the clinical pregnancy rate was not different between the two treatment groups, only the later live-birth rate. This was largely due to the tendency for more postimplantation losses in the HSA-supplemented treatment group. Furthermore, the live-birth rate was not higher when transferring on day 2 or 3, only when transferring at the blastocyst stage. The embryo's ability to implant and give rise to a live birth was preferentially impaired by suboptimum culture medium protein supplementation during days 3–5 of in vitro culture. In support, there were no differences observed in day-3 embryo morphological quality when fertilized and cultured in HSA or SSS® (40). It appears then that the beneficial effect of SSS® supplementation in this study was realized only after activation of the embryonic genome (day 3) which implicates a possible role of growth factors. When considering a standardization or simplification of macromolecular embryo culture media supplementation with more defined preparations such as hyaluronate (28), PVP, PVA (25), or recombinant human albumin (27), the results observed after the first 3 days of culture, may not apply to extended culture to the blastocyst stage.

Upon further investigation of the babies born after embryo culture in media supplemented with HSA or with HSA + SSS®, it was observed that not only were the live birth rate lower when supplementing with HSA alone, but also that the sex ratio of the babies were skewed towards males (Fig. 2). We know that the addition of d-glucose or glucosamine to the culture medium at the 8-cell stage increases the male:female sex ratio of IVF bovine embryos, but had no effect when added earlier before the maternal-zygotic transition (59). A recent meta-analysis of four studies and 2,587 offspring demonstrated a distinct increase in the male:female sex ratio when transferring blastocysts, but not when transferring cleavage-stage embryos (60). The observation of a more even sex ratio when transferring cleavage-stage embryos and the results from studies in the cow suggest that supplementation with HSA alone similarly might have led to the selective loss of female embryos after the 8-cell stage, implying a role of X-linked

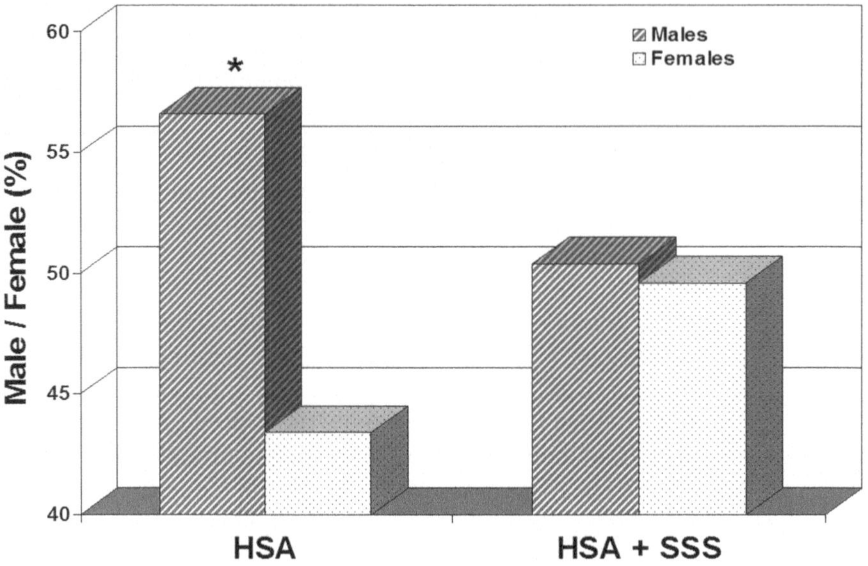

Fig. 2. Male:female sex ratio for live births from embryo culture media supplemented with HSA or HSA+SSS®. The sex ratio was different for live births when embryos were cultured in media supplemented with HSA-only. χ^2, $P<0.009$.

gene products. This condition was corrected by supplementation with a more complex macromolecular source such as SSS®. There is increasing evidence for a major role of sex chromosome dosage (61). Sex-dependent transcriptional differences may affect many different molecular pathways such as protein and glucose metabolism, DNA methylation and epigenetic regulation with developmental consequences such as sex-selective embryo loss under suboptimal in vitro culture conditions (62). Therefore, the increased live-birth rate when supplementing with SSS® may be a direct result of the survival of more female embryos in culture.

When deciding on a macromolecular medium supplement, one must consider not only the supplement, but also the interaction between the medium and the supplement. The combination of supplement and culture medium should be considered rather than selecting the medium and macromolecular supplement solely on their individual merits (23). A relative simple medium such as P-1 (Irvine Scientific, Santa Anna, California) may benefit more from a complex macromolecular supplement such as SSS® (51) than a more complex medium would such as the G-series (Vitrolife, Gotheburg, Sweden) where simple HSA supplementation may yield acceptable clinical results. Even though a specific batch of more complex protein supplement such as SSS® may prove to be of significant clinical benefit (33), the increased complexity also carries with it an increased risk of disease transmission, variability, and the risk of containing embryo toxins such as endotoxins, lipoproteins, or inappropriate cytokines and/or growth factors (8). Supplementation with an inappropriate protein may be expressed as altered embryo metabolism which can be sex-specific, altered gene expression and embryonic imprinting, observable in vitro morphological changes, premature or delayed blastulation, ultrastructural impairments or effects on the fetal weight, especially in ruminant species. After supplementation of in vitro maturation and embryo culture medium with serum in cattle (35) and sheep (63), inappropriate gene imprinting resulted in increased gestational lengths and birth weights. Rigorous batch-specific screening, quality management, and batch shelf-life optimization as discussed previously are essential to prevent or limit the risk associated with macromolecular culture medium supplementation. Using a simple macromolecular culture medium supplement that is safe, consistent, and has been well tested is preferable over the use of an unconfirmed, incompletely screened, even though a potentially superior complex supplement (64).

1.5. Conclusion and Notes

Currently, it appears that most embryo culture media are still supplemented with proteins rather than with nonprotein macromolecules or recombinant protein products. HSA is probably the most common supplement followed by globulin-enriched preparations such as SSS® (Irvine Scientific), SPS® (SAGE), LGPS® (Life Global), and others. Clinical experience and increased awareness of the safety and regulation of embryo culture media in a rapidly expanding global IVF market, suggest that serum supplementation and the Co-Culture of embryos and somatic cells belong to the past. It appears that defined nonprotein or recombinant protein supplements is becoming a viable alternative during gamete and embryo manipulation procedures, but that biological protein supplements are still preferred for any extended period of embryo culture. Understanding the goals and purpose of supplemented macromolecules in embryo culture media during each step of the laboratory IVF process should assist us in choosing the safest and most consistent macromolecule for each step, but also selecting a product that has the capability of delivering the best clinical outcome.

Each batch of biological protein supplement is unique, even if supplied by the same manufacturer. The embryotrophic performance of a batch of protein supplement may not be repeated by a new lot, as protein preparations cannot be processed identically. Each lot of protein supplement typically contains many lot-specific, potentially harmful and unintended hormone and protein contaminants. Macromolecular embryo culture medium supplements should be identified as one of the highest risk factors in an IVF laboratory that may contribute towards clinical compromise and; therefore, it is essential to develop a comprehensive quality management program, specifically tailored towards identifying safe macromolecules for embryo culture. When evaluating a potential source of a macromolecular embryo culture medium supplement, the presence of embryotrophic capabilities as well as the absence of embryotoxic tendencies should be confirmed. With each batch of macromolecular supplement posing a new risk, all efforts should be made to use a proven batch of supplement for as long as the expiration date will allow.

It appears that the beneficial effect of more complex protein supplements such as SSS are pronounced after the activation of the embryonic genome and may be related to the presence of beneficial growth factors. Any conclusions based on current and future studies investigating the use of defined macromolecules or recombinant proteins only during the first 3 days of in vitro culture must be questioned, as the observations may not apply to extended culture to the blastocyst stage. With increasing evidence to support in vitro sex-chromosome dosage and possible sex-dependent in vitro culture sensitivity, lower live-birth rates due to suboptimum embryo culture medium supplementation after day 3 of embryo culture may be a direct result of the preferential loss of female embryos. When deciding on a culture system, thought should be given not only to the individual components contributing to the culture system, but specifically also to the interaction between the chosen media and macromolecular supplement. A simple low-risk macromolecular supplement or ready-to-use pre-supplemented culture media should be preferred over a higher risk, more complex product, if a comprehensive macromolecular quality management program is not feasible. The question remains if the increasingly simplification of embryo culture media supplements is ready for large-scale, clinical use. Instead, are we just starting the quest to define the specific temporal growth factors that should be carefully added to complex, but defined embryo culture media, to be able to compete with, or surpass our current clinical results using growth-factor contaminated, complex protein supplements.

Acknowledgments

I thank the dedicated physicians, embryologists, nurses, and administrative personnel working in our program who are always willing to support new studies and investigational endeavors. I gratefully acknowledge Vitrolife Inc. (Inglewood, Colorado) for their sponsorship and technical support of several investigations and studies in our quest to continually improve culture conditions and embryo culture medium formulations.

References

1. Steptoe PC, Edwards RG (1978) Birth after the reimplantation of a human embryo. Lancet 2(8085):366
2. Rock J, Menkin MF (1944) In vitro fertilization and cleavage of human ovarian eggs. Science 100(2588):105–107
3. McLaughlin L (1982) The Pill, John Rock and the Church. A biography of a revolution, 1st edn. Little, Brown and Company, Boston
4. Steptoe PC, Edwards RG, Purdy JM (1971) Human blastocysts grown in culture. Nature 229(5280):132–133
5. Dokras A, Sargent IL, Redman CWG (1993) Sera from woman with unexplained infertility inhibit both mouse and human embryo growth in vitro. Fertil Steril 60:285–292
6. Kemeter P, Feightinger W (1984) Pregnancy following in vitro fertilization and embryo transfer using pure human serum as culture and transfer medium. Fertil Steril 41(6): 936–937
7. Staessen C, Van Den Abbeel E, Carle M et al (1990) Comparison between human serum and Albuminar-20™ supplement for in vitro fertilization. Hum Reprod 5(3):336–341
8. Leveille MC, Carnegie J, Tanphaichitr N (1992) Effects of human sera and human serum albumin on mouse embryo culture. J Assist Reprod Genet 9(1):45–52
9. Weathersbee PS, Pool TB, Ord T (1995) Synthetic serum substitute (SSS): a globulin-enriched protein supplement for human embryo culture. J Assist Reprod Genet 12(6): 354–360
10. Pool TB (2004) Development of culture media for human assisted reproductive technology. Fertil Steril 81:287–289
11. Alberda AT, Van Os HC, Zeilmaker GH (1989) Hepatitis B virus infectie bij vrouwen behandeld met in vitro fertilisatie. Neederlands Tijdskrift Voor Geneeskunde 133:20–25
12. Klein R, Dumble LJ (1993) Transmission of Creutzfeldt-Jakob disease by blood transfusion. Lancet 341:768
13. Otani T, McDonough PG (1995) Earthquakes and prions. Fertil Steril 63:1137–1139
14. Wiemer KE, Cohen J, Amborskir GF et al (1989) In-vitro development and implantation of human embryos following culture on fetal bovine uterine fibroblast cells. Hum Reprod 4(5):595–600
15. Bongso A, Ng SC, Ratnam S (1990) Co-cultures: their relevance to assisted reproduction. Hum Reprod 5(8):893–900
16. Wiemer KE, Hoffman DI, Maxson WS et al (1993) Embryonic morphology and rate of implantation of human embryos following co-culture on bovine oviductal epithelial cells. Hum Reprod 8(1):97–101
17. Ménézo YJR, Sakkas D, Janny L (1995) Co-culture of the early human embryo: factors affecting human blastocyst formation in vitro. Microsc Res Tech 52(1):50–56
18. Rieger D, Grisart B, Semple E et al (1995) Comparison of the effects of oviductal cell co-culture and oviductal cell-conditioned medium on the development and metabolic activity of cattle embryos. J Reprod Fertil 105:91–98
19. Forsdahl F, Bertheussen K, Bungum LJ et al (1994) A study on extended culture time with embryo replacement at the morula and blastocyst stage. Hum Reprod 9:142–143
20. Barnes FL, Crombie A, Gardner DK et al (1995) Blastocyst development and pregnancy after in vitro maturation of human primary oocytes, intracytoplasmic sperm injection and assisted hatching. Hum Reprod 10:3243–3247
21. Behr B, Pool TB, Milki A et al (1999) Preliminary clinical experience with human blastocyst development in vitro without co-culture. Hum Reprod 14(2):454–457
22. Dugan KJ, Shalika S, Smith RD et al (1997) Comparison of synthetic serum substitute and fetal cord serum as media supplements for in vitro fertilization: a prospective, randomized study. Fertil Steril 67(1):166–168
23. Laverge H, De Sutter P, Desmet R, et al (1997) Prospective randomized study comparing human serum albumin with fetal cord serum as protein supplement in culture medium for in-vitro fertilization (IVF). Hum Reprod 12 (10):2263–2266
24. Tucker KE, Hurst BS, Guadagnoli S et al (1996) Evaluation of synthetic serum substitute versus serum as protein supplementation for mouse and human embryo culture. J Assist Reprod Genet 13(1):32–37
25. Blake D, Svalander P, Jin M et al (2002) Protein supplementation of human IVF culture media. J Assist Reprod Genet 19(3):137–143
26. Bungum M, Humaidan P, Bungum L (2002) Recombinant human albumin as protein source in culture media used for IVF: a prospective randomized study. Reprod Biomed Online 4(3):233–236
27. Takenaka M, Horiuchi T (2007) Recombinant human albumin supports mouse blastocyst development and improves fetal development. Reprod Med Biol 6(4):195–201

28. Gardner DK, Lane M (1998) Culture of viable human blastocysts in defined sequential serum-free media. Hum Reprod 13(Suppl 3): 148–159
29. Ali J, Shahata MA, Al-Natsha SD (2000) Formulation of a protein-free medium for human assisted reproduction. Hum Reprod 15(1):145–156
30. Friedler S, Schachter M, Strassburger D et al (2007) A randomized clinical trial comparing recombinant hyaluronan/recombinant albumin versus human tubal fluid for cleavage stage embryo transfer in patients with multiple IVF-embryo transfer failure. Hum Reprod 22(9): 2444–2448
31. Simon A, Safran A, Revel A et al (2003) hyaluronic acid can successfully replace albumin as the sole macromolecule in a human embryo transfer medium. Fertil Steril 79(6): 1434–1438
32. Tanikawa T, Harada T, Ito M et al (1999) Globulins in protein supplements promote the development of preimplantation embryos. J Assist Reprod Genet 16:555–557
33. Meintjes M, Chantilis SJ, Ward DC et al (2009) A randomized controlled study of human serum albumin and serum substitute supplement as protein supplements for IVF culture and the effect on live birth rates. Hum Reprod 24(4):782–789
34. Richter KS (2008) The importance of growth factors for preimplantation embryo development and in vitro culture. Curr Opin Obstet Gynecol 20(3):292–304
35. Behboodi E, Anderson GB, BonDurant RH et al (1994) Birth of large calves that developed from in-vitro derived bovine embryos. Theriogenology 44:227–232
36. Hardy K, Spanos S (2002) Growth factor expression and function in the human and mouse preimplantation embryo. J Endocrinol 172:221–236
37. Lane M, Hooper K, Gardner DK (2001) Effect of essential amino acids on mouse embryo viability and ammonium production. J Assist Reprod Genet 18:519–525
38. Alvarez JG, Storey BT (1995) Differential incorporation of fatty acids into and peroxidative loss of fatty acids from phospholipids of human spermatozoa. Mol Reprod Dev 42:334–346
39. Hambiliki F, Ljunger E, Olof Karlström P et al (2009) Hyaluronan-enriched transfer medium in cleavage-stage frozen-thawed embryo transfers increases implantation rate without improvement of delivery rate. Fertil Steril 94(5):1669–1673
40. Graham MC, Partridge AB, Lewis V et al (1995) A prospective comparison of synthetic serum substitute and human serum albumin in culture for in vitro fertilization-embryo transfer. Fertil Steril 64(5):1036–1038
41. Rienzi L, Ubaldi F, Anniballo R et al (1998) Preincubation of human oocytes may improve fertilization and embryo quality after intracytoplasmic sperm injection. Hum Reprod 13(4):1014–1019
42. Ho JY, Chen MJ, Yi YC et al (2003) The effect of preincubation period of oocytes on nuclear maturity, fertilization rate, embryo quality and pregnancy outcome in IVF and ICSI. J Assist Reprod Genet 20(9):358–364
43. Guerif F, Cadoret V, Poindron J et al (2003) Overnight incubation improves selection of frozen-thawed blastocysts for transfer: preliminary study using supernumary embryos. Theriogenology 60(8):1457–1466
44. Gorrill MJ, Rinehart JS, Tamhane AC et al (1991) Comparison of the hamster sperm motility assay to the mouse one-cell and two-cell embryo bioassays as quality control tests for in vitro fertilization. Fertil Steril 55(2): 345–354
45. Claasens OE, Harrison KL (1999) Optimizing sensitivity of human sperm motility assay for embryology toxicity testing. In: Proceedings of the 11 th world congress on IVF and human reproductive genetics. Sydney, Australia
46. Pemble LB, Kaye PL (1976) Whole protein uptake and metabolism by mouse blastocysts. J Reprod Fertil 78:149–157
47. O'Neil C (2005) The role of PAF in embryo physiology. Hum Reprod Update 11(3): 215–228
48. Ammit AJ, O'Neill C (1997) Studies of the nature of the binding by albumin of platelet-activating factor released from cells. J Biol Chem 272:18772–18778
49. O'Neil C (1997) Evidence for the requirement of autocrine growth factors for development of mouse preimplantation embryos in vitro. Biol Reprod 56:229–237
50. Paria BC, Dey SK (1990) Preimplantation embryo development in vitro: cooperative interactions among embryos and role of growth factors. Proc Natl Acad Sci USA 87(12):4756–4760
51. Hink JH, Hidalgo J, Seeberg VP et al (1957) Preparation and properties of a heat-treated human plasma protein fraction. Vox Sanguinis 2(3):174–186
52. Ben-Yosef D, Yovel I, Schwartz T et al (2001) Increasing synthetic serum substitute (SSS) concentrations in P1 glucose/phosphate-free

medium improves implantation rate: a comparative study. J Assist Reprod Genet 18(11): 588–592

53. Mosley AK, Brouwer KL (1997) Heat treatment of human serum to inactivate HIV does not alter protein binding of selected drugs. Ther Drug Monit 19:477–479
54. Snyman E, Van der Merwe JV (1986) Endotoxin-polluted medium in a human *in vitro* fertilization program. J In Vitro Fertil Embryo Transfer 5:335–342
55. Mahani IM, Davar R (2007) Hyaluronic acid versus albumin in human embryo transfer medium. East Mediterr Health J 13(4): 876–880
56. Gray CW, Morgan PM, Kane MT (1992) Purification of an embryotrophic factor from commercial bovine serum albumin and its identification s citrate. J Reprod Fertil 94: 471–480
57. Fishel S, Jackson P, Webster J et al (1988) Endotoxins in culture medium for human *in vitro* fertilization. Fertil Steril 49:108–111
58. Nagata Y, Shirakawa K (1996) Setting standards for the levels of endotoxin in the embryo culture media of human *in vitro* fertilization and embryo transfer. Fertil Steril 65:614–619
59. Kimura K, Spate LD, Green MP et al (2005) Effects of d-glucose concentration, d-fructose, and inhibitors of enzymes of the pentose phosphate pathway on the development and sex ratio of bovine blastocysts. Mol Reprod Dev 72:201–207
60. Chang HJ, Lee JR, Jee BC et al (2009) Impact of blastocyst transfer on offspring sex ratio and the monozygotic twinning rate: a systematic review and meta-analysis. Fertil Steril 91(6): 2381–2390
61. Bermejo-Alvarez P, Rizos D, Lonergan P et al (2011) Transcriptional sexual dimorphism during preimplantation embryo development and its consequences for developmental competence and adult health and disease. Reproduction 141:563–570
62. Gutierrez-Adan A, Perez-Crispo M, Fernandez-Gonzalez R et al (2006) Developmental consequences of sexual dimorphism during preimplantation embryonic development. Reprod Domest Anim 41(Suppl 2):54–62
63. Holm P, Walker SK, Seamark RF (1996) Embryo viability, duration of gestation and birth weight in sheep after transfer of in vitro matured and in vitro fertilized zygotes cultured in vitro or in vivo. J Reprod Fertil 107:175–181
64. Pool TB, Martin JE (1994) High continuing pregnancy rates after in vitro fertilization-embryo transfer using medium supplemented with a plasma protein fraction containing α and β globulins. Fertil Steril 61:714–719

Chapter 9

Media Composition: Antioxidants/Chelators and Cellular Function

Catherine M.H. Combelles and Margo L. Hennet

Abstract

Protection of embryos against oxidative insults during culture is necessary to maintain viability. Generation of excessive levels of reactive oxygen species (ROS) is triggered by various components of the in vitro environment, most of which embryos do not normally encounter in vivo. To compensate for these deficiencies in the culture environment, antioxidants and chelators are often used to control or suppress ROS levels as embryos develop. However, there is no consensus regarding dosage, time of exposure, or appropriate combinations of antioxidants and chelators in embryo culture. In order to elucidate this aspect of an embryo's chemical surroundings in vitro, we present the current knowledge on the function and effect of each antioxidant or chelator that is often included in an embryo culture medium.

Key words: Antioxidants, Chelators, Embryo culture, Reactive oxygen species, Medium formulation, EDTA, Thiol compounds, Vitamins, Citrate, Taurine

1. Introduction

Oxidative stress is an important concern for embryo development. It occurs when unstable, reactive oxygen-containing molecules called reactive oxygen species (ROS) reach levels high enough to cause structural and functional damage to DNA, proteins, and lipids (1, 2). ROS contribute to embryonic block and cell death in vitro (3–6), as their production during in vitro culture is known to surpass that of in vivo conditions due to several prooxidant factors present in the culture environment (7).

The ROS commonly found in the reproductive tract are superoxide anion, hydrogen peroxide, and hydroxyl radicals (8, 9). In vitro, production of these ROS and others is aggravated in the embryo and its culture medium by exposure to atmospheric oxygen concentration (20 % opposed to the 2–6 % in vivo) (10–12), light (13, 14), metal ions (15), and media deficiencies (16, 17).

Gary D. Smith et al. (eds.), *Embryo Culture: Methods and Protocols*, Methods in Molecular Biology, vol. 912,
DOI 10.1007/978-1-61779-971-6_9,

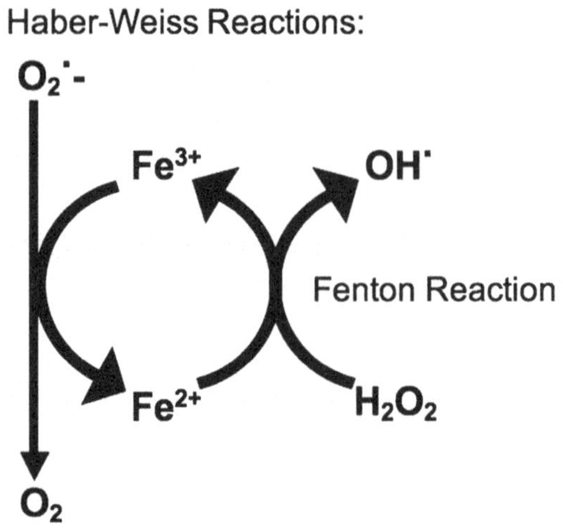

Fig. 1. The Haber–Weiss reactions generate hydroxyl radicals from hydrogen peroxide and iron. These reactions are catalyzed by ferrous and ferric ions. The first step in these reactions is the reduction of ferric ions to ferrous ions by superoxide (generating oxygen). The Fenton Reaction, shown on *right*, is the second step in Haber–Weiss reactions in which hydrogen peroxide is converted to hydroxyl radicals, while ferrous ion is oxidized to ferric ion.

In particular, metal ions such as iron and copper in culture medium accelerate the Haber–Weiss and Fenton reactions to form the most reactive ROS, hydroxyl radicals (Fig. 1) (18). Additionally, free ferrous ions can propagate lipid oxidation directly once it has been initiated by ROS (19). Most laboratory reagents and chemicals contain trace amounts of iron and other metals at levels high enough to catalyze de novo ROS generation (1.6–19.4 μM) (18, 20). When added to embryo culture media, iron leads to increased levels of hydrogen peroxide and higher rates of block or early mortality in mouse embryos (21). Addition of ROS or ROS-generating compounds directly to culture media also has this effect (4). A direct relationship exists between hydrogen peroxide concentration and apoptosis or DNA fragmentation in mouse embryos cultured in vitro (22), and in vitro cultured mouse embryos produce more hydrogen peroxide than do in vivo derived ones (6, 15). It is widely agreed that oxidative stress during in vitro culture impairs embryo development.

Two types of molecules, antioxidants and chelators, protect against ROS-mediated damage. Antioxidants are broadly defined as molecules capable of inhibiting the oxidation of other molecules. With respect to embryo development, antioxidant activities are found in some enzymes, amino acids, vitamins, energy substrates, and thiol compounds. Antioxidants act either enzymatically or nonenzymatically, the latter by serving as an oxidizable substrate for ROS. Enzymatic antioxidants include superoxide dismutase (SOD), catalase, and glutathione peroxidases (GPx), while some

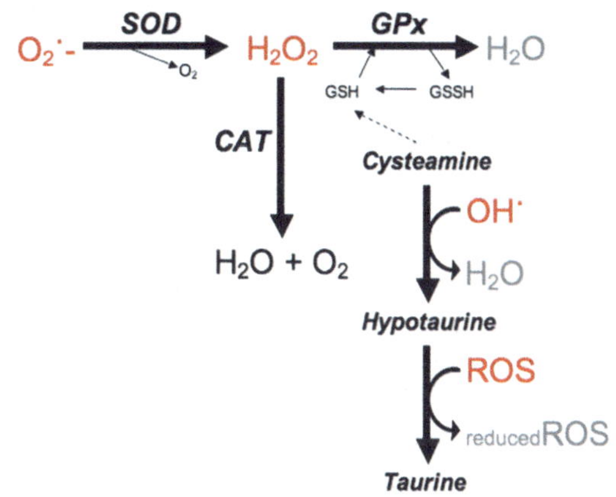

Fig. 2. Many antioxidants act cooperatively to reduce or neutralize ROS. Shown here is one such web of interactions between SOD (superoxide dismutase), GPx (glutathione peroxidase), CAT (catalase), cysteamine, hypotaurine, and taurine. ROS are shown in *red*. The *dashed arrow* from cysteamine indicates that this molecule contributes to glutathione (GSH) synthesis.

nonenzymatic antioxidants are glutathione (GSH), cysteamine (CSH), taurine, hypotaurine, thioredoxin, vitamins, and pyruvate.

In many cases the function of one antioxidant is intertwined with that of several others (Fig. 2). SOD reacts with superoxide radicals to produce hydrogen peroxide, which is subsequently degraded into nontoxic products by catalase and GPx. GSH serves as a necessary substrate for GPx reactions in addition to being a central nonenzymatic defense against ROS in embryos (23–26). CSH reduces hydroxyl radicals and increases GSH synthesis, in part by converting oxidized GSH back to reduced GSH (27). In turn, CSH is converted into hypotaurine when it is oxidized by hydroxyl radicals (28), and hypotaurine is converted to taurine after it reacts with ROS. Taurine neutralizes aldehydes, which are cytotoxic end products of lipid oxidation by ROS (29). Thioredoxin is a small redox protein that reduces the disulphide of substrate proteins (30). Lastly, Vitamin C modifies lipoproteins to increase their resistance to metal ion-mediated oxidation and protects against DNA damage, Vitamin E is a lipid-soluble ROS scavenger, and pyruvate reduces hydrogen peroxide.

Chelators are molecules with at least two negatively charged groups that allow them to form complexes, or chelates, with metal ions that have multiple positive charges. The negatively charged groups—usually containing oxygen, sulfur, or nitrogen—are spaced along the molecule such that binding to a metal creates a ring structure with at least one covalent bond (Fig. 3). The most stable chelate rings are those with 5–7 bonds, or points of contact, between chelator and metal ion (31). The word chelate derives from the Greek word "chele," meaning claw.

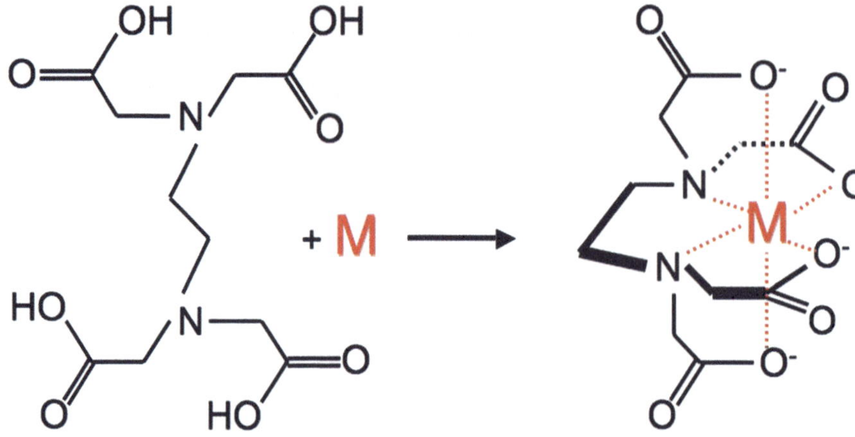

Fig. 3. The chelating activity and molecular structure of EDTA is shown to demonstrate the mechanism of action of chelators. EDTA has four carboxylic acid groups and two amine groups with lone pair electrons. These six groups allow it to form six points of contact with the metal. Note the ring structure formed when EDTA binds the metal—this is typical of all chelates. The *red* "M" represents a metal, and *red dashed lines* signify bonds between the chelator and metal.

Chelation serves many physiological purposes and always causes alterations in the metal ion involved. One alteration with particular relevance to this chapter is that chelation sequesters metal ions, to a certain extent obstructing their ability to participate in Fenton and Haber–Weiss reactions that propagate ROS. Additionally, chelates facilitate the formation and function of many enzymes and are involved in the intracellular transport of metals. Numerous chelators exist as either naturally derived or synthetic molecules. Some with relevance to embryo culture include ethylenediaminetetraacetate (EDTA), citrate, desferrioxamine (DFO), transferrin, ethylenediaminediacetate (EDDA), ethyleneglycolbistetraacetate (EGTA), nitrilotriacetate (NTA), and diethylenetriaminepentaacetic acid (DTPA). It should also be noted that all amino acids are, to some degree, capable of chelation (31, 32).

Antioxidants and chelators are naturally present in the mammalian embryo, oviduct and uterus. It is in this environment that fertilized eggs develop to blastocysts, the period of development that in vitro embryo culture seeks to accommodate. Therefore, although the in vivo environment cannot be replicated in vitro, it offers insights as to the requirements of early embryo development. GPx has been identified as the predominant enzymatic antioxidant in the bovine oviduct, with different isoforms prevailing in various subregions of the oviduct (33). Catalase and SOD activities have also been reported in the oviductal fluid of cow, pig, human (34) and in mouse and rabbit (35, 36). Vitamin C and pyruvate are present in human and mouse oviductal fluids, respectively (37, 38). Lastly, hypotaurine and taurine are the most prevalent amino acids in oviduct fluids of cows, sheep, pigs, rabbits, and mice (39).

In uterine fluids, taurine is the predominant amino acid present in humans, cows, and rabbits (40). At least 14 other amino acids including glutamine, glycine, and alanine have been detected in mammalian uterine and oviduct fluids (40). Cysteamine has also been reported in genital secretions and follicular fluid of cows, pigs and dogs, but not in oviduct fluids (41). Glutathione is known to be a pivotal intracellular nonenzymatic antioxidant in mammalian embryos (23–26), and early embryos also contain transcripts for SOD, catalase, and GPx enzymes (42).

In addition to amino acids, chelators found in the mammalian female reproductive tract include transferrin and citrate. Specifically, transferrin is an abundant protein in oviduct fluid (43). Citrate is also found in embryos, with blastocysts containing up to 12 times more citrate than earlier stage embryos (44, 45). Metals are also present in the physiological surroundings of growing embryos. Potassium, calcium, magnesium, and zinc ions are reported in the oviduct fluids of cow, sheep, rabbit, humans, and horses (46). Trace amounts of other heavy metals are likely to exist in the reproductive tract in the form of protein nucleotides (ATP, GTP) or citrate complexes (47).

Limiting ROS levels during in vitro embryo culture is widely regarded as a necessary protection against the in vitro environment. A survey of commonly used human embryo culture media revealed similarities, yet some differences, in the inclusion of components with established or suspected antioxidant and chelating effects. Table 1 summarizes the incidence of inclusion for a total of 9, 9, and 2 D1-D3, D3-D5, and D1-D5 culture media, respectively. The presence of each of the following components is identified with their primary suspected activity noted in parenthesis: taurine (antioxidant), citrate (chelator), vitamins (antioxidant), glutathione (antioxidant), cysteine/cystine (precursors to antioxidant synthesis), pyruvate (antioxidant), serum proteins (antioxidant), amino acids (chelator), and EDTA (chelator). It is important to note that not all of these compounds are added with the specific

Table 1
Proportions of human embryo culture media containing components with antioxidant or chelating activity

	Taurine[a]	Vitamins[b]	Glutathione	Cysteine cystine	EDTA	Sodium citrate
D1-D3	7/8	3/8	1/9	3/8	7/9	4/8
D3-D5	5/8	7/9	2/9	7/7	1/9	2/8
D1-D5	0/2	0/2	0/2	2/2	2/2	0/2

[a]For both D1-D3 and D3-D5 media, 1/8 contains both taurine and hypotaurine
[b]For D1-D3 media, one includes Vitamin C only, one Vitamin B5 only, and one Vitamins B5, B6, B9. For D3-D5 media, four contain Vitamins B5, B6, B9, one Vitamin C, and one Vitamin B5 only

intent to serve as antioxidants or chelators, with some of them rather included for other effects that they are known to elicit. This is notably the case for pyruvate, vitamins, amino acids, and serum supplements. The presence of each of these compounds will be discussed below, reporting on the historical basis and evidence for their inclusion as well as mechanisms of antioxidant/chelating action whenever known. This will be done first for antioxidants, followed by chelators. There are other antioxidant and chelating agents that are not currently included in human embryo culture media but that they have been investigated for use in animal embryo cultures; these will also be discussed so as to represent the state of knowledge and their probable or improbable inclusions in future human embryo media.

1.1. Antioxidants

1.1.1. Taurine and Hypotaurine

Taurine is currently included in a large proportion of human embryo culture media for the duration of preimplantation development; 88 % (D1-D3) and 63 % (D3-D5) of media contain taurine, with the proportion rising to 88 % (D3-D5) when including media with cysteine/cystine (given these amino acids are metabolic precursors of taurine). It is relevant to note that two commercial media include hypotaurine in addition to taurine; further, hypotaurine can be synthesized from cysteine, and it is also a metabolic precursor of taurine (48). Also pertinent is the ability of embryos to take up taurine (49).

Taurine supplementation during embryo culture resulted in improved development to blastocyst and increased cell numbers in the mouse (50, 51) and in the rabbit (52), with a similar outcome for hypotaurine on bovine embryos (41). Hypotaurine was also established as being essential to permit in vitro development of hamster embryos (53). However, a study in 2–4 cell stage human embryos showed comparable development to the blastocyst stage with supplementation of the medium with either taurine or glutamine, with no further benefits with both taurine and glutamine (54). Similarly, addition of taurine (with or without glutamine) did not affect embryonic development from the zygote to the blastocyst stage in the mouse (55). It is perhaps of interest to note that the two studies which showed no effects of taurine both deviated from other reports in the source or time of exposure of embryos: in the mouse study, in vivo fertilized zygotes were subsequently exposed in vitro (55), and in the human, exposure did not start until 45 h after fertilization (54). While most studies have focused on supplementation from the zygote through blastocyst stages, the temporal effects of taurine on preimplantation development in vitro were also directly tested in the mouse. When examining the influence of taurine supplementation on blastocyst formation, exposure in the first 2 days after insemination, notably around the 2 cell stage, proved most beneficial (51). Further, hypotaurine supplementation was shown to be more beneficial to embryonic development for in vitro versus in vivo fertilized embryos in the hamster (53). Evidence to date thus suggests the importance of taurine and hypotaurine particularly during fertilization and the first cleavage division (17).

While hypotaurine and taurine can serve as oxygen radical scavengers (56, 57), there is only indirect evidence supporting such mechanism of action in embryos. Two studies in the bovine demonstrated improved blastocyst formation for embryos cultured with taurine or hypotaurine under what were presumably increased oxidative stress conditions (exposure to a free radical generator, or 20 % versus 5 % O_2, from the 4–8 cell to blastocyst stage) (58, 59). Of note are other proposed protective functions of taurine on embryonic development, among which is its action as an osmolyte in face of osmotic imbalances (60). Furthermore, it is relevant to consider that taurine can also act as a chelator (52). Future studies on the exact mechanism of action of taurine and hypotaurine (notably with respect to antioxidant activity) will shed additional light into their needs; whether hypotaurine is a superior supplement over taurine also merits consideration. This is particularly relevant since hypotaurine is a more effective antioxidant than taurine (56).

In light of the high concentrations of hypotaurine and taurine in the oviductal and uterine fluids and their overall positive effects in experimental studies, their inclusion in current media appears warranted although not all studies reported benefits (perhaps due to interactions with other culture components). Also, whether supplemented cysteine precursors permit sufficient taurine synthesis, and the protection afforded by it, in the embryo merits direct evaluation. It is relevant to point to the range of concentrations that have been tested, namely between 0.1–50 mM hypotaurine and 0.1–14 mM taurine. While the measured concentrations of taurine and hypotaurine in oviductal fluids (mouse, cow, ewe, sow, rabbit) ranged from 0.05–0.3 mM and 0.15 and 0.28 mM respectively (39, 50), effects were only apparent in vitro when using more than 0.5 mM hypotaurine or more than 1 mM taurine. Further, some studies showed maximal benefits at intermediate concentrations within tested ranges (e.g., 0.5–1 mM hypotaurine in the bovine by Guyader-Joly et al. (41), and 1 mM hypotaurine in the hamster by Barnett and Bavister (53)). For taurine, a comparable effect was reported for 2.5–10 mM concentrations in the rabbit (52), maximum benefits for 5 and 10 mM in the mouse (50) and similar changes at the tested 7 and 14 mM concentrations in the bovine (58). However, the only commercial media with information available on the concentration of supplemented taurine contain 0.05 or 0.1 mM, doses which are both lower than what has proved beneficial with in vitro testing, but yet within reported physiological ranges in the oviduct. Interestingly, taurine concentrations are particularly elevated in human uterine fluid (4–12 mM depending on the cycle stage) (40). Taken together, there is a need to further evaluate the optimal concentrations of taurine and/or hypotaurine for use in vitro and at different times during preimplantation development.

1.1.2. Vitamins C and E

Several media include vitamins, from a single type to a mixture of them. Of direct relevance to this chapter is the inclusion of Vitamin C (or ascorbic acid), a powerful antioxidant. When stressed with an exogenous source of ROS for a timed period, mouse embryos benefited from vitamin supplementation, with Vitamin C protecting against the induced reduction in blastocyst formation and Vitamin E doing so to a much lesser extent (61). In human embryos, Tarin et al. (62) found no effects of ascorbate supplementation on fertilization and embryonic development to day 3 (whether in a simple or complex medium with iron and copper). However, the study only examined a single concentration (similar to body fluid levels) and one pre-compaction developmental time point. In the hamster, there was also no effect of Vitamin C (10 μM) on cleavage and blastocyst formation, when cultured under 5 % oxygen in a defined protein-free medium (63).

Vitamin E (α-tocopherol) is not routinely added to current human embryo culture formulations; however, it also has antioxidant activity with several animal studies reporting on its effects. In the bovine, blastocyst development and embryo size following transfer was increased with Vitamin E supplementation at 100 μM throughout in vitro culture (64). During ovine embryo culture, α-tocopherol (particularly at 200 μM) resulted in improved cleavages, morula, and blastocyst formation along with increased blastocyst cell numbers. Interestingly, the effect was apparent with culture in 20 % and not 5 % O_2 (65). Further, DNA damage was decreased in embryos cultured with 200 μM Vitamin E when compared to no vitamin (20 % O_2). In porcine embryos cultured under 20 % O_2, Vitamin E (50 μM) yielded more blastocysts, increased blastocyst cell numbers, decreased hydrogen peroxide content of 2–4 cell embryos, and reduced DNA damage in 8–16 cell stage embryos when compared with controls (66). Similar positive effects of Vitamin E were obtained during the in vitro preimplantation development of buffalo blastocysts (67). For embryos that were exogenously stressed with a chemical prooxidant, bovine morulae showed improved development to blastocyst (with increased formation, hatching, and cell numbers) when exposed to 400 μM Trolox, a hydrosoluble analogue of Vitamin E (68). Taken together, experimental studies in animal models support some benefits of culturing embryos in the presence of Vitamins C or E. That said, there are several notes of caution specific to the use of these two vitamins in culture media.

First, a paradox exists when supplementing media with Vitamins C or E, given that under certain conditions (e.g., the presence of a transition metal such as iron) these vitamins can act as prooxidant agents (69–71). In this light, Olson and Seidel (64) tested the inclusion of the chelator EDTA in the medium together with Vitamins C and E; while it was hypothesized that EDTA would further the benefits of vitamin supplementation (by perhaps avoiding the risk of any prooxidant activity by the vitamins), the data showed no improvements on blastocyst formation in the bovine. This is in contrast to supplementation with EDTA without vitamins (72), thus suggesting interactions between Vitamins C, E, and EDTA. Understanding the effects of Vitamin C supplementation is likely very complex, particularly in light of recent findings in cultured somatic cell indicating that the dual activity of Vitamin C (as antioxidant or prooxidant) varies with concentration, time, and cellular location (71).

Second, there are known interactions between Vitamins C and E, with Vitamin C regenerating tocopherol from tocopheroxyl radicals (formed when Vitamin E scavenges the oxygen radicals); several studies have thus tested synergism when supplementing the embryo culture medium with both vitamins. Notably, Olson and Seidel (64) documented the detrimental effects of supplementing the bovine embryo culture medium with both Vitamins C and E (at 100 μM each). An increase in apoptotic cells in blastocysts was also obtained after supplementation with Vitamins C and E (at 100 μM each) in porcine embryos (73). In human sperm, cosupplementation with both vitamins induced DNA damage as well as increased DNA damage after H_2O_2 exposure, but yet less ROS was produced upon H_2O_2 exposure (74). The lack of additional benefits when supplementing with both vitamins may thus not necessarily be due to their actions as prooxidants, but rather by other mechanisms. Regardless, a note of caution resides when combining the use of these two vitamins, particularly if not added within optimal levels.

1.1.3. Vitamin B's

Many embryo culture media contain a cocktail of Vitamin B's, and while most studies have reported on their effects when combined, some reports have provided further insight by testing the individual activity of a few. The relevance of discussing Vitamin B supplementation in this chapter lies in the ability of Vitamin B6 (pyridoxine) and Vitamin B9 (folic acid) to scavenge ROS (75–77). Further, Vitamin B5 (panthothenate) can defend indirectly against free radicals by repairing damaged lipids or influencing glutathione synthesis (78, 79). Panthothenate (B5) supplementation showed a significant positive effect on development to blastocyst and term following embryo transfer; this improvement was not surpassed by the already-beneficial effect of select combinations of Vitamin B (including B5). Together, this indicates synergistic interactions between some Vitamin B's (except for B5) and the significant stimulatory effect of B5, even when used alone (63). When omitted from a mixture of vitamin supplements from the 8-cell to blastocyst stage, the lack of Vitamin B5 significantly decreased blastocyst hatching in the hamster (80). When pyridoxine (Vitamin B6) was tested in combination with thiamine, panthothenate, inositol, folic acid, and choline, there was no improvement in blastocyst formation and cell numbers with culture of mouse embryos from the 1 cell stage (81). Further, omission of pyridoxine from a vitamin mixture did not influence blastocyst formation or hatching for in vivo derived 8-cell hamster embryos (80). Confounding interactions between these vitamins cannot be ignored, thereby masking the potential benefits of pyridoxine alone. Lastly, pyridoxal (B6) did not result in changes in cleavage or blastocyst formation of hamster embryos cultured under 5 % oxygen in a protein-free medium (63). As to folic acid (Vitamin B9), it did not promote blastocyst development or cell numbers during embryo culture in the mouse (82), hamster (63, 80) and rabbit (83). A requirement for endogenous folate exists (82), but as long as embryos possess an adequate pool of folic acid, there does not appear to be a need for folic acid as a medium supplement.

Overall, the effects of Vitamin B enrichment are complex since one Vitamin B may be beneficial while others may be inhibitory when used in particular combination. Some caution may thus be required, and the requirements (and potential antioxidant effects) for each vitamin remain to be examined empirically and during defined time windows of preimplantation development. Furthermore, evidence points to species differences when it comes to the influences of vitamin supplementation on embryo culture (80, 81, 83, 84).

1.1.4. Thiol Compounds: Glutathione (GSH), Cysteamine (CSH), Cysteine, Cystine, and β-Mercaptoethanol (BME)

Only 3 out of the 20 surveyed commercial media include GSH (1 for D1-D3 and 2 for D3-D5); however, 12 out of 17 media (3 for D1-D3, 2 for D1-D5, and 7 for D3-D5) contain the amino acids cystine and/or cysteine, itself a precursor of GSH (directly or via CSH synthesis). Cystine results from the oxidation of two cysteine amino acids together, but in the presence of a thiol (such as CSH or BME) cystine can be reduced to cysteine. Of pertinence is a rather large body of evidence on the beneficial effects of thiol compounds during mammalian embryo culture. Also, any changes in the thiol-redox status of embryos may lead to developmental arrest and/or cell death (85), and the redox state could influence embryonic gene expression and function (86).

Several studies have examined direct supplementation with GSH during embryo culture. An early study tested the addition of 0.5–4 μM GSH on development past the 2 cell block in the mouse, with no improvement at these low concentrations (87). During bovine in vitro culture (under 5 % O_2), 1 mM GSH resulted in improved cleavage and blastocyst development (11). Under defined conditions (SOF with BSA, citrate, and amino acids), improvement was observed when added 6 days post-insemination as opposed to between days 2 and 10 or days 2 and 5, time periods of exposure that resulted in a reduction in expanded blastocysts. Conversely, cleavage, morula, and blastocyst formation were promoted with a continuous treatment of embryos with GSH under undefined conditions (with cumulus cell coculture and serum) (11). In a pig model (cultured under 20 % O_2 in defined conditions with BSA), continuous exposure to 0.125–0.5 mM GSH increased the frequency of blastocyst formation and cell numbers when cultured in the absence of BSA; with BSA, these concentrations of GSH further improved blastocyst yield but with no additive effects of GSH beyond BSA on cell numbers (88). Also during porcine embryo culture (under 5 % oxygen in a defined medium with BSA and BME), 0.5–1.0 μM GSH increased blastocyst formation and total cell counts, with a decrease in the number of apoptotic cells after 6 days of culture (89). Interestingly, intracellular GSH levels were increased with exposure to 1.0 μM GSH at the 4 cell and blastocyst stage, but not at the 8–12 cell stage (89). Both pig studies only examined continuous exposure to GSH, yet no negative effects of early exposure was observed, unlike in the bovine experiments. Intrinsic differences in the lipid composition (targets of peroxidation) across species should be considered; given its high lipid content (90), the porcine embryo may particularly benefit from a reducing environment.

Glutathione itself is not effectively transported in cells (91), and thus its utility when provided exogenously may be limited. However, aforementioned studies support a role for GSH when present extracellularly, perhaps by preventing lipid peroxidation of cellular membranes by extracellular ROS. Other compounds may provide a superior alternative to increase intracellular GSH levels, such as the low molecular weight thiols, CSH and BME.

The use of CSH appears particularly relevant in embryos that cannot utilize extracellular cystine for GSH synthesis, as shown in the bovine (23); in addition, the cysteine present in culture medium is rapidly autooxidized into cystine. CSH can reduce cystine to cysteine, and its inclusion in culture medium would in turn permit increased GSH synthesis, even in the absence of a cell monolayer. After continuous culture in 50 μM CSH from the 6–8 cell stage (in the presence of serum), an increased proportion of blastocysts and hatched blastocysts formed in the bovine (23). de Matos et al. (92) tested the effects of CSH supplementation during day 2 to day 4 of development in bovine embryos; 50 μM CSH resulted in improved frequencies of morula and blastocyst formation. Also in bovine embryos, a small study involving the continuous exposure to 100 μM CSH from the 1-cell to blastocyst stage resulted in reduced development to morula and a complete absence of blastocyst when compared to controls (41). Of likely relevance in reconciling these seemingly conflicting results is the use of serum in a complex medium and a coculture monolayer (that may compete for the thiol source) by Guyader-Joly et al. (41). With a web of intersecting pathways in the metabolism of thiol compounds, and potential interactions with several molecules, the culture system can clearly change the effects of CSH supplementation during embryo culture. As CSH may prove beneficial or detrimental, careful attention should be given to not only the timing of its inclusion in embryo culture medium, but also the types of medium and supplements used. Further adding to the complexity of the issue is the potential conversion of CSH into hypotaurine and taurine (with their specific effects discussed above), in turn shifting the balance of these three compounds in the culture medium.

The basis for enriching the medium with the thiol compound BME stems from its ability to promote cystine uptake into cells, and in turn the de novo synthesis of GSH. During bovine embryo culture in a chemically defined system, BME supplementation (10 or 100 μM) improved blastocyst formation and cell numbers (93, 94), notably more significantly in 8–16 cell embryos when compared to <8 cell embryos at the start of exposure (93). Also, BME yielded further benefits when supplemented with serum, presumably by rendering GSH synthesis precursors (present in serum) available to the embryo (93). Upon exposure of 6–8 cell bovine embryos to 50 μM BME, development to blastocyst and hatching was increased with a concomitant rise in intracellular GSH levels (23). In another bovine study, development was promoted only with BME exposures (6.25–12.5 μM) during the first three cleavages, with no detectable benefits when 16-cell embryos were treated, a time when endogenous GSH levels were increasing (95). All of these early bovine studies were performed under 20 % O_2, with a subsequent study indicating that the promoting effect of 50 μM BME supplementation permitted embryonic development to blastocysts at rates similar to when culturing bovine embryos under 5 % O_2 (96). In a coculture system, also supplemented with serum, BME only enhanced blastocyst development when bovine embryos were cultured under 20 % O_2 (97), once again pointing to the importance of pertinent culture conditions when considering the inclusion of any antioxidant supplement. Importantly, addition of BME augmented cystine uptake in the embryo, thus providing a mechanism for the beneficial effects of BME (96). Reduced H_2O_2 levels and decreased DNA fragmentation was observed in bovine embryos cultured with 50 μM BME under 20 % O_2, together with improved blastocyst development (66). The ability of 100 μM BME to protect bovine embryos from prooxidant induced damage was further buttressed by a reduction in cell death, increase in cell numbers, and stimulation of GSH synthesis (68). Lastly, the use of synthetic BME raises a potential issue since its safety for development to term is not yet known.

The amino acid cysteine is routinely added to many embryo culture media, and given that it is a substrate for GSH synthesis, it could modulate the reductive environment of the embryo. In the bovine, 0.6 mM cysteine supplementation during the first 72 h of culture (under 7 % O_2, in SOF with BSA, EDTA, and low glucose) improved morula and blastocyst formation, albeit not significantly (98). When present throughout the culture period (in SOF with serum), 0.6 mM cysteine did not affect cleavage rates but fewer blastocysts formed, thus cautioning the use of cysteine supplementation. Further, oxygen tension influenced blastocyst rates since culture under 5 % O_2 without cysteine resulted in the highest rates with the lowest proportion of apoptotic cells (99). But under 5 % O_2, inner cell mass numbers were higher with than without cysteine (99). Interestingly, the use of *N*-acetyl-l-cysteine (with increased stability over cysteine) did not improve bovine embryonic development while cysteine did (98). Overall, further interpretation of cysteine's effects necessitates a better understanding of cystine uptake and the metabolism of cysteine in embryos. Of relevance are studies on cysteine supplementation during in vitro maturation showing that in bovine oocytes cysteine supplementation results in increased intracellular GSH synthesis (27), but much work still awaits for the embryo in which cystine uptake appears minimal unless BME or CSH is also added (96).

There is a limited pool of thiol compounds within the embryo. Specifically, intracellular GSH levels decrease ten-fold from the time of fertilization through the blastocyst stage during in vitro mouse development (24). This would reflect a progressive depletion in maternal stores of GSH present in the oocyte and in face of the oxidizing environment of in vitro culture. Further, it is at the blastocyst stage that mouse embryos first demonstrate the ability to synthesize GSH de novo (100). Also, intracellular GSH levels are lowest at the 2–8 cell, higher at the 9–16 cell, and highest at the hatched blastocyst stage in cultured bovine embryos (95). The need to provide exogenous sources of GSH would thus vary during preimplantation development (with de Matos et al. (101) confirming a beneficial window of exposure during days 2–4 alone). In light of the existing data, one must consider species differences in the developmental timeline for the biosynthesis of GSH by the embryo. Regardless, enrichment of the culture medium with compounds that increase GSH synthesis would help alleviate GSH depletion in the developing embryo, in turn keeping the thiol-redox status of the embryo adequate. This is particularly relevant during culture, given that GSH levels are higher in embryos derived in vivo versus in vitro (24).

1.1.5. Pyruvate

Although pyruvate is included in all embryo culture media as an obligatory energy substrate for early cleavage divisions, it can also protect embryos from ROS-mediated damage. In the presence of hydrogen peroxide, pyruvate decarboxylates to produce acetate, carbon dioxide, and water (102). Studies of somatic cells have shown that pyruvate is able to protect against hydrogen peroxide-mediated injury and apoptosis (103–105), and two studies of cell culture media reported that the presence of pyruvate accounted for differences in hydrogen peroxide levels in otherwise identical media (106, 107). Similarly, investigations of pyruvate's roles in embryo culture have revealed protective effects against ROS. In bovine embryos, the addition of 0.3 mM pyruvate to modified SOF containing 10^{-5} M hydrogen peroxide significantly reduced hydrogen peroxide levels and protected embryos from hydrogen peroxide-induced injury (108). A study of mouse embryos reports a similar effect in which pyruvate, but not glucose or a test control, allows normal blastocyst development after the addition of 10^{-4} M hydrogen peroxide to culture media (109). Interestingly, pyruvate has been reported to fully protect zygotes from hydrogen peroxide-induced damage, while only partially protecting blastocysts in the cow (108). This indicates that embryos might vary in their sensitivity to ROS, and in their requirements for protection, in a developmental stage-dependent manner.

Nonenzymatic antioxidants are depleted as they react with and neutralize ROS. This concept is particularly relevant to embryo culture, as media deficiencies can cause metabolic stress that results in oxidative stress and cellular damage. Pyruvate's antioxidant function is nonenzymatic, and therefore its presence decreases as it lowers hydrogen peroxide levels. This dynamic has been demonstrated experimentally (106). Interestingly, one study investigating the effect of pyruvate on early mouse embryo development in the presence or absence of other antioxidants and chelators found no difference in pyruvate depletion between groups. However, as hydrogen peroxide levels were not artificially induced in the experiment, the authors suggest that pyruvate scavenging of hydrogen peroxide may have been occurring at levels too low to detect with the tests used (47). If excess hydrogen peroxide or other ROS are generated in embryo culture media, depletion of essential media components such as pyruvate may occur.

1.1.6. Protein Supplements

Most embryo culture media contain a protein source, as human serum albumin (HSA) or serum substitute supplement or replacement (SSS or SSR, itself made up of 84 % HSA and 16 % human globulins) (110). Besides the potential presence of unknown factors with antioxidant or chelating activity in these preparations, there is some evidence pointing to the antioxidant activity of albumin. In mouse embryos, inclusion of SSS results in decreased intracellular H_2O_2 levels when compared to culture in human tubal fluid alone at 24, 48, and 72 h (111), thus suggesting an ROS scavenging activity for SSS. This is not surprising given that albumin may influence the rates of free radical reactions in the culture medium by serving as a reservoir of steroids, vitamins, and metals. Specifically, albumin will bind copper, rendering it less available to accelerate radical reactions; and while albumin itself may become damaged (from the formation of bound hydroxyl radicals), other essential molecular targets will in turn be protected (112, 113). Albumin is thus not defined as a chain-breaking or preventative antioxidant, but rather, like pyruvate, a "sacrificial" one and as such it must be considered whenever evaluating the antioxidant defenses present in the embryo culture medium.

1.1.7. Enzymatic Antioxidants and Thioredoxin

While not currently included in human embryo culture media, exogenously provided enzymatic antioxidants and thioredoxin (TRX) have been tested during animal embryo cultures.

In mouse strains with a 2 cell block, supplementation with SOD (50 μg/ml) during mouse embryo culture (from the pronuclear to blastocyst stage) resulted in improved formation of blastocysts (35), especially with culture in 5 % O_2, thus indicating an additive effect of SOD and low oxygen (114). Chun et al. (36) also showed the positive effects of SOD (370 U/50 μl microdrop) on blastocyst formation in the mouse. A follow-up study by the same group also indicated that SOD (500 μg/ml) exerted the greatest developmental benefits when supplemented throughout the culture period rather than during limited periods within the first 30 h post-insemination (115). Also in the mouse, exogenous SOD benefited blastocyst formation at 100 U/ml while a higher concentration at 1,000 U/ml resulted in detrimental effects. However, the positive influence of SOD supplementation was only evident under 5 % O_2 (and not under 20 %) and it varied greatly across mouse strains (116). When cultured under protein-free conditions, rabbit zygotes exposed to SOD for 72 h demonstrated a significant increase in expanding blastocysts and cell numbers at 600 U/ml (with no effects at lower concentrations, and no further increase or detriments beyond 600 U/ml) (52).

Not all studies report beneficial effects of SOD supplementation during embryo culture, with Legge and Sellens (117) documenting no changes in mouse embryonic development with SOD (5,000 U/ml) in a serum-free medium. In the bovine, Liu and Foote (58) showed no influences of SOD (300 or 600 U/ml) between the 4 and 8 cell and blastocyst stages, when cultured under either 5 % or 20 % O_2. There was also no effect of SOD (1,500–3,000 U/ml) during bovine embryo culture (under low oxygen) with either undefined (coculture with a monolayer) or defined (with BSA and citrate) conditions (11). Also under defined conditions (with BSA and other supplements, including EDTA), SOD addition (10 and 1,000 U/ml) did not influence cleavage and blastocyst formation after exposure in the first 72 h of culture in the bovine (98). These contradicting results could be accounted for by significant differences in the formulations of the culture medium used in these various studies, among which is the inclusion of a protein source (with antioxidant activity), amino acids, or chelators such as citrate or EDTA (see later sections). Specifically, EDTA-containing KSOM was used in the study by Liu and Foote (58) and with effective metal chelation, ROS generation may be limited and there may not be any benefits of SOD beyond ones already provided by EDTA. This is supported by experimental data testing the effects of combined versus single supplementations with SOD and EDTA in the mouse (47). Interestingly, in a different mouse strain, SOD was able to substitute effectively for EDTA during mouse embryo culture (36). In light of the varied results across studies supplementing the medium with SOD, potential species or even strain differences should be considered carefully, as well as the bioactivity of SOD through culture depending on the enzyme preparation and/or how often supplementation is replaced in the medium.

SOD supplementation resulted in greater beneficial effects under 20 % when compared to 5 % O_2 during mouse embryogenesis in vitro (47). In bovine embryos, SOD promoted the rate of blastocyst formation when supplemented from the 4 to 6 cell stage, albeit only under 20 % O_2 (59), further buttressing the existence of interactions between oxygen tension and enzymatic antioxidant supplementation. Therefore, while the use of enzymatic antioxidants may prove beneficial under certain culture conditions, their inclusion may not be warranted as long as other measures are taken to reduce free radical formation, notably in the absence of transition metal ions and under low O_2. However, conflicting results exist since there are augmented benefits of SOD enrichment under low O_2 (114), or even the lack of SOD effects when cultured at high oxygen (116). Another culture medium supplement that may mask the effects of SOD is glucose. Indeed, high glucose concentrations result in increased ROS generation (118, 119), with thus an increased likelihood to obtain an effect of SOD supplementation. Oxygen tension and glucose levels may help explain some of the discordant results, with for instance no effect under low glucose (1.5 mM) and low O_2 (98) and beneficial effects under high glucose (5.56–8.3 mM) and high O_2 (52, 114).

Catalase supplementation (1,000 or 5,400 U/ml) was also tested during mouse embryo culture with no detectable effects (116, 117); similarly, there was no influence of catalase (250–1,000 U/ml) on the development of rabbit zygotes cultured for 72 h in a protein-free medium under 20 % O_2 (52), or on the cleavage and blastocyst formation rates after exposure in the first 72 h of culture (at 5 and 127 U/ml, together with EDTA, under 7 % O_2) in the bovine (98). In another study, catalase supplementation (100 U/ml) promoted mouse blastocyst formation, but this effect occurred only in the absence of EDTA and under 20 % O_2 (47). Nonetheless the effect was minimal when compared to other manipulations, namely SOD, EDTA, or percent oxygen manipulations. Of relevance may be the ineffectiveness of exogenous catalase in reducing intracellular H_2O_2 levels in the embryo (6) perhaps reflecting the inability of catalase to enter the embryo. Future studies should consider the utility of exogenous catalase, as well as the catalase defenses already present in the embryo; if catalase stores are adequate, added catalase may not prove effective. The same considerations apply for SOD, although these enzymatic antioxidants could still modulate extracellular levels of ROS. Lastly, another difficulty in drawing conclusions on the benefits or detriments of enzymatic antioxidant supplementation lies not only in the studies including several antioxidants (even if inadvertently), or conversely focusing on a single one. The latter is a limitation since these antioxidants act cooperatively (Fig. 9.2), and the former may not be targeting the optimal combinations (and/or the right concentrations) of antioxidants.

TRX (20–2,000 μg/ml, from *E. coli*) was also tested during animal embryo culture, particularly given its involvement in post-implantation events in mouse Trx knock-out (120) and transgenic (121) models. In the mouse, TRX promoted blastocyst formation (122), and diminished the 2 cell block in vitro with its influence limited to development from the pronuclear to the 2 cell stage (115, 123). However, Nasr-Esfahani and Johnson (87) showed no improvement in the ability of mouse embryos to develop beyond a 2 cell block upon supplementation with TRX (up to 500 μg/ml). All these cultures were under 20 % O_2; so in light of these conflicting results, one must consider differences in culture conditions that may impact the endogenous protection present in the embryo or the generation of free radicals, and in turn effects of TRX supplementation may vary.

During porcine embryo cultures at 5 % O_2, TRX (1 mg/ml) improved blastocyst formation and total cell numbers; there were also fewer apoptotic cells and the intracellular redox status (GSH content/H_2O_2 levels) of the embryo was increased at the 4 cell and blastocyst stages (89). When examined in the bovine model for which a developmental arrest exists around the 8 cell stage, reduced recombinant TRX (0.5 μg/ml) promoted blastocyst formation (124). In line with the time-sensitive effect reported in the mouse, TRX proved beneficial only when added between 24 and 44 h after insemination; surprisingly, this effect was only apparent under 5 % O_2, with even a decrease in blastocyst cell numbers under 20 % O_2 (124). The potential oxidation, and thus inactivation, of TRX in an oxidizing environment of 20 % O_2 merits further investigation, notably in the presence of the embryo. Further, TRX can also act as an oxidant with the promotion of disulfide bond formation under certain conditions (30).

TRX can be taken up by lens epithelial cells (125), and perhaps also in embryos although direct evidence is currently lacking. All studies to date (with the exception of Bing et al. (124)) used the oxidized form of TRX for supplementation; however upon cell entry, oxidized TRX may be reduced by thioredoxin reductase. Lastly, additive benefits of TRX supplementation with either SOD or a low oxygen tension were documented in mouse blastocyst formation (123). Evidence to date thus warrants further experimental testing the potential benefits of supplementation with enzymatic antioxidants or thioredoxin. Lastly, no studies have yet examined medium enrichment with glutathione peroxidase, in spite of its abundance in the oviductal fluid (33).

The beneficial influences of SOD supplementation likely result from a compromised ability of the embryo to neutralize superoxide radicals, which may exert detrimental effects on development past the 2 cell stage in the mouse (35, 117). This is further supported by detection of a comparable in vitro peak in H_2O_2 at the 2 cell stage between arresting and nonarresting mouse strains (6). Thioredoxin may prevent the oxidation of thiol groups on maternal proteins that would play a functional role in developmental progression; this is based on the ability of thioredoxin to attenuate the mouse 2 cell block, when supplemented in the culture medium during the early stages of development (115, 122, 123). Natsuyama et al. (126) further demonstrated the activation of $p34^{cdc2}$, a key regulator of the cell cycle when SOD or TRX was added during mouse embryo culture. However, not all cleavage divisions may be promoted by TRX (124), and protein(s) targeted for reduction by TRX remain to be identified. Lastly, TRX may also play a role in modulating embryonic cell death, with its supplementation during porcine embryo culture diminishing the proportion of apoptotic cells in blastocysts (89).

1.2. Chelators

1.2.1. EDTA

EDTA is the most common chelator in commercial human embryo culture medium (Table 1), included in 78 % (D1-D3), 100 % (D1-D5), and 11 % (D3-D5) of embryo culture media. It is a synthetic, polyprotic acid with four carboxylic acid groups and two amine groups with lone pair electrons, allowing it to sequester metal ions in a 1:1 metal-to-EDTA complex with six points of contact (Fig. 3).

The first beneficial effects of EDTA for embryo culture were reported in 1977, when the addition of EDTA to Whitten's medium (127) overcame the mouse 2-cell block in vitro (128). Subsequent studies in the mouse and cow confirmed this effect, and have shown that EDTA positively contributes to embryo development in a variety of media (21, 129–134). However, despite its current acceptance as a component in human embryo culture media, the mechanisms of EDTA's effect on embryo development are not fully established and appropriate exposure of embryos to EDTA is still debated.

No consensus exists for the optimal concentration of EDTA, but it most likely depends on the degree of free metal contamination in the culture medium. In one study, 3–5 μM EDTA enhanced bovine embryo development in a chemically defined medium, while 27, 81 and 125 μM EDTA had increasingly detrimental effects (72). Other studies have had success with 10 or 100 μM EDTA in KSOM, SOF, and other chemically defined media (132–137). It is relevant to consider that trace levels of metal ions may have been different in each study's media, thus influencing the amount of EDTA needed for best embryo development. In excess, EDTA may deprive embryos of necessary metal ions (64, 138), but at the right concentration EDTA can protect embryos from toxic amounts of metal ions, and by extension ROS, in the culture environment. For instance, the addition of EDTA to cell-free culture medium diminished levels of ROS compared to EDTA-free medium (139), and in one study EDTA protected developing mouse embryos from toxic effects of impurities—suspected to be zinc—in silicone oil used as overlay (140). To maintain appropriate, steady levels of free metal ions in protein-free culture media, the use of an optimized metal ion buffer has been proposed. This buffer is based on a ligand mixture of 40 μM citrate (see section below) and 4.3 μM EDTA (141).

EDTA may have a developmental stage-dependent effect on embryos. Several in vitro studies suggest that embryos continually exposed to EDTA have impaired development compared to those exposed to EDTA during pre-compaction development only. Proponents of limited EDTA exposure report that treatment with EDTA beyond the 8 cell stage is detrimental, causing decreased cell numbers in the ICM of bovine blastocysts and lowered fetal body weights in mice after transfer of embryos to surrogate mothers (133, 137, 142). Various amounts of time (24, 48, 72 h post-insemination, or until compaction) have been proposed as the best length of EDTA treatment in the mouse and bovine (133, 137, 143). Currently, media exist for pre- or post-compaction human embryo culture, with only the pre-compaction culture media containing EDTA.

Other researchers point out that the detrimental effects of EDTA post-compaction may simply be a question of concentration. Using sequential simplex optimization, the optimal concentration of EDTA in KSOM was reported to be 10 μM (144). A study comparing embryo culture protocols found that mouse embryos cultured continuously in KSOM showed normal development, while those cultured continuously in DM2 experienced impaired development as measured by blastocyst cell counts and hatching rates (135). While KSOM contains 10 μM EDTA, DM2 and the media used in aforementioned studies reporting detrimental effects of continuous EDTA exposure contain 100 μM EDTA (133, 137, 142). In a study of human embryos, five out of nine blastocysts cultured continuously in KSOM led to live births (145). Consequently, it has been proposed that lower concentrations of EDTA (10 μM) may not cause the same detrimental effects in post-compaction embryos as higher concentrations (100 μM). A recent study using mouse embryos reported that optimal blastocyst development was achieved using 100 μM EDTA until 2 days post-fertilization, followed by 10 μM EDTA (141).

At least one mechanism has been proposed to explain why EDTA might affect pre-compaction embryos differently than post-compaction embryos. In mouse embryos, EDTA has been shown to regulate glycolysis by inhibiting cytosolic kinases such as 3-phosphoglycerate kinase (137). Embryos become dependent on the glycolytic pathway for energy only after compaction. Before then, premature utilization of glycolysis leads to reduced development via inhibition of mitochondrial respiratory function (16, 146). Therefore, by limiting glycolysis, EDTA may alleviate the 2-cell block (in mice) and promote cleavage-stage development of embryos, but interfere with post-compaction development. Investigation of intracellular magnesium levels and glycolytic activity during EDTA treatment of embryos led to the proposal that EDTA inhibits 3-phosphoglycerate kinase by reducing the availability, through chelation, of magnesium ions required for normal activity of the enzyme (147). However, other evidence suggests that EDTA may inhibit 3-phosphoglycerate kinase via another, unknown mechanism, but not through chelation (148).

Interestingly, some studies have observed that EDTA significantly reduces blastocyst cell counts in mouse embryos while simultaneously enhancing or not affecting blastocyst formation rates (36, 47, 135). It is therefore possible that an increase in rates of blastocyst formation due to EDTA might belie a second, negative effect of the chelator on embryo quality under some circumstances. Furthermore, it should be considered whether EDTA is an absolute requirement for human embryo culture when sufficient and appropriate nutrients and antioxidants are present. Compared to embryos cultured in media with no EDTA, the rates of blastocyst formation were not significantly different in embryos cultured in KSOM (145). In another human study, rates of blastocyst formation, implantation, and pregnancy were not significantly different between embryos cultured without EDTA and those cultured with EDTA until the 6–8 cell stage in sequential media (149). In both of these studies EDTA was present at 10 μM and not 100 μM.

1.2.2. Citrate

Citrate is a component of several commercial media (33 %) used for human embryo culture. It was first identified as an embryotrophic component of BSA in 1992, and was shown to stimulate rabbit blastocyst growth (150). Citrate is thought to have several roles in promoting embryo development, such as serving as an energy substrate, stimulating fatty acid synthesis (151, 152), acting as an allosteric activator of acetyl-coenzyme A carboxylase, and functioning as a chelator (138, 150, 151, 153). As a chelator, citrate has a high affinity for ferric, calcium, strontium, manganese, and magnesium ions (138). In particular, chelation of calcium ions by citrate has been proposed to be an important event in maintaining junctional integrity in embryos (150, 151, 154). This mechanism is discussed later, as it may apply to other chelators as well. In embryo culture, citrate has been reported to have variable effects on embryo development, possibly due in part to its multiple activities and the differing compositions of culture media. Citrate added to G1 and SOF media improved blastocyst development in the bovine (153, 155, 156). A study in the mouse reported that citric acid improved embryo development over controls, but was not as effective as EDTA (130). Other studies in the bovine found that citrate in a chemically defined medium had no significant influence on embryo development (157–159), while another found that a beneficial effect was evident only when citrate was used in concert with nonessential amino acids in SOF (160). Lastly, one study reports that citrate is beneficial to bovine embryos during coculture (152). The authors suggest that the chelating activity of citrate removes potential metabolites and toxicants secreted by somatic cells in coculture.

1.2.3. Experimental Chelators

Other chelators used experimentally in studies of animal embryo culture include transferrin, DFO, DTPA, EDDA, EGTA and NTA. Although they are generally not included in human embryo culture media, studies of these chelators offer new information and bring to light concerns about the general use of chelators in embryo culture. Transferrin is a protein with a high affinity for iron, with each transferrin molecule capable of binding two atoms of ferric ion (161). In studies of mouse embryos, bovine transferrin promoted development through the 2-cell block after 22 h of treatment following zygote retrieval (21, 162). Interestingly, human transferrin had no beneficial effect on mouse embryos and was even embryotoxic at some concentrations (21). These results demonstrate that species differences are important to consider when adding chelators to embryo culture media. In another mouse study, the absence of transferrin when EDTA and amino acids were present had no effect on embryo development (137). Such a result demonstrates that the effects of some chelators may be eclipsed by others, rendering their presence unnecessary under some conditions.

DFO is a bacterial siderophore—a category of molecules that are among the strongest soluble ferric ion chelators known—and binds iron in a one-to-one ratio with six points of contact. At high levels, DFO is embryotoxic to mouse embryos and at low levels does not overcome the 2 cell block (21). Concentrations of 1 or 10 μM DFO beginning at the 1 cell stage leads to lower cleavage rates than those of control embryos in mice (148). DFO is also a hypoxia-induced factor (HIF) protein stabilizer. When bovine morulae are cultured under 7 % oxygen (but not 2 %) with up to 1 μM DFO, HIF-regulated gene expression increases in developing blastocysts (163), suggesting that, from a genetic perspective, DFO might mimic a reduced oxygen environment. It is relevant to consider, then, that other chelators or chelates may also modulate gene expression.

DTPA is structurally similar to EDTA, and was introduced in 1978 (164). In concentrations of 10 and 100–500 μM, DTPA enhances development of certain strains of mouse embryos susceptible to the 2-cell block in vitro (21, 148). In the bovine, 125 μM DTPA was embryotoxic, while 3–5 μM DTPA positively contributed to embryo development (72). The extreme differences in reported concentrations may be a consequence of species or culture media differences. In the mouse study, BSA was present in the culture medium and may have buffered the embryos from the chelator, allowing for higher concentrations of DTPA to be tolerated (72). Lastly, other chelators such as EDDA, EGTA dipicolinic acid and NTA have also been shown by one study to alleviate the 2-cell block in mice (148).

1.2.4. Notes on Chelator Use in Embryo Culture

Varying degrees of success with different chelators in embryo culture are most likely partially attributable to their individual molecular structures. It is relevant to consider that chelators such as EDTA, DTPA and citrate can induce the catalytic conversion of hydrogen peroxide into hydroxyl radicals under certain circumstances. If the chelator to metal ratio is low, EDTA or DTPA may increase the rate of these oxidative processes and effectively act as prooxidants, but if the chelator is present in excess relative to metals and hydrogen peroxide (~2:1 chelator to metal ratio), it may decrease the rate of such oxidative processes, protecting embryos from ROS generation (165). However, one must keep in mind that, as described earlier, most chelators can also be embryotoxic beyond a threshold. Furthermore, the molecular structure of a chelator, for instance the number of points of contact between chelator and metal, affects how the metal ion is altered during chelation. It is proposed that the absence of positive effects with DFO in embryo culture is due to its exceptionally high affinity for iron. DFO-bound iron cannot participate in catalytic reactions, a trait which, if a positive requirement for intracellular iron exists during early embryo development, may interfere with necessary cell processes (21, 148). Conversely, chelators with lower binding affinities may function as iron transporters, or "chaperones," and thus stimulate development while simultaneously limiting the generation of ROS. The molecular structure and charge of a chelator also determines whether it can function extracellularly or intracellularly. In mice, transferrin uptake occurs in blastocysts but not in earlier stage embryos (162), suggesting that beneficial effects of transferrin early in development, such as overcoming the 2 cell block, are due to chelation, and later effects are due in part to iron transport. On the other hand, EDTA is negatively charged, meaning it cannot permeate the cell membrane. Therefore, it is a chelator of extracellular metal ions (148). In fact, if embryos are injected with EDTA either in perivitelline space or intracellularly, only the former develop normally (130).

Each chelator has different binding affinities for specific metal ions. For instance, transferrin binds ferric ions, while DTPA and EDTA bind both ferric and ferrous ions. Dipicolinic acid is used most often as a selective zinc chelator (166). EDTA, DTPA, and transferrin can each act as ferroxidases under some conditions, converting ferrous ions to ferric ions. This conversion might inhibit the creation of dangerous hydroxyl radicals by limiting ferrous ion availability in Haber–Weiss reactions (Fig. 1). Additionally, EDTA, DTPA, and transferrin, but not DFO, will also bind other metal ions such as copper. As copper can also participate in Haber–Weiss reactions, this characteristic may confer additional protection against ROS generation in embryo culture. Lastly, it has been proposed that chelation of calcium ions may open tight junctions to facilitate solute transport (154). Such a function would be important for maintaining junctional integrity and, therefore, for compaction and blastocoel formation in embryos as well (150, 151). In particular, this function has been investigated and proposed for citrate and EGTA (151, 154).

1.3. Conclusion

In this chapter we have attempted to summarize the current use of, and knowledge about, antioxidants and chelators in embryo culture. Although many uncertainties still exist regarding dosage, time of exposure, and even effect of various molecules intended for embryo protection against ROS, it is clear that a careful balance must be found between sufficient protection and harmful excess. Furthermore, the chemical nature of each antioxidant or chelator is highly relevant to its function and how it will interact with all other media components. It is unusual to find an antioxidant or chelator whose exclusive effect is to lower ROS concentrations. In addition to their protective roles, these molecules are also potential modulators of cell function and they may influence the functions of other antioxidants or chelators.

The influence of the chemical milieu of culture media should not be ignored when evaluating the impact of antioxidants or chelators. As molecules are continually altered, generated or depleted, it is a dynamic system that remains in constant flux. Therefore, one cannot assume that the balance and function of antioxidants and chelators remain the same throughout the culture period. For instance, whether or not the medium is renewed during culture will affect the presence of potentially important supplements, simply due to their stability in that particular chemical environment. The half-lives of these antioxidants and chelators, such as Vitamin C, are influenced by the availability of free oxygen and metals (71, 138). Additionally, another consideration is that various endogenous compounds may readily oxidize in embryo culture media. The ROS produced by these reactions could be "hidden" by the antioxidant function of other media components such as pyruvate. If so, pyruvate levels may decrease while the prooxidant compounds remain unidentified (106).

The redox status of an embryo, involved in the expression of many genes and normal cell function, may determine the directionality of an antioxidant or chelator's effect on development. At normal physiological levels, ROS play essential roles in cell signaling (167), and the importance of their presence has been suggested in embryos (111). Therefore, the goal of antioxidant and chelator addition to culture medium is to maintain safe levels of ROS, rather than to eliminate them altogether. Disproportionate concentrations of antioxidants or chelators with respect to ROS may interfere with normal cell processes. Optimal concentrations for such supplements need to be determined for each instance of use, and for the cultured human embryo specifically.

The sites of excess free radical production and sites of action for supplemented antioxidants and chelators merit further examination. An understanding of whether protection against ROS is needed extracellularly or intracellularly (or both) would allow for a more targeted approach to culture media optimization. Furthermore, more work is needed to distinguish between the direct effects of protective supplements and the consequences of their interactions with other components of the culture medium. Lastly, there may be relevant differences in the responses to antioxidants and chelators of embryos that originate from maturation and/or fertilization in vitro versus in vivo.

Finally, there is some support for the preferential use of "physiological" rather than artificial antioxidants and chelators in embryo culture. Such compounds would include proteins, amino acids, and vitamins instead of synthetic molecules such as EDTA. It has been suggested that, if cultured under 5 % oxygen, physiological antioxidants and chelators would suffice as supplemental protection against ROS (47). Although antioxidants and chelators are routinely included in embryo culture media, there are many important aspects of their function that remain to be elucidated. In the meantime, caution should be used when supplementing embryo culture media, and an effort should be made to understand how each added component may affect the chemical environment of the embryo.

References

1. Bertout J, Mahutte N, Preston S, Behrman H (2004) Reactive oxygen species and ovarian function. In: Leung PCK, Adashi EY (eds) The ovary, 2nd edn. Elsevier, London, pp 353–368
2. Halliwell B (2007) Biochemistry of oxidative stress. Biochem Soc Trans 35:1147–1150
3. Hashimoto S, Minami N, Yamada M, Imai H (2000) Excessive concentration of glucose during in vitro maturation impairs the developmental competence of bovine oocytes after in vitro fertilization: relevance to intracellular reactive oxygen species and glutathione contents. Mol Reprod Dev 56:520–526

4. Cebral E, Carrasco I, Vantman D, Smith R (2007) Preimplantation embryotoxicity after mouse embryo exposition to reactive oxygen species. Biocell 31:51–59
5. Betts DH, Madan P (2008) Permanent embryo arrest: molecular and cellular concepts. Mol Hum Reprod 14:445–453
6. Nasr-Esfahani MH, Aitken JR, Johnson MH (1990) Hydrogen peroxide levels in mouse oocytes and early cleavage stage embryos developed in vitro or in vivo. Development 109:501–507
7. Bedaiwy MA, Falcone T, Mohamed MS, Aleem AA, Sharma RK, Worley SE, Thornton J, Agarwal A (2004) Differential growth of human embryos in vitro: role of reactive oxygen species. Fertil Steril 82:593–600
8. Krajcir N, Chowdary H, Gupta S, Agarwal A (2008) Female infertility and assisted reproduction: impact of oxidative stress. Curr Women's Health Rev 4:7
9. Agarwal A, Gupta S, Sharma RK (2005) Oxidative stress and its implications in female infertility—a clinician's perspective. Reprod Biomed Online 11:10
10. Fischer B, Bavister BD (1993) Oxygen tension in the oviduct and uterus of rhesus monkeys, hamsters and rabbits. J Reprod Fertil 99:673–679
11. Luvoni GC, Keskintepe L, Brackett BG (1996) Improvement in bovine embryo production in vitro by glutathione-containing culture media. Mol Reprod Dev 43:437–443
12. Pabon JE Jr, Findley WE, Gibbons WE (1989) The toxic effect of short exposures to the atmospheric oxygen concentration on early mouse embryonic development. Fertil Steril 51:896–900
13. Grzelak A, Rychlik B, Bartosz G (2001) Light-dependent generation of reactive oxygen species in cell culture media. Free Radic Biol Med 30:8
14. Nakayama T, Noda Y, Goto Y, Mori T (1994) Effects of visible light and other environmental factors on the production of oxygen radicals by hamster embryos. Theriogenology 41:499–510
15. Goto Y, Noda Y, Mori T, Nakano M (1993) Increased generation of reactive oxygen species in embryos cultured in vitro. Free Radic Biol Med 15:69–75
16. Catt JW, Henman M (2000) Toxic effects of oxygen on human embryo development. Hum Reprod 15(Suppl 2):199–206
17. Bavister BD (1995) Culture of preimplantation embryos: facts and artifacts. Hum Reprod Update 1:91–148
18. Halliwell B, Gutteridge JM (1989) Free radicals in biology and medicine, 2nd edn. Oxford University Press, New York, NY
19. Minotti G (1993) Sources and role of iron in lipid peroxidation. Chem Res Toxicol 6:134–146
20. Gutteridge JM (1987) A method for removal of trace iron contamination from biological buffers. FEBS Lett 214:362–364
21. Nasr-Esfahani M, Johnson MH, Aitken RJ (1990) The effect of iron and iron chelators on the in-vitro block to development of the mouse preimplantation embryo: BAT6 a new medium for improved culture of mouse embryos in vitro. Hum Reprod 5:997–1003
22. Yang HW, Hwang KJ, Kwon HC, Kim HS, Choi KW, Oh KS (1998) Detection of reactive oxygen species (ROS) and apoptosis in human fragmented embryos. Hum Reprod 13:998–1002
23. Takahashi M, Nagai T, Hamano S, Kuwayama M, Okamura N, Okano A (1993) Effect of thiol compounds on in vitro development and intracellular glutathione content of bovine embryos. Biol Reprod 49:228–232
24. Gardiner CS, Reed DJ (1994) Status of glutathione during oxidant-induced oxidative stress in the preimplantation mouse embryo. Biol Reprod 51:1307–1314
25. Gardiner CS, Reed DJ (1995) Synthesis of glutathione in the preimplantation mouse embryo. Arch Biochem Biophys 318:30–36
26. Gardiner CS, Salmen JJ, Brandt CJ, Stover SK (1998) Glutathione is present in reproductive tract secretions and improves development of mouse embryos after chemically induced glutathione depletion. Biol Reprod 59:431–436
27. de Matos DG, Furnus CC (2000) The importance of having high glutathione (GSH) level after bovine in vitro maturation on embryo development effect of beta-mercaptoethanol, cysteine and cystine. Theriogenology 53:761–771
28. Harvey MB, Arcellana-Panlilio MY, Zhang X, Schultz GA, Watson AJ (1995) Expression of genes encoding antioxidant enzymes in preimplantation mouse and cow embryos and primary bovine oviduct cultures employed for embryo coculture. Biol Reprod 53:532–540
29. Ogasawara M, Nakamura T, Koyama I, Nemoto M, Yoshida T (1993) Reactivity of taurine with aldehydes and its physiological role. Chem Pharm Bull (Tokyo) 41:2172–2175
30. Debarbieux L, Beckwith J (2000) On the functional interchangeability, oxidant versus

reductant, of members of the thioredoxin superfamily. J Bacteriol 182:723–727
31. Lindenbaum A (1973) A survey of naturally occurring chelating ligands. Adv Exp Med Biol 40:67–77
32. Jeppsen RB (2001) Toxicology and safety of Ferrochel and other iron amino acid chelates. Arch Latinoam Nutr 51:26–34
33. Lapointe J, Bilodeau JF (2003) Antioxidant defenses are modulated in the cow oviduct during the estrous cycle. Biol Reprod 68:1157–1164
34. Lapointe S, Sullivan R, Sirard MA (1998) Binding of a bovine oviductal fluid catalase to mammalian spermatozoa. Biol Reprod 58:747–753
35. Noda Y, Matsumoto H, Umaoka Y, Tatsumi K, Kishi J, Mori T (1991) Involvement of superoxide radicals in the mouse two-cell block. Mol Reprod Dev 28:356–360
36. Chun YS, Kim JH, Lee HT, Chung KS (1994) Effect of superoxide dismutase on the development of preimplantation mouse embryos. Theriogenology 41:511–520
37. Paszkowski T, Clarke RN (1999) The Graafian follicle is a site of l-ascorbate accumulation. J Assist Reprod Genet 16:41–45
38. Gardner DK, Leese HJ (1990) Concentrations of nutrients in mouse oviduct fluid and their effects on embryo development and metabolism in vitro. J Reprod Fertil 88:361–368
39. Guerin P, Guillaud J, Menezo Y (1995) Hypotaurine in spermatozoa and genital secretions and its production by oviduct epithelial cells in vitro. Hum Reprod 10:866–872
40. Casslen BG (1987) Free amino acids in human uterine fluid. Possible role of high taurine concentration. J Reprod Med 32:181–184
41. Guyader-Joly C, Guerin P, Renard JP, Guillaud J, Ponchon S, Menezo Y (1998) Precursors of taurine in female genital tract: effects on developmental capacity of bovine embryo produced in vitro. Amino Acids 15:27–42
42. El Mouatassim S, Guerin P, Menezo Y (2000) Mammalian oviduct and protection against free oxygen radicals: expression of genes encoding antioxidant enzymes in human and mouse. Eur J Obstet Gynecol Reprod Biol 89:1–6
43. Virji N, Phillips DM, Dunbar BS (1990) Identification of extracellular proteins in the rat cumulus oophorus. Mol Reprod Dev 25:339–344
44. Pemble LB, Kaye PL (1986) Whole protein uptake and metabolism by mouse blastocysts. J Reprod Fertil 78:149–157
45. Barbehenn EK, Wales RG, Lowry OH (1974) The explanation for the blockade of glycolysis in early mouse embryos. Proc Natl Acad Sci USA 71:1056–1060
46. Aguilar J, Reyley M (2005) The uterine tubal fluid: secretion, composition and biological effects. Anim Reprod 2:15
47. Orsi NM, Leese HJ (2001) Protection against reactive oxygen species during mouse preimplantation embryo development: role of EDTA, oxygen tension, catalase, superoxide dismutase and pyruvate. Mol Reprod Dev 59:44–53
48. Fellman JH, Green TR, Eicher AL (1987) The oxidation of hypotaurine to taurine: bis-aminoethyl-alpha-disulfone, a metabolic intermediate in mammalian tissue. Adv Exp Med Biol 217:39–48
49. Van Winkle LJ, Dickinson HR (1995) Differences in amino acid content of preimplantation mouse embryos that develop in vitro versus in vivo: in vitro effects of five amino acids that are abundant in oviductal secretions. Biol Reprod 52:96–104
50. Dumoulin JC, Evers JL, Bras M, Pieters MH, Geraedts JP (1992) Positive effect of taurine on preimplantation development of mouse embryos in vitro. J Reprod Fertil 94:373–380
51. Dumoulin JC, Evers JL, Bakker JA, Bras M, Pieters MH, Geraedts JP (1992) Temporal effects of taurine on mouse preimplantation development in vitro. Hum Reprod 7:403–407
52. Li J, Foote RH, Simkin M (1993) Development of rabbit zygotes cultured in protein-free medium with catalase, taurine, or superoxide dismutase. Biol Reprod 49:33–37
53. Barnett DK, Bavister BD (1992) Hypotaurine requirement for in vitro development of golden hamster one-cell embryos into morulae and blastocysts, and production of term offspring from in vitro-fertilized ova. Biol Reprod 47:297–304
54. Devreker F, Van den Bergh M, Biramane J, Winston RL, Englert Y, Hardy K (1999) Effects of taurine on human embryo development in vitro. Hum Reprod 14:2350–2356
55. Devreker F, Hardy K (1997) Effects of glutamine and taurine on preimplantation development and cleavage of mouse embryos in vitro. Biol Reprod 57:921–928
56. Aruoma OI, Halliwell B, Hoey BM, Butler J (1988) The antioxidant action of taurine, hypotaurine and their metabolic precursors. Biochem J 256:251–255
57. Huxtable RJ (1992) Physiological actions of taurine. Physiol Rev 72:101–163

58. Liu Z, Foote RH (1995) Development of bovine embryos in KSOM with added superoxide dismutase and taurine and with five and twenty percent O_2. Biol Reprod 53:786–790
59. Fujitani Y, Kasai K, Ohtani S, Nishimura K, Yamada M, Utsumi K (1997) Effect of oxygen concentration and free radicals on in vitro development of in vitro-produced bovine embryos. J Anim Sci 75:483–489
60. Dumoulin JC, van Wissen LC, Menheere PP, Michiels AH, Geraedts JP, Evers JL (1997) Taurine acts as an osmolyte in human and mouse oocytes and embryos. Biol Reprod 56:739–744
61. Wang X, Falcone T, Attaran M, Goldberg JM, Agarwal A, Sharma RK (2002) Vitamin C and vitamin E supplementation reduce oxidative stress-induced embryo toxicity and improve the blastocyst development rate. Fertil Steril 78:1272–1277
62. Tarin JJ, de los Santos MJ, de Oliveira MN, Pellicer A, Bonilla-Musoles F (1994) Ascorbate-supplemented media in short-term cultures of human embryos. Hum Reprod 9:1717–1722
63. McKiernan SH, Bavister BD (2000) Culture of one-cell hamster embryos with water soluble vitamins: pantothenate stimulates blastocyst production. Hum Reprod 15:157–164
64. Olson SE, Seidel GE Jr (2000) Culture of in vitro-produced bovine embryos with vitamin E improves development in vitro and after transfer to recipients. Biol Reprod 62:248–252
65. Natarajan R, Shankar MB, Munuswamy D (2010) Effect of alpha-tocopherol supplementation on in vitro maturation of sheep oocytes and in vitro development of preimplantation sheep embryos to the blastocyst stage. J Assist Reprod Genet 27.483–490
66. Kitagawa Y, Suzuki K, Yoneda A, Watanabe T (2004) Effects of oxygen concentration and antioxidants on the in vitro developmental ability, production of reactive oxygen species (ROS), and DNA fragmentation in porcine embryos. Theriogenology 62:1186–1197
67. Thiyagarajan B, Valivittan K (2009) Ameliorating effect of vitamin E on in vitro development of preimplantation buffalo embryos. J Assist Reprod Genet 26:217–225
68. Feugang JM, de Roover R, Moens A, Leonard S, Dessy F, Donnay I (2004) Addition of beta-mercaptoethanol or Trolox at the morula/blastocyst stage improves the quality of bovine blastocysts and prevents induction of apoptosis and degeneration by prooxidant agents. Theriogenology 61:71–90
69. Halliwell B (1996) Vitamin C: antioxidant or pro-oxidant in vivo? Free Radic Res 25:439–454
70. Yamamoto K, Niki E (1988) Interaction of alpha-tocopherol with iron: antioxidant and prooxidant effects of alpha-tocopherol in the oxidation of lipids in aqueous dispersions in the presence of iron. Biochim Biophys Acta 958:19–23
71. Osiecki M, Ghanavi P, Atkinson K, Nielsen LK, Doran MR (2010) The ascorbic acid paradox. Biochem Biophys Res Commun 400:466–470
72. Olson SE, Seidel GE Jr (2000) Reduced oxygen tension and EDTA improve bovine zygote development in a chemically defined medium. J Anim Sci 78:152–157
73. Jeong YW, Park SW, Hossein MS, Kim S, Kim JH, Lee SH, Kang SK, Lee BC, Hwang WS (2006) Antiapoptotic and embryotrophic effects of alpha-tocopherol and l-ascorbic acid on porcine embryos derived from in vitro fertilization and somatic cell nuclear transfer. Theriogenology 66:2104–2112
74. Donnelly ET, McClure N, Lewis SE (1999) The effect of ascorbate and alpha-tocopherol supplementation in vitro on DNA integrity and hydrogen peroxide-induced DNA damage in human spermatozoa. Mutagenesis 14:505–512
75. Matxain JM, Ristila M, Strid A, Eriksson LA (2006) Theoretical study of the antioxidant properties of pyridoxine. J Phys Chem A 110:13068–13072
76. Hellmann H, Mooney S (2010) Vitamin B6: a molecule for human health? Molecules 15:442–459
77. Stocker P, Lesgards JF, Vidal N, Chalier F, Prost M (2003) ESR study of a biological assay on whole blood: antioxidant efficiency of various vitamins. Biochim Biophys Acta 1621:1–8
78. Slyshenkov VS, Rakowska M, Moiseenok AG, Wojtczak L (1995) Pantothenic acid and its derivatives protect Ehrlich ascites tumor cells against lipid peroxidation. Free Radic Biol Med 19:767–772
79. Slyshenkov VS, Dymkowska D, Wojtczak L (2004) Pantothenic acid and pantothenol increase biosynthesis of glutathione by boosting cell energetics. FEBS Lett 569:169–172
80. Kane MT, Bavister BD (1988) Vitamin requirements for development of eight-cell hamster embryos to hatching blastocysts in vitro. Biol Reprod 39:1137–1143
81. Tsai FC, Gardner DK (1994) Nicotinamide, a component of complex culture media, inhibits

mouse embryo development in vitro and reduces subsequent developmental potential after transfer. Fertil Steril 61:376–382
82. O'Neill C (1998) Endogenous folic acid is essential for normal development of preimplantation embryos. Hum Reprod 13:1312–1316
83. Kane MT (1988) The effects of water-soluble vitamins on the expansion of rabbit blastocysts in vitro. J Exp Zool 245:220–223
84. Shirley BA (1989) Inhibition of development of preimplantation mouse embryos in vitro by some water soluble vitamins. Med Sci Res 17:465–466
85. Liu L, Trimarchi JR, Keefe DL (1999) Thiol oxidation-induced embryonic cell death in mice is prevented by the antioxidant dithiothreitol. Biol Reprod 61:1162–1169
86. Harvey AJ, Kind KL, Thompson JG (2002) REDOX regulation of early embryo development. Reproduction 123:479–486
87. Nasr-Esfahani MH, Johnson MH (1992) Quantitative analysis of cellular glutathione in early preimplantation mouse embryos developing in vivo and in vitro. Hum Reprod 7:1281–1290
88. Wang WH, Day BN (2002) Development of porcine embryos produced by IVM/IVF in a medium with or without protein supplementation: effects of extracellular glutathione. Zygote 10:109–115
89. Ozawa M, Nagai T, Fahrudin M, Karja NW, Kaneko H, Noguchi J, Ohnuma K, Kikuchi K (2006) Addition of glutathione or thioredoxin to culture medium reduces intracellular redox status of porcine IVM/IVF embryos, resulting in improved development to the blastocyst stage. Mol Reprod Dev 73:998–1007
90. McEvoy TG, Coull GD, Broadbent PJ, Hutchinson JS, Speake BK (2000) Fatty acid composition of lipids in immature cattle, pig and sheep oocytes with intact zona pellucida. J Reprod Fertil 118:163–170
91. Meister A, Anderson ME, Hwang O (1986) Intracellular cysteine and glutathione delivery systems. J Am Coll Nutr 5:137–151
92. de Matos DG, Gasparrini B, Pasqualini SR, Thompson JG (2002) Effect of glutathione synthesis stimulation during in vitro maturation of ovine oocytes on embryo development and intracellular peroxide content. Theriogenology 57:1443–1451
93. Caamano JN, Ryoo ZY, Thomas JA, Youngs CR (1996) Beta-mercaptoethanol enhances blastocyst formation rate of bovine in vitro-matured/in vitro-fertilized embryos. Biol Reprod 55:1179–1184
94. Caamano JN, Ryoo ZY, Youngs CR (1998) Promotion of development of bovine embryos produced in vitro by addition of cysteine and beta-mercaptoethanol to a chemically defined culture system. J Dairy Sci 81:369–374
95. Lim JM, Liou SS, Hansel W (1996) Intracytoplasmic glutathione concentration and the role of beta-mercaptoethanol in preimplantation development of bovine embryos. Theriogenology 46:429–439
96. Takahashi M, Nagai T, Okamura N, Takahashi H, Okano A (2002) Promoting effect of beta-mercaptoethanol on in vitro development under oxidative stress and cystine uptake of bovine embryos. Biol Reprod 66:562–567
97. Lim JM, Reggio BC, Godke RA, Hansel W (1999) Development of in-vitro-derived bovine embryos cultured in 5 % CO_2 in air or in 5 % O_2, 5 % CO_2 and 90 % N_2. Hum Reprod 14:458–464
98. Ali AA, Bilodeau JF, Sirard MA (2003) Antioxidant requirements for bovine oocytes varies during in vitro maturation, fertilization and development. Theriogenology 59:939–949
99. Van Soom A, Yuan YQ, Peelman LJ, de Matos DG, Dewulf J, Laevens H, de Kruif A (2002) Prevalence of apoptosis and inner cell allocation in bovine embryos cultured under different oxygen tensions with or without cysteine addition. Theriogenology 57:1453–1465
100. Gardiner CS, Reed DJ (1995) Glutathione redox cycle-driven recovery of reduced glutathione after oxidation by tertiary-butyl hydroperoxide in preimplantation mouse embryos. Arch Biochem Biophys 321:6–12
101. de Matos DG, Herrera C, Cortvrindt R, Smitz J, Van Soom A, Nogueira D, Pasqualini RS (2002) Cysteamine supplementation during in vitro maturation and embryo culture: a useful tool for increasing the efficiency of bovine in vitro embryo production. Mol Reprod Dev 62:203–209
102. Holleman MAF (1904) Notice sur l'action de l'eau oxygenee sur les acides alpha-cetoniques et sur les dicetones. Recl Trav Chim Belg 23:4
103. Andrae U, Singh J, Ziegler-Skylakakis K (1985) Pyruvate and related alpha-ketoacids protect mammalian cells in culture against hydrogen peroxide-induced cytotoxicity. Toxicol Lett 28:93–98
104. O'Donnell-Tormey J, Nathan CF, Lanks K, DeBoer CJ, de la Harpe J (1987) Secretion of pyruvate. An antioxidant defense of mammalian cells. J Exp Med 165:500–514
105. Salahudeen AK, Clark EC, Nath KA (1991) Hydrogen peroxide-induced renal injury.

A protective role for pyruvate in vitro and in vivo. J Clin Invest 88:1886–1893
106. Long LH, Halliwell B (2009) Artefacts in cell culture: pyruvate as a scavenger of hydrogen peroxide generated by ascorbate or epigallocatechin gallate in cell culture media. Biochem Biophys Res Commun 388:700–704
107. Babich H, Liebling EJ, Burger RF, Zuckerbraun HL, Schuck AG (2009) Choice of DMEM, formulated with or without pyruvate, plays an important role in assessing the in vitro cytotoxicity of oxidants and prooxidant nutraceuticals. In Vitro Cell Dev Biol Anim 45:226–233
108. Morales H, Tilquin P, Rees JF, Massip A, Dessy F, Van Langendonckt A (1999) Pyruvate prevents peroxide-induced injury of in vitro preimplantation bovine embryos. Mol Reprod Dev 52:149–157
109. O'Fallon JV, Wright RW (1995) Pyruvate revisited: a non-metabolic role for pyruvate in preimplantation embryo development. Theriogenology 43:1
110. Weathersbee PS, Pool TB, Ord T (1995) Synthetic serum substitute (SSS): a globulin-enriched protein supplement for human embryo culture. J Assist Reprod Genet 12:354–360
111. Esfandiari N, Falcone T, Agarwal A, Attaran M, Nelson DR, Sharma RK (2005) Protein supplementation and the incidence of apoptosis and oxidative stress in mouse embryos. Obstet Gynecol 105:653–660
112. Halliwell B (1988) Albumin—an important extracellular antioxidant? Biochem Pharmacol 37:569–571
113. Halliwell B, Gutteridge JM (1990) The antioxidants of human extracellular fluids. Arch Biochem Biophys 280:1–8
114. Umaoka Y, Noda Y, Narimoto K, Mori T (1992) Effects of oxygen toxicity on early development of mouse embryos. Mol Reprod Dev 31:28–33
115. Nonogaki T, Noda Y, Narimoto K, Umaoka Y, Mori T (1991) Protection from oxidative stress by thioredoxin and superoxide dismutase of mouse embryos fertilized in vitro. Hum Reprod 6:1305–1310
116. Payne SR, Munday R, Thompson JG (1992) Addition of superoxide dismutase and catalase does not necessarily overcome developmental retardation of one-cell mouse embryos during in-vitro culture. Reprod Fertil Dev 4:167–174
117. Legge M, Sellens MH (1991) Free radical scavengers ameliorate the 2-cell block in mouse embryo culture. Hum Reprod 6:867–871
118. Karja NW, Kikuchi K, Fahrudin M, Ozawa M, Somfai T, Ohnuma K, Noguchi J, Kaneko H, Nagai T (2006) Development to the blastocyst stage, the oxidative state, and the quality of early developmental stage of porcine embryos cultured in alteration of glucose concentrations in vitro under different oxygen tensions. Reprod Biol Endocrinol 4:54
119. Iwata H, Akamatsu S, Minami N, Yamada M (1999) Allopurinol, an inhibitor of xanthine oxidase, improves the development of IVM/IVF bovine embryos (>4 cell) in vitro under certain culture conditions. Theriogenology 51:613–622
120. Matsui M, Oshima M, Oshima H, Takaku K, Maruyama T, Yodoi J, Taketo MM (1996) Early embryonic lethality caused by targeted disruption of the mouse thioredoxin gene. Dev Biol 178:179–185
121. Kobayashi-Miura M, Nakamura H, Yodoi J, Shiota K (2002) Thioredoxin, an anti-oxidant protein, protects mouse embryos from oxidative stress-induced developmental anomalies. Free Radic Res 36:949–956
122. Natsuyama S, Noda Y, Narimoto K, Umaoka Y, Mori T (1992) Release of two-cell block by reduction of protein disulfide with thioredoxin from *Escherichia coli* in mice. J Reprod Fertil 95:649–656
123. Goto Y, Noda Y, Narimoto K, Umaoka Y, Mori T (1992) Oxidative stress on mouse embryo development in vitro. Free Radic Biol Med 13:47–53
124. Bing YZ, Hirao Y, Takenouchi N, Che LM, Nakamura H, Yodoi J, Nagai T (2003) Effects of thioredoxin on the preimplantation development of bovine embryos. Theriogenology 59:863–873
125. Spector A, Yan GZ, Huang RR, McDermott MJ, Gascoyne PR, Pigiet V (1988) The effect of H_2O_2 upon thioredoxin-enriched lens epithelial cells. J Biol Chem 263:4984–4990
126. Natsuyama S, Noda Y, Yamashita M, Nagahama Y, Mori T (1993) Superoxide dismutase and thioredoxin restore defective p34cdc2 kinase activation in mouse two-cell block. Biochim Biophys Acta 1176:90–94
127. Whitten WK (1971) Nutrient requirements for the culture of preimplantation embryos in vitro. In: Raspa G (ed) Advances in the Biosciences, Vol 6. Pergamon Press, Oxford, pp 129–140
128. Abramczuk J, Solter D, Koprowski H (1977) The beneficial effect EDTA on development of mouse one-cell embryos in chemically defined medium. Dev Biol 61:378–383

129. Loutradis D, John D, Kiessling AA (1987) Hypoxanthine causes a 2-cell block in random-bred mouse embryos. Biol Reprod 37:311–316
130. Fissore RA, Jackson KV, Kiessling AA (1989) Mouse zygote development in culture medium without protein in the presence of ethylenediaminetetraacetic acid. Biol Reprod 41:835–841
131. Nasr-Esfahani MH, Winston NJ, Johnson MH (1992) Effects of glucose, glutamine, ethylenediaminetetraacetic acid and oxygen tension on the concentration of reactive oxygen species and on development of the mouse preimplantation embryo in vitro. J Reprod Fertil 96:219–231
132. Nakao H, Nakatsuji N (1990) Effects of co-culture, medium components and gas phase on in vitro culture of in vitro matured and in vitro fertilized bovine embryos. Theriogenology 33:591–600
133. Gardner DK, Lane MW, Lane M (2000) EDTA stimulates cleavage stage bovine embryo development in culture but inhibits blastocyst development and differentiation. Mol Reprod Dev 57:256–261
134. Mehta TS, Kiessling AA (1993) The developmental potential of mouse embryos conceived in Ham's F-10 medium containing ethylenediaminetetraacetic acid. Fertil Steril 60:1088–1093
135. Biggers JD, McGinnis LK, Lawitts JA (2005) One-step versus two-step culture of mouse preimplantation embryos: is there a difference? Hum Reprod 20:3376–3384
136. Biggers JD, McGinnis LK, Summers MC (2006) One-step versus two-step culture of mouse preimplantation embryos. Hum Reprod 21:4
137. Gardner DK, Lane M (1996) Alleviation of the '2-cell block' and development to the blastocyst of CF1 mouse embryos: role of amino acids, EDTA and physical parameters. Hum Reprod 11:2703–2712
138. Chaberek S, Martell AE (1959) Organic sequestering agents. Wiley, New York, NY
139. Martin-Romero FJ, Miguel-Lasobras EM, Dominguez-Arroyo JA, Gonzalez-Carrera E, Alvarez IS (2008) Contribution of culture media to oxidative stress and its effect on human oocytes. Reprod Biomed Online 17:652–661
140. Erbach GT, Bhatnagar P, Baltz JM, Biggers JD (1995) Zinc is a possible toxic contaminant of silicone oil in microdrop cultures of preimplantation mouse embryos. Hum Reprod 10:3248–3254
141. Hentemann M, Bertheussen K (2009) New media for culture to blastocyst. Fertil Steril 91:878–883
142. Lane M, Gardner DK (2007) Embryo culture medium: which is the best? Best Pract Res Clin Obstet Gynaecol 21:83–100
143. Gardner DK, Vella P, Lane M, Wagley L, Schlenker T, Schoolcraft WB (1998) Culture and transfer of human blastocysts increases implantation rates and reduces the need for multiple embryo transfers. Fertil Steril 69:84–88
144. Lawitts JA, Biggers JD (1991) Optimization of mouse embryo culture media using simplex methods. J Reprod Fertil 91:543–556
145. Biggers JD, Racowsky C (2002) The development of fertilized human ova to the blastocyst stage in KSOM(AA) medium: is a two-step protocol necessary? Reprod Biomed Online 5:133–140
146. Crabtree HG (1929) Observations on the carbohydrate metabolism of tumours. Biochem J 23:536–545
147. Lane M, Gardner DK (2001) Inhibiting 3-phosphoglycerate kinase by EDTA stimulates the development of the cleavage stage mouse embryo. Mol Reprod Dev 60:233–240
148. Matsukawa T, Ikeda S, Imai H, Yamada M (2002) Alleviation of the two-cell block of ICR mouse embryos by polyaminocarboxylate metal chelators. Reproduction 124:65–71
149. Macklon NS, Pieters MH, Hassan MA, Jeucken PH, Eijkemans MJ, Fauser BC (2002) A prospective randomized comparison of sequential versus monoculture systems for in-vitro human blastocyst development. Hum Reprod 17:2700–2705
150. Gray CW, Morgan PM, Kane MT (1992) Purification of an embryotrophic factor from commercial bovine serum albumin and its identification as citrate. J Reprod Fertil 94:471–480
151. Holm P, Booth PJ, Schmidt MH, Greve T, Callesen H (1999) High bovine blastocyst development in a static in vitro production system using SOFaa medium supplemented with sodium citrate and myo-inositol with or without serum-proteins. Theriogenology 52:683–700
152. Sung LY, Du F, Xu J, Chang W, Nedambale TL, Zhang J, Jiang S, Tian XC, Yang X (2004) The differential requirement of albumin and sodium citrate on the development of in vitro produced bovine embryos. Reprod Nutr Dev 44:551–564
153. Keskintepe L, Brackett BG, Menezo Y, Kaufmann RA, Servy EJ, Bagis H (1998) Influences of EDTA and sodium citrate on initial bovine embryo development in defined conditions. Theriogenology 49:1

154. Cho MJ, Scieszka JF, Burton PS (1989) Citric acid as an adjuvant for transepithelial transport. Int J Pharm 52:3
155. Kuran M, Robinson JJ, Staines ME, McEvoy TG (2001) Development and de novo protein synthetic activity of bovine embryos produced in vitro in different culture systems. Theriogenology 55:593–606
156. Keskintepe L, Brackett BG (1996) In vitro developmental competence of in vitro-matured bovine oocytes fertilized and cultured in completely defined media. Biol Reprod 55:333–339
157. Rorie RW, Miller GF, Nasti KB, McNew RW (1994) In vitro development of bovine embryos as affected by different lots of bovine serum albumin and citrate. Theriogenology 42:397–403
158. Lim KT, Lee BC, Kang SK, Hwang WS (2003) Effects of protein source and energy substrates on the in vitro development of bovine embryos in a two-step culture system. J Vet Sci 4:73–78
159. Gomez E, Duque P, Diaz E, Diez C (2001) Effects of acetoacetate on in vitro development of bovine embryos in medium containing citrate and myo-inositol. Reprod Domest Anim 36:189–194
160. Keskintepe L, Burnley CA, Brackett BG (1995) Production of viable bovine blastocysts in defined in vitro conditions. Biol Reprod 52:1410–1417
161. Anderson GJ, Vulpe CD (2009) Mammalian iron transport. Cell Mol Life Sci 66: 3241–3261
162. Nasr-Esfahani MH, Johnson MH (1992) How does transferrin overcome the in vitro block to development of the mouse preimplantation embryo? J Reprod Fertil 96: 41–48
163. Harvey AJ, Kind KL, Thompson JG (2007) Regulation of gene expression in bovine blastocysts in response to oxygen and the iron chelator desferrioxamine. Biol Reprod 77: 93–101
164. Buettner GR (1988) DETAPAC (DTPA), aqueous stock solution. University of Iowa, Iowa City, IA
165. Engelmann MD, Bobier RT, Hiatt T, Cheng IF (2003) Variability of the Fenton reaction characteristics of the EDTA, DTPA, and citrate complexes of iron. Biometals 16:519–527
166. Larson AA, Kitto KF (1999) Chelation of zinc in the extracellular area of the spinal cord, using ethylenediaminetetraacetic acid disodium-calcium salt or dipicolinic acid, inhibits the antinociceptive effect of capsaicin in adult mice. J Pharmacol Exp Ther 288: 759–765
167. Kamata H, Hirata H (1999) Redox regulation of cellular signalling. Cell Signal 11:15

Chapter 10

Media Composition: pH and Buffers

Jason E. Swain

Abstract

The proper pH of media is a crucial parameter for optimizing efficacy of gamete and embryo culture. Selecting the right media pH and stabilizing this pH are important variables in minimizing intracellular stress and optimizing development. Regulation of intracellular (pHi) and extracellular pH (pHe) is discussed, as well as methods to prevent harmful pHe oscillations. Furthermore, proper approaches to ensure accurate measurement of media pHe are described.

Key words: pH, Hydrogen ion, Zwitterionic buffer, HEPES, MOPS, Carbon dioxide, Embryo, Oocyte

1. Introduction

Discussions of embryo culture media often entail insights into the roles of supplemental additives, including energy substrates or proteins, and the impact various concentrations or combinations of these substances have on supporting embryo development. However, underlying these types of supplemental additive studies is one of the most basic principles of in vitro cell culture—minimizing stresses imposed on the cells during their manipulations within the in vitro environment. Indeed, minimizing extracellular stressors minimizes intracellular perturbations and helps maintain homeostasis, which is a key component in optimizing the culture system. This is not only accomplished through media formulations but also through precise handling of cells and supplies within the lab, as well as tedious attention to detail. Paramount in this endeavor of minimizing stress is that cells are relegated to the protective confines of the laboratory incubator, and timing of excursions from the incubator is minimized. However, the incubator only provides a protective environment when it is set at the proper parameters to supply the necessary humidity, temperature, and gas concentrations

Gary D. Smith et al. (eds.), *Embryo Culture: Methods and Protocols*, Methods in Molecular Biology, vol. 912, DOI 10.1007/978-1-61779-971-6_10, © Springer Science+Business Media, LLC 2012

required by gametes and embryos. This gas concentration, specifically CO_2, regulates external media pH (pHe). Furthermore, at least for brief periods, incubators must be opened and cells removed for various purposes through the course of their time within the lab.

To be clear, pHe of culture media is an important variable to consider when optimizing the culture system, as an improper pHe can impede embryo development. Because pHe is an endpoint we can directly control within the laboratory, it therefore deserves and demands strict adherence to a defined and narrow range. Furthermore, efforts to maintain a stable pH for procedures performed outside the protective confines of the incubator are also imperative in preventing intracellular perturbations and detrimental effects on gamete function and embryo development. Thus, it is readily apparent that monitoring of pHe and use of protective measures to stabilize pHe, such as buffers, are extremely important factors to consider in optimizing embryo culture systems and further examination and refinement of their use is warranted.

1.1. Defining pH

pH is a measure of the acidity or basicness of a substance, whereby acids are defined as substances that increase H^+ ion concentration and bases as substances that decrease H^+ ion concentration. Thus, pH is the measure of H^+ ion concentration. Notably, these ions exist in very minute amounts in solution, and, as a result, are not easy to denote. Therefore, a negative logarithmic scale has been adopted, which yields a more manageable numbering system.
As an example:

$$0.0000001\,H^+ \text{ ions.}$$

Subjected to a –log transformation becomes:

$$-\log 0.0000001 = 7.$$

Putting this into the form of an equation, one ends up with the familiar notation of:

$$-\log[H^+].$$

As an interesting note, –log can be represented by the letter "p." Others use "p" to signify "power of hydrogen." Regardless of the reason, replacing this in the above equation then yields:

$$p[H^+] \text{ or pH.}$$

Thus, one ends up with the notation of $pH = -\log(H^+)$ and the resulting familiar scale of 1–14, with acids being measured 1–7, bases being measured 7–14, with midpoint neutrality of pH 7.

Importantly, external pH of culture media is generally denoted as pHe, while the internal pH of cells is denoted pHi. The importance of this distinction is discussed below.

1.2. Importance of pHi and pHe

To appreciate the importance of an appropriate and stable pHe, one must first understand the regulation of pHi. Short-term regulation of pHi in cells is achieved by the limited physiochemical buffering capacity of the cytoplasm (1). Additionally, embryos contain transporters to maintain pHi (1–12). These include regulators to combat intracellular acidosis such the sodium dependent HCO_3^-/Cl^- exchanger (HCE) and the Na^+/H^+ antiporter, or the HCO_3^-/Cl^- exchanger to combat alkalosis (Fig. 1). Interestingly, ability to regulate pHi varies with cell stage. Morula and blastocyst stage embryos appear to have more rigorous control over their pHi, possibly due to formation of tight-junctions between cells (9), while denuded mature oocytes lack robust pHi regulatory capacity. The ability to regulate pHi, normally supplied via surrounding cumulus cells (13, 14), does not appear within denuded mature mouse oocytes until several hours after fertilization (5, 15) and bovine and human oocyte appear to have very limited ability to combat alkalosis (6, 7, 10). In the case where when one purposely removes cumulus cells, such as with ICSI, the extremely sensitive nature of the oocyte and importance of a stable pHe becomes evident. However, it should also be noted that though pHi regulators do exist in embryos, pHi initially follows pHe until these mechanisms are activated (10), and that laboratory procedures such as vitrification can impact these regulators and delay their activation (16). Thus, one can now start to see the importance of pHe.

To directly observe the impact of pHe, one needs to simply examine embryo development over a range of pHe (2, 5, 6, 12, 17).

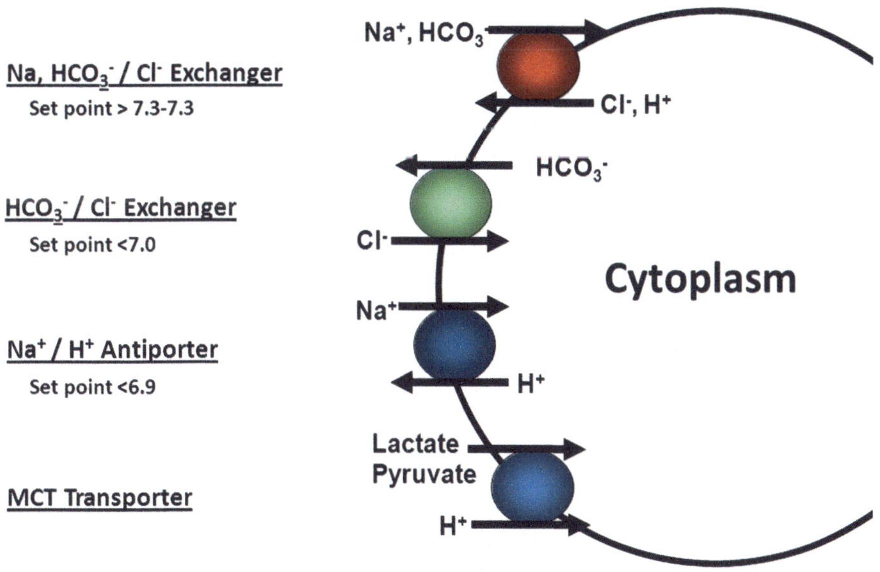

Fig. 1. Regulatory transporters identified in embryos to regulate pHi. Though pHi initially follows pHe in embryos, these regulatory mechanisms allow embryos to develop over a range of pHe.

This impact is likely due to the change in pHi. pHi controls several intracellular processes that can impact embryo, as well as resulting fetal development (18). As an example, raising (~7.4) or lowering (~6.8) pHi in mouse embryos for only 3 h disrupts localization of mitochondrial and actin microfilaments compared to controls (~7.2) (19). Even minor rises in pHi can also dramatically impact embryo metabolism through regulation of various enzymes, such as phosphofructokinase (PFK). Raising pHi ~0.1–0.15 units significantly increased embryo glycolysis and lowered oxidative metabolism (9, 16), which can dramatically impact developmental competence. Alternatively, energy expended maintaining pHi when pHe is not optimal may be an explanation for reduced embryo development. In this respect, studies on pHi have been instrumental in defining the appropriate pHe. pHi of human oocytes and embryos has been repeatedly shown to be approximately ~7.1, which is much lower than the 7.4 of most bodily fluids.

1.3. Practical pHe

Now that the importance of pHi and pHe have been established, and clear relationship defined, it is helpful to discuss pHe from a practical standpoint in an attempt to optimize the culture system. To this end, it may be helpful to view pHe as occurring in three phases within the laboratory: equilibration, set-point, and stabilization (Fig. 2).

1.3.1. pHe Equilibration

Inside the laboratory incubator, pHe is the primarily the result of CO_2 and the amount of bicarbonate in the media. Gaseous CO_2 dissolves in solution to produce carbonic acid, which reaches an equilibrium with the amount of dissolved bicarbonate. Generally the bicarbonate level, supplied as sodium bicarbonate, is set by the media manufacturing company. Thus, to regulate the set-point of our media pHe, CO_2 values on the incubator are adjusted. This is an inverse relationship, with pH decreasing as CO_2 levels increase.

Equilibration then entails the timing needed for diffusion of CO_2 into the media and the timing of the above reaction. Thus, starting pHe and ending pHe are important factors to consider. This is why some laboratories gas their media bottles with ~6 %

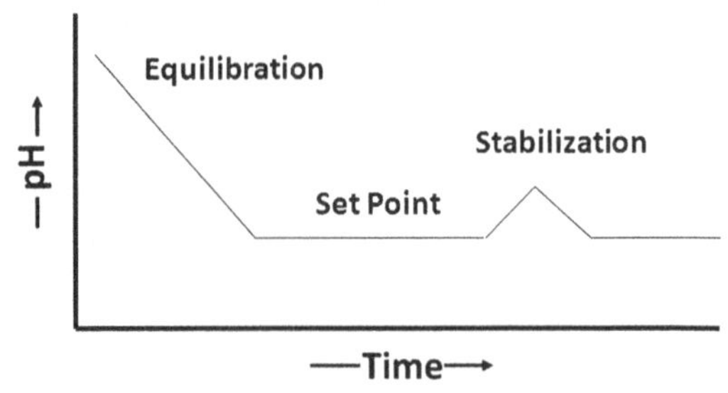

Fig. 2. The three stages of pHe that can be defined for in vitro embryo culture. Each stage is important when considering optimizing the culture system.

CO_2 after opening, to lower the initial pHe and help speed equilibration timing. Furthermore, use of smaller volumes of media in culture dishes, using less oil overlay, and even tilting the lid/dish to increase gas exchange and ensuring oil does not create a gas impermeable seal around the lid can help speed equilibration timing.

1.3.2. pHe Set Point

The question of which CO_2 concentration to use is a difficult one to answer. For the set-point of pHe, though CO_2 and bicarbonate are the major contributors to pHe, they are not the only elements to consider. As an example, protein source and concentration can both affect pHe. Additionally, elevation of the laboratory may also be a factor. Thus, the same basal media may not yield the same set-point pHe from laboratory to laboratory, even if the same CO_2 levels are supplied. We begin to now see the problem with simply setting a specific CO_2 concentration on our incubators or in premixed cylinders of gas, such as 6 %.

In regard to the optimal pHe, again, this is a difficult question and there likely is no "best" pHe. Drifting too far away from the pHi of around ~7.1 likely stresses the embryo, as more resources are required to maintain the proper pHi. Therefore, conventional wisdom tells us that pHe should be slightly higher than pHi to help offset the acidification that occurs as a result of intracellular metabolic processes. Thus, commercial media manufacturers often recommend, and many laboratories culture their embryos in the range of 7.2–7.4, with some labs opting to now culture embryos closer to the 7.2 range. Unfortunately, there likely is no "optimum" pHe, as this will vary from medium to medium based on its ingredients. Amounts of monocarboxylic acids, like lactate and pyruvate, in culture media can lower pHi (9, 20). Additionally, certain amino acids, such as glycine, taurine, and glutamine, act as zwitterions and help in buffering pHi (9). Thus, embryos grown in media with different amounts of these components may have different pHi, though the pHe may be the same. Furthermore, whether early cleavage stage embryos prefer a different pHe than later stages of embryo development is unknown (though as mentioned, it is known that later stages of embryos, like the morula and blastocyst, can regulate their pHi more rigorously than early cleavage stage embryos). That being said, there are some weak data to suggest a slightly more alkaline pHe may benefit fertilization. Dale et al. (7) found higher rates of sperm binding to empty zona pellucidae at pHe 7.5 compared to lower pHe's, though sperm binding to an empty zona and relevance to fertilization may be questionable. This has led some to perform IVF with "high-low-high" paradigm, with oocytes utilizing higher pHe for fertilization, cleavage stage embryo development using a lower pHe, and blastocyst culture using higher pHe. This is often accomplished by the media manufacturer by adjusting bicarbonate levels, so any potential impact on development cannot be attributed to pH alone, and may just as likely be due to changes in bicarbonate levels. It should also be noted that this paradigm may not agree with what the cells would be experiencing in regard to the pHe environment in the reproductive tract in vivo. For example, bovine usually have an alkaline oviduct and a more acidic uterus (21, 22). Thus, it remains unclear if using a single pHe, or changing pHe during embryo culture carries any benefit. Regardless, while various commercial companies list a wide range of acceptable pHe values (Table 1) for their media, and some labs may argue for a benefit of a slightly lower pHe for embryo culture, there is no argument that tight regulation and a narrow acceptable pHe range is a critical component of a rigorous quality control program aimed at optimizing the culture system.

Table 1
Recommended pHe ranges of various commercial embryo culture media

Irvine		Origio	
P1	7.27–7.32	Universal IVF	7.3–7.4
ECM	7.2–7.25	ISM1	7.2–7.3
Single-step (SSM)	7.28–7.32	ISM2	7.35–7.45
Multi-blast	7.3–7.4	EmbryoAssist	7.2–7.3
HTF	7.2–7.3	BlastAssist	7.35–7.45
SAGE		**Life Global**	
Fert Media	7.3 ± 0.1	Global	7.2–7.4*
Cleavage Media	7.2 ± 0.1	Global Fert	7.2–7.4*
Blastocyst Media	7.3 ± 0.1	Blastocyst	7.2–7.4*
IVM	7.2 ± 0.1	HTF	7.2–7.4*
		HTFxtra	7.2–7.4*
Cook		**Vitrolife**	
Sydney IVF Cleavage	7.3–7.5	G5 Series	7.27 ± 0.07
Sydney IVF Blastocyst	7.3–7.5		
Sydney IVF Fert	7.3–7.5		

Recommended ranges may vary based on cell type and stage of development. Optimal pHe will vary because different media composition will impact pHi. Furthermore, amount of protein can alter pHe and should be considered (*asterisk* indicates company recommends pH be adjusted in the laboratory through manipulation of incubator CO_2 levels to an optimum of 7.3)

1.3.3. pHe Stabilization

Perhaps as important as the set-point of pHe is stabilization of that value. Most notable with emerging low-volume culture approaches is that cell culture in high density may lead to a harmful decrease in pHe. However, of more concern are damaging rises in pHe due to laboratory incubator environmental perturbations, such as incubator door openings/closings. This is one reason why patient numbers are limited per incubator, to limit door openings/closing, and why dishes are only observed outside the incubator for very brief periods of time. Furthermore, once a shift in pHe has occurred, the issue of reequilibration becomes a concern as the pHe must return to its set-point, a process that can take several hours.

Though use of inner incubator doors and oil overlay can reduce gas escape, and use of smaller incubators can aid in gas recovery, these approaches are not perfect. Thus, it becomes increasingly clear that for procedures performed outside the incubator, such as cryopreservation, ICSI, and embryo transfer, use of pH buffers within the media aid in pHe stabilization and they are valuable tools in minimizing environmental stress.

1.4. pHe Buffers

For procedures performed outside the laboratory incubator, use of biological pH buffers has offered a significant advantage in stabilizing detrimental deviations in pHe. Approximately 20 synthetic organic buffers, derived largely from zwitterionic amino acids, were formulated to maintain pHe within a desired range (23–25). These buffers, commonly referred to as Good's buffers for the scientist who developed them, provide supplemental buffering capacity over physiologic pH range of approximately 6.1–8.3. This is often accomplished by lowering the media bicarbonate concentration (to ~4 mM) and inclusion of a particular buffer (~21 mM) to maintain pHe without the need for the elevated CO_2 concentrations of the incubator. Importantly, different cell types display varying sensitivity to individual zwitterionic buffers and their concentrations (25). Thus, determining the suitability of specific buffers and concentrations needed to maintain pHe for use with mammalian gametes and embryos in IVF is crucial. Two buffers commonly used in commercially available handling media for ART are 4-(2-hydroxyethyl)-1-piperazineethanesulfonic acid (HEPES) and 3-(*N*-morpholino)-propanesulfonic acid (MOPS), and are selected based on their pKa values, an indication of their optimal buffering capacity (Table 2).

Several studies have examined efficacy of HEPES in various cell types, and in regard to pKa and pH buffering, it is almost an ideal buffer for gametes and embryo. Keeping in mind the impact temperature plays in pH buffering capacity, when working with cells at 37 °C, HEPES has a pKa value of 7.31 (very near to the pHe many labs culture embryos). More recently, at least two commercial companies also now include the less well-studied buffer MOPS in their IVF handling media, which has a lower pKa value of 6.95 at 37 °C (25), near the pHi of embryos. Importantly, both

Table 2
Various zwitterionic buffers exist that can be used to stabilize pHe of IVF handling media

Buffer	pHe range	pKa 25 °C	pKa 37 °C
MOPS	6.5–7.9	7.20	7.02
TES	6.8–8.2	7.40	7.16
HEPES	6.8–8.2	7.48	7.31
DIPSO	7.0–8.2	7.60	7.35
MOBS	6.9–8.3	7.60	na
TAPSO	7.0–8.2	7.60	7.39

An important consideration, in addition to potential toxicity, is the optimal pH buffering capacity (pKa) value at the working temperature. Optimal pHe buffering is obtained when the buffer pKa value is closest to the desired working pHe of the media. In the case of IVF, common pHe values are slightly above the pHi of oocytes and embryos and range from ~7.2 to 7.4 (data obtained from refs. (21–23))

buffers are used with great success, and each has its own potential downside (26). As a result, the literature needs to be carefully considered before condemning or promoting a particular buffer. Importantly, there may still be room for improvement when it comes to buffers and efficacy of IVF. Many culture habits are based on results from older somatic cell studies and use of different buffer types, concentrations and combinations may be useful for gametes and embryos. For example, historically, HEPES or MOPS at ~21 mM has been the industry standard for IVF handling media. Rationale for selection of this buffer concentration is not entirely clear, but was probably used to maintain media osmolality as bicarbonate levels are reduced to 4 mM in most handling media. Thus, 21 mM of Na-conjugated buffer would maintain a similar osmolality to the original 25 mM Na-bicarbonate buffers used in incubator media. However, utilization of lower buffer concentrations below 21 mM will still maintain a stable pHe at room atmosphere (Fig. 3). Additionally, combination of buffers may offer a means to improve upon current monobuffered practices, as combination buffered media have been successfully used to reduced individual buffer concentration, to adjust and optimize pKa, and to culture embryos (26, 27) (Fig. 4).

Though likely media and cell dependent, we still do not know the ideal pHe to culture our gametes and embryos, or what buffering system may be optimal. What is known is that it is readily apparent that pHe is an important variable for consideration in minimizing external stress and optimizing the IVF culture system and that careful and correct monitoring of the endpoint is required to improve embryo growth.

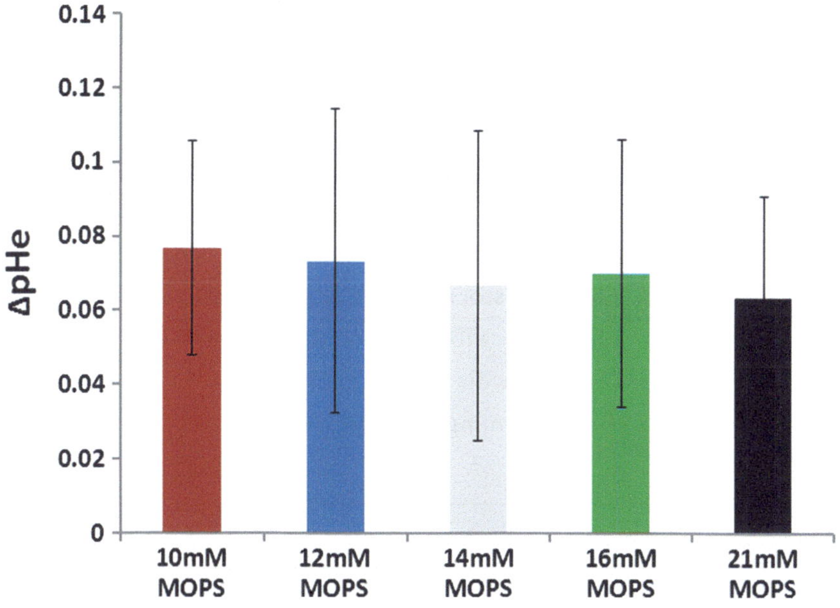

Fig. 3. Change in pHe over ~1 month of MOPS buffered media with varying concentrations of buffer. Use of lower buffer concentrations maintains pHe of IVF handling media that contains 4 mM sodium bicarbonate. Lower buffer concentration may help improve current handling media by reducing potential toxicity. All measurements were taken at ~37 °C. Data is presented as a mean ± SEM and is based on three separate media/measurements.

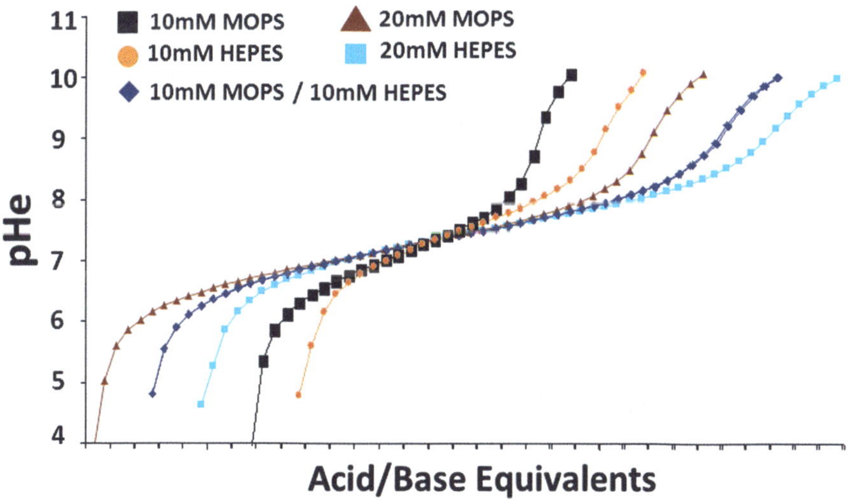

Fig. 4. Combination of various zwitterionic buffers allows for the adjustment of pKa, or optimal buffering, value to coincide with the desired pHe of the culture media. Additionally, concentration of individual buffers is reduced, thereby reducing potential dose-dependent toxicity. The graph shows pHe changes of various buffered media in response to addition of acid or base. The midpoint of the slopes is the ~pKa value, which differs based on type and concentration of buffers used (adapted from ref. (27)).

2. Materials

Components for measuring pHe

1. Appropriate pH electrode.
2. Functioning pH meter with temperature compensation or an ATC probe.
3. Electrode storage and filling solutions.
4. Electrode calibration standards (7 and 10).
5. Culture media for testing.
6. 37 °C warming block (optional).

3. Methods

3.1. Meter Calibration

1. Immediately prior to testing a sample, calibrate the pH meter according to the manufacturer's protocol. It is important to ensure temperature is compensated for by setting the pH meter to the working temperature (~37 °C) or using an ATC probe placed into a 37 °C water bath or warming block (Fig. 5) (see Notes 1–3).

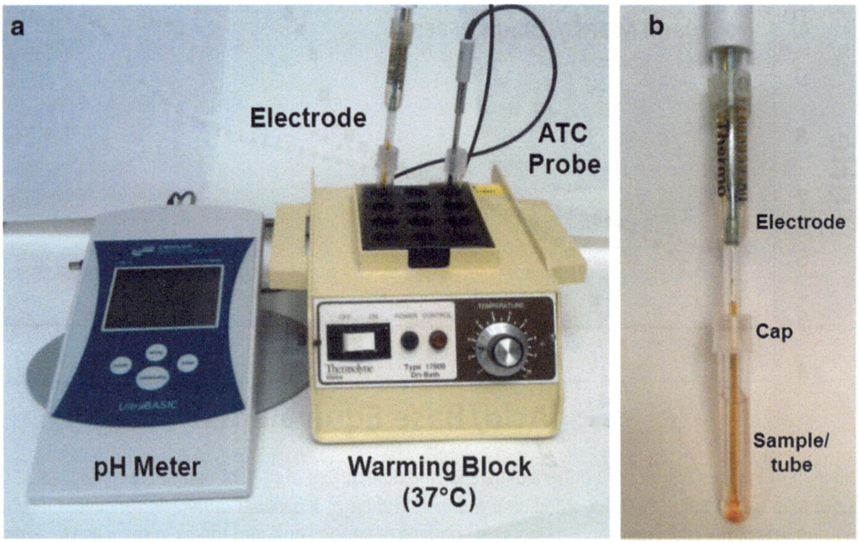

Fig. 5. (**a**) Photograph of a set-up used to measure pH of culture media. The pH meter is outfitted with both a double junction glass electrode, as well as an ATC probe to compensate for temperature. A warming block is used to maintain temperature of samples for accurate calibration and readings. (**b**) The electrode has been fitted through the cap of a test tube to ensure a tight seal and to help stabilize pH readings when samples are removed from the incubator.

2. Aliquot appropriate amount of calibration standards into test tubes immediately prior to use and cap. pH standards should bracket your desired pHe value (see Note 4).
3. Warm standards to working temperature immediately before use (~37 °C) (see Note 5).
4. Once warmed, place pH electrode into standard pH 7. Allow reading to stabilize and calibrate (~6.98 at 37 °). Rinse the electrode with deionized H_2O and gently blot dry. Wiping the probe can affect the pH reading and should be avoided. Place electrode into pH standard 10 and repeat.
5. Once calibrated, place the electrode back into standard pH 7 and verify reading (~6.98 ± 0.02). If the reading is out of range, repeat calibration (see Note 6).

3.2. pHe Measurement

1. For culture media used in the incubator, aliquot media into a loosely capped test tube and place into incubator. Volume of media will vary depending on tube used and size of the pH probe/device. Allowing ~18–24 h incubation ensures adequate temperature and gas equilibration, though less time may be used for smaller volumes of media if desired (see Note 7).

   For handling media with HEPES or MOPS, warm to 37 °C in a water bath or warming block for ~30 min in a capped test tube (see Note 8).
2. Following equilibration of media and calibration of the meter, quickly remove a test sample tube from incubator and cap. Quickly move the tube to the pH meter and place pH electrode into media. This is made easier if the pH meter is setup near the incubators. Allow the reading to stabilize and record the reading. Repeat for each incubator and each medium used in that particular incubator (see Notes 9 and 10).

3.3. pHe Adjustment

1. If pHe readings in each incubator are in range (acceptable range set by lab), no further action is required. However, if pHe of a particular incubator is out of range, the incubator CO_2 levels should be adjusted according to manufacturer's instructions to raise or lower pHe. Raise CO_2 levels to lower pHe and lower CO_2 levels to raise pHe to fall within the specified range (see Notes 11–13). If pH readings are slow and pH meter is sluggish, replacement of the probe may be needed (see Note 14).

4. Notes

1. With new pH electrode and meter technology, some setups have specialized instructions for use. This protocol details use of a standard benchtop pH meter with a normal KCl filled

double junction electrode. Variations may be acceptable. These are simply recommendations based on experience.

2. A double junction glass KCl-filled electrode is recommended for use due to protein content of culture media and for use with organic buffers. This helps prevent clogging of the junction and ensures rapid and efficient readings. However, other electrodes will suffice in many cases, though care may be needed in cleaning/restoring electrodes to ensure rapid response and accurate readings. Regardless of the electrode, proper storage conditions and cleaning of electrodes is required for optimal performance. In this case, electrodes should be stored in a 3 M KCl solution, not water, and the electrode should remain filled with 3 M KCl.
3. Ensure electrode is filled with proper solution before use and that sample gate is open if present. It is recommended accuracy of any new electrode be verified against a known standard or electrode before clinical implementation. Some electrodes will yield different pH readings (28, 29), which could be problematic if these readings are used for adjusting atmosphere of incubators.
4. It is important to use freshly aliquoted calibration standards each time to ensure accurate calibration, as pH of standards can drift following extended exposure to air. Use of a 2 point calibration with standards of pH 7 and 10 is sufficient, though a three point calibration using pH 4 can also be used if desired.
5. Though this step is not always necessary, doing so ensures more accurate pH readings due to the temperature effects on pH and pH electrodes.
6. When done calibrating, or when electrode is not in use, store the electrode properly according to manufacturer's protocols. This usually entails placing the electrode into a 3 M KCl electrode storage solution.
7. Media tested should be exactly what would be used in the laboratory for normal gamete and embryo culture. It is important to use the same basal media, protein source, and concentration to ensure appropriate comparison, as different media may yield a different pHe due to variations in media composition. It is also important to test the pHe of handling media to ensure that it falls within the correct/expected range.
8. An important practical consideration regarding pH buffers and their use in IVF handling media entails verifying working pHe. Though many labs measure pHe of their media used within the incubator, most do not verify pHe of their handling media utilized for procedures outside the incubator. This is unwise, as a comparison of various commercial handling medium demonstrates a varying range of pHe (Table 3). To this end, one must

Table 3
Measurement of pHe should be performed on all media utilized within the IVF lab, including that of handling media containing zwitterionic buffers

Commercial medium (HEPES-HTF based)	pHe (mean ± SEM)
Medium #1	7.28 ± 0.005
Medium #2	7.27 ± 0.003
Medium #3	7.26 ± 0.003
Medium #4	7.08 ± 0.007
Medium #5	7.08 ± 0.005

Temperature affects pHe of these handling media with higher temperatures resulting in a lower pHe. Analysis of pHe of various media indicate that not all maintain the same pHe when measurements are taken at 37 °C

consider the impact of temperature on pHe, as mentioned above. Temperature not only affects the reading of the pH electrode but also causes a change in actual pHe. This is especially apparent when working with zwitterionic buffers. Raising the working temperature of media containing a buffer like HEPES or MOPS results in a corresponding decrease in pHe. Thus, IVF laboratories need to validate pHe of all culture media in their own hands, both that used inside and outside the incubator, and ensure that pHe is measured correctly to ensure proper readings.

9. It is helpful if there is a tight fit or seal between the electrode and the walls of the test tube. This can be accomplished by fitting a rubber gasket around the electrode to fill the gap when placed into the test tube to help stabilized the reading. Alternatively, a small hole can be made in a test tube cap and the electrode slipped through this hole (Fig. 10.5). The cap/electrode can then be fitted onto the test tube to create a seal. This setup can be used during electrode storage, and if the same size test tubes are used, the cap/electrode can be easily fitted over the test sample test tubes to create a seal and help stabilize pH readings.

10. Two test two tubes per incubator per media can also be used and the average the two readings recorded if increased accuracy is desired. Weekly pH reading is likely sufficient, though daily monitoring is also acceptable. Daily pH monitoring does entail use of more culture media, so cost may become a factor.

11. The first step in monitoring pHe of media within the IVF laboratory is to first set an acceptable target range for pHe in your lab. This range should be narrow to help ensure optimal quality control. For reasons described above, there is likely not an optimal pHe, but certainly a narrow range is beneficial in reducing lab variability. A range of 7.27–7.32 at 37 °C is a good starting point, but guidance can be offered by a particular media manufacturer.
12. Readings can be performed on a weekly or daily basis based on lab protocols. Weekly measurement may be more practical when one considers cost of culture media and other variables.
13. Importantly, if testing a new lot of media, perform the pH reading in a side by side fashion with the old verified lot. If the pHe of the new lot is out of range and cannot be easily brought within range by a small adjustment of CO_2 levels, repeat to verify before contacting media company to request a new lot of media. Additionally, all new lots of culture media should have pHe verified before clinical implementation.
14. Annual replacement of the pH probe is recommended, though earlier replacement may be needed, especially in labs who perform daily pH measurements.

References

1. Barr KJ, Garrill A, Jones DH, Orlowski J, Kidder GM (1998) Contributions of Na+/H+ exchanger isoforms to preimplantation development of the mouse. Mol Reprod Dev 50(2): 146–153
2. Zhao Y, Chauvet PJ, Alper SL, Baltz JM (1995) Expression and function of bicarbonate/chloride exchangers in the preimplantation mouse embryo. J Biol Chem 270(41):24428–24434
3. Lane M, Baltz JM, Bavister BD (1998) Regulation of intracellular pH in hamster preimplantation embryos by the sodium hydrogen (Na+/H+) antiporter. Biol Reprod 59(6): 1483–1490
4. Lane M, Baltz JM, Bavister BD (1999) Bicarbonate/chloride exchange regulates intracellular pH of embryos but not oocytes of the hamster. Biol Reprod 61(2):452–457
5. Lane M, Baltz JM, Bavister BD (1999) Na+/H+ antiporter activity in hamster embryos is activated during fertilization. Dev Biol 208(1):244–252
6. Lane M, Bavister BD (1999) Regulation of intracellular pH in bovine oocytes and cleavage stage embryos. Mol Reprod Dev 54(4): 396–401
7. Dale B, Menezo Y, Cohen J, DiMatteo L, Wilding M (1998) Intracellular pH regulation in the human oocyte. Hum Reprod 13(4): 964–970
8. Edwards LJ, Williams DA, Gardner DK (1998) Intracellular pH of the mouse preimplantation embryo: amino acids act as buffers of intracellular pH. Hum Reprod 13(12):3441–3448
9. Edwards LJ, Williams DA, Gardner DK (1998) Intracellular pH of the preimplantation mouse embryo: effects of extracellular pH and weak acids. Mol Reprod Dev 50(4):434–442
10. Phillips KP, Leveille MC, Claman P, Baltz JM (2000) Intracellular pH regulation in human preimplantation embryos. Hum Reprod 15(4):896–904
11. Phillips KP, Petrunewich MA, Collins JL, Baltz JM (2002) The intracellular pH-regulatory HCO_3^-/Cl^- exchanger in the mouse oocyte is inactivated during first meiotic metaphase and reactivated after egg activation via the MAP kinase pathway. Mol Biol Cell 13(11):3800–3810
12. Zhao Y, Baltz JM (1996) Bicarbonate/chloride exchange and intracellular pH throughout preimplantation mouse embryo development. Am J Physiol 271(5 Pt 1):1512–1520
13. Fitzharris G, Baltz JM (2006) Granulosa cells regulate intracellular pH of the murine growing oocyte via gap junctions: development of

independent homeostasis during oocyte growth. Development 133(4):591–599

14. FitzHarris G, Siyanov V, Baltz JM (2007) Granulosa cells regulate oocyte intracellular pH against acidosis in preantral follicles by multiple mechanisms. Development 134(23): 4283–4295
15. Phillips KP, Baltz JM (1999) Intracellular pH regulation by HCO_3^-/Cl^- exchange is activated during early mouse zygote development. Dev Biol 208(2):392–405
16. Lane M, Lyons EA, Bavister BD (2000) Cryopreservation reduces the ability of hamster 2-cell embryos to regulate intracellular pH. Hum Reprod 15(2):389–394
17. Leclerc C, Becker D, Buehr M, Warner A (1994) Low intracellular pH is involved in the early embryonic death of DDK mouse eggs fertilized by alien sperm. Dev Dyn 200(3):257–267
18. Zander-Fox DL, Mitchell M, Thompson JG, Lane M (2010) Alterations in mouse embryo intracellular pH by DMO during culture impair implantation and fetal growth. Reprod Biomed Online 21(2):219–229
19. Squirrell JM, Lane M, Bavister BD (2001) Altering intracellular pH disrupts development and cellular organization in preimplantation hamster embryos. Biol Reprod 64(6): 1845–1854
20. Gibb CA, Poronnik P, Day ML, Cook DI (1997) Control of cytosolic pH in two-cell mouse embryos: roles of H(+)-lactate cotransport and Na+/H+ exchange. Am J Physiol 273(2 Pt 1):C404–419
21. Hugentobler S, Morris DG, Kane MT, Sreenan JM (2004) In situ oviduct and uterine pH in cattle. Theriogenology 61(7–8):1419–27
22. Elrod CC, Butler WR (1993) Reduction of fertility and alteration of uterine pH in heifers fed excess ruminally degradable protein. J Anim Sci 71(3):694–701
23. Good NE, Winget GD, Winter W et al (1966) Hydrogen ion buffers for biological research. Biochemistry 5(2):467–477
24. Good NE, Izawa S (1972) Hydrogen ion buffers. Methods Enzymol 24:53–68
25. Ferguson WJ, Braunschweiger KI, Braunschweiger WR et al (1980) Hydrogen ion buffers for biological research. Anal Biochem 104(2):300–310
26. Swain JE (2010) Optimizing the culture environment in the IVF laboratory: impact of pH and buffer capacity on gamete and embryo quality. Reprod Biomed Online 21(1):6–16
27. Swain JE, Pool TB (2009) New pH-buffering system for media utilized during gamete and embryo manipulations for assisted reproduction. Reprod Biomed Online 18(6):799–810
28. Quinn P, Cooke S (2004) Equivalency of culture media for human in vitro fertilization formulated to have the same pH under an atmosphere containing 5 % or 6 % carbon dioxide. Fertil Steril 81(6):1502–1506
29. Pooler PM, Wahl ML, Rabinowitz AB, Owen CS (1998) pH measurement: a comparison of results using glass membrane electrodes vs solid-state electrodes as a function of solution composition. Anal Biochem 256(2):238–240

Chapter 11

Media Composition: Growth Factors

Aparna Hegde and Barry Behr

Abstract

Despite the fact that the fundamental principle underlying the most common method of culture media constitution is that of mimicking the natural environment of the preimplantation embryo, one major difference that remains between current embryo culture media and in vivo conditions is the absence of growth factors in vitro. Numerous growth factors are known to be present in the in vivo environment of human and nonhuman preimplantation embryos, often with peak concentrations corresponding to when fertilization and preimplantation embryo growth would occur. Although these growth factors are found in very small concentrations, they have a profound effect on tissue growth and differentiation through attachment to factor-specific receptors on cell surfaces. Receptors for many different growth factors have also been detected in human preimplantation embryos. Preimplantation embryos themselves express many growth factors. The growth factors and receptors are metabolically costly to produce, and thus their presence in the environment of the preimplantation embryo and in the embryo respectively strongly implies that embryos are designed to encounter and respond to the corresponding factors. Studies of embryo coculture also indirectly suggest that growth factors can improve in vitro development. Several animal and human studies attest to a probable beneficial effect of addition of growth factors to culture media. However, there is still ambiguity regarding the exact role of growth factors in embryonic development, the optimal dose of growth factors to be added to culture media, the combinatorial effect and endocrine of growth factors in embryonic development.

Key words: Growth factors, Embryo culture, Growth factor receptors, Coculture

1. Introduction

Culture media formulations have seen various improvements in the last two decades that have substantially enhanced the ability to maintain viable embryos in culture until the blastocyst stage consistently (1). Formulations have more or less been crystallized around one of the two dominant philosophies underlying culture media, namely, the sequential formulations and the monoculture ones. However, it would be complacent to assume that culture conditions have been optimized. Of prime concern is the need for

Gary D. Smith et al. (eds.), *Embryo Culture: Methods and Protocols*, Methods in Molecular Biology, vol. 912,
DOI 10.1007/978-1-61779-971-6_11,

culture media to evolve continually towards the end goal of providing an environment for the embryos that is as close to the one it is exposed to in vivo as it travels down the fallopian tube to the uterus. One potentially important consideration for improving culture media in the future could be the need for culture media to provide embryos with the various factors that it is exposed to in its natural environment that promote cell division, proliferation, and differentiation, namely, the growth and immune factors. Simply put, cells do not thrive in nature without these factors. Similarly there is increasing interest in recent years on the potential role of growth and immune factors produced by embryos while in culture media as markers of embryonic viability.

1.1. Culture Media: Why Is There a Need for Further Improvements?

There has been a long-held view that the choice of culture media for the cleavage stage embryo is not of primary importance due to the fact that many different media have been used successfully with dramatically different formulations. Historically, human embryos have been cultured in formulae from both ends of the spectrum of complexity: simple salt solutions and complex tissue culture media.

1.1.1. Biochemical Stress on the Embryo

It is testament to the nutritional tolerance and plasticity of the human embryo that embryos can adapt and grow within a plethora of conditions and widely varying formulae. However, the degree of adaptation to environmental conditions and/or the rate of adaptation to changing conditions will also place a varying degree of biochemical stress on the embryo (2). Embryos themselves have differing abilities to develop in any one particular culture system, depending on their genetic background (3, 4). Evidence is emerging now that adaptive responses by the embryos during culture can have important effects on subsequent developmental profiles. The culture environment per se or changes to the culture environment result in quantitative differences in the level of mRNA for a variety of genes (5–12) performed microarray studies to show that culture conditions cause a small but significant change in the transcriptome. Several animal studies have also demonstrated that morphological and biochemical abnormalities resulting from in vitro fertilization and embryo culture may persist long after exposure to the in vitro environment, through fetal development and into adulthood. Studies have reported distorted body size and altered proportion of tissues (skeletal, muscle, adipose tissue) and organs (heart, liver, kidneys, brain, thyroid) in fetal (13–20) and newborn (21, 22) cattle, mice, and sheep. Placental abnormalities are common and may lead to adverse effects in even subsequent generations of naturally conceived offspring (14, 23). Disruption of normal genomic imprinting may contribute to some of the developmental abnormalities observed in animals. Studies of mouse, cattle, and sheep IVF have demonstrated an increased risk of imprinting errors due to altered DNA and histone methylation

(18–20, 24–26), and these imprinting errors have been linked to "large offspring syndrome (LOS)", the most well-known perinatal abnormality resulting from IVF in animals such as sheep and cattle (18–20).

Although direct evidence comparing human embryo development in vivo versus in vitro is very limited, it is generally accepted that embryo growth in vitro is suboptimal even with current sequential media, much like the more outwardly visible differences, such as slower growth rate and fewer cells, observed in nonhuman mammals (27). Available studies regarding developmental risks among human children resulting from IVF are mostly reassuring and there is little evidence that imprinting disorders other than Beckwith–Weidemann syndrome and Angelman syndrome are more common with assisted reproduction (25–31). Nonetheless, IVF births are at increased relative risk (perhaps as much as doubled) of neonatal mortality, preterm delivery, low birth weight and a variety of major and minor congenital malformations, including cardiovascular, musculoskeletal, urogenital, neurological, and gastrointestinal defects, and are more likely to require neonatal intensive care and hospitalization or physiotherapy and speech therapy in infancy (28, 32–42). One well-documented abnormality associated with human IVF is small size for gestational age (43–47). A recent study of prepubescent children aged 4–10 years concluded that permanent metabolic alterations result from IVF after finding significantly different IGF expression, lipid profiles, body mass index, and height compared to naturally conceived children (48).

Thus a complete understanding of the effects of culture media conditions on subsequent embryonic development and long-term effects during adult life is still far from our grasp. Most likely, there is no adaptation by an embryo to its environment that has no consequence (2). However, there is enough experimental evidence to support the contention that culture conditions that are closer to physiological conditions produce good quality embryos that are more aligned with in vivo-derived embryos and cause fewer deviations in the transcriptome and proteome compared with embryos derived from culture under reportedly more stressful conditions (49–51). And physiological conditions would certainly involve the presence of growth factors. Also such deviation away from in vivo-derived embryo gene expression and/or protein synthesis profiles is most likely to result in reduced subsequent embryo viability and/or perturbed fetal development and possibly have far reaching long-term effects (2).

1.2. Growth Factors and Embryo Development

Though the principle hypothesis underlying sequential media is that of mimicking the natural environment of the preimplantation embryo, there is still a major limitation, the absence in vitro of growth factors that the embryo is naturally exposed to in vivo. Admittedly, there are no studies that have established a direct causal relationship between the absence of growth factors and any detrimental effect of culture media on transcriptome, genome, and later development. However, in vitro growth has been found to slow down growth and reduce cell numbers and survival (52–61). Also Giritharan et al. (61) found in a microarray analysis of global gene expression patterns in mouse blastocysts conceived naturally as compared to those conceived with IVF found significantly different expression of 1912 genes, including (29) that exhibited differences greater than four-fold, with proliferation apoptosis and morphogenetic pathways being most affected by in vitro culture (61). Studies have also found changes in mitochondrial and nuclear reduction (62, 63), intercellular connections and communication reduction (59, 64), cytoskeletal deficiencies (54), and increased vacuoles and intercellular debris (59, 62, 65). Given the crucial role that growth factors play in the growth pathways as well as in cell division, proliferation and differentiation, it would not be specious to assume that some of these effects could be due the absence of growth factors. Significantly, altered expression of growth factors like IGF-1, IGF-2, LIF, and FGF has been found with culture of mammalian embryos (66–71). Evidence for a role of growth factors in the development of embryos is provided by four important established facts, namely, the expression of growth factors in the reproductive tract, expression of receptors for growth factors on embryonic cell surfaces, the enhancement of embryo growth by coculture regardless of the cell type used, and expression of growth factors by the embryos themselves.

1.2.1. Expression of Growth Factors in the Reproductive Tract

There is increasing evidence in recent years of the presence of growth factors in the reproductive tract of humans and nonhuman primates, often with peak concentrations corresponding to when fertilization and preimplantation growth would occur (27). Insulin-like growth factor-1,72 IGF-2 (73), granulocyte–macrophage colony stimulating factor (GM-CSF), heparin-binding epidermal growth factor (HB-EGF), vascular endothelial growth factor (VEGF) (74–80), leukemia inhibitory factor (LIF), epidermal growth factor (EGF), transforming growth factor alpha (TGF-α), fibroblast growth factor (FGF) (81–83), macrophage colony stimulating factor (M-CSF) (84), platelet derived growth factor (PDGF) (85, 86), colony stimulating factor (CSF) (87), inhibin (88–91), fibronectin, and activin A (92) are produced by various parts of the reproductive tract at various times during the menstrual cycle and are likely to have important roles to play in embryo growth and development.

Significantly, expression of IGF-2 (93) and activin A (94) by granulosa cells, and expression of VEGF (95, 96) and inhibin (89, 90) by both granulosa cells and the corpus luteum are stimulated by hCG. This suggests that they may have a role to play in the developing embryo. The follicular fluid collected during oocyte retrieval has been found to contain FGF, IGF-1, IGF-2, and MCSF (73, 83, 84).

Table 1 describes the various growth factors that are produced by the different parts of the reproductive tract at various times in the menstrual cycle.

It almost seems self-evident that the growth factors are present in the reproductive tract fluids that the embryo is exposed to as it travels down the fallopian tube for a particular reason. And that reason most likely would be to fulfill the role that they normally play in normal cellular growth, proliferation and differentiation pathways. While these pathways that the growth factors present in the reproductive tract act upon could very well be those of maternal tissue, it is very plausible that they act on embryonic tissue. Growth factors are metabolically costly to produce and would not be expressed without a purpose (27).

1.2.2. Receptor Expression by Preimplantation Embryos

Additional compelling evidence for the addition of growth factors to culture media is the presence of growth factor receptors on embryonic cell surfaces. Receptors too are metabolically costly to produce and thus their presence implies that embryos are designed to encounter and respond to the corresponding factors (27). Table 2 describes the distribution of receptors of various growth factors that have been detected in human preimplantation embryos. Receptors for HB-EGF, EGF, TGF-α, LIF, GM-CSF, CSF, platelet activating factor (PAF), activin, PGDF, IGF-1, and IGF-2 are expressed either by unfertilized oocytes or the embryos or by both (72, 90, 96–101).

1.3. Embryo Coculture

The past experience with coculture offers a snapshot of the advantages that may be accrued from addition of growth factors to culture media. Various studies have demonstrated the enhancement of embryo growth that occurs with coculture regardless of the particular cell type used (102 108). Coculture was the only way that embryonic development till the blastocyst stage could be supported prior to the development of sequential media. Addition of Vero cells as coculture along with sequential media has been shown to increase the number of cells at cleavage stage by almost 32 % and at the blastocyst stage by almost 50 % as compared to sequential media alone (109). The cocultured cells release a variety of different growth factors that are believed to be responsible for the improved embryo development and this improvement is seen to occur despite the different mix of growth factors produced by different cultured cells. Table 3 describes the enhancement caused by various different cell types when added as coculture cells. Menezo and Sakkas significantly noted that no monozygotic twinning was reported in over 800 deliveries resulting from cocultured blastocyst transfers (110). This suggests that the growth factor supplements secreted due to coculture may have made the culture

Table 1
Expression of growth factors within the reproductive tract in relation to the menstrual cycle

Various cell types and fluids within the reproductive tract	Growth factors	Phase of the menstrual cycle
Endometrium	Leukemia inhibitory factor (LIF) by the luminal epithelial cells (137–139)	Peak in the luteal phase (137–139)
	Activin A by the epithelial cells (92)	Throughout the menstrual cycle
	Colony stimulating factor (CSF) (87)	Higher in luteal phase (87)
	Epidermal growth factor (EGF) (140)	Throughout the menstrual cycle except during early follicular phase (140)
		Throughout the menstrual cycle (78)
	Fibronectin	Peak in mid-luteal phase (141)
	Granulocyte–macrophage colony stimulating factor (GM-CSF) primarily by the luminal and glandular epithelial cells (141)	Peak just before implantation (142–144)
	Heparin-binding epidermal growth factor (HB-EGF) by the apical surface of the luminal epithelium (142–144)	
	Insulin-like growth factor 1 primarily in the luminal and glandular epithelium (72)	Peak from late follicular to mid-luteal phase (145)
	Insulin-like growth factor 2 (146, 147)	During the luteal phase primarily (146, 147)
	Inhibin (92)	
	Platelet derived growth factor (PDGF) (85, 86, 111)	Primarily during luteal phase (77, 79, 80)
	Vascular endothelial growth factor (VEGF) on the apical surface of the epithelium (77, 79, 80)	
Oviduct	EIF (148–150)	Peak from late follicular to mid-luteal phase
	GM-CSF	Peak in late follicular through mid-luteal phase (141)
	HB-EGF	Peak from late follicular to mid-luteal phase (151)
	LIF	Throughout the cycle (152)
	TGF-α	Peak from late follicular to mid-luteal phase (82, 97, 148–150, 154)
	VEGF	Peak during periovulatory period (75, 153)

(continued)

Table 1 (continued)

Various cell types and fluids within the reproductive tract	Growth factors	Phase of the menstrual cycle
Follicular fluid	FGF and fibronectin (83) Insulin like growth factor-1 (73) Insulin like growth factor-2 (73) M-CSF (84)	Throughout the menstrual cycle
Corpus luteum	Inhibin (stimulated by hCG) (88, 89) VEGF (stimulated by hCG) (150)	
Cumulus/granulosa cells	Activin A (stimulated by hCG) (94) FGF (81, 82) Fibronectin primarily in the corona cells (155) Insulin-like growth factor 1 (97) Insulin-like growth factor 2 (stimulated by hCG) (93) Inhibin (He et al. (90)), stimulated by hCG (91) LIF (156) M-CSF (84) TGF-α (97)	
	VEGF	In mature follicles and subsequent to ovulation (74) stimulated by hCG (95, 157)

conditions more robust such that it counteracted the weakening of intercellular connections observed with in vitro culture, and thereby reducing, if not eliminating the increased risk of monozygotic twinning associated with extended embryo culture (111–113).

Coculture is however associated with a couple of major drawbacks, first, there is an unregulated alteration in culture conditions depending upon the type of cells that are used in coculture. Coculture leads to highly unstable and dynamic environmental conditions (114). Second the addition of nonautologous and especially nonhuman cells may be associated with pathogenic contamination. Use of xenotransplants, i.e., use of nonhuman cells is hence illegal. However, the use of autologous cells is also fraught with problems. It is technically challenging and if autologous endometrial cells are used in coculture, there is a possibility of introducing the very same deleterious factors that is the cause of infertility in the first place. For example, use of autologous endometrial cells from an endometriosis patient could possibly lead to introduction of modified immune factors.

Table 2
Expression of receptors for the growth factors by oocytes and the embryos

Factor	Oocyte	2- to 4-cell stage	6- to 8-cell stage	Blastocyst	Comments	References
Activin A		+	+	+	Increased with endometrial cell coculture	(90)
CSF-1		+	+			(158)
EGF	+	+	+	+		(96–98)
GM-CSF		+	+	+		(159)
HB-EGF	+	+	+	+		(98, 99)
Insulin-like growth factor-1	+	+	+	+		(97, 118)
Insulin-like growth factor-2	+	+	+	+		(118)
LIF	+	+	+	+		(161–163)
PAF		+	+	+		(100)
PDGF		+	+	+		(101)
TGF-α		+	+	+		(96–98)

Table 3
Enhancement of growth by different cells added in coculture

Cell type	Enhancement of growth
Autologous cumulus cell coculture	Increased cell numbers (103), reduced fragmentation (103, 102), and increased blastocyst formation (163)
Human oviduct cells	Increased fertilization rates, cell numbers, and implantation rates (105, 106)
Autologous endometrial cells	Reduce fragmentation (107)
Ovarian cancer cell coculture	Increased blastocyst formation (108)
Fetal bovine uterine fibroblasts	Reduced embryo fragmentation, increased cell-to-cell adherence, blastomere expansion, zona thickness variation, and improved hatching and increased ongoing pregnancy rates (164)
Bovine oviductal epithelium	Increased zona thickness variation and implantation and pregnancy rates (165)
Buffalo rat liver cells	Increased pregnancy and implantation (166)
Vero (African green monkey kidney) cells	Improved embryo quality, increased cleavage rates, blastocyst formation, and pregnancy rates (167–169)

It is a bit surprising that coculture as a technique gained credence prior to the addition to specific growth factors to the culture media despite the fact the that the growth factors added to culture media in coculture and their concentrations are unknown. It is not surprising that coculture as a technique has been all but abandoned, but the experience with coculture does provide evidence for the value of addition of known quantities of specific growth factors to culture media to support embryonic development.

1.4. Production of Growth Factors by the Embryos in Culture Media

Group culture is routinely practiced in IVF labs and the prime reason for the improved embryonic development that has been observed is believed to be the production of growth factors by the embryos that act through receptors on embryonic cell surfaces (115–117). Minimization of media volume is also found to be beneficial for the same reasons (115–117). Various studies have documented the production of growth factors by preimplantation embryos including EGF, TGF-α, IGF-1, IGF-2, VEGF, PDGF, PAF, and fibronectin (96, 97, 100, 101, 118–121).

There is increasing evidence currently on the potential use of the various growth factors produced by culture media as prognostic biomarkers of embryo viability and implantation potential. Juriscova et al. first reported on the potential biomarker role for sHLA-G in 1996 (122). In 1999, Menicucci et al. reported on the presence of sHLA-G molecules in the supernatant cultures of early human embryos obtained by in vitro fertilization and also on its possible prognostic role in implantation (123). Various studies have reported that presence of sHLA-G in the culture supernatant was associated with improved implantation rates and pregnancy rates (124–127). Rebmann et al. (128) in a multicenter trial in Germany found that sHLA-G testing could improve pregnancy rate from 30 to 40 % and the sHLA-G status of the embryos could be associated with pregnancy after single embryo transfer doubling the probability of pregnancy rate to 26 %. However, they found that the morphological scoring system is still the best strategy for the selection of embryos but that sHLA-G might be considered as a second parameter if a choice has to be made between embryos of morphologically equal quality.

Similarly Roudebush et al. (100) have showed that patients who became pregnant had a higher PAF level in their culture mediums than patients who did not become pregnant following IVF.

1.5. Current Evidence Regarding Growth Factors

1.5.1. Animal Studies

Numerous animal studies have demonstrated the benefits of adding a variety of individual growth factors to embryo culture media (see Table 4). Desai et al. (129, 130) compared nearly a dozen individual factors, including GM-CSF, LIF, TGF-α, IGF-1, IGF-2, PDGF, and EGF to Vero cell coculture and controls. They found increased and more rapid blastocyst formation, increased hatching rates, increased cell numbers, increased zona thickness variation, and improved inner cell mass quality with nearly all individual

Table 4
Beneficial effects of growth factors seen in animal studies

Growth factor	Beneficial effects seen in animal studies
IGF-1	Increased blastocyst formation and cell numbers in mice (117, 170, 171) pigs, (172, 173), rabbits (174), and cows (175–179). Also reduced apoptosis, fragmentation and apoptosis in pig, rabbit, cow (173–176); increased blastocyst formation rates, improved blastocyst quality, increased zona pellucida thickness variation, and increased hatching in mice (129, 130, 171) and increased pregnancy/live birth in cows (180)
LIF	Decreased mouse embryo fragmentation and increased blastocyst formation, hatching, and trophoblast outgrowth (181–183), while in sheep it reduced degeneration and increased fertilization, ICM cells number, hatching, and implantation rates (184, 185)
GM-CSF	Increased cattle blastocyst formation rates (186). In mice GM-CSF increased blastocyst formation rates, accelerated blastocyst formation and hatching, increased glucose uptake, increased cell numbers, increased the viable implantation rate, decreased apoptosis, greatly enhanced blastocoele reexpansion following cryopreservation (14, 187–189) and has also been shown to partially alleviate numerous fetal and adult developmental abnormalities associated with IVF (14). GM-CSF has also been shown to reverse the effect of culture on labyrinthine growth and increase the surface area of placental trophoblast available for nutrient exchange (14)
M-CSF	Accelerated blastocyst formation and increased cell numbers in mice (190)
HB-EGF	Increased blastocyst formation in rats (191) and increased hatching, trophoblastic growth, and implantation rates in mice (192)
EGF, fibronectin	Increased cell numbers, blastocyst formation, rate of blastocyst formation, blastocyst quality, and hatching in mice (129, 130), increased blastocyst formation and pregnancy/live birth in cows (193), and doubled rat implantation rates (194)
PAF	Increased blastocyst formation, fertilization rates, enhanced preimplantation development, reduced degeneration, fragmentation and apoptosis, and increased birth rates in mice (115, 116, 195, 196)
TGF-α	Reduced apoptosis in preimplantation ovine embryos (197); increased blastocyst formation, rate of blastocyst formation, blastocyst quality, zona pellucida thickness variation, hatching, and trophoblastic growth in mice (129, 130, 198)

growth factors versus controls, although no individual growth factor performed as well as Vero cell coculture. This suggests that a complex mixture would replicate the natural environment much more accurately than individual factors.

1.5.2. Human Studies

Many randomized controlled trials report beneficial effects of specific growth factor additions to human embryo culture media as well (see Table 5).

There have been very few clinical trials in which human embryos grown in the presence of deliberately added growth

Table 5
Beneficial effects seen in human studies

Growth factors	Beneficial effects seen in human studies
LIF	More than doubled blastocyst formation rates and more than tripled the percentage of top-quality blastocysts (199, 200)
IGF-1	Decreased apoptosis, and increased blastocyst formation rates by more than 50 % and ICM cell numbers by nearly 60 %, while IGF-1 antibody inhibited blastocyst formation (72, 201)
GM-CSF	Doubled blastocyst formation rates and increased the speed of blastocyst development, increased hatching rates, reduced apoptosis by 50 % and increased the number of viable ICM cells by 30 % (160, 202)
HB-EGF	Improved both blastocyst formation and hatching by more than 50 % (203)
EGF	Increased secretion of plasminogen activators (which are associated with implantation) by preimplantation embryos (204)
Laminin or fibronectin	Enhanced embryo hatching (120)

factors have actually been transferred back to IVF patients (27). Results have been very favorable for the few studies available. PAF increased pregnancy rates following IVF by 70 % (131). Addition of LIF to transfer media was found to increase implantation and pregnancy rates, but the sample size of 37 patients was too small to confirm statistical significance (132). GM-CSF supplementation resulted in a 50 % increase in clinical pregnancy rates following day 3 embryo transfer (133). Shapiro et al. (134) found that the addition of GM-CSF to sequential blastocyst culture media resulted in greater cell numbers at the cleavage stage and a 50 % increase in the numbers of expanded blastocysts available for transfer, while implantation rates and pregnancy rates per transfer were similar with or without exposure to GM-CSF.

These studies undoubtedly attest to the fact that optimal benefits with minimal potential for developmental disruption, could be achieved with many different growth factors or a preferably a combinations of factors that accurately reproduces the composition of the natural environment. The most obvious factors to include in culture media would be those proven to be present in the in vivo environment, and for which embryos have receptors, i.e., LIF, GM-CSF, HB-EGF, IGF-1, IGF-2, TGF-α, CSF, EGF, PDGF, and activin (27). Other factors present in the reproductive tract, such as VEGF, FGF, fibronectin, laminin, and inhibin, or those with proven benefits such as PAF, are also good candidates for inclusion (27).

2. Conclusion

It would be prudent to conclude that culture media currently in use are suboptimal. Undoubtedly studies attest to a probable beneficial effect of addition of growth factors to culture media. But a real world application of the same is still a long way off. It may be that the reluctance to include growth factors in culture media is due to complacency regarding the adequacy of current culture systems to produce "viable" embryos as stated by Richer et al. (27). However, it is equally true that there is still ambiguity regarding the exact proportion of detrimental effects of culture conditions on embryonic development that can be attributed to lack of growth factors. A direct causal relationship has not yet been established. Growth factors have been shown to have beneficial effects but it has not yet been proven that their absence is associated with particular deleterious effects. Even if a direct causal relationship were to be established, a very important issue that then would need to be answered is about the exact dose of growth factors that needs to be added. Estimating the concentration of growth factors in physiological fluids is difficult given current technological applications. Yes it cannot be denied that emerging technologies like proteomic microarray analysis (135, 136), applied to follicular, tubal, and endometrial fluid, may soon enable a much more comprehensive assay of growth factor composition in the in vivo environment (27). But obtaining uncontaminated in vivo fluids for laboratory analysis is difficult especially oviductal fluid. If a dose that is more than that present physiologically is added, there is a very real possibility of inducing developmental abnormalities. It cannot be denied that the full knowledge of the developmental effects of growth factors is yet not available. However, estimating the concentration of growth factors in physiological fluids is just part of the solution. Various growth factors act as autocrine factors and there is a possibility that their effects are distorted in the in vitro environment. Also several growth factors are produced elsewhere and act on the embryo in the vivo environment through their receptors on the embryonic surface. It would be next to impossible to understand these endocrine effects. There is also the issue of what cocktail of growth factors to add. Growth factors do not act in isolation and various combinations of growth factors are present in the in vivo environment. However, most studies have examined individual growth factors and not the effects of a combination of them. It is likely that a combination of factors will create an environment that mimics the natural one more closely than individual factors. An accurate estimation of the dose effect, endocrine effect, and combinatorial effect is thus difficult.

Thus, in conclusion, the addition of growth factors to culture media is the most promising next step in the path towards improving culture media constitution in order to increase IVF success rates. The most obvious factors to include in culture media would be those that have been demonstrated to be present in vivo and for which embryos have receptors, namely, activin, CSF, EGF, GM-CSF, HB-EGF, IGF-1, IGF-2, LIF, PDGF, and TGF-α. But more studies are necessary regarding the dose and combination of growth factors that should be used before addition of growth factors to culture media can become an everyday practice.

References

1. Behr B, Wang H (2004) Effect of culture conditions on IVF outcome. Eur J Obstet Gynecol Reprod Biol 115S:S72–S76
2. Thompson J, Mitchell M, Kind K (2007) Embryo culture and long-term consequences. Reprod Fertil Dev 19(1):43–52
3. Whittens W, Biggers J (1968) Complete development in vitro of the pre-implantation stages of the mouse in a simple chemically defined medium. J Reprod Fertil 17(2):399–401
4. Payne S, Munday R, Thompson J (1992) Addition of superoxide dismutase and catalase does not necessarily overcome developmental retardation of one-cell mouse embryos during in-vitro culture. Reprod Fertil Dev 4:167–174
5. Ho Y, Doherty A, Schultz R (1994) Mouse preimplantation embryo development in vitro, effect of sodium concentration in culture media onRNAsynthesis and accumulation and gene expression. Mol Reprod Dev 38:131–141
6. Wrenzycki C, Herrman D, Carnwarth J, Niemann H (1999) Alterations in the relative abundance of gene transcripts in preimplantation bovine embryos cultured in medium supplemented with either serum or PVA. Mol Reprod Dev 53:8–18
7. Wrenzycki C, Herrmann D, Keskintepe L, Martins A Jr, Sirisathien S, Brackett B, Niemann H (2001) Effects of culture system and protein supplementation on mRNA expression in pre-implantation bovine embryos. Hum Reprod 16:893–901
8. Niemann H, Wrenzycki C (2000) Alterations of expression of developmentally important genes in preimplantation bovine embryos by in vitro culture conditions, implications for subsequent development. Theriogenology 53:21–34
9. Rizos D, Lonergan P, Boland M, Arroyo-Garcia R, Pintado B, De La Fuente J, Gutierrez-Adan A (2002) Analysis of differential messenger RNA expression between bovine blastocysts produced in different culture systems, implications for blastocyst quality. Biol Reprod 66:589–595
10. Lonergan P, Rizos D, Gutierrez-Adan A, Fair T, Boland M (2003) Oocyte and embryo quality, effect of origin, culture condition and gene expression patterns. Reprod Domest Anim 38:259–267
11. Lonergan P, Pedersen H, Rizos D, Greve T, Thomsen P, Fair T, Evans A, Boland M (2004) Effect of the postfertilization culture environment on the incidence of chromosome aberrations in bovine blastocysts. Biol Reprod 71:1096–1100
12. Kind K, Collett R, Harvey A, Thompson J (2005) Oxygen-regulated expression of GLUT-1, GLUT-3, and VEGF in the mouse blastocyst. Mol Reprod Dev 70:37–44
13. Bertolini M, Moyer A, Mason J, Batchelder C, Hoffert K et al (2004) Evidence of increased substrate availability to in vitro-derived bovine foetuses and association with accelerated conceptus growth. Reproduction 128:341–354
14. Sjoblom C, Roberts C, Wikland M, Robertson S (2005) Granulocyte–macrophage colony-stimulating factor alleviates adverse consequences of embryo culture on fetal growth trajectory and placental morphogenesis. Endocrinology 146:2142–2153
15. Farin P, Farin C (1995) Transfer of bovine embryos produced in vivo or in vitro, survival and fetal development. Biol Reprod 52:676–682
16. Maxfield E, Sinclair K, Broadbent P et al (1998) Short-term culture of ovine embryos modifies fetal myogenesis. Am J Physiol 274:E1121–E1123
17. Crosier A, Farin C, Rodriguez K et al (2002) Development of skeletal muscle and expression of candidate genes in bovine fetuses from embryos produced in vivo or in vitro. Biol Reprod 67:401–408
18. Young L, Fernandes K, McEvoy T et al (2001) Epigenetic change in IGF2R is associated

with fetal overgrowth after sheep embryo culture. Nat Genet 27:153–154

19. Hiendleder S, Mund C, Reichenbach H et al (2004) Tissue-specific elevated genomic cytosine methylation levels are associated with an overgrowth phenotype of bovine fetuses derived by in vitro techniques. Biol Reprod 71:217–223
20. Hiendleder S, Wirtz M, Mund C et al (2006) Tissue-specific effects of in vitro fertilization procedures on genomic cytosine methylation levels in overgrown and normal sized bovine fetuses. Biol Reprod 75:17–23
21. Sangild P, Schmidt M, Jacobsen H et al (2000) Blood chemistry, nutrient metabolism, and organ weights in fetal and newborn calves derived from in vitro-produced bovine embryos. Biol Reprod 62:1495–1504
22. van Wagtendonk-de Leeuw A, Mullaart E, de Roos A et al (2000) Effects of different reproduction techniques, AI MOET or IVP, on health and welfare of bovine offspring. Theriogenology 53:575–597
23. McEvoy T, Sinclair K, Broadbent P et al (1998) Postnatal growth and development of Simmental calves derived from in vivo or in vitro embryos. Reprod Fertil Dev 10:459–464
24. Rivera R, Stein P, Weaver J et al (2008) Manipulations of mouse embryos prior to implantation result in aberrant expression of imprinted genes on day 9.5 of development. Hum Mol Genet 17:1–14
25. Li T, Vu T, Ulaner G, Littman E, Ling J, Chen H, Hu J, Behr B, Giudice L, Hoffman A (2005) IVF results in de novo DNA methylation and histone methylation at an Igf2-H19 imprinting epigenetic switch. Mol Hum Reprod 11(9):631–640
26. Fauque P, Jouannet P, Lesaffre C et al (2007) Assisted reproductive technology affects developmental kinetics, H19 imprinting control region methylation and H19 gene expression in individual mouse embryos. BMC Dev Biol 7:116
27. Richter K (2008) The importance of growth factors for preimplantation embryo development and in-vitro culture. Curr Obstet Gynecol 20(3):292–304
28. Lidegaard O, Pinborg A, Andersen AN (2005) Imprinting diseases and IVF, Danish National IVF cohort study. Hum Reprod 20:950–954
29. DeBaun MR, Niemitz EL, Feinberg AP (2003) Association of in vitro fertilization with Beckwith–Wiedemann syndrome and epigenetic alterations of LIT1 and H19. Am J Hum Genet 72:156–160
30. Gicquel C, Gaston V, Mandelbaum J et al (2003) In vitro fertilization may increase the risk of Beckwith–Wiedemann syndrome related to the abnormal imprinting of the KCN1OT gene. Am J Hum Genet 72:1338–1341
31. Halliday J, Oke K, Breheny S et al (2004) Beckwith–Wiedemann syndrome and IVF, a case–control study. Am J Hum Genet 75:526–528
32. Chang A, Moley K, Feinberg A, Wangler M, DeBaun M (2005) The association between Beckwith–Weidemann syndrome and assisted reproductive technology, a case series of nineteen patients. Fertil Steril 83:349–354
33. Doornbos M, Maas S, McDonnell J, Vermeiden J, Hennekam R (2007) Infertility, assisted reproduction technologies and imprinting disturbances, a Dutch study. Hum Reprod 22(9):2476–2480
34. Sutcliffe A, Peters C, Bowdin S, Temple K, Reardon W, Wilson L, Clayton-Smith J, Brueton L, Bannister W, Maher E (2006) Assisted reproductive therapies and imprinting disorders – a preliminary British survey. Hum Reprod 21(4):1009–1011
35. Hansen M, Kurinczuk J, Bower C, Webb S (2002) The risk of major birth defects after intracytoplasmic sperm injection and in vitro fertilization. N Engl J Med 346:725–730
36. Stromberg B, Dahlquist G, Ericson A et al (2002) Neurological sequelae in children born after in-vitro fertilisation, a population-based study. Lancet 359:461–465
37. Helmerhorst F, Perquin D, Donker D, Keirse M (2004) Perinatal outcome of singletons and twins after assisted conception, a systematic review of controlled studies. BMJ 328:261
38. Jackson RA, Gibson KA, Wu YW, Croughan MS (2004) Perinatal outcomes in singletons following in vitro fertilization, a meta-analysis. Obstet Gynecol 103:551–563
39. Bonduelle M, Wennerholm U, Loft A et al (2005) A multicentre cohort study of the physical health of 5-year-old children conceived after intracytoplasmic sperm injection, in vitro fertilization and natural conception. Hum Reprod 20:413–419
40. Klemetti R, Gissler M, Sevon T et al (2005) Children born after assisted fertilization have an increased rate of major congenital anomalies. Fertil Steril 84:1300–1307
41. Olson C, Keppler-Noreuil K, Romitti P et al (2005) In vitro fertilization is associated with an increase in major birth defects. Fertil Steril 84:1308–1315
42. Shevell T, Malone F, Vidaver J et al (2005) Assisted reproductive technology and

pregnancy outcome. Obstet Gynecol 106: 1039–1045
43. Helmerhorst F, Perquin D, Donker D, Keirse M (2004) Perinatal outcome of singletons and twins after assisted conception, a systematic review of controlled studies. BMJ 328:261
44. Jackson R, Gibson K, Wu Y, Croughan M (2004) Perinatal outcomes in singletons following in vitro fertilization, a meta-analysis. Obstet Gynecol 103:551–563
45. Doyle P, Beral V, Maconochie N (1992) Preterm delivery, low birthweight and small-for-gestational-age in liveborn singleton babies resulting from in-vitro fertilization. Hum Reprod 7:425–428
46. Wang J, Clark A, Kirby C et al (1994) The obstetric outcome of singleton pregnancies following in-vitro fertilization/gamete intrafallopian transfer. Hum Reprod 9:141–146
47. Koudstaal J, Braat DD, Bruinse HW et al (2000) Obstetric outcome of singleton pregnancies after IVF, a matched control study in four Dutch university hospitals. Hum Reprod 15:1819–1825
48. Miles HL, Hofman PL, Peek J et al (2007) In vitro fertilization improves childhood growth and metabolism. J Clin Endocrinol Metab 92:3441–3445
49. Rinaudo P, Schultz R (2004) Effects of embryo culture on global pattern of gene expression in preimplantation mouse embryos. Reproduction 128:301–311
50. Katz-Jaffe M, Linck D, Schoolcraft W, Gardner D (2005) A proteomic analysis of mammalian preimplantation embryonic development. Reproduction 130:899–905
51. Corcoran D, Fair T, Park S, Rizos D, Patel O et al (2005) Suppressed expression of genes involved in transcription and translation in in vitro compared with in vivo cultured bovine embryos. Reproduction 131:651–660
52. Machaty Z, Day BN, Prather RS (1998) Development of early porcine embryos in vitro and in vivo. Biol Reprod 59:451–455
53. Farin P, Slenning B, Britt J (1999) Estimates of pregnancy outcomes based on selection of bovine embryos produced in vivo or in vitro. Theriogenology 52:659–670
54. Wang W, Abeydeera L, Han Y et al (1999) Morphologic evaluation and actin filament distribution in porcine embryos produced in vitro and in vivo. Biol Reprod 60:1020–1028
55. Khurana N, Niemann H (2000) Effects of cryopreservation on glucose metabolism and survival of bovine morulae and blastocysts derived in vitro or in vivo. Theriogenology 54:313–326
56. Montag M, Koll B, Holmes P, van der Ven H (2000) Significance of the number of embryonic cells and the state of the zona pellucida for hatching of mouse blastocysts in vitro versus in vivo. Biol Reprod 62:1738–1744
57. Holm P, Booth P, Callesen H (2002) Kinetics of early in vitro development of bovine in vivo- and in vitro-derived zygotes produced and/or cultured in chemically defined or serum-containing media. Reproduction 123:553–565
58. Papadopoulos S, Rizos D, Duffy P et al (2002) Embryo survival and recipient pregnancy rates after transfer of fresh or vitrified, in vivo or in vitro produced ovine blastocysts. Anim Reprod Sci 74:35–44
59. Rizos D, Ward F, Duffy P et al (2002) Consequences of bovine oocyte maturation, fertilization or early embryo development in vitro versus in vivo, implications for blastocyst yield and blastocyst quality. Mol Reprod Dev 61:234–248
60. Watkins A, Platt D, Papenbrock T et al (2007) Mouse embryo culture induces changes in postnatal phenotype including raised systolic blood pressure. Proc Natl Acad Sci USA 104:5449–5454
61. Giritharan G, Talbi S, Donjacour A et al (2007) Effect of in vitro fertilization on gene expression and development of mouse preimplantation embryos. Reproduction 134:63–72
62. Crosier AE, Farin PW, Dykstra MJ et al (2000) Ultrastructural morphometry of bovine compact morulae produced in vivo or in vitro. Biol Reprod 62:1459–1465
63. Crosier AE, Farin PW, Dykstra MJ et al (2001) Ultrastructural morphometry of bovine blastocysts produced in vivo or in vitro. Biol Reprod 64:1375–1385
64. Boni R, Tosti E, Roviello S, Dale B (1999) Intercellular communication in in vivo- and in vitro-produced bovine embryos. Biol Reprod 61:1050–1055
65. Abe H, Otoi T, Tachikawa S et al (1999) Fine structure of bovine morulae and blastocysts in vivo and in vitro. Anat Embryol (Berl) 199:519–527
66. Wrenzycki C, Herrmann D, Carnwath JW, Niemann H (1996) Expression of the gap junction gene connexin43 (Cx43) in preimplantation bovine embryos derived in vitro or in vivo. J Reprod Fertil 108:17–24
67. Eckert J, Niemann H (1998) mRNA expression of leukaemia inhibitory factor (LIF) and

its receptor subunits glycoprotein 130 and LIF-receptor-beta in bovine embryos derived in vitro or in vivo. Mol Hum Reprod 4:957–965
68. Stojanov T, Alechna S, O'Neill C (1999) In-vitro fertilization and culture of mouse embryos in vitro significantly retards the onset of insulin-like growth factor-II expression from the zygotic genome. Mol Hum Reprod 5:116–124
69. Stojanov T, O'Neill C (2001) In vitro fertilization causes epigenetic modifications to the onset of gene expression from the zygotic genome in mice. Biol Reprod 64:696–705
70. Bertolini M, Beam SW, Shim H et al (2002) Growth, development, and gene expression by in vivo- and in vitro-produced day 7 and 16 bovine embryos. Mol Reprod Dev 63:318–328
71. Lazzari G, Wrenzycki C, Herrmann D et al (2002) Cellular and molecular deviations in bovine in vitro-produced embryos are related to the large offspring syndrome. Biol Reprod 67:767–775
72. Lighten AD, Moore GE, Winston RM, Hardy K (1998) Routine addition of human insulin-like growth factor-I ligand could benefit clinical in-vitro fertilization culture. Hum Reprod 13:3144–3150
73. Wang TH, Chang CL, Wu HM et al (2006) Insulin-like growth factor-II (IGF-II), IGF-binding protein-3 (IGFBP-3), and IGFBP-4 in follicular fluid are associated with oocyte maturation and embryo development. Fertil Steril 86:1392–1401
74. Kamat BR, Brown LF, Manseau EJ et al (1995) Expression of vascular permeability factor/vascular endothelial growth factor by human granulosa and theca lutein cells. Role in corpus luteum development. Am J Pathol 146:157–165
75. Gordon JD, Mesiano S, Zaloudek CJ, Jaffe RB (1996) Vascular endothelial growth factor localization in human ovary and fallopian tubes, possible role in reproductive function and ovarian cyst formation. J Clin Endocrinol Metab 81:353–359
76. Lam PM, Briton-Jones C, Cheung CK et al (2003) Vascular endothelial growth factor in the human oviduct, localization and regulation of messenger RNA expression in vivo. Biol Reprod 68:1870–1876
77. Torry D, Holt V, Keenan J, Harris G, Caudle M, Torry R (1996) Vascular endothelial growth factor expression in cycling human endometrium. Fertil Steril 66(1):72–80
78. Hornung D, Lebovic DI, Shifren JL et al (1998) Vectorial secretion of vascular endothelial growth factor by polarized human endometrial epithelial cells. Fertil Steril 69:909–915
79. Zhang L, Scott P, Turley H et al (1998) Validation of antivascular endothelial growth factor (anti-VEGF) antibodies for immunohistochemical localization of VEGF in tissue sections, expression of VEGF in the human endometrium. J Pathol 185:402–408
80. Sugino N, Kashida S, Karube-Harada A et al (2002) Expression of vascular endothelial growth factor (VEGF) and its receptors in human endometrium throughout the menstrual cycle and in early pregnancy. Reproduction 123:379–387
81. Watson R, Anthony F, Pickett M et al (1992) Reverse transcription with nested polymerase chain reaction shows expression of basic fibroblast growth factor transcripts in human granulosa and cumulus cells from in vitro fertilisation patients. Biochem Biophys Res Commun 187:1227–1231
82. Di Blasio AM, Vigano P, Cremonesi L et al (1993) Expression of the genes encoding basic fibroblast growth factor and its receptor in human granulosa cells. Mol Cell Endocrinol 96:R7–R11
83. Seli E, Zeyneloglu HB, Senturk LM et al (1998) Basic fibroblast growth factor, peritoneal and follicular fluid levels and its effect on early embryonic development. Fertil Steril 69:1145–1148
84. Salmassi A, Zhang Z, Schmutzler A et al (2005) Expression of mRNA and protein of macrophage colony-stimulating factor and its receptor in human follicular luteinized granulosa cells. Fertil Steril 83:419–425
85. Boehm K, Daimon M, Gorodeski I et al (1990) Expression of the insulin-like and platelet-derived growth factor genes in human uterine tissues. Mol Reprod Dev 27:93–101
86. Desai N, Goldfarb J (1996) Growth factor/cytokine secretion by a permanent human endometrial cell line with embryotrophic properties. J Assist Reprod Genet 13:546–550
87. Kauma S, Aukerman S, Eierman D, Turner T (1991) Colony-stimulating factor-1 and C-FMS expression in human endometrial tissues and placenta during the menstrual cycle and early pregnancy. J Clin Endocrinol Metab 73:746–751
88. Davis S, Krozowski Z, McLachlan R, Burger H (1987) Inhibin gene expression in the

human corpus luteum. J Endocrinol 115:R21–R23
89. Wang H, Lu S, Han X et al (1992) Control of inhibin production by dispersed human luteal cells in vitro. Reprod Fertil Dev 4:67–75
90. He Z, Liu H, Mele C et al (1999) Expression of inhibin/activin subunits and their receptors and binding proteins in human preimplantation embryos. J Assist Reprod Genet 16:73–80
91. Vanttinen T, Liu J, Hyden-Granskog C et al (2000) Regulation of immunoreactive inhibin A and B secretion in cultured human granulosa-luteal cells by gonadotropins, activin A and insulin-like growth factor type-1 receptor. J Endocrinol 167:289–294
92. Petraglia F, Florio P, Luisi S et al (1998) Expression and secretion of inhibin and activin in normal and neoplastic uterine tissues. High levels of serum activin A in women with endometrial and cervical carcinoma. J Clin Endocrinol Metab 83:1194–1200
93. Ramasharma K, Li C (1987) Human pituitary and placental hormones control human insulin-like growth factor II secretion in human granulosa cells. Proc Natl Acad Sci USA 84:2643–2647
94. Muttukrishna S, Groome N, Ledger W (1997) Gonadotropic control of secretion of dimeric inhibins and activin A by human granulosa-luteal cells in vitro. J Assist Reprod Genet 14:566–574
95. Neulen J, Raczek S, Pogorzelski M et al (1998) Secretion of vascular endothelial growth factor/vascular permeability factor from human luteinized granulosa cells is human chorionic gonadotrophin dependent. Mol Hum Reprod 4:203–206
96. Chia C, Winston R, Handyside A (1995) EG, TGF-alpha and EGFR expression in human preimplantation embryos. Development 121:299–307
97. Smotrich D, Stillman R, Widra E et al (1996) Immunocytochemical localization of growth factors and their receptors in human preembryos and fallopian tubes. Hum Reprod 11:184–190
98. Jiang Q, Chen S, Xing F (2000) Expression of epidermal growth factor receptor in human preimplantation embryos. Zhonghua Fu Chan Ke Za Zhi 35:468–469
99. Chobotova K, Chobotova K, Spyropoulou I, Çarver J et al (2002) Heparin-binding epidermal growth factor and its receptor ErbB4 mediate implantation of the human blastocyst. Mech Dev 119:137–144
100. Roudebush W, Purnell E, Stoddart N, Fleming S (2002) Embryonic platelet-activating factor, temporal expression of the ligand and receptor. J Assist Reprod Genet 19: 72–78
101. Osterlund C, Wramsby H, Pousette A (1996) Temporal expression of platelet-derived growth factor (PGDF)-A and its receptor in human preimplantation embryos. Mol Hum Reprod 2(7):507–512
102. Freeman M, Whitworth C, Hill G (1995) Granulosa cell co-culture enhances human embryo development and pregnancy rate following in-vitro fertilization. Hum Reprod 10:408–414
103. Mansour R, Aboulghar M, Serour G, Abbass A (1994) Co-culture of human pronucleate oocytes with their cumulus cells. Hum Reprod 9:1727–1729
104. Quinn P, Margalit R (1996) Beneficial effects of coculture with cumulus cells on blastocyst formation in a prospective trial with supernumerary human embryos. J Assist Reprod Genet 13:9–14
105. Yeung W, Lau E, Chan S, Ho P (1996) Coculture with homologous oviductal cells improved the implantation of human embryos, a prospective randomized control trial. J Assist Reprod Genet 13:762–767
106. Kervancioglu M, Saridogan E, Atasu T et al (1997) Human fallopian tube epithelial cell co-culture increases fertilization rates in male factor infertility but not in tubal or unexplained infertility. Hum Reprod 12:1253–1258
107. Nieto F, Watkins W, Lopata A et al (1996) The effects of coculture with autologous cryopreserved endometrial cells on human in vitro fertilization and early embryo morphology, a randomized study. J Assist Reprod Genet 13:386–389
108. Ben-Chetrit A, Jurisicova A, Casper R (1996) Coculture with ovarian cancer cell enhances human blastocyst formation in vitro. Fertil Steril 65:664–666
109. Shapiro B, Richter K, Harris D, Daneshmand S (2001) A randomized controlled study comparing human embryo growth in sequential blastocyst media with or without Vero cell coculture. Fertil Steril 76:s3–s4
110. Menezo Y, Sakkas D (2002) Monozygotic twinning, is it related to apoptosis in the embryo? Hum Reprod 17:247–248
111. Behr B, Fisch J, Racowsky C, Miller K, Pool T, Milki A (2000) Blastocyst-ET and monozygotic twinning. J Assit Reprod Genet 17(6):349–351

112. Milki A, Jun S, Hinckley M, Behr B, Giudice L, Westphal L (2003) Incidence of monozygotic twinning with blastocyst transfer compared to cleavage-stage transfer. Fertil Steril 79(3):503–506
113. Wright V, Schieve L, Vahratian A, Reynolds M (2004) Monozygotic twinning associated with day 5 embryo transfer in pregnancies conceived after IVF. Hum Reprod 19(8):1831–1836
114. Desai N, Goldfarb J (1998) Co-cultured human embryos may be subjected to widely different microenvironments, pattern of growth factor/cytokine release by Vero cells during the co-culture interval. Hum Reprod 13:1600–1605
115. O'Neill C (1998) Autocrine mediators are required to act on the embryo by the 2-cell stage to promote normal development and survival of mouse preimplantation embryos in vitro. Biol Reprod 58:1303–1309
116. O'Neill C (1997) Evidence for the requirement of autocrine growth factors for development of mouse preimplantation embryos in vitro. Biol Reprod 56:229–237
117. Jin Y, Guo X, Li L et al (2001) The effect of autocrine factors on development of early embryos of mouse. Shi Yan Sheng Wu Xue Bao 34:77–80
118. Lighten A, Hardy K, Winston R, Moore G (1997) Expression of mRNA for the insulin-like growth factors and their receptors in human preimplantation embryos. Mol Reprod Dev 47:134–139
119. Punjabi U, Vereecken A, Delbeke L et al (1990) Embryo-derived platelet activating factor, a marker of embryo quality and viability following ovarian stimulation for in vitro fertilization. J In Vitro Fert Embryo Transf 7:321–326
120. Turpeenniemi-Hujanen T, Feinberg R, Kauppila A, Puistola U (1995) Extracellular matrix interactions in early human embryos, implications for normal implantation events. Fertil Steril 64:132–138
121. Hwu Y, Chen C, Li S et al (2006) Expression of vascular endothelial growth factor messenger ribonucleic acid and protein in human preimplantation embryos. Fertil Steril 85:1830–1832
122. Juriscova A, Casper R, MacLusky N, Mills G, Librach C (1996) HLA-G expression during preimplantation human embryo development. Proc Natl Acad Sci USA 93:161–165
123. Menicucci A, Noci I, Fuzzi B, Criscuoli L, Scarselli G, Barricordi O (1991) Nonclassic sHLA class I in human oocyte culture medium. Hum Immunol 60:1054–1057
124. Fuzzi B, Rizzo R, Criscuoli L, Noci I, Melchiorri L, Scarselli B et al (2002) HLA-G expression in early embryos is a fundamental prerequisite for the obtainment of pregnancy. Eur J Immunol 32(2):311–315
125. Gardner D, Sakkas D (2003) Assessment of embryo viability, the ability to select a single embryo for transfer – a review. Placenta 24:5–12
126. Noci I, Fuzzi B, Rizzo R, Melchiorri L, Criscuoli L, Diabizzi S et al (2005) Embryonic soluble HLA-G as a marker of developmental potential in embryos. Hum Reprod 20(1):138–146
127. Sher G, Keskintepe L, Nouriani M, Roussev R, Batzofin J (2004) Expression of sHLA-G in supernatants of individually cultured 46-h embryos, a potentially valuable indicator of "embryo competency" and IVF outcome. Reprod Biomed Online 9:74–78
128. Rebmann V, Switala M, Eue I, Grosse-Wilde H (2010) Soluble HLA-G is an independent factor for the prediction of pregnancy outcome after ART, a German multi-centre study. Hum Reprod 25(7):1691–1698
129. Desai N, Lawson J, Goldfarb J (2000) Assessment of growth factor effects on post-thaw development of cryopreserved mouse morulae to the blastocyst stage. Hum Reprod 15:410–418
130. Desai N, Kattal N, AbdelHafez F et al (2007) Granulocyte–macrophage colony stimulating factor (GM-CSF) and co-culture can affect post thaw development and apoptosis in cryopreserved embryos. J Assist Reprod Genet 24:215–222
131. O'Neill C, Ryan J, Collier M et al (1989) Supplementation of in-vitro fertilisation culture medium with platelet activating factor. Lancet 2:769–772
132. Borini A, Bulletti C, Cattoli M et al (1997) Use of recombinant leukemia inhibitory factor in embryo implantation. Ann N Y Acad Sci 828:157–161
133. Kim D, Kim M, Kang H et al (2001) The supplementation of granulocyte–macrophage colony-stimulating factor (GM-CSF) in culture medium improves the pregnancy rate in human ART programs. Fertil Steril 76:s6
134. Shapiro B, Richter K, Daneshmand S et al (2003) Granulocyte–macrophage colony-stimulating factor enhances human embryo development to the blastocyst stage, a randomized study. Fertil Steril 79:15–16
135. Huang J, Mehrens D, Wiese R et al (2001) High-throughput genomic and proteomic analysis using microarray technology. Clin Chem 47:1912–1916

136. Soulet D, Rivest S (2002) Perspective, how to make microarray, serial analysis of gene expression, and proteomic relevant to day-to-day endocrine problems and physiological systems. Endocrinology 143:1995–2001
137. Arici A, Engin O, Attar E, Olive D (1995) Modulation of leukemia inhibitory factor gene expression and protein biosynthesis in human endometrium. J Clin Endocrinol Metab 80:1908–1915
138. Cullinan E, Abbondanzo S, Anderson P et al (1996) Leukemia inhibitory factor (LIF) and LIF receptor expression in human endometrium suggests a potential autocrine/paracrine function in regulating embryo implantation. Proc Natl Acad Sci USA 93:3115–3120
139. Tsai H, Chang C, Hsieh Y, Lo H (2000) Leukemia inhibitory factor expression in different endometrial locations between fertile and infertile women throughout different menstrual phases. J Assist Reprod Genet 17:415–418
140. Sato A, Bo M, Otani T, Mochizuki M (1995) Expression of epidermal growth factor, transforming growth factor-alpha and epidermal growth factor receptor messenger RNA in human endometrium and endometrial carcinoma. Nippon Sanka Fujinka Gakkai Zasshi 47:473–478
141. Zhao Y, Chegini N (1999) The expression of granulocyte–macrophage-colony stimulating factor (GM-CSF) and receptors in human endometrium. Am J Reprod Immunol 42:303–311
142. Yoo H, Barlow D, Mardon H (1997) Temporal and spatial regulation of expression of heparin-binding epidermal growth factor-like growth factor in the human endometrium, a possible role in blastocyst implantation. Dev Genet 21:102–108
143. Leach R, Khalifa R, Ramirez N et al (1999) Multiple roles for heparin-binding epidermal growth factor-like growth factor are suggested by its cell-specific expression during the human endometrial cycle and early placentation. J Clin Endocrinol Metab 84:3355–3363
144. Stavreus-Evers A, Aghajanova L, Brismar H et al (2002) Co-existence of heparin-binding epidermal growth factor-like growth factor and pinopodes in human endometrium at the time of implantation. Mol Hum Reprod 8:765–769
145. Tang X, Rossi M, Masterson B, Chegini N (1994) Insulin-like growth factor I (IGF-I), IGF-I receptors, and IGF binding proteins 1–4 in human uterine tissue, tissue localization and IGF-I action in endometrial stromal and myometrial smooth muscle cells in vitro. Biol Reprod 50:1113–1125
146. Giudice L, Dsupin B, Jin I et al (1993) Differential expression of messenger ribonucleic acids encoding insulin-like growth factors and their receptors in human uterine endometrium and decidua. J Clin Endocrinol Metab 76:1115–1122
147. Giudice L, Lamson G, Rosenfeld R, Irwin J (1991) Insulin-like growth factor-II (IGF-II) and IGF binding proteins in human endometrium. Ann N Y Acad Sci 626:295–307
148. Morishige K, Kurachi H, Amemiya K et al (1993) Menstrual stage-specific expression of epidermal growth factor and transforming growth factor-alpha in human oviduct epithelium and their role in early embryogenesis. Endocrinology 133:199–207
149. Chegini N, Zhao Y, McLean F (1994) Expression of messenger ribonucleic acid and presence of immunoreactive proteins for epidermal growth factor (EGF), transforming growth factor alpha (TGF alpha) and EGF/TGF alpha receptors and 125I-EGF binding sites in human fallopian tube. Biol Reprod 50:1049–1058
150. Kurachi H, Morishige K, Imai T et al (1994) Expression of epidermal growth factor and transforming growth factor-alpha in fallopian tube epithelium and their role in embryogenesis. Horm Res 41(Suppl 1):48–154
151. Pfeifer T, Chegini N (1994) Immunohistochemical localization of insulin-like growth factor (IGF-I), IGF-I receptor, and IGF binding proteins 1–4 in human fallopian tube at various reproductive stages. Biol Reprod 50: 281–289
152. Keltz M, Attar E, Buradagunta S et al (1996) Modulation of leukemia inhibitory factor gene expression and protein biosynthesis in the human fallopian tube. Am J Obstet Gynecol 175:1611–1619
153. Lam P, Briton-Jones C, Cheung C et al (2003) Vascular endothelial growth factor in the human oviduct, localization and regulation of messenger RNA expression in vivo. Biol Reprod 68:1870–1876
154. Sugino N, Kashida S, Takiguchi S et al (2000) Expression of vascular endothelial growth factor and its receptors in the human corpus luteum during the menstrual cycle and in early pregnancy. J Clin Endocrinol Metab 85:3919–3924
155. Familiari G, Verlengia C, Nottola S et al (1996) Heterogeneous distribution of fibronectin, tenascin-C, and laminin immunoreactive material in the cumulus-corona

cells surrounding mature human oocytes from IVF-ET protocols – evidence that they are composed of different subpopulations, an immunohis-tochemical study using scanning confocal laser and fluorescence microscopy. Mol Reprod Dev 43:392–402
156. Piccinni M, Scaletti C, Mavilia C et al (2001) Production of IL-4 and leukemia inhibitory factor by T cells of the cumulus oophorus, a favorable microenvironment for preimplantation embryo development. Eur J Immunol 31:2431–2437
157. Neulen J, Yan Z, Raczek S et al (1995) Human chorionic gonadotropin-dependent expression of vascular endothelial growth factor/vascular permeability factor in human granulosa cells, importance in ovarian hyperstimulation syndrome. J Clin Endocrinol Metab 80:1967–1971
158. Zolti M, Ben-Rafael Z, Meirom R, Shemesh M, Bider D, Mashiach S, Apte R (1991) Cytokine involvement in occytes and early embryos. Fertil Steril 56(2):265–272
159. Sjoblom C, Wikland M, Robertson S (2002) Granulocyte–macrophage colony-stimulating factor (GM-CSF) acts independently of the beta common subunit of the GM-CSF receptor to prevent inner cell mass apoptosis in human embryos. Biol Reprod 67: 1817–1823
160. Sharkey A, Dellow K, Blayney M et al (1995) Stage-specific expression of cytokine and receptor messenger ribonucleic acids in human preimplantation embryos. Biol Reprod 53:974–981
161. van Eijk M, Mandelbaum J, Salat-Baroux J et al (1996) Expression of leukaemia inhibitory factor receptor subunits LIFR beta and gp130 in human oocytes and preimplantation embryos. Mol Hum Reprod 2:355–360
162. Wanggren K, Lalitkumar P, Hambiliki F et al (2007) Leukaemia inhibitory factor receptor and gp130 in the human fallopian tube and endometrium before and after mifepristone treatment and in the human preimplantation embryo. Mol Hum Reprod 13:391–397
163. Quinn P, Margalit R (1996) Beneficial effects of coculture with cumulus cells on blastocyst formation in a prospective trial with supernumerary human embryos. J Assist Reprod Genet 13:9–14
164. Wiemer K, Cohen J, Wiker S et al (1989) Coculture of human zygotes on fetal bovine uterine fibroblasts, embryonic morphology and implantation. Fertil Steril 52:503–508
165. Tucker M, Ingargiola P, Massey J et al (1994) Assisted hatching with or without bovine oviductal epithelial cell co-culture for poor prognosis in-vitro fertilization patients. Hum Reprod 9:1528–1531
166. Hu Y, Maxson W, Hoffman D et al (1998) Co-culture with assisted hatching of human embryos using buffalo rat liver cells. Hum Reprod 13:165–168
167. Schillaci R, Ciriminna R, Cefalu E (1994) Vero cell effect on in-vitro human blastocyst development, preliminary results. Hum Reprod 9:1131–1135
168. Magli M, Gianaroli L, Ferraretti A et al (1995) Human embryo co-culture, results of a randomized prospective study. Int J Fertil Menopausal Stud 40:254–259
169. Turner K, Lenton E (1996) The influence of Vero cell culture on human embryo development and chorionic gonadotrophin production in vitro. Hum Reprod 11: 1966–1974
170. Harvey M, Kaye P (1992) Insulin-like growth factor-1 stimulates growth of mouse preimplantation embryos in vitro. Mol Reprod Dev 31:195–199
171. Lin T, Yen J, Gong K et al (2003) IGF-1/IGFBP-1 increases blastocyst formation and total blastocyst cell number in mouse embryo culture and facilitates the establishment of a stem-cell line. BMC Cell Biol 4:14
172. Kim S, Lee G, Lee S et al (2005) Embryotropic effect of insulin-like growth factor (IGF)-I and its receptor on development of porcine preimplantation embryos produced by in vitro fertilization and somatic cell nuclear transfer. Mol Reprod Dev 72:88–97
173. Kim S, Lee S, Kim J et al (2006) Antiapoptotic effect of insulin-like growth factor (IGF)-I and its receptor in porcine preimplantation embryos derived from in vitro fertilization and somatic cell nuclear transfer. Mol Reprod Dev 73:1523–1530
174. Herrler A, Krusche C, Beier H (1998) Insulin and insulin-like growth factor-I promote rabbit blastocyst development and prevent apoptosis. Biol Reprod 59:1302–1310
175. Byrne A, Southgate J, Brison D, Leese H (2002) Regulation of apoptosis in the bovine blastocyst by insulin and the insulin-like growth factor (IGF) super-family. Mol Reprod Dev 62:489–495
176. Makarevich A, Markkula M (2002) Apoptosis and cell proliferation potential of bovine embryos stimulated with insulin-like growth factor I during in vitro maturation and culture. Biol Reprod 66:386–392
177. Moreira F, Paula-Lopes F, Hansen P et al (2002) Effects of growth hormone and insulin-like growth factor-I on development of

in vitro derived bovine embryos. Theriogenology 57:895–907

178. Sirisathien S, Hernandez-Fonseca H, Brackett B (2003) Influences of epidermal growth factor and insulin-like growth factor-I on bovine blastocyst development in vitro. Anim Reprod Sci 77:21–32
179. Palma G, Muller M, Brem G (1997) Effect of insulin-like growth factor I (IGF-I) at high concentrations on blastocyst development of bovine embryos produced in vitro. J Reprod Fertil 110:347–353
180. Block J, Drost M, Monson R et al (2003) Use of insulin-like growth factor-I during embryo culture and treatment of recipients with gonadotropin-releasing hormone to increase pregnancy rates following the transfer of in vitro-produced embryos to heat-stressed, lactating cows. J Anim Sci 81:1590–1602
181. Mitchell M, Swanson R, Hodgen G, Oehninger S (1994) Enhancement of in vitro murine embryo development by recombinant leukemia inhibitory factor. J Soc Gynecol Investig 1:215–219
182. Lavranos T, Rathjen P, Seamark R (1995) Trophic effects of myeloid leukaemia inhibitory factor (LIF) on mouse embryos. J Reprod Fertil 105:331–338
183. Mezhevikina L, Fedorova V, Kapralova I, Fesenko E (2006) Increased survival of preimplantation mouse embryos in medium with recombinant cytokine LIF. Ontogenez 37:55–62
184. Fry R, Batt P, Fairclough R, Parr R (1992) Human leukemia inhibitory factor improves the viability of cultured ovine embryos. Biol Reprod 46:470–474
185. Ptak G, Lopes F, Matsukawa K et al (2006) Leukaemia inhibitory factor enhances sheep fertilization in vitro via an influence on the oocyte. Theriogenology 65:1891–1899
186. de Moraes A, Hansen P (1997) Granulocyte–macrophage colony-stimulating factor promotes development of in vitro produced bovine embryos. Biol Reprod 57:1060–1065
187. Robertson S, Sjoblom C, Jasper M et al (2001) Granulocyte–macrophage colony-stimulating factor promotes glucose transport and blastomere viability in murine preimplantation embryos. Biol Reprod 64:1206–1215
188. Papayannis M, Eyheremendy V, Sanjurjo C, Blaquier J, Raffo F (2007) Effect of granulocyte–macrophage colony stimulating factor on growth, resistance to freezing and thawing and re-expansion of murine blastocysts. Reprod Biomed Online 14(1):96–101
189. Behr B, Mooney S, Wen Y, Polan M, Wang H (2005) Preliminary experience with low concentration of granulocyte–macrophage colony-stimulating factor, a potential regulator in preimplantation mouse embryo development and apoptosis. J Assist Reprod Genet 22(1):25–32
190. Bhatnagar P, Papaioannou V, Biggers J (1995) CSF-1 and mouse preimplantation development in vitro. Development 121: 1333–1339
191. Tamada H, Higashiyama C, Takano H et al (1999) The effects of heparin-binding epidermal growth factor-like growth factor on preimplantation-embryo development and implantation in the rat. Life Sci 64: 1967–1973
192. Lim J, Lee D, Song H et al (2006) Heparin-binding epidermal growth factor (HB-EGF) may improve embryonic development and implantation by increasing vitronectin receptor (integrin alphanubeta3) expression in peri-implantation mouse embryos. J Assist Reprod Genet 23:111–119
193. Lim K, Jang G, Ko K et al (2007) Improved in vitro bovine embryo development and increased efficiency in producing viable calves using defined media. Theriogenology 67:293–302
194. Aflafo E, Sod-Moriah U, Potashnik G, Har-Vardi I (2007) EGF increases expression and activity of Pas in perimplantation rat embryos and their implantation rate. Reprod Biol Endocrinol 29(5):4
195. Shi H, Miller F, Miller K, Kim M (1992) The effect of platelet activating factor on different phases of murine in vitro fertilization. J Assist Reprod Genet 9:373–377
196. Roudebush W, Duralia D, Butler W (1996) Effect of platelet-activating factor (PAF) on preimplantation mouse B6D2F1/J embryo formation. Am J Reprod Immunol 35:272–276
197. Zhou P, Liu D, Cang M et al (2008) TGFalpha and EGFR in ovine preimplantation embryos and effects on development. Anim Reprod Sci 104:370–381
198. Taniguchi F, Harada T, Yoshida S et al (1998) Paracrine effects of bFGF and KGF on the process of mouse blastocyst implantation. Mol Reprod Dev 50:54–62
199. Dunglison G, Barlow D, Sargent I (1996) Leukaemia inhibitory factor significantly enhances the blastocyst formation rates of human embryos cultured in serum-free medium. Hum Reprod 11:191–196

200. Hsieh Y, Tsai H, Chang C, Hsu L, Chang S, Lo H (2000) Prolonged culture of human cryopreserved embryos with recombinant human leukemia inhibitory factor. J Assist Reprod Genet 17(3):131–134
201. Spanos S, Becker D, Winston R, Hardy K (2000) Anti-apoptotic action of insulin-like growth factor-I during human preimplantation embryo development. Biol Reprod 63(5):1413–1420
202. Sjoblom C, Wikland M, Robertson S (1999) Granulocyte–macrophage colony-stimulating factor promotes human blastocyst development in vitro. Hum Reprod 14: 3069–3076
203. Martin K, Barlow D, Sargent I (1998) Heparin-binding epidermal growth factor significantly improves human blastocyst development and hatching in serum-free medium. Hum Reprod 13:1645–1652
204. Khamsi F, Armstrong D, Zhang X (1996) Expression of urokinase-type plasminogen activator in human preimplantation embryos. Mol Hum Reprod 2:273–276

Chapter 12

Culture Systems: Single Step

Ronit Machtinger and Catherine Racowsky

Abstract

Culture media to support development of zygotes to the blastocyst stage is based either on a single medium or sequential (two-step) media. Single medium culture either with or without day 3 renewal is associated with simplified laboratory protocols and lower costs compared with sequential medium. There are currently insufficient clinical data to conclude that one system, either single or sequential, is superior to the other with regard to clinical performance. This chapter summarizes the rationale for use of a single medium, along with supporting animal and human data for culturing embryos in a single medium, renewed or not renewed.

Key words: Single medium, Embryo culture, Blastocyst

1. Introduction

The human oocyte fertilizes in the ampullary region of the fallopian tube in vivo, and the newly formed zygote migrates through the isthmus, traverses the uterotubal junction around day 4 at the morula stage, and subsequently implants in the uterus around day 6, at the stage of blastocyst (1–3). Analyses of energy substrate concentrations in fluid collected from nonpregnant women at various stages of the menstrual cycle have shown that these concentrations differ between the fallopian tube and uterine environments (4, 5). The fluid in the fallopian tube has a relatively high concentration of lactate, and a lower concentration of glucose (5–7). The uterine fluid contains a relatively low concentration of lactate but higher concentrations of glucose; the concentration of pyruvate is unchanged between these regions of the reproductive tract.

Over the last 15 years, increasing attention has been paid to extended culture of human embryos from the zygote to the blastocyst stage (4), with a particular focus being given to development of various culture media. The rationale of media development

Gary D. Smith et al. (eds.), *Embryo Culture: Methods and Protocols*, Methods in Molecular Biology, vol. 912,
DOI 10.1007/978-1-61779-971-6_12, © Springer Science+Business Media, LLC 2012

for the culture of preimplantation embryos has been influenced by two concepts: single vs. sequential culture media. In the case of single culture media, the embryos are cultured in one medium from zygote to blastocyst, either with uninterrupted culture throughout, or with a change to fresh medium of the same type on day 3; in either case, the embryo is cultured in a medium developed from the simplex optimization approach in which the appropriate concentrations of constituents are determined by bioassay ("let the embryo choose" philosophy (reviewed by ref. (8))). Using sequential media, embryos are cultured in one type of medium from zygote to early cleavage stage. On day 3, the embryos are transferred to another medium of different composition, and are cultured in this medium until day 5. This two-step culture media approach is based on the assumption that embryos are naturally exposed to different nutrients in the fallopian tube and the uterus (5), and therefore different components are needed to support the development of zygotes to early cleavage stage embryos vs. early cleavage stage embryos to blastocysts (the "back to nature" philosophy) (9–11).

The scope of this chapter is single culture media. We focus on the history of single medium development and then consider data from animal and human embryos cultured using this medium strategy, as compared with sequential media.

1.1. The Development of Media for Embryo Culture

In 1956, Whitten first showed that 8-cell mouse embryos would develop to the blastocyst stage when cultured in a simple solution, based on a physiological saline (12). Two years later, in 1958, McLaren and Biggers published a landmark paper reporting that blastocysts produced in this way could produce live young when transferred to recipient mothers (13). This pioneer work was done when culture medium, previously supplemented with biological fluids such as serum, was replaced by a chemically defined composition (chemically defined medium, by definition, is a solution of known composition) (4).

Following Whitten's first study in 1956, he then reported that addition of lactate to his medium enabled the two-cell stage mouse embryo, but not the zygote, to develop into a blastocyst (the concept of the "two cell block") (14). These observations raised the possibility that the oviduct provides a unique microenvironment for the preimplantation embryo. This idea was supported in part by Biggers, Gwatkin, and Brinster who showed that the preimplantation mouse embryo would develop from the zygote to the blastocyst in explants of the fallopian tube (15). Brinster later modified Whitten's medium by reducing the calcium concentration and adding pyruvate to produce BMOC2 which supported the development of two-cell embryos to blastocysts at high rates (16). Brinster's medium was further modified by Whittingham by decreasing the lactate concentration and increasing the concentration of pyruvate, resulting in the M16 medium (17). Both media were significant advances for embryo culture and were widely used although, with the exception of certain strains of mice, they could not overcome the two cell block.

In 1989, the two-cell block in mice was finally overcome with the production of CZB medium, a modification of BMOC2 medium which had an increased lactate–pyruvate ratio, and glutamine as a replacement for glucose and EDTA. These modifications suggested that glucose might be inhibitory to early embryo development and that a chelating agent was needed to remove any deleterious trace elements from the medium used for cultivation. CZB allowed the development of the zygote to the blastocyst of several strains of mice that had previously blocked at the two-cell stage (18). Only when the embryos had developed to the morula stage was it necessary to add glucose to ensure a high yield of blastocysts.

1.2. Important Components of Culture Media

The majority of culture media include key components that have important roles in embryo development. Those who advocate use of sequential media have stated that some of the components might be harmful to the embryo at specific stages of development.

1.2.1. Glucose

A number of studies in animal models (mouse, hamster, sheep and cattle) (7, 17–21), as well as in human (22) showed that high levels of glucose in culture may result in retardation or developmental arrest of cleavage-stage embryos. Lack of glucose in the media for the postcompaction stage embryo resulted in impaired development to the blastocyst stage and a loss of viability (7). However, in a carefully controlled systematic design, Biggers and McGinnis observed no inhibitory effect of glucose, even at higher concentrations, on the development of mouse zygotes to the blastocyst stage (23). This may be dependent upon the presence of other media components.

1.2.2. Amino Acids and Ammonium

Embryo culture media contain amino acids, as fallopian tube and uterine fluids contain significant amounts of free amino acids (24). Despite their benefit for embryo development, one concern that exists is the breakdown of amino acids in culture and production of ammonia. Of particular concern in regard to ammonia production is the labile amino acid glutamine. The use of a medium containing glutamine might result in significant levels of ammonium buildup, even when the medium is changed every 24 h (25). Lane and Gardner have reported on the ammonium accumulation in culture medium, which induces aberrant blastocyst differentiation, metabolism, pH regulation, gene expression, and subsequently alters fetal development in the mouse (26). In addition, Orsi and Leese have found that ammonium exposure negatively affects the amino acid metabolism of cattle blastocysts in vitro (27).

Zander et al. showed that exposure of mouse embryos to ammonium at the precompaction stage after culture from either the zygote to two-cell stage or from the two-cell to the eight-cell stage did not affect the ability of development to the blastocyst stage (28). However, the blastocysts had decreased cell numbers and increased levels of cellular apoptosis. Although the implantation rate was not affected, the number of fetuses was reduced by culture with ammonium at the precompaction stage and fetal development was delayed, as observed by reduced crown-rump length and maturity. In contrast, the later stage embryo was more resistant to the negative effects of ammonium (28). Virant-Klun investigated the effect of exposure of human embryos cultured in sequential medium to blastocysts. They reported that exposure of early cleavage human embryos to ammonium resulted in inhibition of the development of human blastocysts in vitro (29).

The above concerns regarding inclusion of glutamine in the medium led to replacing glutamine with L-alanyl-L-glutamine or glycyl-L-glutamine, both stable dipeptides of glutamine, in order to prevent ammonium buildup.

1.2.3. EDTA

Ethylenediaminetetraacetic acid (EDTA) is a chelating agent. The addition of EDTA to culture media has been shown to be beneficial for the development of the embryo from the zygote through cleavage stage (18, 30, 31). On the other hand, exposure of postcompaction stage embryos appears to result in a reduced number of inner cell mass cells and fetal development after transfer (5). This may be due to the ability of EDTA to inhibit glycolysis through impairing 3-phosphoglycerate kinase activity.

1.3. Use of Single Medium

One of the limitations in the development of culture media for the human embryo is the inability to perform empirical studies on the human embryos or to perform large controlled studies on individual medium components. Thus, the majority of data underpinning the development of culture media for clinical work derives from animal models. Nevertheless, based on the differences in energy substrate concentrations found between fallopian tube and uterine fluids, as well as results of experiments with embryos from various animal models, Gardner and Lane suggested that in order to optimize the developmental of embryos grown in culture, two media used in sequence (two step or sequential), are required (32). However, the evidence used to support this assumption is not strong due to problems with fluid sampling. Notably, data is primarily based on analysis of fallopian and uterine fluids collected from non-pregnant women, and at various stages of the menstrual cycle (5), so correlation to concentrations found during the luteal phase when embryos are present in the reproductive tract is unknown.

As discussed by Summers and Biggers (33), these measurements were highly variable, and reflect the overall composition of the female tract fluids, and not necessarily the microenvironment around the embryo. The in vitro environment is different from the in vivo environment. In vitro, the embryos are cultured in an aqueous phase whereas in vivo, the embryo is surrounded by a thin layer of fluid and presumably located very close to the maternal tissue, allowing rapid exchange of gases, nutrients, and wastes. On the other hand, the cultured embryo in an aqueous phase is in a microenvironment in which the concentrations of nutrients become depleted and those of waste products inevitably increase.

Two types of single medium protocol can be used for the culture of human embryos from the zygote to blastocyst stage (34):

1. Interrupted single medium culture: The medium is renewed on the third day of embryo culture (embryos typically at the 6–8 cell stage). Renewing the medium enables replenishment of compounds that have been depleted during the initial culture period. On the other hand, this approach may remove compounds secreted by the embryo during the initial period, such as growth factors.
2. Uninterrupted single embryo culture: The same single medium is used throughout the 5–6 days of culture with no replenishment or replacement on day 3. In this approach, if glutamine is present, use of a dipeptide is especially important.

There are several theoretical and practical advantages to using a single culture medium for culturing human embryos to the blastocyst stage. Theoretical advantages include reducing the possibility of stress to embryos associated with exposing them to a different medium composition on day 3. Switching of the embryos from one medium to another may cause additional osmotic or other shock and/or deprive them of any paracrine or autocrine factor they may have produced during the first culture period. Practical advantages include reducing the number of media required in the IVF lab, which means less inventory control, fewer contact control assays and a lower chance for error in the clinical ART laboratory. Overall, the two step procedure is more labor intensive and expensive than the one step protocol. Comparisons of several characteristics of a single medium (with or without renewal) with those of sequential medium reveal benefits and weaknesses (Table 1).

1.4. Animal Data for Embryos Cultured in Single vs. Sequential Media

Biggers et al. compared the developmental potential of mouse embryos after culture in a single medium (potassium simplex optimized medium supplemented with glucose and amino acids, $KSOM^{AA}$) without renewal, $KSOM^{AA}$ renewed after 48 h, or sequential media (G1/G2) (35). There were no significant differences among the three groups for the proportion of blastocysts, rates of hatching, or the number of cells in either the inner cell mass or trophectoderm. Moreover, no differences among groups were observed for incidence of gross anomalies in the fetuses formed.

Table 1
Comparison of various characteristics of a single medium, with and without renewal, and sequential media

Characteristic	Single medium (nonrenewed)	Single medium (renewed)	Sequential media
Leaves embryos undisturbed	Yes	No	No
Accumulated endogenous growth factors	Left in place	Lost	Lost
Replacement of essential nutrients	No	Yes	Yes
Accumulated toxins	Left in place	Removed	Removed
Relative (potential) environmental stress to embryos	Low	Moderate	High
Requires quality control	One medium	One medium	Two media
Relative labor intensity	Low	Moderate	High
Relative cost	Low	Low	High

Perin et al. examined the effects of single medium ($KSOM^{AA}$) with renewal on day 3, vs. sequential medium (G1/G2) on the ability of cultured mouse zygotes to progress to the blastocyst stage (36). The percentage of zygotes that developed to the blastocyst stage was significantly higher in the single medium group as well as the percentage of blastocysts that partially or completely hatched by day 5. The authors concluded that commercially available $KSOM^{AA}$ medium is superior to sequential G1/G2 media for culturing one-cell embryos up to the blastocyst stage in the mouse IVF model.

1.5. Human Embryo Studies

In 2002, Biggers and Racowsky published a study that evaluated the efficacy of using a one-step system involving $KSOM^{AA}$, the predecessor of Global medium (IVF Online, Guelph, ON) (11). Embryos were cultured with $KSOM^{AA}$ from day 1 to 3, from day 3 to 5 or from day 1 to 5, and compared with the laboratory's standard sequential medium system at that time (P1: Irvine Scientific, Santa Ana, CA; and CCM: Vitrolife, Gothenberg, Sweden). The results showed that the cell numbers and fragmentation scores in day 3 embryos as well as the blastocyst formation rates on day 5 were similar among all groups.

Macklon et al. cultured 158 randomized human embryos to one of three systems: culture in their own single medium (Rotterdam medium) from day 1 to 5 with or without renewal on day 3 or culture in a commercially available sequential culture medium. Embryo quality on day 3 and day 5, and implantation rates and ongoing pregnancies were similar among the three groups (37).

Reed et al. compared the outcome of 893 sibling embryos from 80 IVF cycles cultured side by side in a single medium continuously, without medium renewal on day 3 (Global medium), versus sequential media (G5 series sequential media, Vitrolife, Gothenberg, Sweden) (38). Day 3 embryo quality was similar for both groups but for day 5, more blastocysts were selected for transfer from the single medium culture compared with the sequential media.

Sepulveda et al. (39) compared the development of human embryos (from zygotes to blastocysts) in a single medium (with medium renewal on day 3) with a sequential media, using a donor oocyte model. The development of 287 zygotes in renewed single medium was compared with 322 zygotes that were cultured in sequential medium. The authors found that the rates of embryo development on days 3, 4, and 5, as well as the implantation rates, were significantly greater for embryos cultured in the renewed single medium compared with those cultured in sequential media.

The above observations remain to be verified in prospective, randomized trials. However, the collective available evidence indicates that embryo utilization rates and pregnancy rates are at least equivalent when embryos are cultured up to day 5/6 in a single versus a sequential system (reviewed by ref. 34).

1.6. Available Commercial Single Medium

Although there are a number of sequential media commercially available, as far as we are aware, there are only three single medium formulations on the market: Global (IVF Online, Guelph, ON), GM 501 (Gynemed, Lenshan, Germany), and SSM (Irvine Scientific; Santa Ana, CA).

As shown in Table 2, though concentrations are unknown, the constituents of the two more recent media, GM 501 and SSM, are very similar to Global, which is a modified $KSOM^{AA}$ medium (Table 2).

1.7. Summary

The development of culture media to support development of zygotes to the blastocyst stage is based on two philosophies: the "let the embryo choose" single medium philosophy, versus the "back to nature" sequential (two step) media philosophy. Those who advocate the single medium system support the premise that changing the media might cause environmental stress to the embryo and/or deprive it of necessary substrates, whether they are components of a more complex single medium or autocrine/paracrine factors produced by the embryo during the culture period. On the other hand, those who believe in the sequential media approach assume that single culture may have a harmful effect on the embryo due to build up of ammonium and exposure of the

Table 2
Constituents of single step media

	Constituent	Global	SSM	GM 501
Salts	Sodium chloride	+	+	+
	Potassium chloride	+	+	+
	Potassium phosphate	+	+	+
	Calcium chloride	+	+	+
	Magnesium sulfate	+	+	+
Buffer	Sodium bicarbonate	+	+	+
Energy substrates	Sodium pyruvate	+	+	+
	Glucose	+	+	+
	Sodium lactate/lactate Na salt	+	+	+
Chelating agent	EDTA	+	+	+
Dipeptide	Alanyl-glutamine	–	+	+
	Glycyl-glutamine	+	–	–
pH Indicator	Phenol Red	+	+	–
Antibiotic	Gentamicin	+	+	–
Amino acids	L-Alanine	+	+	+
	L-Asparagine	+	+	+
	L-Aspartic acid	+	+	+
	L-Glutamic acid	+	+	+
	Glycine	–	+	+
	L-Proline	+	+	+
	L-Serine	+	+	+
	L-Arginine	–	+	+
	L-Arginine HCl	+	–	–
	L-Cystine	+	+	+
	L-Histidine	+	+	+
	L-Isoleucine	+	+	+
	L-Leucine	+	+	+
	L-Lysine	–	+	–
	L-Lysine HCl	+	–	+
	L-Methionine	+	+	+
	L-Phenylalanine	+	+	+
	L-Threonine	+	+	+
	L-Tryptophan	+	+	+
	L-Tyrosine	+	+	+
	L-Valine	+	+	+

embryos to concentrations of some components, such as glucose, that are not physiological. There are currently insufficient clinical data to conclude that one system, either single or sequential, is superior to the other. However, simplified laboratory protocols and lower costs are associated with use of a single medium.

2. Materials

For single step human embryo culture, a tested commercially available single step media should be selected and handled in the appropriate sterile fashion as normally indicted within the IVF laboratory.

2.1. Media Preparation

1. Commercially available single step media for use in human IVF must be selected. These may include Global (IVF Online, Guelph, ON), GM 501 (Gynemed, Lenshan, Germany), and SSM (Irvine Scientific; Santa Ana, CA) or additional products.
2. Appropriate protein source should be selected and added using sterile technique to the culture medium. Choices include, but are not limited to human serum albumin (HSA), recombinant HSA or HSA with globulins.

3. Methods

Carry out all procedures using sterile technique.

1. Once sterile media with protein has been prepared (see Note 1), culture dishes are prepared according to laboratory protocol and placed into the incubator the day before required use for embryo culture (see Notes 2 and 3).
2. Embryos should be placed into the single step medium according to lab protocol. This medium can be for fertilization if desired, or for embryo culture following ICSI/fertilization check (day 1) (see Notes 4 and 5).
3. Embryos are cultured for the appropriate time until used for transfer or cryopreserved. Media can be changed around day 2 or 3 for cleavage embryos according to laboratory protocol. Alternatively, the same single step medium can be used throughout the entire culture period through days 5–6 (see Note 6).

4. Notes

1. Protein source and supplier is ultimately up to the lab. Some media may be supplied with protein already present. Concentration is also laboratory dependent, but may vary depending on procedure. For example, some laboratories may use higher protein content when culturing frozen/thawed embryos.

2. Type of dish, volume of media, and use of oil overlay are all dependent upon individual laboratory preference and protocol.
3. It is recommended dishes be placed into the appropriate incubator with proper temperature and atmosphere at least 8 h, but preferably greater than 12 h before embryos are to be added to the dish. This allows temperature and pH equilibration, which can be impacted by media volume and oil overlay.
4. If a separate fertilization media is not used, single step media can be used for insemination during IVF cases. Zygotes can then be transferred to a fresh dish/drop of the single step media the following morning after fertilization check. Following ICSI, presumptive zygotes can be placed directly into the single step media following injection.
5. Embryos can be cultured in groups or individually as desired. The goal is to be able to identify and track embryos in accordance with standard laboratory protocol.
6. If medium is to be changed on day 2 or 3, ensure new media and dishes for the final 2–3 days of culture are made and placed into the incubator to equilibrate before use.

References

1. Pauerstein CJ, Eddy CA (1979) The role of the oviduct in reproduction; our knowledge and our ignorance. J Reprod Fertil 55:223–229
2. Diaz S, Ortiz ME, Croxatto HB (1980) Studies on the duration of ovum transport by the human oviduct. III. Time interval between the luteinizing hormone peak and recovery of ova by transcervical flushing of the uterus in normal women. Am J Obstet Gynecol 137:116–121
3. Lyons RA, Saridogan E, Djahanbakhch O (2006) The reproductive significance of human fallopian tube cilia. Hum Reprod Update 12:363–372
4. Biggers JD (1998) Reflection of the culture of preimplantation embryo. Int J Dev Biol 42: 879–884
5. Gardner DK, Lane M (1996) Alleviation of the '2-cell block' and development to the blastocyst of CF1 mouse embryos: role of amino acids, EDTA and physical parameters. Hum Reprod 11:2703–2712
6. Leese HJ, Tay JI, Reischl J, Downing SJ (2001) Formation of fallopian tubal fluid: role of a neglected epithelium. Reproduction 121: 339–346
7. Lane M, Gardner DK (2007) Embryo culture medium: which is the best? Best Pract Res Clin Obstet Gynaecol 21(1):83–100
8. Biggers JD (2000) Thoughts on embryo culture conditions. Reprod BioMed Online 4(1):30–38
9. Leese HJ (1991) Metabolism of the preimplantation mammalian embryo. Oxford Rev Reprod Biol 13:35–72
10. Leese HJ (1998) Human embryo culture: back to nature. J Assist Reprod Gene 15:466–468
11. Biggers JD, Racowsky C (2002) The development of fertilized human ova to the blastocyst stage in KSOM^{AA} medium: is a two-step protocol necessary? Reprod Biomed Online 5(2): 133–140
12. Whitten WK (1956) Culture of tubal ova. Nature 177:96
13. McLaren A, Biggers JD (1958) Successful development and birth of mice cultivated in vitro as early as early embryos. Nature 182(4639):877–878
14. Whitten WK (1957) Culture of tubal ova. Nature 179:1081–1082
15. Biggers JD, Gwatkin RB, Brinster RL (1962) Development of mouse embryos in organ cultures of fallopian tubes on a chemically defined medium. Nature 194:747–749
16. Brinster RL (1972) Cultivation of the mammalian embryo. In: Rothblat G, Cristofalo V (eds)

Growth, nutrition and metabolism of cells in culture. Academic, New York
17. Whittingham DG (1971) Culture of mouse ova. J Reprod Fertil 14:7–21
18. Chatot CL, Ziomek CA, Bavister BD, Lewis JL, Torres I (1989) An improved culture medium supports development of Random-bred 1-cell mouse embryos in vitro. J Reprod Fertil 86:679–688
19. Schini SA, Bavister BD (1988) Two-cell block to development of cultured hamster embryos is caused by phosphate and glucose. Biol Reprod 39:1183–1192
20. Thompson JG, Simpson AC, Pugh PA, Tervit HR (1992) Requirement for glucose during in vitro culture of sheep preimplantation embryos. Mol Reprod Dev 31:253–257
21. Takahashi Y, First NL (1992) In vitro development of bovine one-cell embryos: influence of glucose, lactate, pyruvate, amino acids and vitamins. Theriogenology 37:963–978
22. Quinn P (1995) Enhanced results in mouse and human embryo culture using a modified human tubal fluid medium lacking glucose and phosphate. J Assist Reprod Genet 12:97–105
23. Biggers JD, McGinnis L (2001) Evidence that glucose is not always an inhibitor of mouse preimplantation development in vitro. Hum Reprod 16(1):153–163
24. Miller JF, Schulz GA (1987) Amino acid content of preimplantation rabbit embryos and fluids of the reproductive tract. Biol Reprod 36:125–129
25. Gardner DK, Lane M (1993) Amino acids and ammonium regulate mouse embryo development in culture. Biol Reprod 48:377–385
26. Lane M, Gardner DK (2003) Ammonium induces aberrant blastocyst differentiation, metabolism, pH regulation, gene expression and subsequently alters fetal development in the mouse. Biol Reprod 69(4):1109–1117
27. Orsi NM, Leese HJ (2004) Ammonium exposure and pyruvate affect the amino acid metabolism of bovine blastocysts in vitro. Reproduction 127(1):131–140
28. Zander DL, Thompson JG, Lane M (2006) Perturbations in mouse embryo development and viability caused by ammonium are more severe after exposure at the cleavage stages. Biol Reprod 74(2):288–294
29. Virant-Klun I, Tomazevic T, Vrtacnik-Bokal E, Vogler A, Krsnik M, Meden-Vrtovec H (2006) Increased ammonium in culture medium reduces the development of human embryos to the blastocyst stage. Fertil Steril 85: 526–528
30. Erbach GT, Bhatnagar P, Baltz JM, Biggers JD (1995) Zinc is a possible toxic contaminant of silicone oil in microdrop cultures of preimplantation mouse embryos. Hum Reprod 10:3248–3254
31. Nasr-Esfahani MH (1992) Effects of glucose, glutamine, ethylenediaminetetraacetic acid and oxygen tension on the concentration of reactive oxygen species and on development of the mouse preimplantation embryo in vitro. J Reprod Fertil 96:219–231
32. Gardner DK, Lane M, Schoolcraft WB (2002) Physiology and culture of the human blastocyst. J Reprod Immunol 55(1–2):85–100
33. Summers MC, Biggers JD (2003) Chemically defined media and the culture of mammalian preimplantation embryos: historical perspective and current issues. Hum Reprod Update 9(6):557–82
34. Biggers JD, Summers MC (2008) Choosing a culture medium: making informed choices. Fertil Steril 90(3):473–483
35. Biggers JD, McGinnis LK, Lawitts JA (2005) One-step versus two-step culture of mouse preimplantation embryos: is there a difference? Hum Reprod 20(12):3376–3384
36. Perin PM, Maluf M, Nicolosi Foltran Januário DA, Nascimento Saldiva PH (2008) Comparison of the efficacy of two commercially available media for culturing one-cell embryos in the in vitro fertilization mouse model. Fertil Steril 90:1503–1510
37. Macklon NS, Pieters MH, Hassan MA, Jeucken PH, Eijkemans MJ, Fauser BC (2002) A prospective randomized comparison of sequential versus monoculture systems for in-vitro human blastocyst development. Hum Reprod 17(10):2700–2705
38. Reed ML, Hamic A, Thompson DJ, Caperton CL (2009) Continuous uninterrupted single medium culture without medium renewal versus sequential media culture: a sibling embryo study. Fertil Steril 92(5):1783–1786
39. Sepulveda S, Garcia J, Arriaga E, Diaz J, Noriega-Portella L, Noriega-Hoces L (2009) In vitro development and pregnancy outcomes for human embryos cultured in either a single medium or in a sequential media system. Fertil Steril 91:1765–1770

Chapter 13

Culture Systems: Sequential

Patrick Quinn

Abstract

Methods for the culture of preimplantation human embryos evolved primarily from those used for mouse embryos. The initial method was usually culture in a single medium in microdrops of medium under oil for 2–3 days before transfer. Subsequently, extended culture over the whole preimplantation period was used. The debate at present is which system is best, a sequential series of media to accommodate changes in physiology and metabolism of the embryo from a 1-cell zygote to the differentiated blastocyst stage or a single-step culture regime using the same culture medium throughout the preimplantation period. Aspects of the advantages and disadvantages of these two culture systems will be discussed.

Key words: Embryo, Culture media, Metabolism, Morphology

1. Introduction

A possible and somewhat logical concept is to culture mammalian embryos in media that represent the fluid environment they would normally encounter in vivo. This concept has been referred to as the "back to nature" approach as opposed to an alternative method labeled "let the embryo choose" where the embryo is placed in a medium containing all possible, to a limited degree, metabolites it may possibly interact with (1, 2). The main emphasis of this second approach has been in regard to amino acids (see Subheading 1.2.4).

At the beginning of clinical human-assisted reproductive technology (ART) in the 1980s, embryos were replaced in the uterus at any time from day (D) one to D3, with the initial time being D2. Many clinics, depending on the number and quality of embryos, extended embryo transfer to D3 (reviewed in (3)). In the late 1990s embryo culture was extended further to D5/6 (4) and the Gardner team emphasized the need for a specific medium for culture from D3 to D5/6 to meet the specific needs of the embryo during this period. Again, the decision to undertake this extended

Gary D. Smith et al. (eds.), *Embryo Culture: Methods and Protocols*, Methods in Molecular Biology, vol. 912, DOI 10.1007/978-1-61779-971-6_13, © Springer Science+Business Media, LLC 2012

culture was very dependent on the number and quality of embryos on D3, when transfer of embryos from medium designed for the first phase of embryonic development (D1–D3) occurred with their placement in a somewhat more complex medium designed for development in the second phase of preimplantation embryogenesis from D3 to D5/6.

The media in a sequential series are often formulated for fertilization, cleavage, and blastocyst stages by commercial media vendors (5). Cleavage and blastocyst media are those used for the first and second phase of embryo development, namely from D1 to D3 and then D3 to D5/6, respectively. It makes some sense to make a change in media conditions around D3 because of various changes that occur in the human embryo at this time. Included in these changes is the process called compaction and activation of the embryonic genome.

Compaction is the process wherein tight junctional complexes form between the plasma membranes of peripheral blastomeres, binding them tightly together and flattening the outer surface of these blastomeres into a smoother appearance. This flattened appearance in which the view of the individual blastomeres becomes obscured makes the blastomeres appear as a mulberry, hence the embryo at this stage is referred to as a morula. Calcium and magnesium ions are an essential part of the compaction process, and, in their absence, tight junctions do not form. Thus, use of a calcium/magnesium-free medium is often used during D3 biopsy of blastomeres for preimplantation genetic diagnosis to permit easy removal of cells that may not be feasible if compaction has already occurred (6). Furthermore, the occurrence of compaction is a prerequisite for the subsequent development of the embryo into a blastocyst, as the tight junctions between the outer blastomeres allow for the fluid-filled blastocoel cavity to maintain a barrier to the leakage of blastocoel fluid back into the fluid environment surrounding the embryo. The tightly joined outer blastomeres surrounding the blastocoel cavity are referred to as the trophectoderm layer and they act as a transporting epithelium that gives the embryo more control over the regulation of transport of molecules into and out of the blastocoel cavity and subsequently within the blastomeres themselves. This is particularly important with regard to the internal pH (pH_i) in the embryonic cells. Prior to compaction, small increases in pH_i cause substantial increases in the activity of the enzyme phosphofructokinase (PFK) that plays a key role in regulating metabolism (7), namely an increase in glycolytic activity that has a negative effect on the early precompaction preimplantation embryo.

As mentioned, the other major event occurring around D3 in human embryos that may justify changing of media is activation of the embryonic genome (8). It is likely that this event is associated with many changes in metabolic activity, some of which can be reflected in changes in the uptake and metabolism of energy substrates such as pyruvate, lactate, and glucose (7), amino acids (9), fatty acids, purines, and pyrimidines to provide for the large increases in nuclear and mitochondrial DNA and cytoplasmic RNA.

It is yet to be determined whether a single-step medium for culture from D1 to D5/6, a "let the embryo choose" approach, or the use of sequential culture media ("back to nature") provides for better embryo development during the transition from D1–3 to D3–5/6. Current evidence would indicate that both culture strategies work equally as well (10). However, as discussed below, there may be differential requirements between early and late stage embryos that require modulation of pH whether either a sequential media or single-step medium protocol is used.

1.1. Components of Sequential Culture Media and Their Interaction with Embryos

The topic of sequential media components has been reviewed many times (see (3)). As a brief summary, many of the constituents in ART media are found throughout the body. These include various salts and energy substrates and will be discussed in more detail. Interestingly, uniquely defined female reproductive tract components are often not included in modern ART media, although we are aware that some of these components, for example, human oviductin-I/human alpha-fetoprotein (11), are involved with cell proliferation (12). Conversely, some components of current ART media are artificial, for example, EDTA. This is important to point out because the main objective of some in regard to media composition is to follow the "back to nature" strategy, and only include components in the medium at a concentration that is found in the female reproductive tract at the time of preimplantation embryo development. Additionally, it should be pointed out that not all common media components of culture medium are obligatory; however, their presence may enhance the viability of the cultured embryos, e.g., amino acids (13). Similarly, though not unique to the female reproductive tract, the presence of compounds like albumin and hyaluronan in medium has significant positive effects on embryo development (13). A useful concept to keep in mind is that inclusion of components in culture medium that help minimize intracellular stress and maintain cellular homeostasis of the embryo while it is in vitro are often beneficial (7).

1.1.1. Inorganic Ions

Nearly all current ART media have followed the original strategy of Whitten (14) and Brinster (15) and have an ionic composition based on Krebs-Ringer solution (16). The ionic composition of oviduct and uterine fluid was one of the first parameters in these fluids measured and the information was used to formulate some of the early media used for the culture of human (17) and domestic animal (18) preimplantation embryos. Interpretation of the ionic composition of female reproductive tract fluids has been difficult. Analysis of human oviductal fluid by X-ray spectrometry using electron probe excitation (19) indicated high concentrations of potassium compared to similar studies by other investigators; the concentration of phosphate was also high. The authors suggested these anomalies may have been due to differences in patient selection and fluid sampling techniques. There are no more recent analyses of the ionic composition of human reproductive tract fluids, which is unfortunate, as ions such as potassium, phosphate, calcium, and magnesium influence various functions in embryonic cells such as metabolism and cation exchange (17, 20), as discussed below.

It would seem unnecessary to change ionic composition of culture medium between Phase I (D1–D3) and Phase II (D3–D5/6) of human preimplantation embryo development as the female reproductive tract fluid is based on the composition of the transudate from the arterial blood flow that supplies the fallopian tube and uterus and this is likely to remain fairly constant during the preimplantation period. There are several aspects of the ionic composition of embryo culture medium that have to be addressed however to optimize media performance and outcomes. The effect of changes to the ionic composition of ART media has become more evident from the work of Lane, Bavister, and Gardner (7, 20), amongst others and has been incorporated into ART media by these and other investigators. Some of these aspects include:

Phosphate Ions

Excessive levels have been shown to have a negative metabolic effect by phosphorylating glucose that leads to increased glycolysis and a concomitant decrease in oxidation, resulting in inadequate production of energy, the so-called Crabtree effect (4). The absence or very low levels of phosphate ions in ART media does not have any negative effects on embryo development and has been shown to actually increase success rates (21, 22). The early preimplantation embryo also has high levels of ATP (23) which would act as an endogenous source of phosphate. The high ATP:ADP ratio in the early embryo (24) also reduces glycolytic flux in the early embryo by allosteric inhibition of the enzyme PFK which is a flux-generating enzyme of glycolysis (7). The utilization of glucose increases after compaction on D3 of human embryo development (4) as the ATP:ADP ratio falls and the levels of phosphate in the medium should be increased to ensure efficient metabolism of this metabolite.

Ratios of Sodium to Potassium and Calcium to Magnesium

In work involving the creation of Human Tubal Fluid (HTF) medium (17), it was found that there was nearly a four-fold lower sodium to potassium ion ratio in HTF compared to the control medium T6. Interestingly, significantly fewer mouse zygotes developed to expanded blastocysts in T6 medium compared to HTF medium and human clinical pregnancy rates were higher in the HTF compared to the T6 medium. Other media commonly used at that time for human ART, for example Hoppe and Pitts medium

(25) and Ham's F10 also had sodium to potassium ratios closer to HTF compared to T6 medium. The high content of potassium ions in human fallopian tube fluid (19) may be the result of prior mentioned factors including the type of patient from which the sample is collected as well as the method of collection, which may impact the nature of the fluid specimen collected (26). Biggers (26) has also stated that it is very difficult to collect oviduct fluid and most of the analyses are questionable. The overall end-point is that multiple ART media work very well and give equivalent results, keeping in mind that there are very few well-designed prospective randomized trials of ART media comparisons.

Concerning the ratios of calcium to magnesium, the work of Lane (see (7, 27)) using hamster embryos has shown that within the cells of embryos there is a very important interplay between pH_i and calcium and magnesium ions. Calcium is considered to be a universal regulator in all cells and abnormal changes in its concentration can alter protein synthesis, DNA regulation, mitochondrial functions such as oxidative metabolism and cell-to-cell communication. Normally intracellular calcium ion (Ca_i^{2+}) levels are controlled by storage in sites such as endoplasmic reticulum, mitochondria, and the nucleus or by transporting it through the cell membrane or through L-gated calcium channels as needed. All of this handling of Ca_i^{2+} requires energy and can easily be upset by inappropriate embryo handling and metabolic perturbations leading to cellular malfunctions (7). The influx of Ca^{2+} through the calcium channels can be lessened by increasing the magnesium concentration in the medium and this approach has been adopted in various embryo culture media to reduce the uptake of exogenous calcium. Indeed, this strategy improves the development of hamster embryos in vitro (27). Magnesium also has important roles in energy metabolism (see (7)). However, it must be remembered that a calcium spike occurs in spermatozoa by the release of Ca_i^{2+} during fusion with the oocyte membrane (28) and so a lower concentration of magnesium should be present in fertilization medium. To contend with these opposing requirements between fertilization and embryo culture, it is recommended that a lower magnesium concentration (~0.2 mM) be present in fertilization medium and higher levels (~2 mM) in embryo culture media.

Another Inorganic Ionic Component of Media is Sodium Bicarbonate

This will be discussed below in Subheading 1.5.

1.2. Energy Substrates

This subject was one of the first investigated in relation to the development of preimplantation mammalian embryos and has been extensively reviewed (3–5, 26, 29). The triad of pyruvate, lactate, and glucose at concentrations similar to those measured in human reproductive fluids (30) are almost universally used in commercial ART media. It must also be remembered that all 20 amino acids can interact with the metabolic derivatives of these energy substrates and be converted to a Krebs cycle intermediate either directly or through an intermediate conversion to pyruvate, acetyl CoA, or acetoacetyl CoA (31). Indeed, it has been shown that certain amino acids can substitute for pyruvate for culture of 1-cell hamster embryos to the morula/blastocyst stage (32). Additional roles other than substrates for amino acids in embryo culture are discussed in detail later.

1.2.1. Glucose

Gardner's work (30) suggested that as the embryo passes from the ampullary region of the oviduct to the uterus, it is exposed to lower concentrations of pyruvate and lactate, and, importantly, an increased concentration of glucose. This pattern of exposure is reflected in the metabolic activity in embryos during the preimplantation period. Most of these metabolic studies have been done with mouse embryos and have been extrapolated to the human situation, though there are several studies with human embryos that confirm these findings. To summarize, the preimplantation embryo has abundance of glycolytic enzymes and under certain conditions, such as the presence of enough glucose and phosphate, glycolysis (the conversion of glucose to lactate) is increased to the detriment of oxidative metabolism. This is known as the Crabtree effect and invariably leads to the blockage or reduction in embryo development, such as the 2-cell block in certain strains of mice (4). This phenomena can be alleviated in several ways. One is to reduce or remove phosphate altogether in cleavage medium. Another approach is to add amino acids to the medium. A third method is to add EDTA to cleavage stage medium which inhibits glycolysis by chelating intracellular magnesium which is a cofactor for metabolic kinases in the Embden-Meyerhof glycolytic pathway (7, 33). The addition of EDTA and amino acids to medium for the culture of zygotes from the mouse CF1 strain that blocks at the 2-cell stage in the absence of EDTA and amino acids, overcomes the 2-cell block and produces a higher ratio of ATP:ADP at the 2-cell stage equivalent to that in F1 hybrid 2-cells that do not block (4). Gardner and Lane (33) also showed that the presence of amino acids and EDTA overcame the 2-cell block in the presence of relatively high concentrations of glucose and phosphate. After compaction of the embryo, glycolysis becomes a more important energy generating pathway, and hence EDTA is removed from the medium and glucose concentration is increased. This changed metabolism in the embryo as it develops toward the blastocyst stage has been related to a period of greater anoxia it is exposed to in the uterus (34) which would start to limit the supply of energy from oxidative phosphorylation. It should be mentioned that there is some debate as to whether EDTA has to be taken out of culture medium after compaction. Biggers et al. (35) found that reducing the EDTA concentration from 0.1 mM, the concentration of EDTA used by Gardner and Lane (33) in their earlier studies, to 0.01 mM had no detrimental effect on the continued growth and development of mouse embryos after the 8-cell stage. These parameters were no different from the results Biggers et al. (35) obtained with Gardner and Lane's G2 medium used after compaction, so the general trend has been to omit EDTA from medium for this stage of development in sequential series of media.

1.2.2. Pyruvate

Overall, pyruvate is a very important energy component for the early cleaving embryo whereas glucose becomes more predominant after compaction. Gardner et al. (36) have recently reported that increased glucose uptake is closely related to increased implantation and live birth rates of cultured human embryos on both D4 and D5 of development. Increased uptake of glucose and decreased glycolysis appear to be the important parameters involved (36, 37).

1.2.3. Lactate

Lactate is important for cleavage stage embryos. In relation to lactate in ART media, sodium-DL-lactate was the source initially used in embryo media. Although the L-lactate isomer is the metabolically consumed isomer, both D- and L-lactate influence pH_i and only 5 mM D-L-lactate can significantly reduce pH_i in mouse embryos by 0.15 pH units (38). Therefore, some commercial media now include only the L-lactate form, thus providing the bioactive form of lactate and avoiding excessive levels of lactate by eliminating its D-isomer. It has been recommended that ART media still contain lactate so that there is a suitable intracellular ratio of lactate/pyruvate which is in equilibration with the cytoplasmic ratio of NAD^+/NADH (29).

1.2.4. Amino Acids

The role of amino acids during early mammalian embryo development are many and varied, ranging from biosynthetic precursors and energy sources to osmolytes, intracellular pH buffers, antioxidants, chelators, and regulators of differentiation (13). Amino acids are present in reproductive tract fluids (39) and so it is logical that they be included in embryo culture media. In the beginning of human ART one of the media used by Edwards and Steptoe (40) was Ham's F10 medium, which contains a range of amino acids. However, in the 1980s, simple culture media such as Tyrode's T6 and HTF that lacked amino acids became popular. Extensive studies by Gardner, Lane and colleagues in the 1990s in both mouse and domestic species emphasized the beneficial effects of amino acids on embryo development (13) and all commercial sources of human ART media now contain amino acids. In addition to amino acids being present in the female reproductive tract, the oocyte and embryo have specific transport systems for amino acids (41) and they have an endogenous pool of amino acids (42). Importantly, the presence of amino acids in culture and handling media prevents the efflux of endogenous amino acids from the embryo (43).

The concentration and necessity for all amino acids during all of the preimplantation period are debatable. Gardner and Lane (13, 44) advocated that based on mouse studies, only nonessential amino acids were necessary for precompaction human embryos during culture from D1 through D3 but that after compaction, both essential and nonessential amino acids should be present. In fact, when mouse zygotes were cultured with essential amino acids, their development was impaired (13). This has been verified by others. Similarly in the hamster, Bavister and his group (32) have shown that when tested in medium containing single amino acids, the nonessential amino acids asparagine, aspartate, glycine, serine as well as taurine, which are all present at high concentrations in female reproductive tract fluid, stimulated blastocyst development from the zygote stage but the essential amino acids cysteine, isoleucine, leucine, phenylalanine, threonine, and valine were inhibitory. The essential amino acid histidine was the only one that was stimulatory.

The overall conclusions from these studies are that nonessential amino acids stimulate early cleavage, mitosis rates in the trophectoderm and blastocoel formation. The essential amino acids as a whole are inhibitory to early cleavage but after compaction, they stimulate cleavage rates and formation of the inner cell mass (ICM). A general statement is that nonessential amino acids are present at high and that essential amino acids are at low concentrations in the fallopian tube (13), the site for the early cleaving embryo.

In addition to the effects of amino acids on embryo development in culture media, the consumption and production of amino acids by preimplantation embryos in vitro has been studied. Most of this work has come from Henry Leese's laboratory and has involved several mammalian species, including humans. In the human embryo, leucine was consumed at all stages from D2 to D3 of development through to the blastocyst stage with the significant uptake of serine, arginine, and several other essential amino acids increasing at later stages (45). Based on the uptake of leucine, it may be reasonable to include it in medium for all periods of development but there is uncertainty as to whether leucine is one of the essential amino acids that may be inhibitory to the early cleaving embryo. The amino acids that were consistently produced by the embryos at all preimplantation stages were alanine and glutamine. Thus, it may be wise to exclude alanine from human ART media for both the precompaction and compaction phases of development. The omission of alanine may help the continued transamination of pyruvate into alanine during the removal of ammonia from the embryo by the enzyme transaminase. Also, in addition to being produced by the embryo, glutamine readily deaminates, producing toxic ammonium so it has been recommended that glutamine be omitted from culture medium (46). Glutamine is synthesized by the enzyme glutamine synthetase in the embryos and embryos lacking this enzyme die (46). Therefore, use of a more stable glutamine stable dipeptide may advantageous, as it has shown that this strategy considerably reduced the accumulation of ammonium in embryo culture medium (47), however, effect of the conjugate amino acid of the dipeptide, either glycine or alanine, should also be considered when adding dipeptide glutamine to culture media. There has been a report that alanyl-glutamine gave better results than glycyl-glutamine when added to medium for the culture of human embryos (48), while another report indicates glycyl-glutamine is superior for mouse embryos (49). Furthermore, as an added means of reducing ammonia production from amino acid breakdown and optimizing conditions for embryo development, concentrations of amino acids commonly utilized in media have been reduced (50, 51).

1.3. Vitamins

Whether the inclusion of vitamins in human ART media is beneficial is as yet to be determined. There are several reports from nonhuman animal studies that vitamins can stimulate or, more precisely, maintain embryo development pre- and postimplantation by lessening the stress of in vitro culture. Lane and Gardner (52) reported that the vitamins in Eagle's Minimal Essential Medium (MEM) when included in mouse tubal fluid medium in combination with the amino acids in MEM maintained more normal metabolic activity in mouse and rat blastocysts. When these blastocysts were transferred, implantation rates and fetal weights were similar to in vivo controls in contrast to the absence of the vitamins and amino acids, which were lower. Only the water soluble vitamin pantothenate has been found to stimulate the development of hamster zygotes to the blastocyst stage (53) whereas in the rabbit, several other water soluble vitamins stimulate the expansion and hatching of blastocysts (54). Based on these reports and the presence of vitamins in media such as Ham's F10 and Eagle's MEM used in the early days of human IVF and sporadically since (55), commercial ART media companies have added vitamins to all or some of their ART media for the entire culture period. The trend was to add all eight vitamins in MEM medium, but subsequently the number of vitamins has been reduced to only pantothenate with or without three other vitamins. It would be fair to say that it is not really known what the absolute vitamin requirement is for human embryos at present.

1.4. Growth Factors

Peptide growth factors are present in the mammalian female reproductive tract and the embryo contains receptors for many of them (reviewed by Kane et al. (56)). Growth factors have been added to human ART media and have promoted the formation of blastocysts (reviewed in (57)) but only one growth factor, granulocyte macrophage colony stimulating factor (GM-CSF) has been clinically tested and has shown a moderate increase in pregnancy rates with no increase in the rate of miscarriages or congenital disorders in the babies born (58). Investigation of the benefit of adding growth factors to human ART media is still required, especially as there is concern that this practice may lead to abnormal development (59). Furthermore, whether a sequential exposure to various growth factors offers any benefit is unknown.

1.5. pH and Temperature Control

The importance of pH in IVF has recently been reviewed (60) and is presented in Chapter 10 of this book.

The use of Krebs-Ringer bicarbonate-based medium under an atmosphere of 5 % CO_2 in air was used by Whitten (61) to culture 2-cell mouse embryos to the blastocyst stage. Whitten (62) was then able to culture 1-cell embryos from F1 hybrid mice to the blastocyst stage in a medium containing pyruvate, calcium lactate, and glucose but with 22.6 mM sodium bicarbonate instead of 25 mM sodium bicarbonate that had become the standard for the culture of mouse embryos based on the work of Brinster (15). Whitten (62) recommended that the medium have an osmolarity between 240 and 310 milliosmols (optimal range was 250–280 milliosmols) and a pH of about 7.2. It was not until the turn of the century that it became recognized that the pH of culture medium for human ART should be lower than the commonly used value of 7.4 and that the optimal pH varied at different stages of development. A pH of around 7.2 was suggested for early cleavage and a higher level (7.3–7.4) for blastocyst formation (63). Findings that the pH_i for early cleaving human embryos was about 7.1 (64) supported the idea that the pH of cleavage medium should be around 7.2 to relieve any excessive stress on the embryo when the pH had varied too far away from the optimum value. Phillips et al. (64) showed that the early cleaving human embryo had specific transport systems for the exchange of H^+ and HCO_3^- ions through the cell membrane so the embryo can maintain pH_i within a somewhat narrow range over somewhat larger excursions in external pH. It was reported that when the pH of medium rose above 7.4 due to inadequate gassing of the atmosphere, embryo morphology and resulting pregnancy rates were poorer than usual (65). pH_i, in addition to being a powerful regulator of many key enzyme pathways and reactions that control both oxidative metabolism and glycolysis, also regulates cellular processes such as cell division, differentiation, and cytoskeletal events (7). pH levels in culture media can drift for various reasons such as altitude, temperature, and the capacity of the culture system to maintain CO_2 levels caused by the frequency of incubator door openings or other leakage problems so it is important to measure pH in each incubator used on a regular basis.

Sodium bicarbonate in the medium in concert with carbon dioxide in the atmosphere has the greatest effect on pH of the medium and it is recognized that measuring the pH of culture medium is more important than to just measure the CO_2 concentration in the culture atmosphere (66). As mentioned, it is hypothesized that the optimal pH may differ for different stages of embryo development. In the early Phase I cleavage stage of development, a pH in the range of 7.20–7.25 appears to give the best development from D1 to D3 (Fig. 1). pH_i measurements for human embryos after compaction in Phase II of development are not available, but observational data indicate that a pH in the range of 7.30–7.35 is better for blastocyst development than the lower pH recommended for cleavage medium (Fig. 2). To avoid having to change CO_2 levels to accommodate these differences in required pH, media companies may instead change the sodium bicarbonate concentration so that the different pH levels are achieved under the same concentration of CO_2, thus simplifying the incubator requirements for the extended culture of embryos through the entire preimplantation period. Such a strategy is not possible when using a single-step medium for the whole of the preimplantation period and one would have to compromise pH levels and use the same pH for the whole period. Alternatively, differences in the pH of the single-step medium could be varied by adjusting CO_2 levels in the atmosphere before and after compaction on D3. Importantly, both bicarbonate levels and CO_2 levels can impact embryo development. Therefore, consideration of their impact while adjusting pH needs to also be considered.

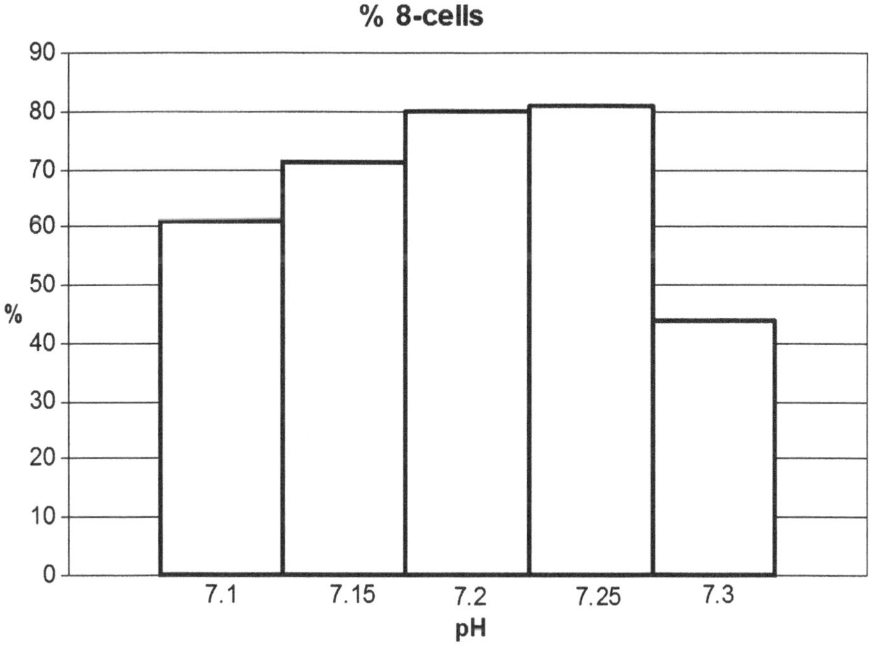

Fig. 1. Effect of pH on percentage of zygotes developing to 8-cells on D3 of culture. Based on 96,431 embryos (K. Miller, unpublished).

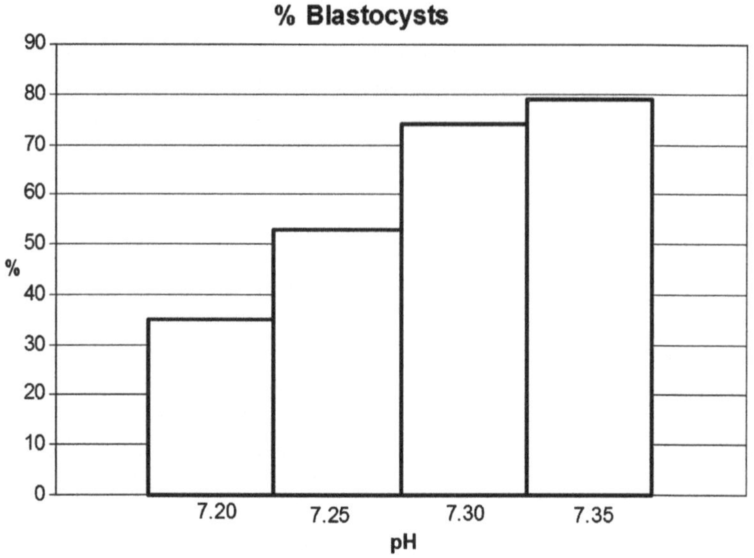

Fig. 2. Effect of pH on percentage of D3 embryos developing to blastocysts on day 5/6 of culture. Based on 5,582 patients (K. Miller, unpublished).

Concerning the temperature of culture systems, it has generally been assumed that the core body temperature of 37 °C would be optimal for the culture of human gametes and embryos. It has been found though that, like scrotal temperature, ovarian follicles and the fallopian tube in several mammalian species are 1–2 °C lower than the core body temperature and that could help in promoting a lower beneficial metabolic rate in the oocyte and early embryo; this is in line with the "Quiet Embryo Hypothesis" of Leese (67). There is some evidence that a temperature in the range of 36.5–36.9 °C may be the optimal range for human IVF (68) and the data in Fig. 3 pertaining to the effects of aspirate temperature show that this parameter has an effect on the subsequent proportion of embryo transfers that occurred on D5 and their associated implantation rates. It is not known if any differential temperature requirements exist for the human preimplantation embryo that may further improve existing sequential culture systems.

1.6. Static or Dynamic Platforms

Classically, a static culture system has been used for in vitro development with increasing success rates in succeeding decades (69). The most frequently used static system is drops or wells of culture medium under oil (15). After a slow beginning, dynamic culture systems in which the culture medium is in motion around the gametes and embryos have been more frequently studied and have culminated in microfluidic devices (70). However, although embryo development has been reported to be enhanced in microfluidic devices, there has not yet been any reported increase in pregnancy rates over static systems. One of the most obvious

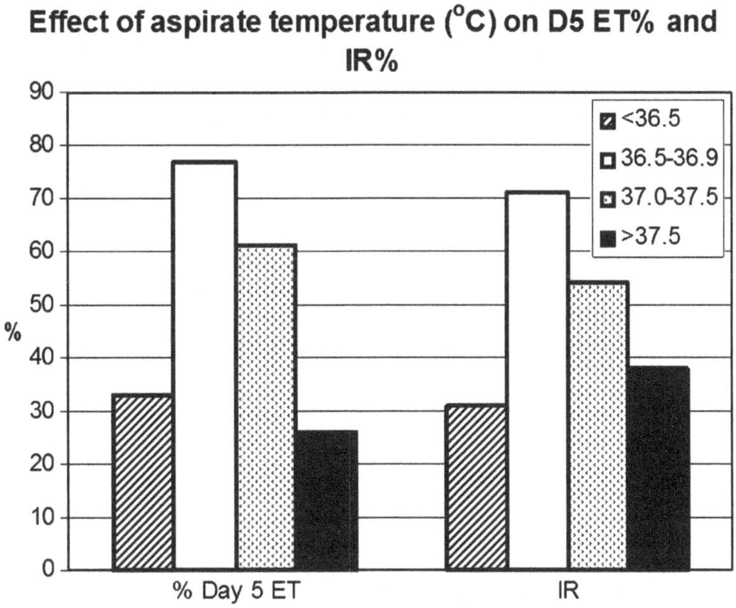

Fig. 3. Effect of aspirate temperature on percentage of ETs on D5 and implantation rates. Based on 1,811 patients (K. Miller, unpublished).

negative effects of dynamic culture systems is the removal of autocrine/paracrine growth factors produced by the embryo (71). If anyone has observed a freshly excised mouse oviduct under a microscope soon after ovulation, they would invariably see the constant to and fro movement of the cumulus masses within the tube. This mechanical activity is both muscular and cellular and mixes gametes and embryos with tubal secretions. The cellular contribution to this movement is mostly due to ciliary beating which is directed toward the uterus. Similar movement of culture fluid during in vitro development has been shown to improve embryonic development. This improvement occurs whether the flow of fluid is unidirectional or oscillatory via mechanical vibration (72). One of the first successes with mechanical vibration was reported by Hoppe and Pitts (25) who used constantly shaking culture dishes for mouse IVF. A static system for mouse IVF was also developed (73) but required higher sperm concentrations than those used by Hoppe and Pitts (25). Recent studies in humans by Isachenko et al. (72), using a mechanical micro-vibration system, showed a significant increase in embryonic development in vitro and significantly increased pregnancy rates with this system. The vibratory actions were pulsatile (5 s once every hour). No studies with constant application of vibration, as used by Hoppe and Pitts (25) for mouse IVF, have been reported. Mechanical vibration also positively stimulates proliferation and secretory activity in cultured somatic cells (74).

Another modification to the culture system is the use of the so-called Well-of-the-Well system (WOW) (75) where small microwells are created in the bottom of a 4-well culture dish so that the embryos can be cultured individually in the microwells while still benefiting from the beneficial effects of group culture. The WOW system produced higher blastocyst rates than conventional culture in a sibling embryo study and gave nearly a 50 % clinical pregnancy rate in women who previously failed at least one IVF cycle associated with medium to low embryo quality and/or impaired embryo development (75).

Perhaps the usefulness of microfluidic or culture with vibration is the removal of toxic components in the vicinity of the embryo whereas the WOW system has the opposite effect and probably enhances the containment of autocrine/paracrine factors in the vicinity of the embryo. Both of these approaches are compatible with sequential culture systems, though dynamic flow through a microfluidic device has more potential by seamlessly allowing changing of the culture media over time. It will be interesting to see if these two systems can be combined to improve embryo development.

1.7. Summary and Conclusions

Embryonic stages at the beginning and end of the preimplantation period, the zygote and the blastocyst, are not only different in their morphology but also differ in their physiology and biochemistry. The question then is will one medium in a single-step culture protocol from the zygote to blastocyst stage (10) be sufficient or will embryo development be better served using a sequential series of media. Inappropriate culture conditions, whether they be due to medium formulations, pH, temperature, physical conditions of the culture system or other factors and/or a combination of all of the above can induce cellular stress in the embryo (4). The use of the single-step culture system has recently been tested on a more regular basis and in comparison to the sequential media appears to be just as effective if not better (10). Some studies (76, 77) however, including that of Reed et al. (10) have not monitored or optimized pH at the different stages of development that may have had a substantial impact on the performance of the different media involved. Some of the differences between a sequential media series and the single-step medium protocol are listed in Table 1. At this stage, there is still no conclusive evidence that one system is better than the other in terms of outcomes.

2. Materials

1. Various size pipette tips.
2. Appropriate embryo tested culture dishes.
3. Fertilization, cleavage, and blastocyst media.

Table 1
Comparison of sequential media and single-step medium culture protocols

Factors	Sequential media	Single-step medium
Change of medium	Require change of medium around D3 of culture	Can be used for complete preimplantation period with no medium change
Energy substrates	Change on D3	No change
Presence of EDTA	Blastocyst medium does not contain EDTA that may reduce glycolysis	Contains EDTA that may compromise blastocyst development by slowing glycolysis
Effect of essential amino acids	Cleavage medium does not contain essential amino acids that may impair development from the zygote stage	Contains essential amino acids that may compromise early development
pH	Cleavage medium has a pH of around 7.20–7.25 which optimizes development from the zygote to 8-cell stage Blastocyst medium has a pH between 7.30 and 7.35 which is optimal for blastocyst development	pH is similar during the whole culture period unless the atmospheric CO_2 concentration is changed on D3. Studies comparing sequential media to single-step medium culture have not optimized pH

4. Handling media (i.e., HEPES or MOPS buffered).
5. Appropriate protein source.
6. Embryo tested mineral oil.
7. Laboratory incubator.
8. Microscope.

3. Methods

1. The day before ovum pickup (OPU), label culture dishes. When making drops of medium, use a sterile pipette tip, rinsing the tip twice with culture medium before making the drops.
2. *For in vitro fertilization (IVF)*. Prepare fertilization media in a sterile manner with appropriate protein and concentration. Place drops of fertilization medium into the dish according to laboratory protocol (see Notes 1 and 2).
3. Immediately cover the drops with oil and place the dish in the CO_2 incubator (see Notes 3–5).
4. *For intracytoplasmic sperm injection (ICSI)*. Prepare and place dishes of cleavage media in the CO_2 incubator as above (see Notes 3–5).
5. When placing the dishes in the incubator, gently remove the lid of the dish and set it at an angle on the side of the dish to

allow for complete gas exchange. Dishes must gas for a minimum of 4 h before use (or overnight).

6. On the day of OPU (D0) for IVF cases, prepare culture dishes with cleavage medium as in step 4.
7. *For both IVF and ICSI on D0 or D1 before fertilization check.* Prepare handling media according to laboratory protocol. Media can be left at room temperature overnight, or prepared early on the morning of D1 and warmed in air to 37 °C on a heating plate. In either case, warm the dishes to 37 °C on the morning of D1 before use.
8. *For IVF late on D0 or early on D1.* Prepare a wash dish with handling media.
9. On D1, or day of fertilization check, gently remove the cumulus cells by stripping oocytes according to laboratory protocol (see Note 6). Gently wash the stripped oocytes well in handling media. Then place fertilized oocytes in dishes of cleavage media. For ICSI'ed oocytes, successfully fertilized eggs can remain in cleavage medium. Zygotes can remain in cleavage media until D3 (see Note 7).
10. D2 zygotes can be examined according to laboratory protocol. Dishes of blastocyst media should be prepared as above and placed into the incubator at least 4 h prior to use on D3.
11. *On D3 after blastocyst medium dishes have equilibrated for at least 4 h.* For embryos that are to be cultured to D5/6, remove the embryos from the cleavage medium culture dishes and place into dishes of blastocyst medium (see Note 8).

4. Notes

1. Protein type and concentration can vary depending on laboratory protocol. HSA is recommended for fertilization at a concentration of 5 mg/mL.
2. Type of dish and volume of media will vary depending on laboratory protocol.
3. Volume of oil will vary depending on lab protocol.
4. Prepare no more than two dishes at a time to minimize evaporation and osmolality shifts, out-gassing of CO_2 and drift in the pH of the medium.
5. Temperature and gas environment should be set according to lab protocol. Temperature should be no >37 °C, though some labs use a slightly lower temperature. Reduced oxygen has been found to be beneficial for embryo culture. Additionally, CO_2 concentrations should be set to ensure the pH of the medium falls into an acceptable range prescribed by the media supplier.

6. Stripping of cumulus can be done using hyaluronidase and small pipettes. Actual procedure may vary from lab to lab.
7. It is recommended that pipettes or strippers of 275–300 μm diameter pipette tips be used to minimize medium transfer between drops; transfer volume should be <1 μL.
8. There is anecdotal data that transferring cleaving embryos from cleavage medium to blastocyst medium on D2 or D4 may result in better results than the more traditional medium exchange on D3. It is the responsibility of each individual laboratory to determine their own protocol for when this embryo exchange from cleavage medium to blastocyst medium should be undertaken. To reach this decision, it should be kept in mind that the optimal day of exchange may be patient dependent to some extent, i.e., some patients may do better if the exchange is on D2, others, if the exchange is on D3, and others again, if the exchange is on D4.

References

1. Biggers JD, Culture techniques in animal and human reproductive biology. Workshop on evidence based assisted reproductive technologies (ART). Food and Drug Administration, Department of Health and Human Services, National Institutes of Health, Wednesday, September 18, 2002. www.fda.gov/.../NewsEvents/WorkshopsMeetingsConferences/TranscriptsMinutes/UCM054465.pdf. Accessed 18 June 2011
2. Biggers JD, Summers MC (2008) Choosing a culture medium: making informed choices. Fertil Steril 90:473–483
3. Quinn P (2012) Short culture: day 1/day 2/day 3 embryo culture. In: Nagy ZP, Varghese AC, Agarwal A (eds) Practical manual of in vitro fertilization. Advanced methods and novel devices. Springer, New York, NY
4. Gardner DK (1998) Changes in requirements and utilization of nutrients during mammalian preimplantation embryo development and their significance in embryo culture. Theriogenology 49:83–102
5. Quinn P (2003) Media used in the assisted reproductive technologies laboratories. In: Patrizio P, Tucker MJ, Guelman V (eds) A color atlas for human assisted reproduction: laboratory and clinical insights. Lippincott Williams & Wilkins, Philadelphia
6. Handyside AH et al (1990) Pregnancies from biopsied human preimplantation embryos sexed by Y-specific DNA amplification. Nature 344:768–770
7. Lane M, Gardner DK (2001) Blastomere homeostasis. In: Gardner DK, Lane M (eds) ART and the human blastocyst. Springer, New York, NY
8. Braude P, Bolton V, Moore S (1988) Human gene expression first occurs between the four- and eight-cell stages of preimplantation development. Nature 332:459–461
9. Brison DR et al (2004) Identification of viable embryos in IVF by non-invasive measurement of amino acid turnover. Hum Reprod 19:2319–2324
10. Reed ML et al (2009) Continuous uninterrupted single medium culture without medium renewal versus sequential medium culture: a sibling embryo study. Fertil Steril 92:1783–1786
11. Wagh PV, Lippes J (1993) Human oviductal fluid proteins. V. Identification of human oviductin-I as alpha-fetoprotein. Fertil Steril 59:148–156
12. Keel BA et al (1991) Synergistic action of purified α-fetoprotein and growth factors on the proliferation of porcine granulose cells in monolayer culture. Endocrinology 129:217–225
13. Gardner DK (2008) Dissection of culture media for embryos: the most important and less important components and characteristics. Reprod Fertil Dev 20:9–18
14. Whitten WK (1956) Culture of tubal mouse ova. Nature 177:96
15. Brinster RL (1963) A method for in vitro cultivation of mouse ova from two-cell to blastocyst. Exp Cell Res 32:205–208
16. Krebs HA, Henseleit K (1932) Untersuchungen uber die harnstoffbildung im tierkorper. Zeirschrift fur Pysikalische Chemie 210:33–66

17. Quinn P, Kerin JF, Warnes GM (1985) Improved pregnancy rate in human in vitro fertilization with the use of a medium based on the composition of human tubal fluid. Fertil Steril 44:493–498
18. Tervit HR, Whittinghan DG, Rowson LEA (1972) Successful culture in vitro of sheep and cattle ova. J Reprod Fertil 30:493–497
19. Borland RM et al (1980) Elemental composition of fluid in the human fallopian tube. J Reprod Fertil 58:479–482
20. Lane M, Gardner DK (2000) Regulation of ionic homeostasis by mammalian embryos. Semin Reprod Med 18:195–204
21. Quinn P et al (1995) Successful human in vitro fertilization using a modified human tubal fluid medium lacking glucose and phosphate ions. Fertil Steril 63:922–924
22. Pool TB, Atiee SH, Martin JE (1998) Oocyte and embryo culture: basic concepts and recent advances. In: May JV (ed) Assisted reproduction: laboratory considerations. Infertil Reprod Med Clin N Am 9:181–203
23. Quinn P, Wales RG (1971) Adenosine triphosphate content of preimplantation mouse embryos. J Reprod Fertil 25:133–135
24. Quinn P, Wales RG (1973) The effect of culture in vitro on the levels of adenosine triphosphate in preimplantation mouse embryos. J Reprod Fertil 32:231–241
25. Hoppe PC, Pitts S (1973) Fertilization in vitro and development of mouse ova. Biol Reprod 8:420–426
26. Biggers JD (2001) Thoughts on embryo culture conditions. Reprod Biomed Online 4(Suppl 1):30–38
27. Lane M et al (1998) Intracellular divalent cation homeostasis and developmental competence in the hamster preimplantation embryo. Mol Reprod Dev 50:443–450
28. Rogers BJ, Yanagimachi R (1976) Competitive effect of magnesium on the calcium-dependent acrosome reaction in guinea pig spermatozoa. Biol Reprod 15:614–619
29. Leese HJ (1998) Human embryo culture: back to nature. J Assist Reprod Genet 15:466–468
30. Gardner DK et al (1996) Environment of the preimplantation human embryo in vivo: metabolite analysis of oviduct and uterine fluids and metabolism of cumulus cells. Fertil Steril 65:349–353
31. Pool TB (2001) Summary: five take-home lessons regarding the optimization of culture media and the use of commercial products for human embryo culture. In: Quinn P (ed) ART media. Commercial and in-house products. Proceedings of the 5th Annual Symposium of the American Association of Bioanalysts College of Reproductive Biology, Las Vegas, AAB, St. Louis, MO, 5th May 2001
32. Bavister BD, McKiernan SH (1993) Regulation of hamster embryo development in vitro by amino acids. In: Bavister BD (ed) Preimplantation embryo development. Springer, New York, NY
33. Gardner DK, Lane M (1996) Alleviation of the '2-cell block' and development to the blastocyst of CF1 mouse embryos: role of amino acids, EDTA and physical parameters. Hum Reprod 11:2703–2712
34. Fischer B, Bavister BD (1993) Oxygen tension in the oviduct and uterus of rhesus monkeys, hamsters and rabbits. J Reprod Fertil 99:673–679
35. Biggers JD, McGinnis LK, Lawitts JA (2005) One-step versus two-step culture of mouse preimplantation embryos: is there a difference? Hum Reprod 20:3376–3384
36. Gardner DK et al (2011) Glucose consumption of single post-compaction human embryos is predictive of embryo sex and live birth outcome. Hum Reprod 26:1981–1986
37. Lane M, Gardner DK (1996) Selection of viable mouse blastocysts prior to transfer using a metabolic criterion. Hum Reprod 11:1975–1978
38. Edwards LJ, Williams DA, Gardner DK (1998) Intracellular pH of the preimplantation mouse embryo: effects of extracellular pH and weak acids. Mol Reprod Dev 50:434–442
39. Tay JI et al (1997) Human tubal fluid: production, nutrient composition and response to adrenergic agents. Hum Reprod 12: 2451–2456
40. Steptoe PC, Edwards RG, Purdy JM (1971) Human blastocysts grown in culture. Nature 229:132–133
41. Van Winkle LJ (1988) Amino acid transport in developing animal oocytes and early conceptuses. Biochim Biophys Acta 947:173–208
42. Schultz GA et al (1981) Endogenous amino acid pool sizes in mouse eggs and preimplantation embryos. J Reprod Fertil 61:387–393
43. Kolajora M, Baltz JM (1999) Volume-regulated anion and organic osmolytes channels in mouse zygotes. Biol Reprod 60:964–972
44. Gardner DK, Lane M (1993) Amino acids and ammonium regulate mouse embryo development in culture. Biol Reprod 48:377–385
45. Houghton FD et al (2002) Non-invasive amino acid turnover predicts human embryo development capacity. Hum Reprod 17:999–1005

46. He Y et al (2007) Glutamine synthetase is essential in early mouse embryogenesis. Dev Dyn 236:1865–1875
47. Lane M, Gardner DK (2003) Ammonium induces aberrant blastocyst differentiation, metabolism, pH regulation, gene expression and subsequently alters fetal development in the mouse. Biol Reprod 69:1109–1117
48. Quintans CJ et al (2005) Human IVF outcome in media containing either alanyl-L-glutamine or glycyl-L-glutamine. Fertil Steril 84(Suppl 1): S456
49. Biggers JD, McGinnis LK, Lawitts JA (2004) Enhanced effect of glycyl-L-glutamine on mouse preimplantation embryos in vitro. Reprod Biomed Online 9(1):59–69
50. Eagle H (1959) Amino acid metabolism in mammalian cell cultures. Science 130:432–437
51. Lane M, Hooper K, Gardner DK (2001) Effect of essential amino acids on mouse embryo viability and ammonium production. J Assist Reprod Genet 18:519–525
52. Lane M, Gardner DK (1998) Amino acids and vitamins prevent culture-induced metabolic perturbations and associated loss of viability of mouse blastocysts. Hum Reprod 13:991–997
53. McKiernan SH, Bavister BD (2000) Culture of one-cell hamster embryos with water soluble vitamins: pantothenate stimulates blastocyst production. Hum Reprod 15:157–164
54. Kane MT (1988) The effects of water-soluble vitamins on the expansion of rabbit blastocysts in vitro. J Exp Zool 245:220–223
55. Behr B et al (1999) Preliminary clinical experience with human blastocyst development in vitro without co-culture. Hum Reprod 14:454–457
56. Kane MT, Morgan PM, Coonan C (1997) Peptide growth factors and preimplantation development. Hum Reprod Update 3:137–157
57. Bavister B, Baltz J (2002) Influence of culture media on embryo development. In: De Jonge CJ, Barratt CLR (eds) Assisted reproductive technology. Accomplishments and new horizons. Cambridge University Press, Cambridge
58. Origio (2011) Embryogen. http://www.origio.com/~/media/Origio/Files/PDS/MEC/EmbryoGen_070611_v1_web.ashx. Accessed 16 Sept 2011
59. Rawlins RG (2001) Modern media for in vitro fertilization—an evolution. In: Quinn P (ed) ART media. Commercial and in-house products. Proceedings of the 5th Annual Symposium of the American Association of Bioanalysts College of Reproductive Biology, Las Vegas, AAB, St. Louis, MO, 5th May 2001, pp 43–65
60. Swain JE (2010) Optimizing the culture environment in the IVF laboratory: impact of pH and buffer capacity on gamete and embryo quality. Reprod Biomed Online 21:6–16
61. Whitten WK (1957) Culture of tubal ova. Nature 179:1081–1082
62. Whitten WK (1970) Nutrient requirements for the culture of preimplantation embryos in vitro. Adv Biosci 6:129–141
63. Ferring Pharmaceuticals (1999) The Assisted Reproductive Global Monitor, May Council Meeting, Special Meeting Reporter, Session 1, Council for the Advancement of Ovulation Induction and Assisted Reproductive Technology, Marrs RP, Steinkampf MP (chairpersons)
64. Phillips KP et al (2000) Intracellular pH regulation in human preimplantation embryos. Hum Reprod 15:896–904
65. Quinn P, Stone BA, Marrs RP (1990) Suboptimal laboratory conditions can affect pregnancy outcome after embryo transfer on day 1 or day 2 after insemination in vitro. Fertil Steril 53:168–170
66. Pool TB (2004) Optimizing pH in clinical embryology. J Clin Embryol 7:1–17
67. Leese HJ et al (2008) Metabolism of the viable mammalian embryo: quietness revisited. Mol Hum Reprod 14:667–672
68. Higdon HL, Blackhurst DW, Boone WR (2008) Incubator management in an assisted reproductive technology laboratory. Fertil Steril 89:703–710
69. Quinn P (2004) The development and impact of assisted reproductive technologies culture media. Fertil Steril 8:27–29
70. Heo YS et al (2010) Dynamic microfunnel culture enhances mouse embryo development and pregnancy rates. Hum Reprod 25:613–622
71. Paria BC, Dey SY (1990) Preimplantation embryo development in vitro: co-operative interactions among embryos and role of growth factors. Proc Natl Acad Sci USA 87:4756–4760
72. Isachenko E et al (2010) Mechanical agitation during the in vitro culture of human pre-implantation embryos drastically increases the pregnancy rate. Clin Lab 56:569–576
73. Stanger JD, Quinn P (1982) Fertilization of cumulus-free, zona-intact mouse ova in vitro at high and low sperm concentrations. Gamete Res 5:61–70
74. Isachenko V (2011) In-vitro culture of human embryos with mechanical micro-vibration increases implantation rates. Reprod Biomed Online 22:536–544

75. Vajta G et al (2008) The well-of-the-well system: an efficient approach to improve embryo development. Reprod Biomed Online 17:73–81
76. Sepulveda S et al (2009) In vitro development and pregnancy outcomes for human embryos cultured in either a single medium or in a sequential media system. Fertil Steril 91: 1765–1770
77. Wirleitner B et al (2010) Individual demands of human embryos on IVF culture medium: influence on blastocyst development and pregnancy outcome. Reprod Biomed Online 21:776–782

Chapter 14

Culture Systems: Embryo Co-Culture

Yves J.R. Ménézo, Edouard Servy, Anna Veiga, André Hazout, and Kay Elder

Abstract

During the 1970s, domestic animal biotechnology, i.e., embryo transfer in farm animals, was confronted with the problem of embryonic developmental arrest observed in vitro, especially during the cycle in which maternal to zygotic transition (MZT) cycle takes place. In farm animals, obtaining blastocysts is mandatory, as transfer at earlier stages results in expulsion of the embryo from the vagina. In humans, the first attempts to obtain blastocysts with classical culture media were disappointing, and the use of a coculture strategy was naturally tempting: the first significant results of successful blastocyst development were obtained in the early 1980s, using trophoblastic tissue as a feeder layer in order to mimic an autocrine embryotrophic system. The next supporting cell systems were based on oviduct epithelial cells and uterine cells in order to achieve a paracrine effect. Non-hormone dependence was then demonstrated with the use of prepubertal cells, and finally with the use of established cell lines of nongenital origin (African Green Monkey Kidney, Vero cells). The embryotrophic properties are linked to features of "transport epithelia." Vero cells have been extensively used in human ART, and most of our knowledge about the human blastocyst was gathered with the use of this technology. Coculture is still in current use, but with systems that employ autologous uterine cells. Results following the use of this technology in human ART are superior to those observed with the use of sequential media. The benefit is linked to the release of free radical scavengers and growth factors by the feeder cells. In animal biotechnology, an important part of the "precious embryos," i.e., those resulting from cloning technology, involves coculture with buffalo rat liver (BRL) cells or Vero cells.

Key words: Coculture, Embryo, Blastocyst, Growth factors, Free radical scavengers

1. Introduction

The first mammalian embryo culture systems used in domestic animal biotechnology were dependent upon coculture with feeder cells in order to achieve embryo transfer. Embryo transfer in farm animals can only be performed at the blastocyst stage: transfer at earlier stages of development (after IVF and short duration of culture), prior to Day 5, results in embryos being expelled into the

Gary D. Smith et al. (eds.), *Embryo Culture: Methods and Protocols*, Methods in Molecular Biology, vol. 912, DOI 10.1007/978-1-61779-971-6_14,

vagina. Embryo culture was confronted with the problem of developmental arrest during the cell cycle that corresponds to genomic activation: maternal to zygotic transition (MZT), the shift that occurs from maternal control of metabolism via mRNAs and proteins stored during oocyte maturation, to activation of the zygote genome to direct further embryonic development. This transition occurs at the 2–4 cell stage in the mouse, 4–8 cells in human, and 8–16 cells in bovine. The transition occurs during an extended cell cycle, the longest during early embryo development, which lasts for 24 h before the first embryonic transcripts appear. The embryo then becomes autonomous. Dynamic in vivo conditions are always difficult to reproduce in vitro, and early culture systems yielded blastocyst formation rates that barely reached 50 %. Therefore, coculture, defined as the simultaneous culture of somatic cells together with the embryo, was a logical intermediary step. Bovine embryo coculture began in the early 1980s, using trophoblastic vesicles (1, 2) to mimic an autocrine embryotrophic system. This protocol was the first to allow culture from the 1-cell stage to blastocyst, with live calves delivered after transfer. The second supporting cell systems used oviduct epithelial cells (3), in order to achieve a paracrine effect. This type of cell layer was first employed in the monkey by Goodeaux et al. (4) using uterine cells and in the human by Bongso (5, 6) using homologous oviduct epithelial cells. We were then able to demonstrate that the embryotrophic effect was not hormone-dependent, by using prepubertal oviduct in mouse and in the bovine (7) and nongenital established cell lines in the human (8). Vero (Green Monkey Kidney) cells have the same embryonic origin (mesonephros) as the uterus and oviduct, and have been extensively used in human cell culture systems. Vero, as well as buffalo rat liver (BRL) are still used for bovine embryo production. We discuss the mechanisms whereby coculture can improve embryo quality.

1.1. Prolonged Culture Time

The transfer of any embryo before it has undergone genomic activation is a "blind process," even when kinetic and morphological features have been assessed. The morula stage corresponds to a full remodeling of cells, so transfer at the blastocyst stage is a logical strategy. Moreover, uterine motility is decreased after D5, once the blastocyst stage has been reached (9). Blastocyst transfer as a selection process has been questioned even within some teams (10–12), but this concept is now fully recognized. In terms of cytology, it allows the selection of healthy embryos in translocation carriers (13). This positive selection effect, together with the relatively poor efficacy of PGD/PGS has led the utility of PGD to be contested (14). Moreover, an embryo's cytogenetic status can evolve in a positive manner, with "self correction" between early cleavage stages and blastocyst. An increasing number of teams are using now blastocyst biopsy for PGD (15, 16), following our design of this technology (17). However, as in any biological process, this selection is not perfect, as can be observed by the fact that trisomies 13, 18, and 21 occur naturally. Nonetheless, when a SET "single embryo transfer" is indicated, the decision must be blastocyst transfer.

1.2. Different Technologies

Early embryonic development is bi-phasic. The first cleavage divisions are completely driven by maternal reserves stored during oocyte growth and maturation, especially for the final "presentation/ preparation" of maternal mRNAs as condensed polyadenylated clusters, ready for translation. This first phase is of necessity the most fragile: any attrition in the storage of protein and mRNA required for housekeeping mechanisms, i.e., cell division homeostasis, redox potential regulation, energy metabolism, protection against external insults (ROS, pH, etc.), or DNA repair will result in either developmental arrest or apoptosis. It is also now evident that very early anomalies may lead to delayed embryonic problems ((18); see also large offspring syndrome). The long period of time required for MZT presents a problem, and every supporting system that can avoid delays in this transition allows a certain level of reserves to be kept until the blastocyst stage. The presence of supporting feeder cells may also partially rescue borderline embryos that would have been arrested in vitro, allowing them to pass this critical stage. However, this view is only schematic, as some of the mRNAs stored during maturation are used at the blastocyst stage.

Somatic cells as feeders can be used as cell suspensions, pieces of tissue (trophoblast) or in monolayers. Coculture is a three-partner relationship: culture medium, cells, and embryo. The first assays of these systems began in the early 1980s with the fantastic development of embryo transfer for animal breeding, especially in bovine, and in particular after in vitro maturation and in vitro fertilization proved to be successful. The use of trophoblastic cells (D18 embryo) to culture 1-cell embryos demonstrated that the trophic factors are not hormone-dependent (1, 2). This hypothesis was then confirmed by transferring embryos into prepuberal oviduct in mouse (in vivo and in vitro) and in bovine (7–19), as well as by coculture with de-synchronized tubal or uterine cells (20). It should be mentioned here that coculture with human endometrial cells represents a de-synchronization, as the uterine biopsies are performed during the luteal phase, when there is an abundance of material (i.e., uterine cells). Although there is apparently no hormonal prerequisite, it is also clear that not all types of cells are equally efficient. Fibroblasts are poor feeder layers for embryo culture, for two basic reasons: fibroblasts continue to divide and overgrow, leading to culture medium depletion, generally associated with a decrease in pH (acidification), which is incompatible with the embryo's needs and capacity to manage an acidic pH (21). The second point is that the quality and specificity of growth factors released by fibroblasts do not have counterpart receptors on the embryo. In fact, the ability of cells to sustain embryonic development is limited to epithelial transporters (22). For this reason, we used Vero cells initially, an established cell line that is easy to handle and manipulate and highly controlled, since these cells are used for vaccine production (23). Vero cells can be bought with a certificate of safety from the American Tissue Culture Collection (ATCC, Rockville, Maryland) or from the WHO library, at a passage of around 121: each batch has a documented passage number. They can be easily frozen and thawed, and can conveniently be used routinely in an IVF unit. Tumorigenicity of the cells has never been described before passage 149 (24). Most of the information related to human blastocyst formation is based on information gathered from Vero cell coculture: from paternal and maternal effects (25, 26) to cryopreservation (27, 28). Autologous coculture, using endometrial biopsies ((29, 30), see Fig. 1) or oviduct cells (5, 6) started a little later in human IVF.

A European laboratory (Laboratoires Genevrier, Sophia Antipôlis, France) is now offering the opportunity for autologous coculture from uterine biopsies to be proposed for patients in European IVF units: endocell, the first commercial ready-to-use kit designed for embryo coculture up to the blastocyst stage. Endometrial biopsies are frozen in a previous cycle, and the plated cells are sent back to the IVF lab at the time of the corresponding patient treatment procedure (Figs. 2 and 3 with permission). Cumulus cells have also been proposed as feeder cells for culturing human embryos (31), but although this technique is relatively easy and attractive in principle, the results obtained were contested,

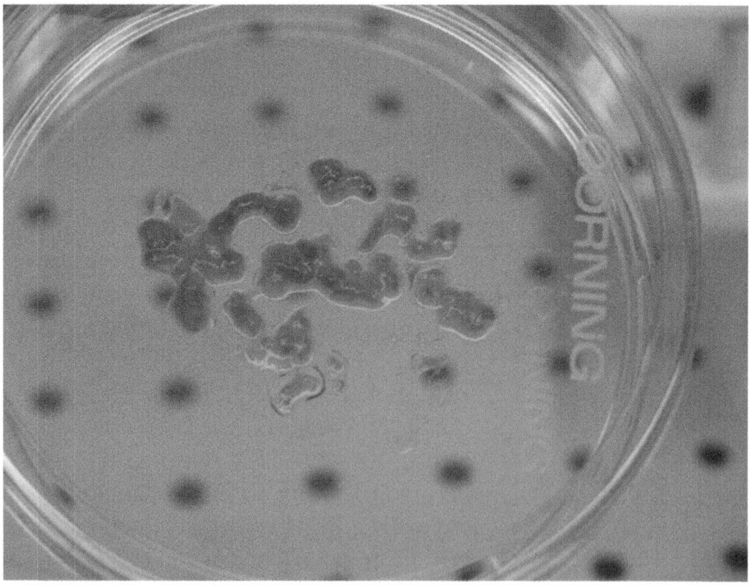

Fig. 1. Fresh biopsy of endometrial cells (25 mG).

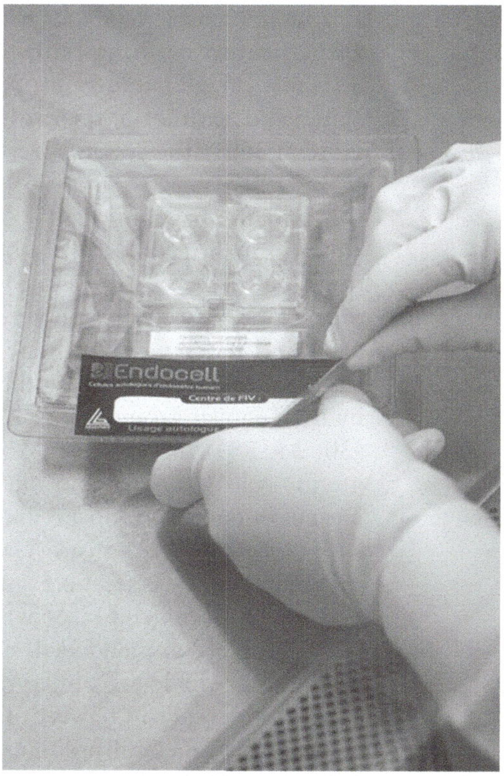

Fig. 2. A 4-well plate in a sterile blister with the autologous uterine cells ready for use (endocell, with permission).

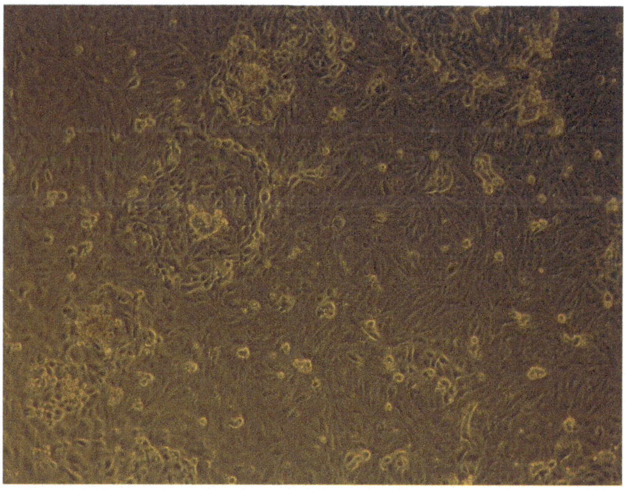

Fig. 3. A well covered with uterine cells (*endocell*, with permission). The uterine biopsy has been sent 1 or 2 months previous to the IVF cycle. The plated wells are sent under a sterile blister "ready for use".

and the technique has not been sustained. Problems encountered with the use of nonestablished cells as feeder layers include difficulty in obtaining sub-passages after trypsinization, rapid aging, and attrition of their trophic capacities is generally observed quite rapidly—the systems are "one shot."

1.3. Why Still Coculture Now

The evidence has confirmed that embryo coculture systems improve embryo quality and result in better pregnancy rates than conventional culture systems (32–35). The blastocyst expansion appears quicker, on Day 5.5 post in vitro insemination (Fig. 4), synthesis of hCG and hatching are initiated earlier (Vero cells: (8–36)). The largest study demonstrating this positive effect has been published by the Spanish group of IVI in Valencia (autologous uterine cells (30, 34, 35)) the group at Cornell University (New York) using uterine cells (32, 33), and our group with the group of institute Dexeus in Barcelona (Spain) and the Servy institute for reproductive endocrinology in Augusta (Ga), all three of us using Vero cells. The impact of this effect is even more important when coculture is performed for difficult cases, when culture with conventional media and SET is recommended. It appears that coculture with Vero or uterine cells yield very similar results. In our experience, we found a major difference in terms of freezing between cocultured blastocysts and the ones cultured in sequential media (37, 38).

Most of the criticism leveled against coculture and blastocyst transfer is not valid: the sex ratio is not modified (1,014 males/945 females: 51.7 %), and the birth weight of babies is not increased (39, 40). The latter could have been a source of concern, considering the large offspring syndrome described in farm animals.

It has been repeatedly said that prolonged culture time increases the incidence of monozygotic twinning. We demonstrated that this consequence is typically due to the effect of sequential media that has insufficient protection from ROS-induced apoptosis (41, 42); monozygotic twinning is not increased in coculture systems.

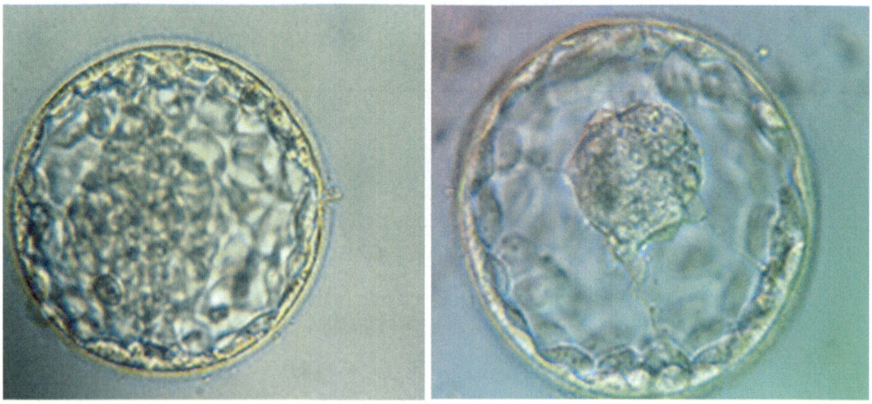

Fig. 4. Two expanded blastocysts obtained on D5 (Vero cells, B2 medium).

The cells of the feeder release sulfur-derived antioxidants such as hypotaurine, the natural free radical scavenger in the embryo environment (43, 44). We observed that therapeutic abortion rates are decreased for cocultured blastocysts (<1 %, 12/1971) and there is also a low incidence of stillbirths and death <3 months (9/1968).

Using a very large cohort of oocytes, Dominguez et al. (35) found that coculture yielded a highly significant difference in blastocyst formation rates, 56.0 % vs. 45.9 %, $p < 0.01$. In their oocyte donation program, where the maternal effect (26) is minimal, the difference is even greater: 70.5 % vs. 56.4 % ($p < 0.0001$). In term of pregnancy and implantation rates, the data follow the same pattern: pregnancy rates 39.1 % vs. 27.5 %, implantation rates 33.3 % vs. 20.9 %. Again, the results obtained with Vero cells are similar ((45): patients with repeated failures of implantation): 39.3 PR per retrieval with cocultured blastocyst transfer.

In fine, a meta-analysis (46) concluded that there is "an overall statistically significant effect of coculture on the implantation rate, clinical pregnancy rate, and ongoing pregnancy rate."

1.4. Coculture and Blastocyst Freezing

In order to develop a universal method of calculating a program's success rate, the overall outcome must be clearly defined. Freezing rate per surplus (non-transferred) embryo is a mandatory consideration, as well as the take home baby rate per frozen embryo, i.e., the percentage of extra births added by the cryopreservation program. This factor is rarely mentioned in the published literature (37, 38). It is usually assumed that prolonged culture time decreases the number of supernumerary embryos available for cryopreservation; as mentioned previously, coculture increases the blastocyst formation rates compared to conventional media. Moreover, the blastocysts obtained are more cryo-resistant, as a result of enhanced protection against ROS-induced damage. Release of sulfur-derived antioxidants by the feeder cells (47) prevents the formation of peroxidized lipids in the membranes. These prostaglandin-like endoperoxides distort the shape of the membrane and create fracture zones during freezing and thawing (48). The increase in cryotolerance may be also related to the higher number of cells per blastocyst (49), related to LIF produced by the feeder layer (50). This means that the take home baby rate per frozen embryo can easily be ten times higher than that observed for early stage embryos (37, 38). In addition to the effect of increased cryo-resistance, there is a selection effect: a very high number of early stage embryos have chromosomal anomalies (51). In our blastocyst program with Vero cells, freezing routinely resulted in exactly 25.4 % extra babies per year. The protocol used has been described elsewhere (28–52). It must be noted here that blastocysts obtained on Day 7 can lead to pregnancy and birth only when they are transferred to recipients prepared for Day 5 transfer. In our experience, fresh transfers on Day 7 never led to pregnancy, as if the "implantation window is closed."

1.5. How Can We Explain the Positive Effect of Coculture?

Coculture systems have been proposed to exert their effect through two different mechanisms: removal of toxic compounds and/or secretion of embryotrophic factors. We attempt to clarify these assertions.

The metabolism of oxygen is very important for the embryo. ROS production by embryos has been repeatedly reported: the three main ROS are produced, i.e., O_2^-, H_2O_2, and OH°. An increased production of these ROS is observed for in vitro produced embryos vs. those conceived in vivo. ROS production is observed to rise in the mouse at the time of fertilization, and they are considered to be at least partly responsible for developmental arrests at the G2/M phase of the second cell cycle, the time of MZT (53). ROS are endogenously produced by various metabolic pathways (e.g., oxidative phosphorylation, glucose in excess, NADPH and xanthine oxidases (44)). ROS have been shown to be deleterious for the preimplantation embryo: they induce DNA strand breaks, mitochondrial alterations, lipid peroxidation, etc. The consequences are numerous, and they lead to developmental arrests, sometimes via apoptosis. H_2O_2 is a strong inducer of apoptosis in the embryo (54), and is the most deleterious embryo "killer." Moreover, it is rarely mentioned that culture media itself can generate ROS when not well designed, and especially when not very well protected against ROS. There are important differences between different types of culture media in term of spontaneous ROS production (55), and protection from ROS effects is not obvious: for example glutathione, the universal ROS scavenger, does not penetrate embryo membranes, and although vitamin C can be protective it can also be pro-oxidant. Clearly, feeder cells are able to generate and release protective sulfur-derived antioxidants (47). Oviduct, uterine, and Vero cells possess the cysteine sulfinic acid pathway (43–47) allowing the synthesis of hypotaurine and taurine, important protective agents in the genital tract in vivo. This exogenous protection reinforces endogenous protection. Vero cells may also counteract unknown detrimental effects, such as those present in a hydrosalpinx environment (56).

1.5.1. Growth Factors

Maternal transcripts and proteins are present in finite quantities in the oocyte, dependent upon the quality of oocyte maturation and especially in relation to the age of the female. The same growth factors that allow the oocyte to gain its competence, i.e., the ability to pass the critical cycle of MZT and allow full development, must be active during the first 3 days of culture (57). It is important to avoid any delay in the initiation of embryonic MZT. Growth factors act in a harmonious and synergistic manner, interacting with metabolic pathways via activation and inhibition. If we consider what is secreted by the different feeder cells used in coculture, a few may be of importance. Using microarray analysis, we have assessed the presence of growth factor receptor transcripts in the

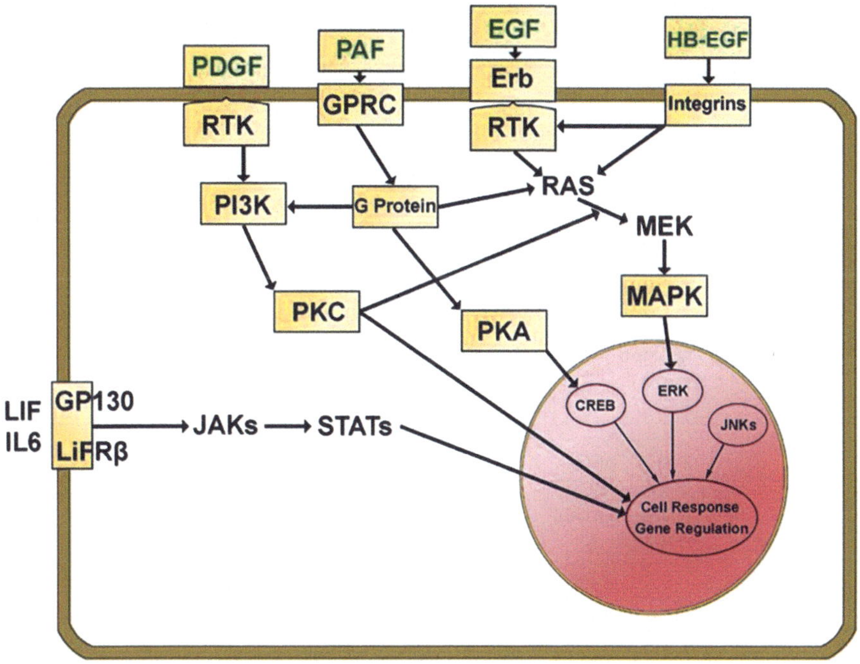

Fig. 5. Growth factors released by feeder cells and impact on early preimplantation development: the diagram shows receptors and their interaction in signaling mechanisms for cell growth, differentiation, and apoptosis. mRNA transcripts for all of the molecules and their signaling pathways have been identified by microarrays on pools of oocytes (for protocol, see ref. (58)). *CREB* cAMP-response element-binding protein, *EGF* epithelial growth factor (TGF alpha), *ERK* extracellular signal-regulated kinase, *GP130* glycoprotein 130, *HB* heparin binding, *IL6* interleukin 6, *JAK* janus kinase, *JNK* c-Jun N-terminal kinase, *LIF* leukemia inhibiting factor, *MAPK* mitogen-activated protein kinase (serine threonine protein kinase), *MEK* MAP-Erk kinase, extracellular-signal-regulated kinase (serine threonine protein kinase), *PAF* platelet activating factor, *PDGF* platelet-derived growth factor, *PI3K* phospho-inositol 3 phosphate, *PKA* PhosphoKinase A, *PKC* PhosphoKinase C, *RAS* (abbreviation of RAt Sarcoma), small GTPases, *RTK* receptor tyrosine kinase, *SMAD* homologue of drosophila protein, mothers against decapentaplegic (MAD), *STAT* signal transducer and activator of transcription, *TGF beta* transforming growth factor beta.

oocyte (58), and their signaling pathways: these are presented in Fig. 5. Coculture cells that express leukemia inhibitory factor (LIF) enhance blastocyst development in vitro in mouse (59), in the bovine (60) and in the ewe (61) where a peak of LIF is also observed prior to ovulation. Uterine cells and Vero cells secrete LIF and interleukin 6 (62, 63). GP130, the common receptor for LIF and IL6, as well as LIF receptor beta are expressed in the oocyte. The signaling pathways are also expressed. The same observation can be made for platelet activating factor (PAF) and platelet-derived growth factor (PDGF). The mitogenic effect of EGF/TGF beta has been demonstrated in the mouse (64), but although these are secreted by Vero cells, it is not clear that all feeder cells release such growth factors. TGF beta receptors 1 and 3, as well as all the Erbs, members of the EGF receptor family, are highly expressed on the human oocyte. In vivo, other growth factors such as growth hormone are present, and corresponding receptors on the human oocyte can also be found (65). However, maintaining perfect

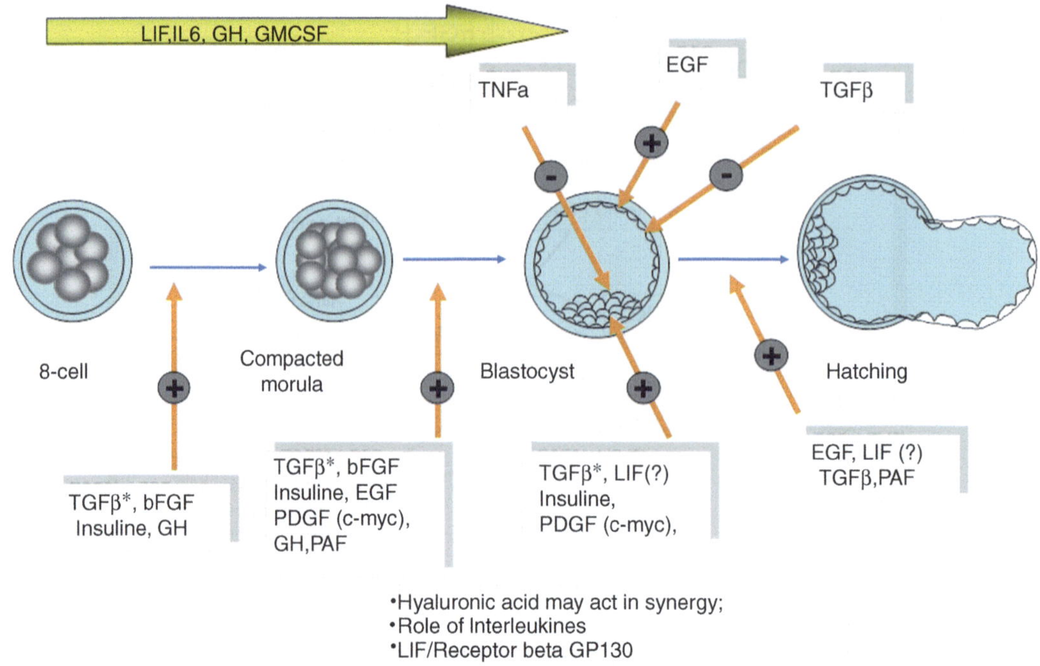

Fig. 6. + Activator, – inhibitor, *MZT* maternal to zygotic transition, *EGF* epithelium growth factor (TGF alpha), *bFGF* basic fibroblast growth factor, *GH* growth hormone, *LIF* leukemia inhibiting factor, *PAF* platelet activating factor, *PDGF* platelet-derived growth factor, *TGF beta* transforming growth factor beta, *TNF alpha* tumor necrosis factor.

synchrony and regulation of differentiation may require both positive and negative effects. Growth factors are redundant and affect gene regulation, but the precise intervention of each is still unknown. For example, one function of apoptosis is to eliminate cells that are damaged by any induced stress. The concept of adding one specific growth factor in order to inhibit apoptosis in culture media can be contested; although the number of blastocysts may be increased, this is accompanied by cytogenetic anomalies (66). The complexity, the cross-linking and the redundancy of all the pathways involved has to be taken into consideration. Most of the growth factors involved in regulating development will be synthesized in the embryo later, post genomic activation, but the exogenous growth factors still regulate embryonic development and differentiation (Fig. 6).

1.6. Conclusions: Coculture Systems, Past and Future

The benefit of coculture systems has been a matter of controversy since the time that they were first used in embryo culture. Irrespective of the type of feeder layer, the main criticisms have been:

1. *The risk of contamination*. Since the time of Pasteur, we have known that there is no spontaneous generation of bacteria and viruses. In a well-equipped laboratory with trained biologists, the risk of contamination is no higher than in classical IVF. The technology is even safer when established cell lines are

used, rather than autologous cells. Established lines have a quality control that is guaranteed by ATCC or WHO. Training of technicians and biologist in cell culture is of course mandatory, but this should in any case be a part of routine IVF training. In the year 2000, the French government mandated that coculture with Vero cells was not recommended, especially because of lack of training within the IVF teams. This may be reversed if the system is developed through a pharmaceutical company that will furnish pre-plated Petri dishes (Nunc). Naturally the supplier will have to maintain a very high level of quality control of the cells (number of sub-passages, stock library, etc.). One laboratory is also preparing autologous uterine cells on request, prepared from a clinical biopsy that is taken in a cycle prior to the IVF treatment cycle; the plated cells are returned ready for use (endocell, Laboratoire Genevrier, Sophia Antipôlis, France: Figs. 4 and 5).

2. *Safety of the procedure for the babies.* The first large-scale blastocyst transfer programs were set up using coculture on Vero cells (67). The first babies were born in France from fresh blastocysts in March 1988 and in November 1988 in the United States (Augusta Reproductive Biology associates, Augusta, Georgia). The first births from frozen blastocysts were in the fall of 1988 in France, and in the spring of 1989 in the United States. The babies are now 24 years of age, well past the age of puberty, and no specific problem has been identified, including overweight or shift in sex ratio (39, 40). Therefore, although the technology of blastocyst culture may have been contested, blastocyst transfer would no longer be challenged.
3. *Time consuming.* Organization of cell culture systems is simply a matter of planning. We have worked with coculture systems for 10 years without problems, and our system has been described in postgraduate courses at American Society for Reproductive Medicine (ASRM) meetings. However, if the cells can be bought ready for use, this aspect is no longer relevant.

To the question "Is there still a place for coculture in IVF?"—the answer is yes: it is not an obsolete concept. All of the careful analyses have demonstrated superiority in terms of results. Compared with sequential media systems, successful coculture has features that are unique. First of all, it is both active and dormant at the same time: it represents a state that is static, without overgrowth that will result in media degradation, but at the same time is an evolving dynamic system, releasing antioxidants and growth factors, a situation that may be considered to mimic in vivo conditions. This is especially important, since some sequential media that lack the so-called essential amino acids (especially methionine) can increase the risk of imprinting anomalies and anomalies linked to DNA damage (68, 69). Saturation/overwhelming of embryo DNA repair capacity leads at best to developmental arrest, or at worst to a tolerance for mutations that might subsequently lead to malignant cell transformation. Coculture may help to provide a better understanding of the embryo's interactions with its environment, and thus help in improving conventional culture media. Early preimplantation development is very far from being understood, and improvements in our various concepts surrounding it are still required.

2. Materials

1. Tissue culture dishes (35 × 10 mm).
2. Tissue culture dishes Falcon 3002 (60 × 15 mm).
3. Culture plates multiwell.
4. Culture flasks.
5. Centrifuge conical graduated tubes.
6. Plastic filters 0.45 μm.
7. B2 medium (with calf serum).
8. HEPES buffered medium 199.
9. Hanks balanced solution w/o Ca and Mg.
10. *Vero cells*. The cells were either obtained from ATCC or WHO collection (UK), at sub-passage 120–134, in a frozen tube containing 25 million cells. A certificate of quality including all the tests for viruses and electron microscopy is given.
11. Trypsin.
12. EDTA.
13. Chlamydia (BioMérieux Kit 5532/1).
14. Mycoplasma (BioMérieux Kit 4240/2).
15. Liquid nitrogen.
16. Liquid nitrogen programmable machine DMSO.
17. Cryo-vials.

3. Methods: Preparation of Vero Cells

3.1. Programmed Handling of the Cells: Careful Organization of the Cell Culture System is Essential, as Shown in Diagram 1

1. From the frozen cells, flasks are seeded with $2–3 \times 10^6$ cells, in B2 medium with 15 % fetal calf serum (FCS) reaching confluence within 4 days ($6–8 \times 10^6$ cells/flask).
2. After trypsinization (0.025 % in calcium-free Hanks balanced salt solution), the cell suspension is split into three parts: one is used to seed a new flask (in B2 + 15 % FCS), one is frozen, and the remaining part is used to seed wells, at a concentration of

100,000 cells per well (seeding in B2 with 7.5 % FCS). Confluence was reached in wells after 3 days.

3. Trypsinization was performed in 0.25 % HEPES buffered medium TC199 containing EDTA (0.11 %). The cells must not be passaged repeatedly: we observed that growth slows down after four sub-passages, and it proved better to freeze fourth passage cells and seed new flasks with a previously frozen aliquot.

3.2. Freezing Vero Cells for Coculture

Methods of cell freezing and thawing have been previously described (47–70).

1. The cells were placed in 2 mL cryo-vials (Nunc) and frozen in B2 medium containing 10 % DMSO.
2. Cells were frozen with the following temperature gradient: −5 °C/min from ambient temperature to −20 °C (hold for 5 min), then −1 °C/min to −30 °C, then −2 °C/min to −60 °C, and then plunge into liquid N_2.
3. After several sub-passages, the cells were tested for Chlamydia and Mycoplasma as an internal quality control. No contamination was ever detected.

3.3. Preparation of Feeder Layers

1. Coculture plates must be incubated at an alkaline (pH 7.4), bicarbonate containing medium. B2, Earle's and TCM 199 are considered as optimal coculture media for human and bovine, goat- and sheep-cloned embryos, suitable for the needs of both the epithelial cells and the embryos.
2. All the plates for the following week are prepared every Friday afternoon, according to the estimated number of cycles scheduled for that week.
3. The cells were seeded in B2 medium alone, at a concentration of approx. 15,000 cells per well.
4. The medium is changed for all the plates on Monday morning: the unattached cells are removed by washing. The medium is then changed again when coculture is initiated, immediately before the embryos are deposited on the layer.
5. The embryos are plated on the cells on Day 2, at the 2–4 cell stage. The coculture medium is not changed until embryo transfer and/or freezing.
6. All plates must be discarded after use, or after 1-week post seeding if not used.

3.4. Coculture with Autologous Uterine Cells

1. A 19–15 mG biopsy is performed during the luteal phase in a previous ovarian cycle (Day 22–24 of the cycle). Two options are then possible:
   (a) Freeze the biopsy with DMSO according to the protocol described, then thaw the sample during the treatment

cycle, and trypsinize (0.25 % in HEPES buffered TCM199, ten times more concentrated than for established cell lines); well seeding is performed 3–5 days before the expected coculture attempt.

(b) The biopsy can be trypsinized (0.25 % in HBSS) immediately using the classical protocol with 0.25 % trypsin.

The primary cell culture is then set up after seeding; when the cells are confluent, the primary culture is trypsinized and the cells are frozen.

Thawing of the cells and seeding of the wells will be performed 3 days before the expected coculture time.

2. Primary cultures cannot be used more than 2–3 times. Currently in France, IVF units are using a commercial system (endocell) instead of an in-house process. The full experimental procedure with details of all reagents and controls has been published elsewhere (70).

4. Notes

1. If required, the purity and the quality of the epithelial cells can be tested via immunochemical techniques, i.e., by testing the keratin content of the cells. The presence of keratin filaments confirms that the cells are epithelial in nature.
2. Detachment of the cells from the plastic support always indicates that there is a problem. This may sometimes happen in primary cultures that have a borderline contamination (e.g., Chlamydia, especially in uterine cells). Detachment can also happen if the culture medium is not well buffered (alkaline: bicarbonate), if a toxic compound is present, or if there is de-regulation of CO_2 in the incubator.

References

1. Camous S, Heyman Y, Meziou W et al (1984) Cleavage beyond the block stage and survival after transfer of early bovine embryos cultured with trophoblastic vesicles. J Reprod Fertil 72:479–485
2. Heyman Y, Ménézo Y, Chesne R et al (1987) In vitro cleavage of bovine and ovine early embryos: improved development using coculture with trophoblastic vesicles. Theriogenology 27:59–68
3. Gandolfi F, Moor RM (1987) Stimulation of early embryonic development in the sheep by co-culture with oviduct epithelial cells. J Reprod Fertil 81:23–28
4. Goodeaux LL, Voelkel SA, Anzalone CA et al (1989) The effect of rhesus epithelial cell monolayers on in vitro growth of Rhesus embryos. Theriogenology 39:187
5. Bongso A, Soon-Chye N, Sathananthan H et al (1989) Improved quality of human embryos when co-cultured with human ampullary cells. Hum Reprod 4:706–713
6. Bongso A, Ng SC, Fong CY et al (1992) Improved pregnancy rate after transfer of embryos grown in human fallopian tubal cell coculture. Fertil Steril 58:569–574
7. Ménézo Y, Hamidi J, Khatchadourian C et al (1989) The murine prepuberal oviduct

supports early embryo development in vitro. Dev Growth Differ 31:551–555

8. Ménézo Y, Guerin JF, Czyba JC (1990) Improvement of human early embryo development in vitro by coculture on monolayers of Vero cells. Biol Reprod 42:301–306
9. Fanchin R, Ayoubi JM, Righini C et al (2001) Uterine contractility decreases at the time of blastocyst transfers. Hum Reprod 16: 1115–1119
10. Kolibianakis EM (2004) Day 5 embryo transfer does not enhance reproductive outcome compared to D3 transfer using the current culture systems. In: ESHRE pre congress course special interest groups "Embryology", Berlin, pp 47–52
11. Kolibianakis EM, Zikopoulos K, Verpoest W et al (2004) Should we advise patients undergoing IVF to start a cycle leading to a day 3 or a day 5 transfer? Hum Reprod 19:2550–2554
12. Papanikolaou EG, Camus M, Kolibianakis EM et al (2006) In vitro fertilization with single blastocyst-stage versus single cleavage-stage embryos. N Engl J Med 354:1139–1146
13. Ménézo Y, Chouteau J, Veiga A (2001) In vitro fertilization and blastocyst transfer for carriers of chromosomal translocation. Eur J Obstet Gynecol Reprod Biol 96:193–195
14. Fauser BC (2008) Preimplantation genetic screening: the end of an affair? Hum Reprod 23:2622–2625
15. Veiga A, Gil Y, Boada M et al (1999) Confirmation of diagnosis in preimplantation genetic diagnosis (PGD) through blastocyst culture: preliminary experience. Prenat Diagn 19:1242–1247
16. McArthur SJ, Leigh D, Marshall JT et al (2008) Blastocyst trophectoderm biopsy and preimplantation genetic diagnosis for familial monogenic disorders and chromosomal translocations. Prenat Diagn 28:434–442
17. Veiga A, Sandalinas M, Benkhalifa M et al (1997) Laser blastocyst biopsy for preimplantation diagnosis in the human. Zygote 5:351–354
18. Fernández-Gonzalez R, Moreira PN, Pérez-Crespo M et al (2008) Long-term effects of mouse intracytoplasmic sperm injection with DNA-fragmented sperm on health and behaviour of adult offspring. Biol Reprod 78: 761–772
19. Papaioannou VE, Ebert KM (1986) Development of fertilized embryos transferred to oviducts of immature mice. J Reprod Fertil 76:603–608
20. Thibodeaux JK, Menezo Y, Roussel DJ et al (1992) Coculture of in vitro fertilized bovine embryos with oviductal epithelial cells originating from different stages of the estrous cycle. J Dairy Sci 75:1448–1455
21. Dale B, Menezo Y, Cohen J et al (1998) Intracellular pH regulation in the human oocyte. Hum Reprod 13:964–970
22. Scodras JM, Pollard JW, Betteridge KJ (1991) Effect of somatic cell type on bovine embryonic development in coculture. Theriogenology 35:269
23. Simizu B, Terasima T (1988) VERO cells, origin, properties and BioMedical applications. In: Chiba university, Japan. pp 1–234
24. Levenbook IS, Petricciani JC, Elisberg BL (1988) Tumorigenicity of Vero cells in VERO cells, origin, properties and BioMedical applications. In: Chiba university, Japan. pp 183–193
25. Janny L, Menezo YJ (1994) Evidence for a strong paternal effect on human preimplantation embryo development and blastocyst formation. Mol Reprod Dev 38:36–42
26. Janny L, Ménézo Y (1996) Maternal age effect on early human embryonic development and blastocyst formation. Mol Reprod Dev 45:31–37
27. Ménézo Y, Nicollet B, Herbaut N et al (1992) Freezing co-cultured human blastocysts. Fertil Steril 58:977–980
28. Ménézo Y, Veiga A (1997) Cryopreservation of blastocysts. In: Gomel V, Leung PCK (eds) In: Vitro fertilization and assisted reproduction, 10th world congress of in vitro fertilization and assisted reproduction, Vancouver, Monduzzi, pp 49–53
29. Jayot S, Parneix I, Verdaguer S et al (1995) Coculture of embryos on homologous endometrial cells in patients with repeated failures of implantation. Fertil Steril 63:109–114
30. Simón C, Mercader A, Garcia-Velasco J et al (1999) Coculture of human embryos with autologous human endometrial epithelial cells in patients with implantation failure. J Clin Endocrinol Metab 84:2638–2646
31. Quinn P, Margalit R (1996) Beneficial effects of coculture with cumulus cells on blastocyst formation in a prospective trial with supernumerary human embryos. J Assist Reprod Genet 13:9–14
32. Barmat LI, Liu HC, Spandorfer SD et al (1998) Human pre-embryo development on autologous endometrial coculture versus conventional medium. Fertil Steril 70:1109–1113
33. Barmat LI, Liu HC, Spandorfer SD et al (1999) Autologous endometrial co-culture in patients with repeated failures of implantation after in vitro fertilization-embryo transfer. J Assist Reprod Genet 16:121–127
34. Mercader A, Garcia-Velasco JA, Escudero E et al (2003) Clinical experience and perinatal outcome of blastocyst transfer after coculture

of human embryos with human endometrial epithelial cells: a 5-year follow-up study. Fertil Steril 80:1162–1168
35. Dominguez F, Gadea B, Mercader A et al (2010) Embryologic outcome and secretome profile of implanted blastocysts obtained after coculture in human endometrial epithelial cells versus the sequential system. Fertil Steril 93:774–782
36. Turner K, Lenton EA (1996) The influence of Vero cell culture on human embryo development and chorionic gonadotrophin production in vitro. Hum Reprod 11:1966–1974
37. Ménézo Y (2004) Cryopreservation of IVF embryos: which stage? Eur J Obstet Gynecol Reprod Biol 113(Suppl 1):S28–S32
38. Ménézo YJ (2004) Blastocyst freezing. Eur J Obstet Gynecol Reprod Biol 115(Suppl 1): S12–S15
39. Ménézo YJ, Chouteau J, Torelló J et al (1999) Birth weight and sex ratio after transfer at the blastocyst stage in humans. Fertil Steril 72:221–224
40. Ménézo Y, Veiga A, Pouly JL (2000) Assisted reproductive technology (ART) in humans: facts and uncertainties. Theriogenology 53:599–610
41. Ménézo Y, Sakkas D (2002) Monozygotic twinning: is it related to apoptosis in the embryo? Hum Reprod 17:247–248
42. Cassuto G, Chavrier M, Menezo Y (2003) Culture conditions and not prolonged culture time are responsible for monozygotic twinning in human in vitro fertilization. Fertil Steril 80:462–463
43. Guérin P, Ménézo Y (1995) Hypotaurine and taurine in gamete and embryo environments: de novo synthesis via the cysteine sulfinic acid pathway in oviduct cells. Zygote 3:333–343
44. Guérin P, El Mouatassim S, Ménézo Y (2001) Oxidative stress and protection against reactive oxygen species in the pre-implantation embryo and its surroundings. Hum Reprod Update 7:175–189
45. Ménézo Y, Sakkas D, Janny L (1995) Coculture of human embryo: factor affecting human blastocyst formation in vitro. Microsc Res Tech 32:50–56
46. Kattal N, Cohen J, Barmat LI (2008) Role of coculture in human in vitro fertilization: a meta-analysis. Fertil Steril 90:1069–1076
47. Ouhibi N, Hamidi J, Guillaud J et al (1990) Co-culture of 1-cell mouse embryos on different cell supports. Hum Reprod 5:737–743
48. Pryor WA, Stanley JP, Blair E (1976) Autoxidation of polyunsaturated fatty acids: II. A suggested mechanism for the formation of TBA-reactive materials from prostaglandin like endoperoxides. Lipids 11:370–379
49. Vlad M, Walker D, Kennedy RC (1996) Nuclei number in human embryos co-cultured with human ampullary cells. Hum Reprod 11: 1678–1686
50. Carnegie JA, Morgan JJ, McDiarmid N et al (1999) Influence of protein supplements on the secretion of leukaemia inhibitory factor by mitomycin-pretreated Vero cells: possible application to the in vitro production of bovine blastocysts with high cryotolerance. J Reprod Fertil 117:41–48
51. Iwarson E, Lundqvist M, Inzunza J et al (1999) A high degree of aneuploidy in frozen-thawed human preimplantation embryos. Hum Genet 104:376–382
52. Kaufman RA, Ménézo Y, Hazout A et al (1995) Cocultured blastocyst cryopreservation: experience of more than 500 transfer cycles. Fertil Steril 64:1125–1129
53. Nasr-Esfahani MM, Johnson MH (1991) The origin of reactive oxygen species in mouse embryos cultured in vitro. Development 113: 551–560
54. Pabon JE Jr, Findley WE, Gibbons WE (1989) The toxic effect of short exposures to the atmospheric oxygen concentration on early mouse embryonic development. Fertil Steril 51: 896–900
55. Martín-Romero FJ, Miguel-Lasobras EM, Domínguez-Arroyo J et al (2008) Contribution of culture media to oxidative stress and its effect on human oocytes. Reprod Biomed Online 17:652–661
56. Kim YB, Ahn SH, Chang DY et al (2002) Vero cell co-culture counteracts the detrimental effects of hydrosalpinx fluid on the development of mouse embryos in vitro. J Korean Med Sci 17:217–219
57. Ménézo Y, Elder K (2011) The enhancers of oocyte competence. In: Tosti E, Boni R (eds) Oocyte maturation and fertilization: a long history for a short event. Bentham Science Publishers Ltd, Sharjah, pp 64–70
58. Ménézo Y Jr, Russo G, Tosti E et al (2007) Expression profile of genes coding for DNA repair in human oocytes using pangenomic microarrays, with a special focus on ROS linked decays. J Assist Reprod Genet 24: 513–520
59. Kauma SW, Matt DW (1995) Coculture cells that express leukemia inhibitory factor (LIF) enhance mouse blastocyst development in vitro. J Assist Reprod Genet 12:153–156
60. Fukui Y, Saito T, Miyamoto A et al (1994) Effect of leukaemia inhibiting factor on in vitro

development of parthenogenetic bovine morulae. Theriogenology 42:1133–1139
61. Ptak G, Lopes F, Matsukawa K et al (2006) Leukaemia inhibitory factor enhances sheep fertilization *in vitro* via an influence on the oocyte. Theriogenology 65:1891–1899
62. Desai NN, Goldfarb JM (1996) Growth factor/cytokine secretion by a permanent human endometrial cell line with embryotrophic properties. J Assist Reprod Genet 13:546–550
63. Desai N, Goldfarb J (1998) Co-cultured human embryos may be subjected to widely different microenvironments: pattern of growth factor/cytokine release by Vero cells during the co-culture interval. Hum Reprod 13:1600–1605
64. Paria BC, Dey SK (1990) Preimplantation embryo development in vitro: cooperative interactions among embryos and role of growth factors. Proc Natl Acad Sci USA 87:4756–4760
65. Ménézo YJ, El Mouatassim S, Chavrier M et al (2003) Human oocytes and preimplantation embryos express mRNA for growth hormone receptor. Zygote 11:293–297
66. Paula-Lopes FF, Hansen PJ (2002) Apoptosis is an adaptive response in bovine preimplantation embryos that facilitates survival after heat shock. Biochem Biophys Res Commun 295:37–42
67. Ménézo Y, Hazout A, Dumont M et al (1992) Coculture of embryos on Vero cells and transfer of blastocysts in humans. Hum Reprod 7(Suppl 1):101–106
68. Ménézo Y, Elder K, Benkhalifa M et al (2010) DNA methylation and gene expression in IVF. Reprod Biomed Online 20:709–710
69. Ménézo Y, Dale B, Cohen M (2010) DNA damage and repair in human oocytes and embryos: a review. Zygote 18:357–365
70. Ouhibi N, Benet G, Ménézo Y (1991) Fetal bovine epithelial cell monolayers: method of culture and identification. J Tissue Culture Meth 13:289–294

Chapter 15

Culture Systems: Low-Oxygen Culture

Borut Kovačič

Abstract

The tension of oxygen measured in the oviducts of several mammals was 5–8.7 %, but this drops in the uterine milieu to <2 % in cows and primates. For embryo culture in human in vitro fertilization (IVF), a non-physiologic level of 20 % oxygen has been used for the past 30 years. However, several animal studies have shown that low levels of oxygen plays an important physiological role in reducing the high levels of detrimental reactive oxygen species within cells, influences the embryonic gene expression, helps with embryo metabolism of glucose, and enhances embryo development to blastocysts. However, clinical studies have given contradictory results. Nevertheless, in nearly all reports, some kind of improvement has been observed, either in embryo development or in implantation and no detriments have been reported. For these reasons, more and more IVF laboratories utilize low oxygen during embryo culture.

Key words: In vitro fertilization, Embryo culture, Atmosphere, Oxygen, Embryo development

1. Introduction

Attention in the early days of the IVF program was focused on providing conditions for embryo culture in vitro which were similar to those in the oviduct and uterus. This led to the equipping of IVF laboratories with CO_2 incubators, which provided a reduced oxygen atmosphere. However, the relatively high pregnancy rate achieved after the transfer of embryos, which have been successfully developed in the atmospheric concentrations (20 %) of oxygen, caused some to abandon low-oxygen equipment and also avoided added cost of supplying of nitrogen. Subsequently, incubators with 5–6 % CO_2 in air atmosphere have become standard in IVF laboratories over the past 30 years. The birth of nearly five million children conceived in the high oxygen culture conditions more than demonstrate that human embryos can adapt well to the atmospheric oxygen concentration. However, more and more experiments

Gary D. Smith et al. (eds.), *Embryo Culture: Methods and Protocols*, Methods in Molecular Biology, vol. 912,
DOI 10.1007/978-1-61779-971-6_15,

examining effects of low and high oxygen on mammalian embryo development suggest that a reduced oxygen culture environment is beneficial.

As will be discussed, numerous animal studies have revealed the beneficial effects of reduced oxygen. Embryos cultured at 5 % oxygen started earlier with first, second, and third cleavages had shorter fourth cell cycle durations, contained less apoptotic cells, expressed lower incidences of mosaicism in each embryonic stage, had a better inner cell mass (ICM) cell proportion, demonstrated a higher proportion of hatching blastocysts and developed in a larger number to the blastocyst stage, compared with embryos cultured at atmospheric oxygen conditions. However, clinical trials involving culture of human embryos at 5 % versus 20 % have given contradictory conclusions, and the reasons for the differences in the results remain unknown. Therefore, further investigations on the low-oxygen effect on human embryos are likely required. However, it is important to note that no report indicates a negative effect of low oxygen on clinical results or embryo development.

In this chapter, arguments for switching routine embryo culture in animal and human IVF from high- to low-oxygen atmospheres are presented. Much of the focus includes animal physiological studies that have investigated oxygen tensions in the oviduct and the uterus, as well as data on the relationships between different metabolic pathways in the embryo and oxygen gradient in the female reproductive tract. Owing to all the arguments provided from animal experiments and due to the caution against any as yet undiscovered adverse environmental impacts on child development, an increasing number of laboratories have already moved to a low-oxygen culture. In the final part of the chapter, several technical details and possible options for building a low-oxygen culture in an IVF laboratory are described.

1.1. Physiologic Environment Oxygen Concentrations

1.1.1. Preimplantation Embryos

The tension of oxygen measured in the oviducts of several mammals ranges from 35 to 60 mmHg (5–8.7 % oxygen). The uterine oxygen concentration in rabbits and hamsters is in a range similar to that in the oviducts but decreases to 3–5 % at the period of implantation. Oxygen concentration in the oviduct of rhesus monkey is 5–8.7 %, while the uterine level is 1.5–2 % (1). The higher level of oxygen in the oviduct compared to the uterus also suggests that the embryo passes through a decreasing oxygen gradient and reaches the lowest oxygen concentration in the morula stage when it commences the compaction process. This coincides with the changes in metabolic pathway preference—from oxidative phosphorylation for energy production in the pre-compaction stages to increased dependence on ATP production by glycolysis in post-compaction embryos (2, 3).

In humans, more attention has been paid to respiratory gases in the uteroplacental tissue during the early pregnancy period, and reports indicate oxygen tension in the coelomic fluid of around

2.2 % (20 mmHg) and below 2 % (15–18 mmHg) in amniotic fluid (4). There are only a few studies which have been conducted on analyzing oxygen concentration in the uterine cavity in nonpregnant women (5, 6). From the first study, an average oxygen concentration in the human uterus was measured below 2 % (15 mmHg). The intrauterine oxygen tension in the second study was measured at the endometrial surface in 21 patients at the time of insemination. The individual oxygen tension varied significantly, with a mean value around 2 % (18.9 mmHg). In some women, time-related variations or rhythmic oscillations were observed, which could be associated with endometrial wave-like activity. To date, no studies have been published on the exact oxygen tension in the human oviduct.

1.1.2. Oocytes

Although the dissolved oxygen content of follicular fluid negatively correlates with the frequencies of chromosomal and cytoplasmic abnormalities of oocytes (7), it is still unclear as to whether low oxygen is beneficial during oocyte maturation. Many authors reported an ovarian venous oxygen concentration of 7 % (41 mmHg) and an intrafollicular concentration of 8 % (51 mmHg) in pigs (8) and as well as in humans (9). However, there are contradictory findings concerning the consumption of oxygen within the follicle. One concluded that most of the oxygen is consumed by outer layers of granulosa cells and the oocyte is in anoxic conditions (10), while others predicted that cumulus cells do not need much oxygen and the majority of the oxygen remains available for oocytes (11).

1.2. Oxygen Utilization

1.2.1. Embryos

The embryo is surrounded by oviductal or uterine fluid, from which it accumulates substrates and oxygen required for growth. The main activity of the early embryos during the initial divisions is in metabolic production of energy and gene activation. During this phase, embryos use aerobic respiration to obtain the oxygen required for the oxidation of substrates such as pyruvate and amino acids. Once an embryo starts growing, it switches to oxidative phosphorylation and aerobic glycolysis (2, 12). The production of energy—ATP molecules—increases at compaction and blastocyst formation and support protein synthesis and ion transport system (13). Byatt-Smith et al. (14) proposed a model in which 5 % oxygen would be able to sustain oxidative phosphorylation in small embryos (from a mouse). However, large (human, bovine) embryos in later stages would establish a gradient with anoxia in the center of the embryo. Potentially supporting this model, studies have reported that 2 % oxygen was too low and a detrimental concentration for post-compact mouse embryos (15, 16) but optimal for bovine blastocysts (17).

The embryo obtains its oxygen via passive diffusion with a rate that is controlled by the oxygen tension in the gas phase, the solubility of oxygen in the medium, the boundary layer around the embryo, and the rate of diffusion through the cytoplasm. The diffusion through a layer of oil appears not to be rate limiting (18). The oxygen is consumed in oxidative phosphorylation within mitochondria and the rate of consumption depends on the surrounding oxygen concentration, the substrates that are available and on mitochondrial enzyme complexes. In the mouse embryo, >70 % of the oxygen is metabolized via oxidative phosphorylation at the blastocyst stage, and <30 % at the 2- to 4-cell stages (19). The trophectoderm (TE) consumes significantly more oxygen, produces more ATP and contained a greater number of mitochondria than the ICM. The major fate of the energy production by TE is likely to be the Na(+), K(+)ATPase—sodium pump enzyme—located in the TE basolateral membrane. The pluripotent cells of the ICM display a quiescent metabolism (20).

1.2.2. Oocytes

The respiration rates for human oocytes vary from 0.5 to 1.1 nl/h and, for human Day-1 to Day-3 embryos, the rates were 0.2–0.7 nl/h (21). Quantitatively, an oxygen consumption rate in bovine early embryos and blastocysts was measured as being 0.38 nl/h versus 1.3 nl/h, respectively, which is 3.4-fold higher in blastocysts compared to those in the early embryo stage (22). In the bovine system, blastocysts with very high or very low respiration rates are nonviable (23). A peak of oxygen consumption was observed at the time of fertilization and prior to cleavages (24).

1.3. Production of Reactive Oxygen Species

Another rationale for use of low oxygen is that it can be potentially toxic for cells through formation of ROS. These ROS can be produced from a variety of sources.

1.3.1. Endogenous Source of ROS

As a side-product of any aerobic metabolism ROS can be produced. ROS may originate from embryo metabolism and/or the surroundings. Endogenous ROS are mostly produced during oxygen reduction in the mitochondria. The electron transfer from the reducing equivalent (NADH-$FADH_2$) to oxygen is carried out with the help of various reductases through the mitochondrial respiratory chain and is often catalyzed by exogenous and endogenous low-molecular-weight molecules (e.g. iron complex). Depending on the number of electrons required for oxygen reduction, different types of reactive oxygen molecules can be generated: the superoxide anion radical ($O_2^{-\bullet}$)—one electron, hydrogen peroxide (H_2O_2)—two electrons, hydroxyl radical ($OH^\bullet$)—three electrons (25). Together with other atoms or molecules, the additional radicals can accidentally occur, e.g. alkoxyl or peroxyl radicals in lipids (26).

1.3.2. Exogenous Source of ROS

The exogenous sources of ROS in physiological conditions are follicular fluid or tubal and uterine milieu. Oocytes and embryos from IVF are cultured in a medium, which can in itself be the source of ROS. It was found that commercial media generate ROS at various rates, depending on the composition, whereas follicular fluid generates ROS at a much lower level (27). In comparison with in vivo produced oocytes, a higher level of membrane damage and cytoplasmic glutathione pool reduction was observed in in vitro-cultured oocytes (27). ROS (O_2^-) can be generated by buffers or molecules in media that are photosensitive at exposure to visible light. Such media components are flavins (28), which fortunately are not common components in current commercial embryo culture media. The ferric or cupric salts in a medium which is exposed to an atmospheric concentration of oxygen can also produce ROS. When H_2O_2 reacts with metal ions (iron or copper), the highly reactive and toxic $OH^•$ is generated (25). In serum or serum supplements, which are frequently used for cell protection, diverse oxidase activity was found and can therefore also produce ROS (25).

1.4. Harmful Effects of ROS on Cells

1.4.1. Mitochondrial Alterations

ROS generation in mitochondria can lead to oxidative damage to mitochondrial proteins, membranes, and DNA (29). The H_2O_2 can pass freely through membranes, whereas $O_2^{-•}$ is not membrane permeable but can pass through ion channels, and alter most types of cellular molecules such as lipids, proteins, and nucleic acids in the cytoplasm (30). The oxidative stress additionally opens the mitochondrial permeability transition pores. As a result, a massive loss of ions and metabolites from mitochondrial matrix occurs, causing mitochondrial alterations, embryo cell block, ATP depletion, and the activation of the cell's apoptotic machinery (31). Additionally, mitochondrial damage can alter their function in the tricarboxylic acid cycle, fatty acid oxidation, the urea cycle, amino acid metabolism, and other reactions (32).

1.4.2. Membrane Lipids and Proteins

Reactive oxygen species like H_2O_2 detrimentally affect membrane lipids via lipid peroxidation. This occurs as a chain reaction and reacts to the surrounding phospholipids (33). This can affect the membrane stability or permeability. The 2-cell block observed in mouse embryos is associated with a rise in lipid peroxides (34). With oxidative stress, the rates of disulphide bonds and disulphide formations in proteins increase within the cell. It has been demonstrated that ROS disturb the ratio of glutathione to glutathione disulphide (35). In addition, ROS frequently cause the inactivation of specific enzymes (36). Other aberrations, including the aggregation of cytoskeletal components and endoplasmic reticulum condensates, have also been reported (37).

1.4.3. DNA

Oxidative stress, such as that caused by hydroxyl radicals ($OH^•$), induces DNA fragmentation (29, 38). The oxidation of DNA deoxyribose causes single-strand breaks in DNA (39). If these single-strand breaks exist at the time of DNA replication, then irreversible double-strand breaks can occur (40). Single-strand breaks also happen where mispairing occurs during DNA polymerization. For example, 8-oxiso-guanine is a product of guanine oxidation and it can mispair with adenine (41). Mitochondrial DNA is particularly sensitive to damage, as it has no protective histones. For this reason, increased rates of mutations are found in mitochondrial DNA and, as a consequence, abnormal phosphorylation enzymes—encoded by mitochondrial DNA—are produced (42).

1.4.4. ROS Effects on Mammalian Embryos In Vitro

ROS have been implicated in the impaired development of mammalian embryos in vitro (43). Some developmental blocks, such as the 2-cell block in mice, are related to ROS action (34). Although, oxygen metabolism is highly active at the blastocyst stage, ROS exposure during oocyte maturation can alter embryo development (44). Besides this, DNA fragmentation, and consequently the irreversible double-strand breaks caused by ROS, may increase in early-stage embryos because of very active DNA replication in these stages. All these types of cellular damage caused by ROS lead to abnormal embryo development, which can be expressed by uneven blastomeres, slower cleavage or an arrest in development (45). Furthermore, the ROS H_2O_2 is the main mediator of apoptosis and related cytoplasmic fragments in blastocysts (46). The relationship between apoptosis and cytoplasmic fragments has already been demonstrated (45).

It should be mentioned that ROS can also be very important for cells. Low nontoxic concentrations of ROS are capable of mediating cellular functions through intracellular signaling via alterations in the redox state (47). A brief pulse of H_2O_2 delivered to bovine embryos influences their capacity to develop to blastocysts (48). Mouse embryo hatching is also associated with a burst of ROS production (49). ROS, like H_2O_2, can influence the regulation of gene expression and may be needed, though careful regulation of its concentration by oxidases is likely important (25). The activity of these regulatory enzymes, like NADPH oxidase and xanthine oxidase, are dependent on the environmental oxygen concentration (29, 50).

1.5. Protection Against ROS

1.5.1. Protective Mechanisms Against ROS Within Oocytes and Embryos

Oocytes and embryos have endogenous protective mechanisms against ROS, which involve the activity of superoxide dismutases (SODs), catalases, and peroxidases, enzymes known as oxygen scavengers. Ovulated oocytes and early embryos are rich in these scavengers, especially superoxide dismutases (51). The protective role of dismutases is to convert $O_2^{-\bullet}$ into H_2O_2. Catalases and peroxidases eliminate H_2O_2 by changing it to H_2O and O_2 (52).

1.5.2. Protective Mechanisms Against ROS in Embryo Culture System

The protection of oocytes and embryos during culture in vitro can be carried out in different ways. Arguably the best approach entails prevention of ROS generation in culture media and embryos.

This can also be achieved by avoiding the coculture of oocytes and embryos with cellular debris, leukocytes, and dead spermatozoa—found to be major ROS generators (53, 54). It was also demonstrated that the aspiration of dead cells and cytoplasmic fragments—a possible source of ROS—from an embryo can improve its developmental ability (55). It may be important to avoid prolonged expose the culture media and paraffin oil to visible light (56), since it can cause elevation of H_2O_2 in the medium. The deleterious effects of visible light and hydrogen peroxide on embryos can be diminished by lowering the oxygen concentration in the incubating atmosphere (57).

Other methods to reduce ROS include scavenging of the harmful species. Many IVF culture media contain various concentrations of different nonenzymatic antioxidants, such as vitamins A, C,and E, pyruvate and sulfur compounds (glutathione, cysteamine, taurine, hypotaurine), and metal chelators (ethylenediaminetetraacetic acid (EDTA)), which can help to protect the cells against lipid peroxidation and DNA damage. As a result, it appears as though commercial media differ a great deal in terms of their protective ability against oxidative stress depending on their composition (27).

Another approach to decrease the harmful effects of oxygen on embryos during cultivation in vitro is to keep them in a lower oxygen concentration, similar to that found in the female reproductive tract. Culture of mammalian embryos under atmospheric oxygen has been shown to result in a higher H_2O_2 concentration per embryo in comparison with a low-oxygen culture (38, 58, 59). The authors reported a reduced H_2O_2 content in embryos associated with a reduction in DNA fragmentation. Similarly, oocytes matured in vitro in a low-oxygen culture produced less H_2O_2 than oocytes matured at atmospheric oxygen (60). It has not yet been possible to conclude that the production of ROS under atmospheric oxygen conditions is the main contributor to poor embryo development.

1.6. Low-Oxygen Control of Gene Expression

Mammalian cells have developed a complex mechanism for sensing oxygen and gene regulation. Rinaudo et al. (61) demonstrated a significant alteration in the gene expression pattern in mouse embryos cultured at 20 % oxygen. However, an increase in antioxidant gene expression at 20 % oxygen has not been observed. Similar results were reported by Harvey et al. (62) for bovine embryos. Oxygen is also known as a regulator of gene expression via hypoxia-inducible factors (HIFs), which are DNA-binding proteins composed of two subunits. One subunit is stable only in low-oxygen conditions. Under low-oxygen conditions, HIFs activate the transcription of many genes, responsible for the regulation of pre- and post-implantation embryonic metabolism and development (61, 63). Furthermore, the oxygen concentration used during embryo culture in vitro may influence embryonic gene expression patterns. It was demonstrated that a relatively small decrease in oxygen from 7 to 2 % can have a significant effect on the expression of genes involved in the regulation of pre- and post-implantation embryonic metabolism and the development of mouse blastocysts (16). Similarly, a reduction in oxygen from 7 % pre-compaction to 2 % post-compaction in bovine embryos resulted in oxygen-regulated changes in gene expression for many proteins involved in ICM constitution and the metabolism of glucose. As a result, a significant increase in the proportion of cells constituting the ICM occurred under 2 % oxygen, rather than under 7 % oxygen (62). The study conducted on global gene expression patterns in mouse embryos showed a similar pattern in embryos developed in vivo and in those developed in vitro under 5 % oxygen. However, several genes involved in cell growth and maintenance and genes responsible for gastrulation were mis-regulated, when embryos were cultured at 20 % oxygen (61). The authors of the study also revealed that the influence of different culture media on the global pattern of gene expression was minor in comparison with the influence of different oxygen concentration (61).

1.7. Animal Embryos Developed In Vitro at Low Versus High Oxygen Tension

Many scientists have tested the effects of changes in oxygen levels on the development of preimplantation embryos in vitro. In one of the first reports in 1978, authors demonstrated that mouse blastocysts cultured at 20 % O_2 had fewer blastomeres than those developing under 5 % O_2 and that the proportion of embryos forming blastocysts decreased linearly with an increase in oxygen concentration from 2.5 to 20 % (64). In a similar experiment, sheep and cow embryos were cultured in different oxygen concentrations of 2–20 % oxygen. The highest proportion of blastocysts from both species developed in 7–8 % oxygen. These blastocysts also contained the largest number of cells compared to embryos from other groups, while 0 % oxygen was deemed detrimental (65). Even a short exposure for 1 h to 20 % oxygen at the beginning of the culture proved detrimental to development beyond the morula stage (65).

Other oxygen studies have analyzed the effect of only 5 and 20 % oxygen concentrations on the development of embryos of different species. Several groups have shown that culture under low oxygen improves blastocyst development in mice (66–68), hamsters (57, 69), rats (70), rabbits (71), cows (72–76), goats (77), and pigs (59). An improvement in the embryo cell number has been observed in several species in response to low oxygen (65, 71, 74, 77, 78). Other studies have revealed several other beneficial effects of reduced oxygen. Embryos cultured at 5 % oxygen started earlier with first, second, and third cleavage (68). They had a shorter fourth cell cycle duration (79), contained less apoptotic cells (78), had a lower incidence of mosaicism in each embryonic stage (80), expressed better ICM cell proportion (81) and ICM cell organization (82), or demonstrated a higher proportion of hatching blastocysts (78) compared with embryos cultured at atmospheric oxygen conditions. In contrast, some studies could not detect a difference in the blastocyst development rate under different oxygen conditions in mice (83), cats (84), and bovine embryos (85, 86). Some of these variations may relate to the strains of mice used, or to differences in media composition, including the content of metal ion chelators, such as BSA or EDTA (87). It was also reported that the oxidative metabolism of glucose was significantly enhanced by reducing the oxygen concentration from 20 to 5 % in mouse morulae and blastocysts (88).

Although most studies demonstrate an enhancement in at least one of the embryo development parameters by using a low-oxygen atmosphere, the optimal oxygen concentration for the fertilization process remains unknown. The authors of two studies on bovine and human IVF observed lower fertilization rate in oocytes at low oxygen levels compared to those fertilized at atmospheric oxygen levels (76, 89), suggesting that the fertilization process probably requires a higher level of oxygen. In contrast with this, Thompson et al. (17) reported that the post-compact stages of bovine embryos developed better by decreasing oxygen concentration from 7 % in the pre-compaction to 2 % in the post-compaction period (17). Such an oxygen gradient during embryo culture could lead to more physiological oxygen conditions, similar to those in the female reproductive tract, with higher oxygen levels in the oviducts and an extremely low concentration of oxygen in the uterus. Further investigation needs to be undertaken in order to confirm this hypothesis.

1.8. Effect of Oxygen Concentration on Animal In Vitro Maturation

Several studies presented a beneficial effect for reduced versus atmospheric (5 % versus 20 %) oxygen tension for oocyte in vitro maturation, while others provided contradictory reports. Low oxygen resulted in better kinetics of the first cleavage of in vitro-matured bovine oocytes in comparison with 20 % oxygen, but development to the blastocyst stage was not enhanced (89). Similarly, oxygen concentration during IVM did not affect mouse oocyte maturation, fertilization, cleavage, blastocyst development, implantation, and a number of viable fetuses. However, the fetal weight was reduced following oocyte maturation at 5 % compared with 20 % oxygen (90). Better blastocyst development was observed for pig (91) and bovine (92) IVM in atmospheric and not in low oxygen. On the contrary, some studies reported beneficial IVM results, including blastocyst development in low-oxygen conditions in porcine (93, 94), mouse (95), and bovine oocytes (60, 96). From all these contradictory results, it can be concluded that a consensus on the optimal oxygen concentration for IVM has not yet been established. To date, no data has been published about the effects of oxygen concentration on human IVM.

1.9. Low Versus High Oxygen Effect on Human IVF Outcome

In contrast with the numerous experiments conducted in animals, which have consistently demonstrated that a low-oxygen culture markedly improves embryo development and IVF outcomes, the 13 similar studies conducted on humans have not so uniformly proven the benefit of a low-oxygen culture for human embryos (53, 76, 97–107). The discrepancy between these results could be explained by some peculiarities between the species. However, an analysis of the published papers undertaken demonstrates that there are only few relevant human data obtained through adequately designed prospectively randomized trials with groups of patients that are of sufficient size. Furthermore, the duration of the human embryo culture is often only 2–3 days long, but animal embryos are usually cultured to the blastocyst stage. There are also differences in the culture media used between laboratories and differences in homogeneity in the patient groups compared. Clinical parameters, such as pregnancy, implantation, or live birth rate, which are usually compared, do not accurately reflect the situation about embryo viability between compared groups, because of the embryo selection for transfer. The real effect of oxygen on the IVF outcome often also passes unnoticed due to alternating and modifying the day of transfer according to the quality of developing embryos.

Two of the 13 studies investigating the oxygen effect on human embryos were retrospective (76, 106). Three studies analyzed embryo development to blastocysts and were not primarily focused on the clinical outcome (99, 102, 103). Only two of them were prospectively randomized, comparing sibling oocytes (102, 103). In both studies, a significant improvement in the quality of early embryos and blastocysts, especially in the ICM were observed. Kovačič and Vlaisavljević (102) noted that 5 % O_2 contributed to the faster development of early and late embryos, resulting in a higher proportion of embryos able to develop into blastocysts. Due to earlier blastocoel formations in embryos from the low-oxygen group, the evaluation of blastocysts—especially ICM morphology—became much easier.

Some of the prospectively randomized clinical studies were focused on early embryo development, analyzing Day-3 transfer outcomes (97, 98, 100). In both studies carried out by Dumoulin et al. (97, 98), no marked improvement was observed in fertilization or cleavage, or in the number of cells per embryo. Neither Dumoulin et al. (97, 98) nor Bahceci et al. (100) found any influence exerted by the different oxygen tensions on the clinical outcome.

Other studies have analyzed extended preimplantation development in either 5 % or 20 % oxygen and compared the clinical results of Day-5 transfers (53, 101, 105). These studies discovered different improvements in the embryo characteristics such as higher embryo utilization rate per cycle (53, 105), or a higher mean embryo score in the early stages (101). Waldenstrom et al. (105) and Catt and Henman (53) were the first to report increasing pregnancy and implantation rates by reducing the oxygen concentration in incubators.

It is interesting that such an influence on clinical results was not observed by Kea et al. (101), Behr et al. (99), and Nanassy et al. (106). In all three studies, the concentration of oxygen was changed during the prolonged embryo culture on Day 3—once from 20 to 5 % and once from 5 to 20 %. Wale and Gardner (68) investigated the temporal responses of preimplantation mouse embryos to oxygen conditions in vitro. They found that a partial (either only in pre- or only in the post-compaction period) or complete culture in atmospheric oxygen resulted in significantly fewer blastocyst cell numbers or blastocysts developed compared to the control low-oxygen group. It can be concluded from all these studies, that embryo culture at reduced oxygen is beneficial only if 5 % oxygen is used during the whole in vitro culture period.

The last two prospective, randomized studies, designed according to the CONSORT statement, analyzed the clinical outcomes of mixed group—Day-3 and Day-5—transfers (104, 107). Meintjes et al. (104) demonstrated the overall increase in live birth implantations (42.9 % versus 30.7 %) in a patient group in which embryos were cultured in a 5 % O_2, compared with rates among women whose embryos were incubated in an atmospheric O_2 environment. However, when transferring embryos on Days 2–3, the implantation rates did not differ between compared groups. The main difference in the oxygen effect on the clinical outcome was thus observed in Day-5 transfer groups. The authors did not find any difference in fertilization, cleavage, or embryo utility rates.

In contrast with the previous study, Kovačič et al. (107) observed marked differences in developmental outcomes with a short and extended culture and in embryo utility rates. Embryos cultured in 20 % oxygen had significant delays in early cleavage divisions and in later cavitation process and, as a result of this, a lower number of blastocysts and optimal blastocysts developed on Day 5, compared to embryos cultured at 5 % O_2. The clinical outcomes in compared groups remained at similar rates, but when the freezing/thawing cycles were included, the cumulative pregnancy rate was higher in the low-oxygen group. Both studies differed by patient ages and quality of cycle, since Kovačič's study was carried out on 50 % of optimal and 50 % of poor responding cycles. The greatest difference in favor of low oxygen was observed precisely in those cycles with a low number of oocytes and Day-3 transfers in which the selection between embryos for transfer was not required.

Regarding the outcomes of all 13 studies, it can be summarized that the lowering of O_2 in the culture atmosphere caused a significant improvement in embryo development and/or quality in as many as nine of them; however, an increase in the pregnancy rate was noted in only four studies. None of the authors have found any worsening related to lower O_2 concentration, either in embryo development or in the pregnancy rate. The contradictory findings point out the need for new clinical trials, which will take into the consideration the cumulative pregnancy rates rather than pregnancy or implantation rates only. It also remains to be determined whether the effect of oxygen concentration is stage specific. This analysis is still being carried out by Cochrane Collaboration (108). Another question that remains is whether 5 % O_2 really is the most optimal oxygen concentration for human embryos in vitro. To date it is not yet known what the most optimal oxygen concentration for human embryo culture is. According to reports from several animal studies, it seems that the in vitro culture of mammalian preimplantation embryos is favored by 5 % O_2 (38 mmHg). This value was found in many studies to be beneficial for mouse embryos, but this conclusion was generalized with a lack of evidence for all mammalian embryos. As the size of the human embryo equates more to ruminants, then it could be speculated that a level around 7 % O_2 is likely to be more advantageous for precompaction embryos than 5 %, as modeled by Byatt-Smith et al. (14). However, the measurements of intrauterine oxygen tension in the Rhesus-monkey reproductive tract showed drops from 8 % in oviducts to only 1.5–2 % in the uterus (1). It is true that very low O_2 concentration stimulates the proliferation of trophectoderm, and ICM favors glycolysis, but at <2 % O_2 in the incubator the anoxic conditions could be created, especially in drops of medium overlaid by oil and in the microenvironment surrounding the embryo. The human embryo may become hypoxic at low oxygen levels in a static culture without stirring (14), while mouse embryos are likely to be able to satisfy their demands for oxygen by diffusion. Thus, to date, it is possible to conclude only that, on the basis of evidence, a reduction of oxygen in atmosphere is likely to benefit the human embryo. However, further physiological, molecular, and clinical studies are needed to evaluate the real effects of oxygen on human embryos in vitro.

2. Materials

There are several options available to provide a reduced oxygen atmosphere for embryo culture. These include use of incubators with oxygen control (see Note 1), or isolated chambers within incubators. Different approaches require different gas sources.

2.1. Incubators

1. *Classic CO_2 incubators.* Almost all types of classic incubators can be found today with the option for reduced oxygen atmosphere (see Note 2). These devices have two gas (CO_2, N_2) or three gas (CO_2, N_2, O_2) connections and two gas sensors, one for CO_2 and one for O_2 (see Note 3).
2. *Single-dish incubators.* These are small table incubators (see Fig. 1). They have several chambers, designed so that single

Fig. 1. System of table incubators for single-dish culture. Each chamber has its own lid.

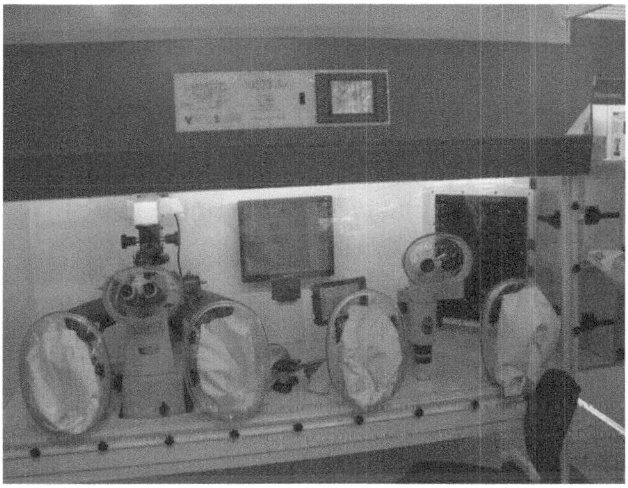

Fig. 2. The incu-workstation has a tightly closed glassy front side with openings for integrated gloves and microscope binoculars. The interior is gassed by CO_2 and N_2 and equipped with microscopes. The culture dishes with embryos can remain in the workstation through the whole culturing period.

culture dish fits to the chamber. Each chamber is usually supplied by a premade gas mixture and is covered by its own lid.

3. *Sealable chambers (desiccators) or gas-impermeable bags.* Once the dish is inserted, the bag or chamber is filled with a preselected gas mixture (5 % O_2, 5–6 % CO_2 and 90 % N_2) and tightly sealed.
4. *Incu-workstations.* Big hoods modified for functioning as a workstation and an incubator at the same time (see Fig. 2).

2.2. Gases

Recommended concentration of gases in the incubators with low-oxygen tension is 5 % O_2, 6 % CO_2 and 89 % N_2 (see Note 4). This can be obtained by connecting the incubator to one of the following gas sources:

1. Separate bottles of CO_2 (laboratory gas of the highest purity 4.5) and N_2 (laboratory gas of the highest purity 5.0).
2. Mixture of 5 % O_2, 5–6 % CO_2 and 89–90 % N_2.
3. Bottle of CO_2 (laboratory gas of the highest purity 4.5) and liquid nitrogen vapor tank to supply N_2 gas.

3. Methods

3.1. How to Establish Low-Oxygen Culture System?

The transition from high- to low-oxygen culture can be a challenging project, both in financial terms and in the organization of work in the laboratory. The major problem occurs in laboratories which are already equipped with a large number of incubators that are not able to regulate oxygen. A frequently asked question in this case is whether it is possible to introduce a low-oxygen culture with existing incubators. In this case, the use of the simple and inexpensive sealed-bag/chamber method (see Note 5) can be used. However, in setting up a new laboratory, it is better to think about low-oxygen culture from the start, when new incubators are purchased and the gas installations in laboratory are installed.

Another common question concerns the number of low-oxygen incubators required to obtain the appropriate low-oxygen culture conditions. The basic rule for good laboratory practice in the IVF is to keep the number of culture dishes per incubator at a minimum and to reduce the opening of incubator to a minimum. In low-oxygen culture, this rule is particularly important because the low-oxygen atmosphere is regulated differently to the atmosphere in CO_2 incubators (see Note 6). The larger the incubator, the larger the amount of N_2 is required to maintain a low-oxygen atmosphere (see Notes 7 and 8). The most economic variant for a N_2 supply for a large number of "big-box" low-oxygen incubators is in using the liquid nitrogen from cryo-laboratory (see Note 9).

It is necessary to verify the recovery time for each gas separately, before using the incubator for clinical purposes (see Note 10). The ability of incubator types in this respect varies from each other (see Note 11). Failures in gas recovery in an incubator can lead to even worse conditions for cultivation with low-oxygen compared to a high-oxygen atmosphere (see Note 12).

A more constant atmosphere can be provided by using small single-dish incubators, because the lid of the chamber is open for individual needs. The consumption of nitrogen is lower due to the small volume of the chamber. The recovery time for temperature and the gases should also be shorter. Despite many advantages, these incubators also have some shortcomings (see Notes 13 and 14).

3.2. Low Oxygen During Manipulation with Oocytes and Embryos

The maintaining of a constant physiological level of reduced oxygen during an IVF/ICSI procedure is difficult. This is hindered by many manipulations with the oocytes and embryos outside the incubator during oocyte pickup, denudation, ICSI, assessment of morphology and embryo transfer. Constant atmosphere throughout the IVF procedure can only be ensured by incu-chambers or isolettes, but use of these approaches is becoming very rare (see Note 15). In order to reduce the exposure time of the biological material to a high concentration of oxygen, special care must be given to the appropriate organization of work with as little burden as possible on low-oxygen incubators.

3.3. Minimal Requirements for Low-Oxygen Culture in an Extensive IVF Program

The main problem in introducing reduced oxygen into the embryo culture is the fact that all classic CO_2 incubators cannot be replaced overnight. Although, it is usually recommended to use one incubator for no more than four patients, we recently demonstrated that the beneficial effects of low oxygen can be achieved also by just using two low-oxygen incubators (and several classic CO_2 incubators) in a busy clinical program (eight oocyte pickups per day, on average, and all blastocyst culture). This was the beginning of the use of the low-oxygen culture in our laboratory. The method has been clinically tested in a sibling oocyte study (1,916 oocytes analyzed) on the effect of reduced versus atmospheric oxygen on embryo development, resulting in higher blastulation (70.1 % versus 58.5 %) rates in low compared to high oxygen groups (102). The following laboratory protocol was used for the low-oxygen group:

1. During the study, two incubators with reduced oxygen are used, but the others are classic CO_2 incubators—all with four inner doors.
2. Low-oxygen incubators are connected to the bottles of laboratory grade CO_2 and bottles of laboratory grade N_2, providing 5 % O_2, 6 % CO_2, and 89 % N_2. Classic incubators contain 6 % CO_2 in the air.
3. Temperature and gas concentrations in the incubators are recorded continuously and checked daily using an independent gas meter.
4. One low-oxygen incubator is chosen for "culturing" (for embryo cultivation from Day 1 until Day 4); another one is a low-oxygen incubator for Day 5 embryos, waiting for transfer and freezing and for the preincubation of media, while others are "working" incubators with atmospheric oxygen (for the incubation of oocytes during oocyte collection, denudation, embryo evaluation, preparation for embryo transfer or freezing and for the preincubation of media). One classic

incubator is used only for semen preparation. The doors of the "culturing" incubator should be opened as rarely as possible, while "working" incubators can be used more frequently.

5. Oocytes and embryos are cultured in drops of media under equilibrated mineral oil in 4-well dishes. Media are aliquoted per culture dishes, covered with mineral oil and placed in the low-oxygen incubator the day before use.
6. The maintenance of constantly low levels of oxygen is not possible on the beginning of the first day of oocyte culture in vitro due to frequent manipulations with the oocytes out of the incubator (isolation from follicular fluid, denudation, and ICSI). From this reason, all oocytes should be kept in the incubator with 20 % of oxygen until insemination.
7. Oocytes which are inseminated by IVF or ICSI are transferred to low-oxygen preincubated media and kept in a reduced oxygen atmosphere.
8. For embryo evaluation, the shelf with all the dishes containing embryos from patients of the same day are taken out of the culturing incubator and transferred into the working incubator with atmospheric oxygen. After the termination of the embryo evaluation of all patients (maximum 15 min), the dishes are returned to the culturing incubator.
9. The doors of the culturing low-oxygen incubator should be opened for a maximum of 15 times per day.

4. Notes

1. Incubators for maintaining the atmosphere with reduced oxygen are often named triple-gas incubators, but this term is a misnomer since 5 % CO_2 in the air is also triple gas (CO_2, O_2, and N_2). Triple-gas incubators have two gas sensors. To ensure more stable conditions, an infra red (IR) sensor is recommended, and not a thermal conductivity (TC) CO_2 sensor. The IR sensor works independently of humidity and temperature. The oxygen sensor is more sensitive and, in comparison with the CO_2 sensor, its shelf life is usually shorter.
2. Physiologists often refer to the tension of a gas in a liquid, rather than its partial pressure, and define the tension of a gas in a solution as the pressure of this gas in an atmosphere with which that liquid sample is in equilibrium. The major reason for using the word tension is that a dissolved gas exerts no measurable pressure.
3. When purchasing an incubator with oxygen control, it should be noted that an oxygen concentration lower than that in the

atmosphere will be used. Although, these types of incubators are usually named "triple-gas" incubators, they are supplied by only two gases—CO_2 and N_2—to ensure 5 % O_2 and 6 % CO_2 in the atmosphere. Therefore, the bottle of oxygen should not be connected and the manometer for oxygen is also not required.

4. When setting the gas values in the incubator menu, the altitude has to be taken into consideration. This is important, since there is a relationship between altitude and atmospheric pressure. At sea level, where atmospheric pressure is 760 mmHg and 20.95 % of this is oxygen, the partial pressure of oxygen is 159 mmHg. At higher altitudes the atmospheric pressure decreases, but the pO_2 must be maintained for proper equilibrium.
5. Culture dishes with a medium can be put in a plastic bag and gassed. The gassing of bags (or tubes with embryos) for a low-oxygen culture for 2 min was sufficient to equilibrate oxygen levels (and pH) in the media. The bags must then be tightly sealed. This approach is simple but the downside to this method is that is rather cumbersome and not suited to a busy laboratory environment.
6. Any prolonged opening of incubator door causes the elevation of oxygen level within incubator. N_2 is used for displacement of air from the chamber to such an extent that it remains only 5 % O_2 inside and nearly all volume of air must be replaced by N_2. When the oxygen sensor detects the set value of 5 %, the N_2 pumping stops.
7. The N_2 consumption rate in large incubators with the frequent opening of doors can be ten times higher than the consumption of CO_2.
8. "Big-box" low-oxygen incubators should have integrated inner split well-sealed doors to prevent the excessive exchanging of chamber atmosphere.
9. For the conversion of liquid nitrogen in the gas phase and using the gaseous nitrogen in the incubators, an evaporator is needed (see Fig. 3). One liter of liquid nitrogen gives around 700 l of gas. Although liquid nitrogen is not sterile, its purity is higher in comparison with bottled nitrogen.
10. The recovery time for O_2 for triple gas incubator can be longer than the recovery time known from classic CO_2 incubators.
11. Incubators regulate both gases in different ways. Due to the displacement of air with N_2, a CO_2 concentration cannot be adjusted faster than the O_2 concentration. The quality of the incubator with reduced oxygen is therefore reflected in the good levels of synchronization between the pushing out of excess O_2 and the recovery of the CO_2 deficit. Therefore, it is recommended that the incubators are equipped with a display that allows for the continuous monitoring of fluctuations in

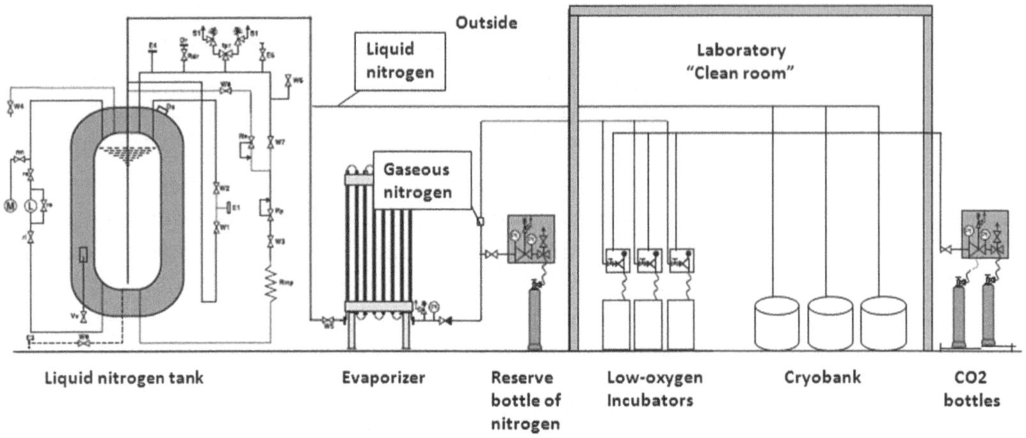

Fig. 3. The scheme of the gas installation in a laboratory with several large low-oxygen incubators. The source of gaseous nitrogen for incubators is the liquid nitrogen. For such an installation, a large tank for liquid nitrogen is installed outside the building. The liquid phase of nitrogen is drawn to containers in the cryobank, but gaseous nitrogen is obtained by vaporizing the liquid nitrogen. Source: Maribor IVF laboratory.

gas concentrations over linear diagrams. Incubators made by different manufacturers vary significantly in the ability to regulate both gases at the same time, which can be seen from the very different types of curves on their displays (see Fig. 4) (personal experience).

12. The repeated opening of the door in triple-gas incubators with a low ability to regulate and synchronize both gases can lead to worse culture conditions compared to that in a classic CO_2 incubator. A long recovery time for CO_2 can change the pH of the medium. It is therefore recommended that additional measurements of pH in the medium are performed.
13. Each incubator should have an independent measurement of temperature, gas concentrations and the pH of media. A drawback to single-dish mini incubators is in the difficulty controlling the physicochemical parameters within the chamber due to its inaccessible interior.
14. Some mini single-dish low-oxygen incubators operate with premixed gas (5 % O_2, 6 % CO_2, 79 % N_2), which is relatively expensive, though amounts used are low due to the low volume of the incubator.
15. The future of IVF laboratories is in equipment which facilitate the maintenance of a constant atmosphere, temperature, and air purity during the IVF procedure. The drawback of incu-chambers is in a more difficult manipulation of material, which can significantly slow the procedure.

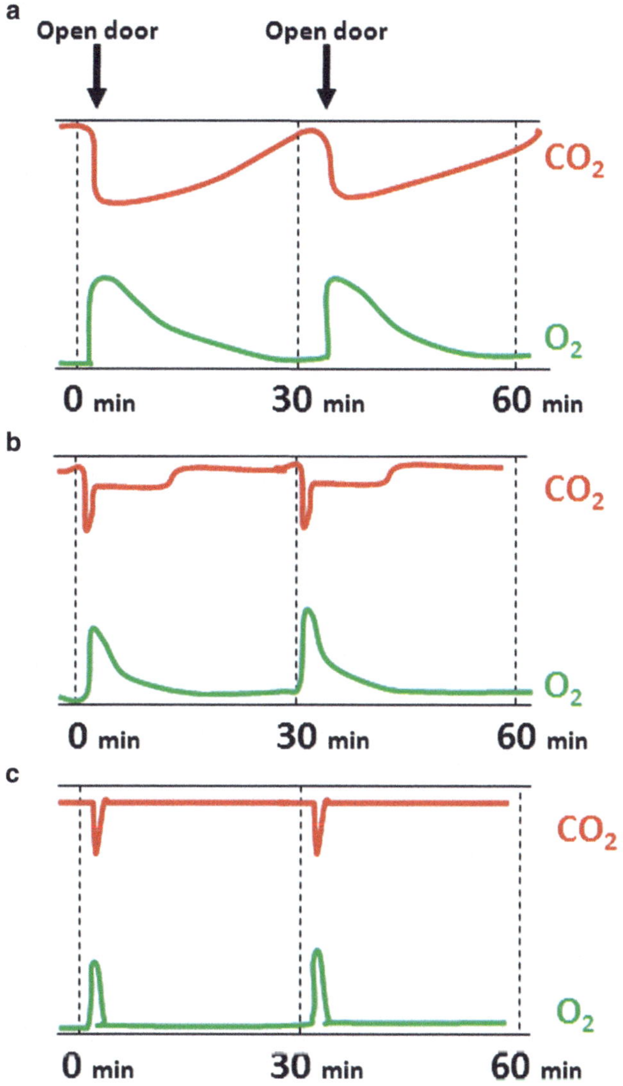

Fig. 4. Different patterns of CO_2 and O_2 recovery after door opening in low-oxygen incubators. (**a**) Recovery rates for O_2 and consequently for CO_2 are too long, since the total air volume in the chamber has to be exchanged by N_2. (**b**) Extensive pumping of N_2 and CO_2 immediately after closing the door causes the rapid initial recovery of both gases. CO_2 is finely adjusted after O_2 is completely recovered. (**c**) Ideal pattern; incubator has the ability to recover the concentrations of both gases very quickly and independently from one another.

References

1. Fischer B, Bavister BD (1993) Oxygen tension in the oviduct and uterus of rhesus monkeys, hamsters and rabbits. J Reprod Fertil 99: 673–679
2. Leese HJ (1995) Metabolic control during preimplantation mammalian development. Hum Reprod Update 1:63–72
3. Thompson JG, Partridge RJ, Houghton FD, Cox CI, Leese HJ (1996) Oxygen uptake and carbohydrate metabolism by *in vitro* derived bovine embryos. J Reprod Fertil 106: 299–306
4. Jauniaux E, Gulbis B, Burton GJ (2003) Physiological implications of the materno-fetal oxygen gradient in human early pregnancy. Reprod Biomed Online 7:250–253
5. Yedwab GA, Paz G, Homonnai TZ, David MP, Kraicer PF (1976) The temperature, pH, and partial pressure of oxygen in the cervix and uterus of women and uterus of rats during the cycle. Fertil Steril 27:304–309
6. Ottosen LD, Hindkaer J, Husth M, Petersen DE, Kirk J, Ingerslev HJ (2006) Observations on intrauterine oxygen tension measured by fibre-optic microsensors. Reprod Biomed Online 13:380–385
7. Van Blerkom J, Antczak M, Schrader R (1997) The developmental potential of the human oocyte is related to the dissolved oxygen content of follicular fluid: association with vascular endothelial growth factor levels and perifollicular blood flow characteristics. Hum Reprod 12:1047–1055
8. Knudsen JF, Litkowski LJ, Wilson TL, Guthrie HD, Batta SK (1978) Concentrations of hydrogen ions, oxygen, carbon dioxide and bicarbonate in porcine follicular fluid. J Endocrinol 79:249–250
9. Shalgi R, Kraicer PF, Sofermanm N (1972) Gases and electrolytes of human follicular fluid. J Reprod Fertil 28:335–340
10. Gosden RG, Byatt-Smith JG (1986) Oxygen concentration gradient across the ovarian follicular epithelium: model, predictions and implications. Hum Reprod 1:65–68
11. Clark AR, Stokes YM, Lane M, Thompson JG (2006) Mathematical modelling of oxygen concentration in bovine and murine cumulus–oocyte complexes. Reproduction 131: 999–1006
12. Martin KL (2000) Nutritional and metabolic requirements of early cleavage stage embryos and blastocysts. Hum Fertil (Camb) 3:247–254
13. Thompson JG, Sherman ANM, Allen NW, McGowan LT, Tervit HR (1998) Total protein content and protein synthesis within pre-elongation stage bovine embryos. Mol Reprod Dev 50:139–145
14. Byatt-Smith JG, Leese HJ, Gosden RG (1991) An investigation by mathematical modelling of whether mouse and human preimplantation embryos in static culture can satisfy their demands for oxygen by diffusion. Hum Reprod 6:52–57
15. Feil D, Lane M, Roberts CT, Kelley RL, Edwards LJ, Thompson JG, Kind KL (2006) Effect of culturing mouse embryos under different oxygen concentrations on subsequent fetal and placental development. J Physiol 572:87–96
16. Kind KL, Collett RA, Harvey AJ, Thompson JG (2005) Oxygen-regulated expression of GLUT-1, GLUT-3, and VEGF in the mouse blastocyst. Mol Reprod Dev 70:37–44
17. Thompson JG, McNaughton C, Gasparrini B, McGowan LT, Tervit HR (2000) Effect of inhibitors and uncouplers of oxidative phosphorylation during compaction and blastulation of bovine embryos cultured *in vitro*. J Reprod Fertil 118:47–55
18. Baltz JM, Biggers JD (1991) Oxygen transport to embryos in microdrop cultures. Mol Reprod Dev 28:351–355
19. Trimarchi JR, Liu L, Porterfield DM, Smith PJ, Keefe DL (2000) Oxidative phosphorylation-dependent and -independent oxygen consumption by individual preimplantation mouse embryos. Biol Reprod 62:1866–1874
20. Houghton FD (2006) Energy metabolism of the inner cell mass and trophectoderm of the mouse blastocyst. Differentiation 74:11–18
21. Scott L, Berntsen J, Davies D, Gundersen J, Hill J, Ramsing N (2008) Human oocyte respiration-rate measurement—potential to improve oocyte and embryo selection? RBM Online 17:461–469
22. Lopes AS, Larsen LH, Ramsing N, Lovendahl P, Raty M, Peippo J, Greve T, Callesen H (2005) Respiration rates of individual bovine *in vitro*-produced embryos measured with a novel, non-invasive and highly sensitive microsensor system. Reproduction 130: 669–679
23. Lopes AS, Madsen SE, Ramsing NB, Løvendahl P, Greve T, Callesen H (2007) Investigation of respiration of individual bovine embryos produced in vivo and *in vitro*

and correlation with viability following transfer. Hum Reprod 22:558–566

24. Lopes AS, Lane M, Thompson JG (2010) Oxygen consumption and ROS production are increased at the time of fertilization and cell cleavage in bovine zygotes. Hum Reprod 25:2762–2773
25. Guerin P, El Mouatassim S, Menezo Y (2001) Oxidative stress and protection against reactive oxygen species in the pre-implantation embryo and its surroundings. Hum Reprod Update 7:175–189
26. Bielski BH, Arudi RL (1983) Preparation and stabilization of aqueous/ethanolic superoxide solutions. Anal Biochem 133:170–178
27. Martín-Romero FJ, Miguel-Lasobras EM, Domínguez-Arroyo JA, González-Carrera E, Alvarez IS (2008) Contribution of culture media to oxidative stress and its effect on human oocytes. Reprod Biomed Online 17: 652–661
28. Grzelak A, Rychlik B, Bartosz G (2001) Light-dependent generation of reactive oxygen species in cell culture media. Free Radic Biol Med 30:1418–1425
29. Halliwell B, Aruoma OI (1991) DNA damage by oxygen-derived species. Its mechanism and measurement in mammalian systems. FEBS Lett 281:9–19
30. Batty E, Jensen K, Freemont P (2009) PML nuclear bodies and their spatial relationships in the mammalian cell nucleus. Front Biosci 14:1182–1196
31. Murphy MP (2009) How mitochondria produce reactive oxygen species. Biochem J 417:1–13
32. Kowaltowski AJ, Vercesi AE (1999) Mitochondrial damage induced by conditions of oxidative stress. Free Radic Biol Med 26:463–471
33. Tappel AL (1980) Vitamin E and selenium protection from in vivo lipid peroxidation. Ann N Y Acad Sci 355:18–31
34. Nasr-Esfahani M, Johnson MH, Aitken RJ (1990) The effect of iron and iron chelators on the *in-vitro* block to development of the mouse preimplantation embryo: BAT6 a new medium for improved culture of mouse embryos *in vitro*. Hum Reprod 5:997–1003
35. Orrenius S, Burkitt MJ, Kass GE, Dypbukt JM, Nicotera P (1992) Calcium ions and oxidative cell injury. Ann Neurol 32(Suppl): S33–42
36. Halliwell B, Gutteridge JM (1990) Role of free radicals and catalytic metal ions in human disease: an overview. Methods Enzymol 186:1–185
37. Tarín JJ (1996) Potential effects of age-associated oxidative stress on mammalian oocytes/ embryos. Mol Hum Reprod 2:717–724
38. Kwon HC, Yang HW, Hwang KJ, Yoo JH, Kim MS, Lee CH, Ryu HS, Oh KS (1999) Effects of low oxygen condition on the generation of reactive oxygen species and the development in mouse embryos cultured *in vitro*. J Obstet Gynaecol Res 25:359–366
39. Salgo MG, Stone K, Squadrito GL, Battista JR, Pryor WA (1995) Peroxynitrite causes DNA nicks in plasmid pBR322. Biochem Biophys Res Commun 210:1025–1030
40. Cox MM, Goodman MF, Kreuzer KN, Sherratt DJ, Sandler SJ, Marians KJ (2000) The importance of repairing stalled replication forks. Nature 404(6773):37–41
41. Marnett LJ (2000) Oxyradicals and DNA damage. Carcinogenesis 21:361–370
42. Taanman JW (1999) The mitochondrial genome: structure, transcription, translation and replication. Biochim Biophys Acta 1410: 102–123
43. Johnson MH, Nasr-Esfahani MH (1994) Radical solutions and cultural problems: could free oxygen radicals be responsible for the impaired development of preimplantation mammalian embryos *in vitro*? Bioessays 16:31–38
44. Blondin P, Coenen K, Sirard MA (1997) The impact of reactive oxygen species on bovine sperm fertilizing ability and oocyte maturation. J Androl 18:454–460
45. Yang HW, Hwang KJ, Kwon HC, Kim HS, Choi KV, Oh KS (1998) Detection of reactive oxygen species (ROS) and apoptosis in human fragmented embryos. Hum Reprod 13:998–1002
46. Pierce GB, Parchment RE, Lewellyn AL (1991) Hydrogen peroxide as a mediator of programmed cell death in the blastocyst. Differentiation 46:181–186
47. Harvey AJ, Kind KL, Thompson JG (2002) REDOX regulation of early embryo development. Reproduction 123:479–486
48. Morales H, Tilquin P, Rees JF, Massip A, Dessy F, Van Langendonckt A (1999) Pyruvate prevents peroxide-induced injury of *in vitro* preimplantation bovine embryos. Mol Reprod Dev 52:149–157
49. Thomas M, Jain S, Kumar GP, Laloraya M (1997) A programmed oxyradical burst causes hatching of mouse blastocysts. J Cell Sci 110:1597–1602
50. Nasr-Esfahani MM, Johnson MH (1991) The origin of reactive oxygen species in mouse embryos cultured *in vitro*. Development 113:551–560

51. El Mouatassim S, Guerin P, Menezo Y (1999) Expression of genes encoding antioxidant enzymes in human and mouse oocytes during the final stages of maturation. Mol Hum Reprod 5:720–725
52. Fridovich I (1995) Superoxide radical and superoxide dismutases. Ann Rev Biochem 64:97–112
53. Catt J, Henman M (2000) Toxic effects of oxygen on human embryo development. Hum Reprod 15(Suppl 2):199–206
54. Enkhmaa D, Kasai T, Hoshi K (2009) Long-time exposure of mouse embryos to the sperm produces high levels of reactive oxygen species in culture medium and relates to poor embryo development. Reprod Domest Anim 44:634–637
55. Alikani M, Cohen J, Tomkin G, Garrisi GJ, Mack C, Scott RT (1999) Human embryo fragmentation *in vitro* and its implications for pregnancy and implantation. Fertil Steril 71:836–842
56. Otsuki J, Nagai Y, Chiba K (2007) Peroxidation of mineral oil used in droplet culture is detrimental to fertilization and embryo development. Fertil Steril 88:741–743
57. Nakayama T, Noda Y, Goto Y, Mori T (1994) Effects of visible light and other environmental factors on the production of oxygen radicals by hamster embryos. Theriogenology 41:499–510
58. Goto Y, Noda Y, Mori T, Nakano M (1993) Increased generation of reactive oxygen species in embryos cultured *in vitro*. Free Radical Biol Med 15:69–75
59. Kitagawa Y, Suzuki K, Yoneda A, Watanabe T (2004) Effects of oxygen concentration and antioxidants on the *in vitro* developmental ability, production of reactive oxygen species (ROS), and DNA fragmentation in porcine embryos. Theriogenology 62:1186–1197
60. Hashimoto S, Minami N, Takakura R, Yamada M, Imai H, Kashima N (2000) Low oxygen tension during *in vitro* maturation is beneficial for supporting the subsequent development of bovine cumulus–oocyte complexes. Mol Reprod Dev 57:353–360
61. Rinaudo PF, Giritharan G, Talbi S, Dobson AT, Schultz RM (2006) Effects of oxygen tension on gene expression in preimplantation mouse embryos. Fertil Steril 86 (4 Suppl):1252–1265
62. Harvey AJ, Kind KL, Thompson JG (2007) Regulation of gene expression in bovine blastocysts in response to oxygen and the iron chelator desferrioxamine. Biol Reprod 77:93–101
63. Semenza GL (2000) Expression of hypoxia-inducible factor 1: mechanisms and consequences. Biochem Pharmacol 59:47–53
64. Quinn P, Harlow GM (1978) The effect of oxygen on the development of preimplantation mouse embryos *in vitro*. J Exp Zool 206:73–80
65. Thompson JG, Simpson AC, Pugh PA, Donnelley PE, Tervit HR (1990) Effect of oxygen concentration on *in-vitro* development of preimplantation sheep and cattle embryos. J Reprod Fertil 89:573–578
66. Pabon JE, Findley WE, Gibbons WE (1989) The toxic effect of short exposures to the atmospheric oxygen concentration on early mouse embryonic development. Fertil Steril 51:896–900
67. Umaoka Y, Noda Y, Narimoto K, Mori T (1992) Effects of oxygen toxicity on early development of mouse embryos. Mol Reprod Dev 31:28–33
68. Wale PL, Gardner DK (2010) Time-lapse analysis of mouse embryo development in oxygen gradients. Reprod Biomed Online 21:402–410
69. McKiernan SH, Bavister BD (1990) Environmental variables influencing *in vitro* development of hamster 2-cell embryos to the blastocyst stage. Biol Reprod 43:404–413
70. Kishi J, Noda Y, Narimoto K, Umaoka Y, Mori T (1991) Block to development in cultured rat 1-cell embryos is overcome using medium HECM-1. Hum Reprod 6:1445–1448
71. Li J, Foote RH, Simkin M (1993) Development of rabbit zygotes cultured in protein-free medium with catalase, taurine, or superoxide dismutase. Biol Reprod 49:33–37
72. Fukui Y, McGowan LT, James RW, Pugh PA, Tervit HR (1991) Factors affecting the in-vitro development to blastocysts of bovine oocytes matured and fertilized *in vitro*. J Reprod Fertil 92:125–131
73. Takahashi Y, Kanagawa H (1998) Effect of oxygen concentration in the gas atmosphere during *in vitro* insemination of bovine oocytes on the subsequent embryonic development *in vitro*. J Vet Med Sci 60:365–367
74. Olson SE, Seidel GE (2000) Reduced oxygen tension and EDTA improve bovine zygote development in a chemically defined medium. J Anim Sci 78:152–157
75. Van Soom A, Yuan YQ, Peelman LJ, de Matos DG, Dewulf J, Laevens H, de Kruif A (2002) Prevalence of apoptosis and inner cell allocation in bovine embryos cultured under different oxygen tensions with or without cysteine addition. Theriogenology 57:1453–1465

76. Higdon HL 3rd, Blackhurst DW, Boone WR (2008) Incubator management in an assisted reproductive technology laboratory. Fertil Steril 89:703–710
77. Batt PA, Gardner DK, Cameron AW (1991) Oxygen concentration and protein source affect the development of preimplantation goat embryos *in vitro*. Reprod Fertil Dev 3:601–607
78. Yuan YQ, Van Soom A, Coopman FO, Mintiens K, Boerjan ML, Van Zeveren A, de Kruif A, Peelman LJ (2003) Influence of oxygen tension on apoptosis and hatching in bovine embryos cultured *in vitro*. Theriogenology 59:1585–1596
79. Lequarre AS, Marchandise J, Moreau B, Massip A, Donnay I (2003) Cell cycle duration at the time of maternal zygotic transition for *in vitro* produced bovine embryos: effect of oxygen tension and transcription inhibition. Biol Reprod 69:1707–1713
80. Bean CJ, Hassold TJ, Judis L, Hunt PA (2002) Fertilization *in vitro* increases nondisjunction during early cleavage divisions in a mouse model system. Hum Reprod 17:2362–2367
81. Harvey AJ, Kind KL, Pantaleon M, Armstrong DT, Thompson JG (2004) Oxygen-regulated gene expression in bovine blastocysts. Biol Reprod 71:1108–1119
82. Enders AC, Boatman D, Morgan P, Bavister BD (1989) Differentiation of blastocysts derived from *in vitro*-fertilized rhesus monkey ova. Biol Reprod 41:715–727
83. Nasr-Esfahani MH, Winston NJ, Johnson MH (1992) Effects of glucose, glutamine, ethylenediaminetetraacetic acid and oxygen tension on the concentration of reactive oxygen species and on development of the mouse preimplantation embryo *in vitro*. J Reprod Fertil 96:219–231
84. Johnston LA, Donoghue AM, O'Brien SJ, Bildt DE (1991) Influence of temperature and gas atmosphere on in-vitro fertilization and embryo development in domestic cats. J Reprod Fertil 92:377–382
85. Betterbed B, Wright RW Jr (1985) Development of one-cell ovine embryos in two culture media under two gas atmospheres. Theriogenology 23:547–53
86. Khurana NK, Niemann H (2000) Effects of oocyte quality, oxygen tension, embryo density, cumulus cells and energy substrates on cleavage and morula/blastocyst formation of bovine embryos. Theriogenology 54:741–756
87. Orsi NM, Leese HJ (2001) Protection against reactive oxygen species during mouse preimplantation embryo development: role of EDTA, oxygen tension, catalase, superoxide dismutase and pyruvate. Mol Reprod Dev 59:44–53
88. Khurana NK, Wales RG (1989) Effects of oxygen concentration on the metabolism of (U-14 C)glucose by mouse morulae and early blastocysts *in vitro*. Reprod Fertil Dev 1:99–106
89. Bermejo-Alvarez P, Lonergan P, Rizos D, Gutiérrez-Adan A (2010) Low oxygen tension during IVM improves bovine oocyte competence and enhances anaerobic glycolysis. Reprod Biomed Online 20:341–349
90. Banwell KM, Lane M, Russell DL, Kind KL, Thompson JG (2007) Oxygen concentration during mouse oocyte *in vitro* maturation affects embryo and fetal development. Hum Reprod 22:2768–2775
91. Park JI, Hong JY, Yong HY, Hwang WS, Lim JM, Lee ES (2005) High oxygen tension during *in vitro* oocyte maturation improves *in vitro* development of porcine oocytes after fertilization. Anim Reprod Sci 87:133–141
92. Oyamada T, Fukui Y (2004) Oxygen tension and medium supplements for *in vitro* maturation of bovine oocytes cultured individually in a chemically defined medium. J Reprod Dev 50:107–117
93. Karja NW, Wongsrikeao P, Murakami M, Agung B, Fahrudin M, Nagai T, Otoi T (2004) Effects of oxygen tension on the development and quality of porcine *in vitro* fertilized embryos. Theriogenology 62:1585–1595
94. Booth PJ, Holm P, Callesen H (2005) The effect of oxygen tension on porcine embryonic development is dependent on embryo type. Theriogenology 63:2040–2052
95. Eppig JJ, Wigglesworth K (1995) Factors affecting the developmental competence of mouse oocytes grown *in vitro*: oxygen concentration. Mol Reprod Dev 42:447–456
96. Corrêa GA, Rumpf R, Mundim TC, Franco MM, Dode MA (2008) Oxygen tension during *in vitro* culture of bovine embryos: effect in production and expression of genes related to oxidative stress. Anim Reprod Sci 104: 132–42
97. Dumoulin J, Vanvuchelen R, Land J, Pieters MH, Geraedts JP, Evers JL (1995) Effect of oxygen concentration on in-vitro fertilization and embryo culture in the human and the mouse. Fertil Steril 63:115–119
98. Dumoulin J, Meijers C, Bras M, Coonen E, Geraedts JP, Evers JL (1999) Effect of oxygen concentration on human in-vitro fertilization

and embryo transfer. Hum Reprod 14: 465–469
99. Behr B, Pool TB, Milki AA, Moore D, Gebhardt J, Dasig D (1999) Preliminary clinical experience with human blastocyst development *in vitro* without co-culture. Hum Reprod 14:454–457
100. Bahceci M, Ciray HN, Karagenc L, Ulug U, Bener F (2005) Effect of oxygen concentration during the incubation of embryos of women undergoing ICSI and embryo transfer: a prospective randomized study. Reprod Biomed Online 11:438–443
101. Kea B, Gebhardt J, Watt J, Westphal LM, Lathi RB, Milki AA, Behr B (2007) Effect of reduced oxygen concentrations on the outcome of *in vitro* fertilization. Fertil Steril 87:213–216
102. Kovačič B, Vlaisavljevič V (2008) Influence of atmospheric versus reduced oxygen concentration on development of human blastocysts *in vitro*: a prospective study on sibling oocytes. Reprod Biomed Online 17:229–236
103. Ciray HN, Aksoy T, Yaramanci K, Karayaka I, Bahceci M (2009) *In vitro* culture under physiologic oxygen concentration improves blastocyst yield on sibling oocytes. Fertil Steril 91(4 Suppl):1459–1461
104. Meintjes M, Chantilis SJ, Douglas JD, Rodriguez AJ, Guerami AR, Bookout DM, Barnett BD, Madden JD (2009) A controlled randomized trial evaluating the effect of lowered incubator oxygen tension on live births in a predominantly blastocyst transfer program. Hum Reprod 24:300–307
105. Waldenstrom U, Engstrom A, Hellberg D, Nilsson S (2009) Low-oxygen compared with high-oxygen atmosphere in blastocyst culture, a prospective randomized study. Fertil Steril 91:2461–2465
106. Nanassy L, Peterson CA, Wilcox AL, Peterson CM, Hammoud A, Carrell DT (2010) Comparison of 5 % and ambient oxygen during days 3–5 of in vitro culture of human embryos. Fertil Steril 93:579–585
107. Kovačič B, Sajko MC, Vlaisavljevič V (2010) A prospective, randomized trial on the effect of atmospheric versus reduced oxygen concentration on the outcome of intracytoplasmic sperm injection cycles. Fertil Steril 94:511–519
108. Bontekoe S, Mantikou E, van Wely M, Seshadri S, Repping S, Mastenbroek S (2011) Low oxygen concentrations for embryo culture in assisted reproductive technologies. Cochrane Database Syst Rev 1. Art. No.: CD008950

Chapter 16

Culture Systems: Embryo Density

Michael L. Reed

Abstract

Embryo density is defined as the embryo-to-volume ratio achieved during in vitro culture; in other words, it is the number of embryos in a defined volume of culture medium. The same density can be achieved by manipulating either the number of embryos in a given volume of medium, or manipulating the volume of the medium for a given number of embryos: for example, a microdrop with five embryos in a 50 μl volume under oil has the same embryo-to-volume ratio (1:10 μl) as a microdrop with one embryo in a 10 μl volume under oil (1:10 μl). Increased embryo density can improve mammalian embryo development in vitro; however, the mechanism(s) responsible for this effect may be different with respect to which method is used to increase embryo density.

Standard, flat sterile plastic petri dishes are the most common, traditional platform for embryo culture. Microdrops under a mineral oil overlay can be prepared to control embryo density, but it is critical that dish preparation is consistent, where appropriate techniques are applied to prevent microdrop dehydration during preparation, and results of any data collection are reliable, and repeatable. There are newer dishes available from several manufacturers that are specifically designed for embryo culture; most are readily available for use with human embryos. The concept behind these newer dishes relies on fabrication of conical and smaller volume wells into the dish design, so that embryos rest at the lowest point in the wells, and where putative embryotrophic factors may concentrate.

Embryo density is not usually considered by the embryologist as a technique in and of itself; rather, the decision to culture embryos in groups or individually is protocol-driven, and is based more on convenience or the need to collect data on individual embryos. Embryo density can be controlled, and as such, it can be utilized as a simple, yet effective tool to improve in vitro development of human embryos.

Key words: Embryo density, Embryo culture, Group culture, Individual culture, Autocrine, Paracrine, Juxtacrine

1. Introduction

During infertility treatment with IVF, embryos are cultured either in groups or singly. How many embryos are cultured together or whether they are cultured individually is often selected by each laboratory for historical reasons or for practical reasons. Some argue that embryos cultured together produce better quality

Gary D. Smith et al. (eds.), *Embryo Culture: Methods and Protocols*, Methods in Molecular Biology, vol. 912,
DOI 10.1007/978-1-61779-971-6_16,

embryos via secretion of embryotrophic, positive signals, others argue that embryos cultured together may either deplete the media of substrates or negatively affect nearby embryos via the transmission of signals. Oftentimes, embryos are cultured singly for the reason that it allows one to continually correlate the morphology or secretome of the embryo throughout culture.

Mammalian embryos are surprisingly durable; they can withstand a variety of culture conditions vastly different from the in vivo environment. Traditional methods for embryo culture can be as simple as placing embryos into a test tube with cell culture medium and maintaining the test tube in an environment with a set temperature and CO_2 concentration; or as complex as placing embryos into agar chips, surgically transferring these chips to oviducts of xenogenous recipients then recovering the embryos at the desired point in time. With a few notable exceptions, embryo culture techniques have remained largely unchanged (1–5).

This chapter focuses on the use of embryo density as a simple, yet practical tool for improving in vitro culture of human embryos, and as such, the chapter includes (1) a review of the literature, to build a knowledge base for understanding *why* embryo density is important, (2) introduction of culture dishes in Subheading 2, that are designed *specifically* for embryo culture, to take advantage of embryo density, and (3) methods for preparing culture dishes for use or study of embryo density, addressing both the traditional, generic petri dishes and the newer generation of culture dishes.

1.1. Embryo Density and Embryo Culture In Vitro

There is evidence that in vitro culture of human embryos can to be improved by increased embryo density; this finding implies (1) that there could be a localized build up of positive factors, e.g., autocrine/paracrine/juxtacrine communication, and/or (2) that increased embryo density limits the diffusion of positive factors away from embryos, compared to culture in larger volumes of culture medium, and/or (3) that increased embryo density facilitates formation of an effective zone of nutrients, metabolites, and oxygen around the embryo or embryos.

Embryo density is defined, for the purposes of this discussion, as the embryo-to-volume ratio achieved during in vitro culture, e.g., the number of embryos in a given volume of culture medium. The ratio can be modified by changing either the number of embryos, or the culture volume. For example, a microdrop with five embryos in a 50 μl volume under oil has the same embryo-to-volume ratio as a microdrop with one embryo in a 10 μl volume under oil. While both methods for increasing embryo density might improve in vitro culture of mammalian embryos, the actual impact, or efficacy of each approach may differ, not so much in terms of obvious morphological metrics, but at the level of transcriptional activity.

It is intriguing to consider that embryo density, like many other facets of IVF that are taken for granted, may have a profound impact on epigenetic normalcy (for additional reading see refs. 6–12). For example, Dumoulin et al. (13) evaluated media from two manufacturers for culture of human embryos after IVF; a simple comparative study that is familiar to every IVF laboratory. Individual embryos were cultured in 5 μl drops of medium (a relatively high embryo density) under oil, for 2–3 days after egg retrieval (embryo transfer occurred on day 2 or 3). The results of the study demonstrated that embryo development, according to select morphological parameters, differed significantly between the two treatments despite the relative similarities in media composition. Interestingly, they found that the birth weights of the infants were significantly different depending upon which culture medium the embryos were exposed to; and for only a very short period of time.

Increasing embryo density is not synonymous with the effects derived from somatic cell coculture, where cell-to-embryo contact may be important, or where the culture medium may be conditioned by factors secreted by various somatic cell types (14, 15). Communal culture of embryos with trophoblast vesicles in the bovine model, for example, is valid for comparative studies on embryo density (16–18) as is culture of individual embryos in the presence of "helper" embryos that have been isolated by embedding them in agarose (19). One of the earliest reports of controlling embryo density, with the intent to improve development (not out of convenience) was the culture of 2-cell hamster embryos through, what was at that time, a virtually impenetrable in vitro 2-cell developmental block (20). Two-cell embryos recovered from oviducts of mated females were grouped in a single microdrop, after which the medium was aspirated until less than a microliter remained, leaving just enough medium to cover the embryos. Division through the 2-cell stage was achieved; however, continued development was not successful; embryos arrested again at 4 cells. Until recently, there has been a paucity of information regarding controlling embryo density for human embryo culture; rather the bulk of the studies have been done with embryos from laboratory, livestock, and domestic animal models (21).

Culturing embryos in groups (also called communal culture) is a common practice; in fact, a recent survey regarding laboratory procedures at ten top-ranked, successful IVF clinics stated that nine of the ten laboratories cultured embryos in groups (22). No specifics regarding the protocols employed by each of these laboratories were provided, but this is not surprising as there are no universally agreed upon protocols to control embryo density, or many other aspects of embryo culture for that matter.

From a practical standpoint, culturing embryos in groups, if data collection on individual embryos is not essential, is logical, convenient, and consistent with managing work load and minimizing costs as fewer disposables may be required. Further, variable embryo density via culturing embryos in groups is an artifact of the need to establish culture dishes well ahead of time, added to not knowing the number of embryos that will be available for culture. There are also many embryologists who are procedurally comfortable with handling embryos in groups, having been trained using poly-ovulatory or superovulated animal models; however, there are added risks to having a larger number of ova/embryos in a single dish.

Culture of embryos individually in defined volumes of medium is also common, but more so for laboratories that strive to maximize data collection, e.g., evaluation of morphological features of individual embryos at specific time points post-ICSI, culture medium collection at specific time points for amino acid or other metabolic evaluations, or for tracking individual embryos after manipulation for genetic analysis. Methods for tracking individual embryos rely, most often, on monitoring embryos in marked positions within a dish, or culture within an individual well, test tube, or capillary tube, though there is a recent notable innovation where individual mouse embryos were tracked by insertion of silicone bar codes into the perivitelline space surrounding the blastomeres (23).

At this point of the discussion, it would be good to address terminology regarding cell-to-cell communication processes; autocrine, paracrine, and juxtacrine represent different modes of communication that can exist between cells, or in the context of preimplantation embryos, between individual blastomeres, between entire embryos, and between cells within embryos. Autocrine communication can be defined as secretion of a chemical or agent by a cell that can be recognized by that same cell that affects a process within that same cell. Paracrine communication can be defined as secretion of a chemical or agent by a cell that affects a process in a different cell that is physically separate but is close in proximity. Juxtacrine communication can be defined as secretion of a chemical or agent by a cell that affects a process in another cell where the two cells are physically attached to each other; the attached cell may or may not be of a different cell type. Autocrine and paracrine communication are also, in general, considered to be receptor-mediated. These definitions may be further confounded by defining each type of communication based on the type of cell secreting or receiving the chemical or agent; specifically autocrine may be used to define communication between the same types of cells that are in close proximity, while paracrine may be used to define communication between different types of cells that are in close proximity.

Putative autocrine/paracrine/juxtacrine factors can be produced by the embryo, and/or the oviduct, and/or the uterine tissues, and for the purposes of a general discussion, the impact that these factors have will depend on the intended target, for example, embryotrophic, oviduct functional modifications, and involvement with implantation processes. For (primarily) monoovulatory species, as example primate, equine, and bovine it is logical that ova would undergo fertilization and development without the benefit of other nearby ova. Embryotrophic factors could be autotrophic, but would also have origins from the maternal environment. For polyovulatory species, as example rodent, porcine, feline, and canine there could be (some) beneficial embryotrophic factors impacting nearby fertilized ova, for a short time at least, but more likely, the majority of the interactions would be between the embryo(s) and the maternal environment. Secretions from fertilized ova can alter the function of the oviduct (24, 25), and secretions of the oviduct can impact development of fertilized ova (26, 27). And like many hormones and growth factors, secretion and circulation may be widespread, where target function depends on the presence of tissue- and/or cell-specific receptors (for additional reading, see refs. (28–43)).

For many human IVF laboratories, the more sophisticated methods for studying embryology, beyond morphological metrics, are simply not available and/or are cost-prohibitive. Ideally, a benchtop analyzer for spent culture medium analysis, if financially feasible, could be placed into each IVF laboratory. There are a number of papers supportive of amino acid and/or nutritional substrate analysis for selection of embryos for transfer where culture medium is sampled, typically, 24 h after embryos are placed into the medium. A profile can be generated to account for changes in the culture medium alone (without embryos), e.g., amino acid degradation, in addition to generation of a protein and amino acid depletion/addition profile for each embryo. By correlating these profiles to previously determined algorithms in combination with morphological data, it is possible to predict, with greater certainty, embryo potential (44–51). Correlating secretome, metabolic, and morphological data for embryo selection will be challenging, however; as there are so many variables associated with in vitro culture of human embryos, it might be difficult to develop a universal embryo selection criterion.

Embryo density may or may not have an obvious impact on the effectiveness of the human embryo culture system; most laboratories, as mentioned above, do not have ready access to newer technologies, and must rely on visible, semi-quantifiable metrics. And the choice of an appropriate metric, or endpoint, is critical. Metrics can be as simple as embryo morphology, speed of development, proportions of good quality embryos at any given time point, or more sophisticated, representing proteomic and metabolomic functions. And so the ability to gather observational data regarding any impact will depend upon the sensitivity of the method used to monitor any given metric, the end-point chosen for any given metric, and the population of patients being studied. For example, the following null hypotheses are presented to contrast the different degrees of sensitivity required for each metric evaluation: "embryo density does not alter the proportion of embryos reaching the blastocyst stage" is insensitive, but typical of simple quality assurance programs, whereas "embryo density does not alter the proportion of blastocysts nor the number of cells (trophectoderm and/or inner cell mass) within embryos that reach the blastocyst stage where the blastocoel volume is at least one-half of the total morphologic volume, at 120 h after intracytoplasmic sperm injection" is better, but can only be achieved in a program where either animal models are available, or there are mechanisms in place for research on spare embryos.

1.1.1. Literature Review

Perhaps the earliest published study demonstrating a *positive* effect of group culture (which is to say, increased embryo density) of human embryos was by Moessner and Dodson (52). Embryos were cultured in 1 ml volumes (organ well dishes) of Ham's F-10 supplemented with processed human serum. Embryos from each of 55 patients were allocated to individual ($n = 156$) or group culture ($n = 179$, with 2–5 embryos per group) with evaluation of the embryos at transfer on day 2 (day 0 = egg retrieval). The authors concluded that group culture enhanced (statistically) cell number and mean embryo score (combination of qualitative morphology score and cell number), though not qualitative morphology score. Almagor et al. (53) reported improved pregnancy outcomes in a selected group of patients when embryos were maintained in groups. Culture of the embryos was in 0.7 ml volumes (organ well dishes) of Ham's F-10 supplemented with processed human cord serum, with random allocation of patients (not embryos within patients) to either individual embryo culture ($n = 42$ patients) or group culture ($n = 49$ patients, 3–6 embryos per group) and evaluation for replacement of embryos on day 2 (day 0 = egg retrieval). Pregnancy rates (sac, embryo, fetal heart) were statistically higher in patients allocated to the group embryo culture treatment, although cell numbers and morphology scores were similar to the control treatment group. These two reports are unique, in that despite the use of very large volumes of culture medium (embryo density ranged from 1:1,000 to 1:116.7 μl) having a larger number of embryos in close proximity improved development.

In contrast, Spyropoulou et al. (54) found no differences between any of the metrics for patients allocated to either group or individual embryo culture. The embryos were cultured in 20 μl drops of EBSS medium supplemented with a serum substitute and human albumin. Allocation of patients (not embryos within patients) to either individual embryo culture ($n = 78$ patients) or group culture ($n = 81$ patients, 3–5 embryos per group) was random, with replacement of embryos on day 2 (day 0 = egg retrieval). No statistically significant differences were found between the two study groups for any of the developmental parameters studied, nor for the proportion of cycles where 2–3 embryos were available for transfer. Additionally, there were no significant differences regarding the proportion of cycles with positive pregnancy tests, clinical pregnancies, or the implantation rates between the two groups. Also, Rijnders and Jansen (55) found no significant differences in blastocyst formation, cleavage arrest and degeneration, or pregnancy and implantation rates between the four culture methods studied; however, the authors did see a trend for a higher proportion of blastocyst formation when embryos were cultured individually in a 5 μl volume. The study compared the effects of group or individual embryo culture, and the effect of incubation volume on development beyond day 3. Embryos were cultured in a mixture of Earle's and Ham's F-10 medium supplemented with a processed plasma solution. Embryos from 25 patients were included ($n = 324$ embryos), cultured in groups (2–4 embryos) in 160–200 μl until day 3, then randomized into (1) small volume group culture (5 μl per embryo, 8–12 embryos per group in a single drop), (2) individual embryo culture in 5 μl volume, (3) individual embryo culture in 160 μl volume, and (4) group embryo culture in 160 μl volume (77 embryos, and the average density was not defined). Embryo evaluation and replacement occurred on day 5 (day 0 = egg retrieval).

Communal culture of human embryos was also studied by Jones et al. (56). Fifty-three patients were recruited over the course of this dynamically changing study; eight culture system variants eventually were evaluated. Zygotes were grouped (2–3) in 10 or 20 μl drops until day 3 (day 0 = egg retrieval), at which time the embryos were either kept in their same grouping, or sorted into groups with similar cell numbers, in 10 or 50 μl drops. Sorting the embryos into groups with similar cell numbers became the standard protocol, as there were no pregnancies in the first protocol with six patient cycles.

Behr et al. (57) utilized group culture for evaluating extended embryo culture without coculture. Embryos were first placed into 150 μl drops of P1 medium supplemented with synthetic serum substitute ($n = 3$–5 embryos per drop; range 1:50–1:30 μl embryo density) followed by embryo replacement on day 3 (day 0 = egg retrieval). Lesser quality, supernumerary embryos ($n = 838$ embryos from 380 patients) were cultured, combining all remaining embryos into a single 150 μl drop of Blastocyst Medium supplemented with synthetic serum substitute. Fifty-four percent of these lesser quality embryos developed to the blastocyst stage by day 5, and were available for cryopreservation. The embryo density for the second phase of culture was not available.

Utilizing a very high embryo density, Ali (58) compared ultra-microdrop (UMD) and continuous ultra-microdrop culture (cUMD) protocols for human embryos to standard microdrop culture. Embryos were individually cultured, under oil in either large drops (20 and 50 μl), or in ultra-microdrops (1.5–2 μl). Embryo transfer occurred on day 2. Also, culture medium renewal (daily) was compared to allowing the medium to remain unchanged during the period of culture. Both the UMD and cUMD methods were superior, in terms of clinical pregnancy and implantation rates in women less than 37 years of age, compared to the routine microdrop culture method. The very high embryo density combined with not changing the culture medium, allowed, in the author's opinion concentration and localization of positive embryotrophic factors.

One possible explanation enhanced embryo development with increased embryo density is a local buildup of positive effectors or factors that are secreted by the embryos, and whose purpose is to communicate with the maternal environment, effectively conditioning the culture medium. Kapiteijn et al. (59) collected and pooled culture medium after exposure to developing embryos (embryo conditioned medium) in an effort to define or identify autocrine/paracrine/juxtacrine factors that could be involved with vascular formation at the time of implantation. Three pooled media (A–C) were identified as having exposure to cleavage stage embryos (2–8 cells, days 1–3, day 0 = egg retrieval) and one pool (Pool D) identified as having exposure to embryos up to the blastocyst stage (day 3–5). The model for the study was a human endometrial microvascular endothelial cell line (hEMVEC) maintained in vitro; this cell line responds, via angiogenic activity, after exposure to VEGF-A and hCG. In the presence of an angiogenic stimulator, the endothelial cells migrate and form capillary structures, or tubes, within a fibrin matrix. The hEMVEC cell system was exposed to the pooled conditioned media. Pool A medium stimulated cell proliferation compared to control medium (not conditioned). Pools B, C, and D elicited a dose response indicating that a soluble factor present in the conditioned media was able to stimulate tube formation, compared to control medium (no tube formation). Challenging the conditioned media with a soluble preparation of VEGF-A receptor eliminated the angiogenic effect of the conditioned media. Finally, the authors demonstrated that VEGF-A was present in the conditioned medium. Taken together, these data indicate that cleaving embryos secrete a cell targeted, biologically active VEGF-A in vitro. And although this study did not attempt to evaluate embryo density, it did demonstrate a "proof of principle", where a positive communication factor can be identified, having been secreted by human embryos in vitro.

To take advantage of localized embryotrophic factors, Vajta et al. (60), employed the WOW format for culture, were small microwells were created in the bottom of a conventional culture well plate for culture of human embryos; in effect creating individual microenvironments, where secretion of any positive factors would be concentrated directly around each embryo. The dimension of a WOW well used in this study was approximately 250 μm wide by 200 μm deep, and concave in shape. The microwells were created by mechanical force, where a sterile round-tip rod was introduced though the mineral oil overlay and culture medium, touching the tip of the rod to the bottom of the well; downward rotational force was applied to the rod, creating small wells in the surface of the dish. There were two studies presented. Study 1 utilized a sibling embryo format where a total of 165 fertilized ova were distributed to either conventional, individual embryo culture or WOW culture. There were no significant differences between the two culture groups with respect to the number of embryos with seven or more cells, nor was there a difference in quality grades on day 3 (day 0 = day of retrieval). Observations on day 5 and 6 indicated that development to blastocyst-stage was significantly better when embryos were maintained in the WOW format. In the second study, all fertilized ova ($n = 347$) were cultured in the WOW format. The blastocyst formation rate was 49.8 % on day 5, and 94 total embryos were transferred to patients, resulting in a clinical pregnancy rate of 48.9 % with a 37 % implantation rate per embryo.

Sequential media are often used for mammalian embryo culture, where the first of the two media are used for culture of embryos from fertilization to day 3, and the second medium is used for culture from day 3 to day 5 or 6. The logic for the use of sequential media is that there are stage-specific nutrient and/or amino acid requirements, and so, by extension, is it possible that embryo density is important only for specific embryonic stage(s) in vitro? Rebollar-Lazaro and Matson (61) evaluated culture of embryos using two formats, (1) group culture up to day 3 (3–5 embryos per group) followed by individual culture to the blastocyst stage on day 5 and 6, and (2) individual culture for entire culture period. The culture medium volume for the study was 15 μl microdrops under oil (1:5 μl up to 1:3 μl, compared to 1:15 μl density for group and individual culture, respectively). The study did not find significant differences between the two treatment groups for clinical pregnancies and implantation rates, and only in women less than 35 years of age were there significantly more blastocysts available from the group culture treatment. The other two possible culture treatments—group culture of all embryos through the entire culture period, and individual culture of embryos up until day 3 following by grouping embryos until day 5 or 6—were unfortunately not included.

Traditional devices for embryo culture include flat-bottom dishes, flat and round-bottom wells, and test tubes; however, there are a few newer products on the market that can be used to take advantage of increased embryo density culture. In a recent prospective study with human embryos, Ebner et al. (62) cultured sibling embryos in three formats using Embryo Corral® dishes. Embryos were cultured individually, grouped but isolated using the fences, and grouped without isolation by fences. Embryo density was based on a 30 μl volume of culture medium; 1:30, 1:7.5, and 1:7.5 μl respectively. Utilizing single blastocyst-stage embryo transfer, where the best embryo was chosen for replacement regardless of culture treatment, it was clearly demonstrated that increasing embryo density via grouping embryos, regardless of whether the embryos were isolated by the plastic posts from each other or not, significantly improved all of the established metrics.

1.2. Maximum Number of Embryos per Volume of Culture Medium Versus Minimum Volume of Culture Medium for Individual Embryos

What is the "best" embryo density for culture of human embryos in vitro? Is communal embryo culture better, worse, or no different from individual embryo culture when embryo density is the same? What volume of medium is ideal for culture of grouped or individual embryos? Could there be consequences to reducing the volume of culture medium below an undefined threshold, on a per-embryo basis? With a very high embryo density, build up of negative factors, e.g., ammonium is one example if the culture medium contains glutamine, or depletion of energy substrates is logical. And when does the balance between positive and negative factors shift?

One recommendation is to not culture human embryos at a density greater than 1:6.25 μl (or in other words, no more than four human embryos per 25 μl volume) to avoid depletion of nutrients and buildup of negative factors in the microdrop (63). These same authors (64) have modified their earlier recommendation to what is now considered by many to be "standard practice", stating that a maximum embryo density of 1:12.5 μl (no more than four human embryos in 50 μl, in addition to changing the culture medium every 48 h) is optimal. Granted, culture media formulations continue to evolve, e.g., replacing glutamine with more stable glutamyl dipeptides, reducing ammonium build-up in the culture medium (1–4) but it can be difficult to determine a correct course of action. A few examples: Rijnders and Jansen (55) utilized 5 μl drops for individual embryos with no detriment, as did Dumoulin et al. (13). Ali (58) utilized drop volumes as low as 1.5–2.0 μl, again without detriment (though embryo transfer occurred on day 2) so it appears that a minimal volume of medium for individual embryo culture has not yet been adequately defined.

1.2.1. Increased Embryo Density and Group Culture

How many embryos can be grouped together in any given volume, and is there a point where embryo density exceeds the ability of the volume to support development, or where negative factors overcome any beneficial group culture effect?

de Oliveira et al. (65) evaluated development of in vitro matured and fertilized bovine ova, where groups of 5, 10, 20, or 30 fertilized ova were cultured in 100 μl of culture medium. Measured endpoints were cleavage rates, development to day 8 (expanded blastocysts) and transcription of HSP 70.1 and Glut-1.

Percent cleavage and development to expanded blastocyst-stage reached a threshold, where these metrics were statistically lower for embryos cultured in groups of 30, compared to 20 or less. Glut-1 expression did not differ between groups, and HSP 70.1 expression was higher with increasing numbers of embryos per microdrop, again until embryo density was over 20 embryos per microdrop; however, the decreased expression of HSP at 30 embryos per microdrop was related to the decreased embryo development in this treatment.

Nagao et al. (66) using bovine embryos, cultured in vitro matured and in vitro fertilized embryos in 50 μl volumes, up to 100 embryos per drop, without detriment to development, and alteration of microdrop volume did not affect development. The study also addressed oxygen concentration, and inclusion/exclusion of inorganic phosphate. For embryos cultured in 50 μl, individually or in groups of five, development was significantly improved by culture at lowered oxygen tension, and without inorganic phosphate. This study also highlights that there can be culture medium by gas atmosphere by embryo density interactions.

With regard to human embryos, a density of 30 embryos per 100 μl or 100 embryos per 50 μl seems improbable. However, in the study presented by Ali (58), the embryo density was as high as nine embryos per 1.5 μl; this would be challenging from a technical standpoint, and equivalent to placing 300 embryos in a 50 μl drop. The culture medium was either replaced daily, or not at all, and it must be pointed out that embryo transfer occurred on day 2. Embryo culture at this density should be undertaken with caution, as nutrient utilization and metabolite buildup is not defined.

Is it possible, for group culture of embryos, that a poor quality or degenerating embryo could have a negative impact on a cohort of healthy embryos? One study addressed this issue directly, using mouse embryos derived via in vitro fertilization. In experiment 1, Salahuddin et al. (67) examined the issue of embryo density (number of embryos per set volume of medium), where 1, 5, 10, or 20 embryos were cultured in groups in 20 μl medium under oil. The beneficial effects of group culture were evident by higher proportions of blastocysts, and reduced time to hatching as embryo density increased. Experiment 2 examined the impact of culturing developing embryos, again in 20 μl drops of medium, with an equal number of unfertilized ova. The proportion of embryos developing to advanced stages (morulae through hatching blastocysts) decreased significantly by the presence of the unfertilized/degenerating ova, compared to controls. Experiment 3 examined development of a single embryo, ten embryos, and ten embryos with ten unfertilized ova, in 20 μl drops. Again, increased embryo density improved development, while coculture of embryos with unfertilized ova negatively affected development of the embryos.

Can there an effect due to differences in developmental stage, e.g., cell number differences when embryos are cultured in groups?

Fertilization and subsequent cell divisions are not synchronous, even with reducing the fertilization interval with ICSI. Jones et al. (56) utilized group culture of embryos, but only after sorting according to developmental stage. Zygotes were grouped (2–3) in 10 or 20 μl drops until day 3 (day 0 = egg retrieval), at which time the embryos were either kept in their same grouping, or sorted into groups with similar cell numbers, in 10 or 50 μl drops. At the end of the study, there were a total of eight protocol variants. Sorting embryos into groups with similar cell numbers became the standard after the first six cycles; no pregnancies resulted in these first cycles, and there was concern that embryos with a low cell number might negatively impact better quality embryos.

Using a large number of in vitro-derived bovine embryos, Larson and Kubisch (68) asked a similar question; do more advanced embryos, e.g., blastocysts, inhibit development of less advanced embryos if cultured together in the same group? In one treatment, embryos advancing to the blastocyst stage were removed to separate drops for continued culture with similarly developing embryos, removing potential influences of an advanced embryo on the slower progressing embryos that remained in the original drop. In a second treatment, all embryos were allowed to remain in group culture, no matter the developmental stage. The authors concluded that the proportions of blastocysts developing in the two groups were not significantly different—there did not appear to be an inhibitory effect of the more advanced embryos on the lesser advanced. Conversely, the presence of the embryos that failed to develop in the groups of embryos did not alter the proportion of blastocysts, nor inhibit hatching of the blastocysts, suggesting that slower progressing embryos also did not inhibit the development of the more advanced embryos. In fact, blastocysts that were allowed to remain in their original groups, and not moved to other drops, exhibited a significantly higher rate of hatching than the counterparts that were moved.

Although not related to embryo density in the traditional sense, there are three references of interest describing (1) the impact of artificially lysed blastomeres on the development of the surviving blastomeres (69), (2) the impact of the pattern of fragmentation, and the benefit of routine removal of fragments on further embryonic development (70), and (3) the benefit of removal of degenerated blastomeres from post-thaw cleavage stage embryos (71). These studies strongly suggest that the close proximity of degenerative material or acellular fragments can negatively affect development of the healthy blastomeres.

1.2.2. Increased Embryo Density and Individual Embryo Culture

What is the minimum volume of culture medium that can be used for a single, individually cultured embryo? Would available nutrients diminish significantly over time, without medium renewal for example? Are embryos cultured individually at greater risk to negative factors?

Ali (58) summarized his work with ultra-microdrop (UMD) and continuous ultra-microdrop culture (cUMD). Embryos were individually cultured, under oil in either large drops (20 and 50 μl), or in ultra-microdrops (1.5–2 μl). Additionally, culture medium renewal (daily) was compared to allowing the medium to remain unchanged during the period of culture. Both the UMD and cUMD methods were superior, resulting in higher clinical pregnancy and implantation rates in women less than 37 years of age, compared to the routine microdrop culture method. The very high embryo density combined with not changing the culture medium, allowed, in the author's opinion concentration and localization of autocrine factors.

Using a microdrop volume of 5 μl, Dumoulin et al. (13) concluded that choice of culture medium (there were two media compared in this study) significantly affected embryo development in vitro to day 2 and 3 (the days for embryo transfer), and birth weight of the infants. While the study did was not designed to specifically study embryo density, the use of the 5 μl volume for individual embryo culture was at least successful; though it is possible that there could be a culture medium formulation by embryo density confounding factor.

A microfluidic platform has recently been described that allows for submicroliter volume embryo culture. Melin et al. (72) designed culture chambers so that valves could be closed on either side, effectively turning the chambers into static culture environments, once the embryos had migrated, or flowed into the chambers. There were four chambers designed; 20, 40, 100, and 500 nl. For this study, two-cell mouse embryos were selected for growth in 100 nl chambers, at a density of two embryos per 100 nl. Controls consisted of mouse embryos cultured in traditional microdrops; two embryos per 5 μl, two embryos per 20 μl, and ten embryos per 20 μl. The development of two embryos in the 100 nl chambers, to the blastocyst stage, was similar to that observed for two embryos cultured in 20 μl (81.8 and 83.3 %, respectively) while development of two embryos in 5 μl was significantly lower, only 3.3 % of all embryos developed to the blastocyst stage. Evaporation of medium during microdrop preparation was suspected for the decreased developmental outcomes in the 5 μl volume.

And recently, Vutyavanich et al. (73) have looked at varying embryo densities by altering (1) the number of embryos per standardized microdrop volume, and (2) altering the volume of medium per standardized number of embryos. Two-cell in vivo-derived embryos from superovulated out-bred ICR mice were used for the three experiments, and metrics were morphological development, differential, and total cell numbers per embryo. For the first experiment 1, 5, 10, and 15 embryos were cultured in 10 μl drops of medium. There were no differences found between embryo density treatments for the basic morphological parameters;

however, there were significantly fewer trophectoderm and inner cell mass cells for embryos cultured individually (the lowest embryo density) compared to the highest embryo density of 15 embryos per 10 μl. Experiment 2 evaluated development and cell numbers of embryos cultured individually in drops of 0.5, 1.0, 2.0, 5.0, and 10.0 μl. Morphological developmental parameters and, surprisingly, inner cell mass cell numbers were not significantly different between culture treatments; however, the number of trophectoderm cells (and hence total number of cells) for embryos cultured in the 0.5 μl volume were significantly lower than the 10 μl volume treatment. Experiment 3 evaluated culture of two embryos per drop, where 0.5, 1.0, and 2.0 μl volumes were utilized. There were no morphological or differential cell number differences observed for embryos cultured, in pairs, in the three volumes of medium. Though not a comparison made by the authors the following is noted, with regard to mean numbers of ICM, Trophectoderm, and total cell numbers, for embryos cultured individually or in pairs, in 0.5, 1.0 μl volumes: (1) 0.5 μl, single embryo; 23.3, 43.1, and 66.4 cells; 0.5 μl, two embryos; 34.1, 59.5, and 93.6, (2) 1.0 μl, single embryo; 23.0, 45.9, and 68.9 cells; 1.0 μl, two embryos; 36.7, 58.1, 94.8 cells, and (3) 2.0 μl, single embryo; 23.3, 50.3, 73.6 cells; 2.0 μl, two embryos; 34.4, 56.4, and 90.9 cells. The pattern appears to indicate that a doubling of embryo density, via communal culture, increased cell numbers per embryo in each of the three volumes. Taking this latter assumption into account, in light of the data as presented, the two different approaches to increasing embryo density may have very different developmental outcomes.

Although embryos cultured individually may benefit from increased embryo density, e.g., reducing the volume of culture medium, there may be instances where individual culture may be less effective than group culture, where embryo toxins for example, override any positive embryo density effect. Hughes et al. (74) demonstrated the protective effect of culturing embryos in groups, effectively reducing the toxicity of cumene hydroperoxide that was added to the culture medium. Embryos cultured individually had significantly reduced development compared to embryos cultured in groups; however, even the protective effect of grouping embryos decreased at higher concentrations of the toxin.

1.3. Dynamic Embryo Culture

Although most of the emphasis in this discussion is, and has been in regard to embryo density and the use of static embryo culture methods, e.g., dishes, wells, test tubes, dynamic embryo culture methods, e.g., microfluidics and tilting-dish culture are briefly introduced here.

At present, microfluidic platform culture is not easily or practically incorporated into the routine IVF laboratory, though in time, the technology may prove to be a practical culture system, where rather than moving ova and sperm to one dish, then moving fertilized ova to another dish or series of dishes with changes in culture medium, it may become feasible to bring capacitated sperm and culture medium to the ova for fertilization and subsequent development, without moving the ova at all (75). The automated culture platforms allow for movement of culture medium around an embryo continuously or in pulses, as well as providing a mechanism for introducing discrete compounds or factors to embryos automatically at any developmental stage or time point. Microfluidic platforms can also be used to study static and dynamic culture methods, without movement of the embryos from the culture device, and can be designed so that spent culture media may be sampled downstream from embryos in culture.

A more practical approach to dynamic embryo culture involves placing culture dishes onto platforms where the rotation and/or degree of tilt of the platform can be carefully controlled. Tilting and/or rotating the culture dish would result in movement of the culture medium and the embryos, eliminating the zone of positive and negative factors that is immediately adjacent to embryos. But also, the movement would expose the embryos to mechanical forces that might be encountered in vivo, including vibration and shear forces (76–78). Microfluidic culture can also be merged with the tilting platform concept to imitate peristaltic movement within the oviduct, as demonstrated by Kim et al. (79) where bovine embryo development was evaluated using two culture methods: (1) a dynamic, tilting microfluid microchannel platform, and (2) standard microdrop culture. Embryo development using the dynamic culture method was superior to the static microdrop culture method.

Using 13–15 embryos per 10 μl volume of culture medium (an embryo density of 1:0.8–1:0.7 μl), Heo et al. (80) compared static microdrop culture to static microfunnel culture (no movement of culture medium in the microfluidic platform) and dynamic microfunnel culture (pulsatile movement of culture medium in the microfluidic platform). Microfunnels, rather than the traditional microchannel design, were built into the microfluidic platform, to physically concentrate the embryos to the center bottom of the funnel, and to limit physical stress or shear forces that might be imposed on embryos if cultured in microchannels. The proportions of embryos developing from in vivo derived mouse embryos to the blastocyst stage (all blastocyst stages combined) were not significantly different between culture treatments; however, the speed of embryo development, based upon the proportion of hatching and hatched blastocysts at an established time point, was improved using microfunnel culture with active movement of culture medium. The number of cells per blastocyst was also significantly higher in the microfunnel with active movement of medium treatment, compared to the static culture methods.

1.4. Static Culture Methods

Devices for embryo culture have evolved away from glass to plastics, though for the most part, these devices have remained relatively unchanged. Test tubes, organ well dishes, watch glasses, vertical fire-polished glass Pasteur pipettes, glass capillary tubes, cell-well plates, and petri dishes of all sizes have been used. A commonality to these devices is that the embryos remain static, in place, inside an incubator of some type and left undisturbed for the most part. With a few exceptions, most laboratories utilize flat, sterile plastic petri dishes, with microdrops of varying volumes and covered in varying volumes of mineral oil. Organ well dishes and cell-well plates, either round or flat bottom, open or covered with oil, are also common.

Several of the newer generation culture dishes designed specifically for embryo culture are introduced; most are readily available commercially, affordable, and have practical applications for embryo density in relation to group and/or individual embryo culture:

GPS Dishware® (Sun IVF, IVFOnline, Guelph, Ontario, Canada), 60 mm petri-style dishes have structured, slanted wells of various sizes, and in one design in particular, the Embryo Corral® dish, is specifically intended for group culture of embryos using a larger well, divided by post, into quadrants of smaller wells that allow sharing of culture medium, but tracking of individual embryos.

The now-commercially available 35 mm WOW dishes (Cryo Innovation Ltd, Budapest, Hungary) accommodate the "well of the well" technique, where there are nine small depressions in the dish for individual embryo culture. More often in the literature, WOW dishes were crafted by the individual embryologist. The commercial WOW dish is designed for traditional embryo culture methods and for use with the Primo Vision time-lapse imaging system (Cryo Innovation Ltd, Budapest, Hungary).

A modified square cell-well plate, the Conical Embryo Culture Plate (Biogenics, Napa California) can be used with either traditional or benchtop incubators, and are uniquely designed with four flat wells and two conical-shape wells, where the conical wells can direct embryos to the center bottom of each well to maximize concentration of any putative embryotrophic factors.

The EmbryoSlide™ was designed specifically for use with the EmbryoScope™ imaging system (Unisense FertiliTech A/S, Denmark), where each device has the shape of a conventional walled microscope slide, with 12 individual wells for embryo culture, per slide. This device is not available separate from the imaging system, but merits mentioning as it is similar, in concept, to the WOW dishes.

The following discussions focus on the physical aspects of embryo density, including communal and individual embryo culture, where "proof of principle" is presented.

1.4.1. Effective "Zone" Surrounding Embryos in Static Culture

Soluble, quantifiable growth factors and other ligands secreted by embryos are purported to be important for embryo development (trophic) and communication between the developing embryos and the maternal environment. And there is direct evidence that culture medium components are modified, or altered to some extent by the embryos themselves, e.g., changes in amino acid profiles across a 24-h period. Regarding static embryo culture, there is good evidence for the presence of a zone, or gradient of dissolved oxygen, potassium, and calcium around the embryo(s), where these gradients, maintained by diffusion, are measurable out to beyond 50 μm from mouse embryos (81, 82). Lopes et al. (83) also demonstrated (using the EmbryoSlide™) that an oxygen gradient, maintained by diffusion, can be measured using bovine embryos cultured individually in 20 μl wells.

Imagine that a static "zone" or concentration gradient will form around each embryo where positive and negative factors would be highest closer to the embryo. Is it possible to take advantage of the different types of culture dishes or systems to limit diffusion of these factors, or to maintain embryos in close proximity, while providing adequate volumes of culture medium? Some comparisons are obvious: micro-wells versus larger wells, flat bottom, curved wells, and macrodrops versus microdrops.

Since the application of microdrop culture with a gas permeable oil overlay by Ralph Brinster in the 1960s (using mouse embryos), this method for culturing embryos has come to dominate the IVF industry, moving beyond organ well and test tube culture. Drops of medium under oil (micro- and macrodrops) tend to form a dome-like structure, where the height of the dome tapers off towards the sides of the drops. The height and shape of the microdrop is limited by the flat surface of the petri dish, the coating on the dish if any, and the weight of the oil used for the overlay. An unintended benefit to culture of embryos in microdrops is that slight motion of the dish, while it is carried from place to place, or gently swirled, tends to move the embryos closer together. There are several new generation culture dishes that are designed to promote maintenance of the embryo(s) in a discreet location at the bottom of a well to take advantage of putative benefits of increased embryo density. The designs of these newer dishes are also useful for applications such as time-lapse computer monitoring of embryo development (84–87), and so there are now several manufacturers of imaging systems declaring support for mammalian embryo development studies, for example: Primo Vision, Cryo Innovation Ltd (Budapest, Hungary); VivaView™, Olympus, Olympus America Inc (Center Valley, PA), USA; EmbryoScope™, Unisense FertiliTech A/S (Aarus N, Denmark).

But how close do embryos need to be to benefit from an autocrine/paracrine/juxtacrine or zone effects, and alternately, how small does a drop or well of medium need to be for individual embryo culture to maintain the beneficial diameter of this zone?

1.4.2. Distance Between Embryos

There are two well-designed studies that looked at the concept that there might be an effective "zone" surrounding an individual embryo, where the circumference of this zone is limited, measurable, and where positive factors might concentrate. The first study involved culture of in vivo and in vitro derived pig embryos. Stokes et al. (88) secured individual embryos at discreet distances from each other in a grid formation—a small culture spot was first prepared on the bottom of the dish with Cell-Tak™, after which the adhesive areas were overlaid with a single 20 μl drop of medium. The endpoints for embryo development were proportion of blastocysts, cell numbers (both inner cell mass and trophectoderm cells), and volume of the blastocysts. In vitro derived embryos developed best when the distance between each embryo was 81–160 μm, and embryos failed to reach blastocyst stages when the distance exceeded 640 μm. In vivo derived embryos developed to the blastocyst stage, individually or in groups, and were less sensitive to the culture distance effect; however, there appeared to be a slight benefit with regard to time to blastocyst formation and maintaining distance between embryos below 481 μm.

In a second similar study, Gopichandran and Leese (89) utilized in vitro derived cattle embryos. Embryos were again secured at measurable, discreet distances in grid formation within drops of medium under oil. The distance between embryos had a significant impact on development, where optimal development occurred when the spacing was no greater than 165 μm, and like the former study results, the beneficial effect diminished with increasing distance. After spacing of more than 540 μm, no blastocysts were formed. Metabolic indicators (glucose uptake, pyruvate uptake, and lactate production) paralleled developmental data, where increasing distance between embryos decreased metabolic efficiency.

In the traditional flat surfaced dish, microdrops of culture medium under an oil overlay form a dome-like structure over the flat surface of the culture dish, and embryos, with some rotational movement of the dish tend move closer together; however, there is no guarantee that embryos will be, or remain close together; and the larger the drop the more likely the embryos are to spread away from each other. One solution to this problem is to place embryos into round bottom or conical wells of culture medium.

Two dish designs promote passive movement of embryos (settling via gravity) into wells, and as a result focus the embryos to a central point where they are in close proximity to each other. The conical well shape, in theory, would also promote a concentration of positive "trophic" factors by limiting diffusion of these factors.

The first example, GPS Dishware® (Sun IVF, IVFOnline, Guelph, Ontario, Canada) consists of three versions of 60 mm petri-style dishes that have structured, slanted wells of various sizes that aid movement of the embryos to the central bottom site, the so-called GPS site (one of these three dish styles is discussed in greater detail in the next section).

The second example is a square, cell-well plate, uniquely suited for use in either traditional or benchtop incubators, has both flat and conical wells. The efficacy of the Conical Embryo Culture Plate (Biogenics, Napa California) was evaluated using bovine embryos derived from in vitro matured and fertilized ova. In three replicates, 565 and 568 fertilized bovine ova were cultured in flat or conical wells, respectively. Embryos were cultured in groups of 10–50 per well, to take advantage of increased embryo density. The proportion of embryos cleaving was not statistically different between the two treatments, 81.9 % cleaved ova and 82.2 % cleaved ova in flat and conical wells, respectively. However, the proportion of transferrable embryos on day 7 of culture (morula and blastocyst stage embryos) was statistically different between the two treatments; 35.2 % of fertilized ova (43.0 % of cleaved ova) developed to morula and blastocyst stages in flat wells, versus 43.1 % of fertilized ova (52.5 % of cleaved ova) developing to morula and blastocyst stages in the conical wells (H. Wei, unpublished data, Trans Ova Genetics, Souix Center, IA). In the context of a commercial bovine embryo production laboratory, the efficiency of production, and hence the number of marketable, viable embryos was increased by almost 19 %, requiring only that embryos are cultured in conical wells rather than flat wells. And given that the embryo density was similar between the two treatments, it appears that the data substantiates the designer's theory that the conical well would benefit embryo development by maintaining embryos in close proximity, and perhaps concentrating embryotrophic substances around the embryos. In the context of a human IVF laboratory, increased efficacy through culture in a conical well as described could result in more embryos of quality sufficient for embryo transfer and/or cryopreservation, with little or no change to existing protocols.

1.4.3 Dishes with Fences

Domestic felines have been studied and modeled extensively, so that assisted reproductive technologies may be better translated or used more effectively in research with endangered species. Two unique studies evaluated the potential benefits of culturing embryos in groups: (1) using embryos of the same species but grouping embryos according to metabolic markers and (2) utilizing group culture of feline embryos with heterospecific embryos, e.g., mouse and cattle embryos. What is unique to these studies is the use of nylon fences secured within the dishes to provide a physical barrier between embryos while allowing free movement of embryo culture medium and soluble components through the fence structure.

In the first study, Spindler and Wildt (90) cultured individual "intermediate" quality, in vitro produced feline embryos in combinations with either (a) groups of low, intermediate, and high quality embryos (as determined at the oocyte stage by glucose uptake) or (b) groups of embryos of different ages. A porous nylon mesh barrier was used to physically isolate the individual intermediate quality embryo from the treatment "companion" embryos, in 20 μl drops of medium. In the first treatment group, a significantly greater proportion of the individual, physically isolated embryos developed to blastocysts, and had greater numbers of cells per embryo, when the companion embryos were derived from the high quality (high glucose uptake) oocytes. Embryos derived from poor quality oocytes did not provide developmental support. In the second treatment group, older companion embryos provided greater support to the individual isolated embryo, compared to age-synchronized embryos and younger embryos, respectively. In the second study, Spindler et al. (91) demonstrated that mouse and cattle embryos could provide a nonspecific factor(s) that improved development of individual, physically isolated feline embryos, maintained in close proximity, but isolated by the fence. In the first treatment group, individual feline embryos were incubated with ten feline, or 10 and 20 mouse embryos. Culture with ten feline or 20 mouse embryos was effective in providing support to the individual embryos. In the second treatment, individual embryos were cultured with ten feline, 20 mouse embryos, or ten cattle embryos. Each variant in this second treatment group was effective at improving development of the individual embryos.

A new, commercially available petri-style dish, available in three versions, has been designed specifically for embryo culture (Universal GPS®, Embryo GPS®, and the Embryo Corral® dishes; Sun IVF, IVFOnline, Guelph, Ontario, Canada). Wells with structured, raised walls were designed onto the bottom of the dish; slanted sides were added to direct the embryo to the center bottom of the well. This center bottom site of each well is called the "GPS" site. The dish can be obtained in three formats; the Universal GPS® dish has three larger off-center wells ringed by smaller wells at the periphery, the embryo GPS® dish has two larger off-center wells ringed by smaller outer wells at the periphery, and the Embryo Corral® dish has two larger off-center wells that are each divided into four equal quadrants by small plastic posts, ringed by smaller wells around the periphery. The larger single wells in each these three designs allows for group culture, but the plastic posts in the Embryo Corral® dish allow for movement of the culture medium and any soluble components between each of the four adjacent wells, while restricting movement of individual embryos between adjacent wells. The distance between the center bottom (the GPS site) of each quadrant well is approximately 4 mm, or 4,000 μm, significantly further than the distances found optimal between embryos, as discussed earlier in this chapter. But do these greater distances impact, or prevent communication between human embryos in adjacent wells?

Neal et al. (92) utilized a sibling embryo culture format for 44 IVF patients, where embryos were randomly allocated, within each patient, to either individual or group culture using the Embryo Corral® dish. Embryo transfer occurred on day 3, where embryos were selected for transfer according to a cumulative embryo score (calculated using cell number and quality score), regardless of culture method. There were no significant differences between culture groups according to the assigned embryo scores; however, the highest pregnancy rates were achieved when embryos were selected exclusively from the group culture treatment. Though the number of patients was limited, this preliminary study demonstrates the concept that increased embryo density can improve embryo viability regardless of no perceived differences in morphological embryo quality.

A recent, prospective well-designed study also compared individual human embryo culture to group embryo culture utilizing the unique properties of the Embryo Corral® dish. Ebner et al. (62) studied sibling embryos, within patients, cultured in three formats; embryos cultured individually, embryos that were grouped but isolated using the fences (plastic posts), and embryos that were grouped without isolation by fences. Embryo density was based on a 30 μl volume of culture medium; 1:30, 1:7.5, and 1:7.5 μl respectively. Outcome measures were embryo development, implantation, and live birth. Utilizing single blastocyst-stage embryo transfer, where the best embryo was chosen for replacement regardless of culture treatment, it was clearly demonstrated that increasing embryo density via grouping embryos, regardless of whether the embryos were isolated from each other or not, significantly improved all of the established metrics. This study demonstrates that not only is increased embryo density, via communal culture of embryos appropriate, but that the dish, as designed, successfully provided a mechanism for individual culture with group culture effects. Also, the individual slant-wall wells may contribute to inhibiting diffusion of positive trophic factors. In consideration for the concerns regarding distance between embryos (see refs. 88, 89) the authors commented on the fact that due to the unique design of the wells, medium in the direct proximity to the embryos at the bottom of each well was moved to the new dish on day 3 along with the embryos, theoretically increasing the opportunity for carrying over any positive factors that might be in the medium.

Another dimension to consider in relation to the aforementioned comments on culture medium carry-over (62) is that renewal of the culture medium, as suggested by Gardner and Lane (64) not only may not be required, but in fact renewal of culture medium, or movement of embryos every 48 h may inadvertently reduce the concentration of any beneficial autocrine/paracrine/juxtacrine, and other positive factors. Continuous culture of human embryos, without culture medium renewal, is not only possible (93), but may be beneficial in light of this discussion.

1.5. Embryo Density; Mircrodrop Size, Shape, and Preparation

Moessner and Dodson (52) and Almagor et al. (53) utilized 1 and 0.7 ml volumes of medium in open cell well dishes, respectively. These volumes, compared to the size of an embryo, are absolutely enormous, and represent an extremely low embryo density per volume. In contrast, O'Neill (94, 95) utilized densities as low as 1:1 μl of medium (ten embryos in 10 μl drops), and Melin et al. (72) created chambers for culture of embryos at nanoliter volumes.

In theory, the shape of a well (conical vs. flat bottom, and round vs. elongated), size of the well (large vs. small vs. microwells), and volume of medium could all contribute to increasing or decreasing diffusion of the mitogenic and autocrine/paracrine/juxtacrine factors.

The "well of the well" or WOW culture system was designed to allow for observation of embryos in extremely small volumes of medium; 0.15 and 0.04 μl of medium (96). An important point regarding the WOW system: the wells are actually small depressions, closely spaced, that are placed onto the bottom of a conventional embryo culture dish or within a cell well of a well dish. This configuration allows for embryos to be placed into the depressions for individual embryo culture. Further, each small depression can be an individually isolated well of its own, where the small volume of medium, e.g., 0.15 μl, is covered by an oil overlay; however, a common format is to form a series of these depressions on the bottom of a dish that is covered by a larger volume of culture medium, for example 500 or 600 μl, which is then covered by the oil overlay creating a contiguous cover of the same culture medium over all of the wells. Each depression allows for physical concentration, perhaps by limiting diffusion, of any autocrine/paracrine/juxtacrine factors and/or localization of metabolic changes, or gradients surrounding the embryo.

The shape of the WOW's can be varied depending on the metal rod used to create the wells. But does the shape of the well have an impact on development? Feltrin et al. (97) modified the WOW technique, to create very narrow microwells for embryo culture. The traditional WOW conical-shaped microwell was compared to the narrow microwells, using cloned bovine embryos. Cleavage and development of the cloned embryos was significantly better in the narrower, modified version of the WOW.

The physical characteristics (volume, diameter, and depth) of WOW-style wells were also evaluated by Hoelker et al. (98), using wells drilled into the bottom of 5-well culture dishes, and culturing individual in vitro derived bovine embryos. The diameter and depth of the wells significantly affected development of the embryos to the blastocyst stages, where deeper and wider wells were beneficial to embryo development; however, cell numbers were not significantly impacted. Culture in the wells was also superior to traditional group culture (16 embryos per culture well, matched to WOW well volumes). The optimal embryo-to-WOW volume ratio was determined to be 1:0.27 μl. With respect to transcriptional activities of bovine embryos cultured at increased embryo density, either by group culture or WOW culture, beneficial effects on both development and abundance of transcripts was improved by both methods of culture, compared to controls (99). Also of interest, transcriptional activity differed according to method of culture, where expression of ATP5A1, PLAC8, and KRT8 by embryos cultured in groups was more similar to expression profiles from matched in vivo derived embryos, and S100A10 and ZP3 expression by embryos cultured individually using the WOW method was more similar to expression profiles of matched in vivo derived embryos; demonstrating that increased embryo density, according to method used, can impact developmental competence.

One proposed alternative to culturing embryos in wells or drops is to utilize small capillary tubes. In an early report on the effects of drop volume and embryo density, Lane and Gardner (100) expressed concern over growing embryos under conditions of the highest surface area to volume ratio, e.g., a 5 μl, dome shaped microdrop. To evaluate this effect, individual embryos were cultured within glass capillary tubes of 5, 10, 20, 40, and 80 μl volumes, contrasted against individual embryo culture in microdrops of 5, 10, 20, 40, 80, 160, and 320 μl. Blastocysts that developed in 5 μl capillary tubes had a significantly higher numbers of cells compared to blastocysts that developed in 5 μl microdrops. Cell numbers per blastocyst were also similar after development in 5, 10, 20, and 40 μl capillary tubes. A similar effect was not found for microdrop culture, where drop sizes of 10 or 20 μl yielded the greatest number of cells per embryo. The authors suggested embryotrophic factors secreted by embryos might be vulnerable to extraction by the lipophilic mineral oil overlay, or that these factors might be denatured more easily, being in close proximity to the medium–oil interface when smaller microdrops, e.g., 5 μl volumes, are used.

In a similar study, mouse embryos were cultured in pairs in small glass capillary tubes, described as "glass oviducts" (hence GO). Thouas et al. (101) cultured embryos in 1 μl volume capillary tubes, open-ended and vertically positioned, for approximately 96 h. Endpoints included development to blastocyst, partial or complete hatching in vitro, number of cells (ICM and trophoblast differentiated), and implantation to day 14.5 p.c. The GO system supported embryonic development in vitro similar to that observed for microdrop "high density" culture (20 embryos per 10 μl), and better than controls (ten embryos per 20 μl); however, implantation rates post culture were similar between treatment groups. A unique aspect to the system is the vertical position of the tube—a micro test tube in reality. The capillaries are described as being 200 μm internal diameter, a little less than twice the diameter of a cleavage stage mouse embryo, and though not commented upon in the text of the reference, one would expect settling of the embryos within the vertical capillary, restricting embryo position within the culture system and minimizing even further the diffusion of positive factors.

The method used to prepare culture dishes is often overlooked, and not evaluated critically, but it is very important to consider the technique, in particular when small volumes of culture medium are pipetted to create microdrops. Swain et al. (102), using a multifactorial design, measured microdrop culture medium osmolality after altering variables associated with culture dish preparation; volume of microdrop (10, 20, and 40 μl), placement of microdrops before or after oil overly, the temperature of the surface where dishes were prepared (room temperature vs. 37 °C), and location relative to air flow (fume hood air flow on vs. fume hood air flow off). Osmolality was measured at 5 min and again at 24 h (osmolality results did not change between these two time intervals). A combination of reduced microdrop volume, increased temperature of the surface where dishes were prepared, and placement of microdrops onto the dish surface before overlaying with mineral oil significantly increased culture medium osmolality by approximately 40 mOsm. Having the air flow on or off in the hood had no significant effect on osmolality alone; however, there was a significant interaction between air flow, reduced culture medium volume, increased temperature of the surface where dishes were prepared, and placing the oil overlay onto the drops after pipetting.

1.6. Concluding Remarks

Mammalian embryo development in vitro can be impacted, positively or negatively, by the density chosen for culture, no matter if the density was derived by choice, or by convenience. There are benefits to culturing embryos individually, in particular for programs that desire to maximize data collection for embryo selection, and culturing embryos in groups also has benefits. But regardless of the style of culture, embryo density should be controlled, if for no other reason than to provide protocol consistency. The evidence in the literature suggests that embryo density impacts development of human, and other mammalian embryos in vitro, where the culture conditions can effect changes in embryonic morphological parameters, and altered cellular processes such as gene transcription. The concept that there are autocrine/paracrine/juxtacrine factors secreted by mammalian embryos in vitro is logical, and demonstrable, though the actual targets for these factors may be oviduct, uterine, and ovarian tissues. In vitro, these putative factors appear to have specific and general positive effects, though negative factors, e.g., cellular toxins, metabolic wastes, reactive oxygen species, and so forth may concentrate, locally, with the use of a static culture system; the challenge then is to provide an in vitro culture environment to maximize localized positive factors, that are not outweighed by the inevitable negative factors.

Specific questions remain unanswered, including, but certainly not limited to: What is the minimum volume of culture medium that can be used for individual embryo transfer? What is the maximum number of embryos that can be placed into group culture in a defined volume of culture medium? Given the same embryo density (embryo-to-volume ratio) are there in vitro developmental differences between embryos cultured individually, or cultured in groups? What are the interactions between the components of the culture medium, the duration of culture without medium renewal, culture medium pH, incubation temperature, incubation atmosphere, and embryo density?

And perhaps the most challenging question, as there is virtually no information for human embryos in vivo: what are the developmental and epigenetic consequences of in vitro culture of human embryos, and what tools, e.g., embryo density, culture medium composition, temperature, atmosphere, and so forth, are appropriate to normalize in vitro embryo culture to the in vivo environment?

2. Materials

2.1. Culture Supplies

Culture media, protein supplementation, mineral oil, pipettes and pipette tips, and other consumables should be selected according to the individual needs of each laboratory. With the exception of some microfluidic platforms and the relatively new generation of plastic dishes, sterile disposable dishes, test tubes, cell well plates, and other devices have all been taken from other applications, adapted for culture of embryos in vitro.

2.2. Specialized Culture Dishes/Devices

The following list of products (and respective manufacturer contact information) is presented for information purposes, and to stimulate interest in the new generation of human embryo culture dishes, some of which are now commercially available; however, this list does not construe endorsement of any product or manufacturer by the author, nor should the list be considered exhaustive, as other similar products may be, or may soon be commercially available.

No manufacturer or brand of "traditional" sterile plastic dishes is mentioned, as there are multiple manufacturers, but note that at least two manufacturers are now offering traditional style dishes that are tested and labeled for use with human embryos.

GPS® Dishware

Sun IVF.

IVFonline.

Guelph, Ontario, Canada.

Conical Embryo Culture Dish
Biogenics.
Napa, California, USA.
Microwell (WOW) Embryo Culture System
Cryo Innovation Ltd.
Budapest, Hungary.
EmbryoSlide™ (Not Available Separate from the EmbryoScope™)
Unisense FertiliTech A/S.
Aarhus N, Denmark.

3. Methods

The purpose of this section is to provide basic, practical information for comparing ease of preparation and unique characteristics of the various dishes, and to encourage embryologists to consider creative alternatives to traditional methods of embryo culture to evaluate embryo density.

There are three embryo culture dishes for which preparation methods are presented in greater detail, as these devices are commercially available and easily obtained: (1) the "traditional" flat sterile plastic petri dish with oil overlay, (2) the Conical Embryo Culture Plate (Biogenics, Napa, California), and (3) the GPS® Dishware dishes (Sun IVF, IVFonline, Guelph, Canada). The Microwell (WOW) Embryo Culture System (Cryo Innovation Ltd, Budapest, Hungary), while very new to the commercial market at the time of this writing (and having limited availability), is discussed in some detail, as is a method for describing preparation of WOW dishes that can be attempted by the investigator. The EmbryoSlide™ (Unisense FertiliTech A/S, Denmark) is not available (yet) separate from the EmbryoScope™; therefore, a description, and limited discussion for preparation of this device are presented.

The respective manufacturer notes and recommendations are current at the time of this writing; however, it is suggested that each laboratory check with the manufacturers for any product updates.

3.1. Good Practices for Preparing Embryo Culture Dishes

Laboratory environments range from actual clean room facilities to "cleaner" room facilities, and so techniques for dish preparation should be practiced, by default, using sterile and aseptic techniques. Do not prepare dishes or culture media under or near ventilation grates even if the incoming air is HEPA filtered. Ideally, prepare dishes in a clean environment, for example in the confines of a HEPA flow hood or cabinet. Running the hood *before* dish preparation is a good idea, but running the hood *during* dish preparation is not advised. Do not use a warm surface for dish preparation, as this may accelerate dehydration of small volumes of medium. Similarly, preparing one dish at a time can help to prevent culture medium dehydration, particularly if small microdrop volumes are used. Prepare at least two dishes per patient to minimize the risk of losing all embryos should there be an incident in the laboratory. Label the dishes clearly (dish and lid) with a minimum of two patient identifier(s), and when in doubt use more identifiers. If you are using dishes that do not have prestamped numbers in the plastic, label the position of each microdrop on the bottom of the dish, in a defined sequence and do not repeat numbers. Alternately, drops may be oriented in a consistent, preplanned pattern, e.g., oriented clockwise compared to patient identifiers on the dish. Inspect the culture medium prior to use for contamination, precipitation, or abnormal color that could indicate that the culture medium is compromised. To pipette culture media, pour or measure a volume of medium into a petri dish, for example, so as not to pipette directly from the culture medium container. Rinse a pipette with a small volume of medium before making drops (or wells), and rinse above the level on the pipette that will be dispensed. For example, if you are dispensing 50 μl drops, rinse the pipette with one or more 100 μl volumes of medium. If an oil overlay is used, select a product that has undergone embryo and/or gamete toxicity testing by the manufacturer. Pipette the oil carefully so as not to disturb the shape of the microdrops, and be consistent with regard to volume of oil used dish to dish; do not pour the oil into the dish, as oil may accumulate on the edge of the dish making dish lid removal more difficult and possibly interfering with gas atmosphere equilibration. The oil may be pipetted before, or after placement of the microdrops (see Note 1). Inspect the culture medium in the dishes before placing embryos into the dish, and during incubation at intervals defined for your laboratory, for contamination, precipitation, or abnormal color that could indicate that the culture medium is compromised.

3.2. Standard 60 mm Flat Sterile Plastic Petri Dishes

Flat sterile plastic petri dishes are available from a number of manufacturers and many if not all of these generic dishes are not tested specifically for embryo toxicity. In recent years, a few manufacturers have offered dishes that are labeled specifically for use in human IVF and as such must meet minimum quality assurance evaluations.

Multiple microdrops per dish are recommended, as embryos should be rinsed through at least one spare microdrop before being placed into the microdrop intended for culture.

The formats listed for the standard 60 mm dishes, using 50 and 15 μl drops are for embryo culture or for use immediately after ICSI. For conventional insemination of oocytes, the same format, with six peripheral and one central drop, can be used; however, the volume of medium for each peripheral microdrop can be increased to 100 μl.

Suggested Procedure for Preparation

Dish format 1—for culturing embryos in groups at a density of 1:16.7–12.5 μl (64).

Six peripheral microdrops—50 μl—for culture of 3–4 embryos per drop.

One central microdrop—50 μl—for rinsing or holding embryos selected for transfer.

1. Rapidly pipette the desired volume of culture medium onto each position on the surface of the labeled dish.
2. Cover the microdrops with 12 ml mineral oil. Pipette the oil carefully so as not to disrupt the structure of the drops.
3. Place the lids onto each labeled culture dish and equilibrate in the incubator overnight.

A variation of this procedure is to pipette the drops of culture medium onto the surface of the dish after the mineral oil has been placed into the dish (see Note 1).

Dish format 2—for culture of individual embryos at a density of 1:15 μl (64).

Six peripheral microdrops—15 μl—for culture of individual embryos.

One central microdrop—50 μl—for rinsing or moving embryos for transfer.

1. Rapidly pipette the desired volume of culture medium onto each position on the surface of the labeled dish.
2. Cover the microdrops with 12 ml mineral oil. Pipette the oil carefully so as not to disrupt the structure of the drops.
3. Place the lids onto each labeled culture dish and equilibrate in the incubator overnight.

A variation of this procedure is to pipette the drops of culture medium onto the surface of the dish after the mineral oil has been placed into the dish (see Note 1).

3.3. Conical Embryo Culture Plate

A volume of 200 μl per conical well is recommended, and use of a mineral oil overlay is also recommended because the volume of medium in the well is small enough to be affected by dehydration (the volume of mineral oil is not defined by the manufacturer; see Note 2). The lid for the dish is unique, with a key and slot format to allow the lid to be more tightly sealed or open to the gas atmosphere. "Open" and "closed" are stamped onto the plastic for "key" placement. Embryos should be rinsed in the equilibrated culture medium in the flat bottom wells, and then gently pipetted onto the center bottom of the conical well(s), which is approximately 1 mm across. The product-insert states that up to 40 embryos can be cultured together in 200 µl of culture medium, which would result in an embryo density of 1:5 µl.

Suggested Procedure for Preparation

1. Prepare the cell well plate according to the manufacturer recommendations.
2. Pipette 1.0 ml of culture medium into each of the flat-bottom wells and the central well.
3. Pipette 200 µl of culture medium into each of two conical wells.
4. Cover each conical well with 1.0 ml of mineral oil. Pipette the oil carefully.
5. Place the lid in the "open" position.
6. Place the dish into the incubator for overnight equilibration.

Rinse the embryos through at least one of the wells containing equilibrated culture medium. Carefully pipette the embryos onto the center bottom of the conical well(s). Replace the lid with the "key" in the "open" position and return the dish to the incubator for culture.

Additional Manufacturer Notes

"Purpose: This plate was designed for culturing embryos from day 1 through the blastocyst stage."

"Species: This plate was designed for embryo culture of all mammalian species, including humans."

"Types of embryos: This plate was designed for culture of embryos derived from in vitro fertilization, nuclear transfer (cloning), and parthenogenesis."

3.4. GPS® Dishware

See the manufacturer package inserts for off-gassing the dishes, preparation of culture medium and oil (not covered in this discussion).

Suggested procedure for preparation:

3.4.1. Universal GPS®

No specific embryo density is suggested for this culture dish; however, in keeping with the theory that embryo density should be no more than four embryos in a 50 µl volume (64) then the larger central wells could each hold 16 embryos, and the outer wells could each hold 12 embryos, although placing fewer embryos per well, and using more dishes is advisable. The larger volumes of medium in each well of this dish make the dish suitable for conventional insemination of oocytes and/or culture of embryos.

1. Pipette up to 150 μl of culture medium into the eight peripheral wells.
2. Pipette up to 200 μl of culture medium into the two central wells.
3. Pipette 13 ml of mineral oil into the dish, carefully so as not to disrupt the medium within the raised wells.
4. Place the lids onto each labeled culture dish and equilibrate in the incubator overnight.

The manufacturer provides a procedure to pipette culture medium into the wells through the oil when using the Universal GPS® dishes; however, the volume of medium may not displace the oil completely, affecting how the embryos settle into the central bottom GPS site of the wells.

3.4.2. Embryo GPS®

No specific embryo density is suggested for this culture dish; however, in keeping with the theory that embryo density should be no more than four embryos in a 50 μl volume (64) then the larger central wells could each hold 12 embryos, and the outer wells could each hold four or five embryos, although placing fewer embryos per well, and using more dishes is advisable. The larger volumes of medium in the central wells of this dish make the dish suitable for conventional insemination of oocytes and/or culture of embryos.

1. Pipette up to 60 μl of culture medium into the eight peripheral wells.
2. Pipette up to 150 μl of culture medium into the three central wells.
3. Pipette 13 ml of mineral oil into the dish, carefully so as not to disrupt the medium within the raised wells.
4. Place the lids onto each labeled culture dish and equilibrate in the incubator overnight.

The manufacturer provides a procedure to pipette culture medium into the wells through the oil when using the embryo GPS® dishes; however, the volume of medium may not displace the oil completely, affecting how the embryos settle into the central bottom GPS site of the wells.

3.4.3. Embryo Corral®

The central wells of this dish are designed so that each quadrant will have an embryo density of 1:30 μl, where four individual embryos, separated by posts, share a total volume of 120 μl culture medium (see Note 3). This dish is less suited towards conventional insemination of oocytes, but better suited for embryo culture.

1. Pipette 6 μl of culture medium into the eight peripheral wells.
2. Pipette 30 μl of culture medium into each quadrant of the two central wells, one well at a time (there are four wells divided

with posts, in each central well). Do not pipette the total volume of 120 μl, as this will result in an uneven distribution of medium in each well.

3. Pipette 13 ml of mineral oil into the dish, carefully so as not to disrupt the medium within the raised wells.
4. Place the lids onto each labeled culture dish and equilibrate in the incubator overnight.

Do *not* pipette culture medium into the wells through the oil when using the Embryo Corral® dishes, as the oil will not be displaced completely by the small volume of culture medium. Improper filling of each well will affect the bridging, or movement of culture medium between quadrants that are separated by the plastic posts (fences), and will affect how the embryos settle into the central bottom GPS site of the wells.

3.5. Microwell (WOW) Embryo Culture System

For many of the studies, WOW dishes are prepared by the investigators. A commercial version of the WOW dish, based on a 35 mm standard flat sterile petri dish, has been developed for use with the Primo Vision imaging system (Cryo Innovation Ltd, Budapest, Hungary). Two methods for dish preparation are presented; the first method is taken from Vajta et al. (60), where the microwells were created by the investigator, and the second method for preparation is taken from Pribenszky et al. (87), a recent case report detailing culture of human embryos in the commercial WOW microwell dishes for time-lapse monitoring. Common to both preparation methods is the use of a contiguous volume of culture medium that covers all of the microwells at once.

The format of the WOW dishes is suited for embryo culture or for use immediately after ICSI; conventional insemination of oocytes should be done in cell wells, test tubes, or microdrops under oil.

3.5.1. Format 1: Preparation of Homemade WOW Dishes by the Investigator

WOW chambers were created manually using Nunc four-well dishes (see Subheadings 2 and 3) (60).

Suggested Procedure for Preparation

1. Pipette 400 μl of culture medium into each well of the culture plate.
2. Pipette 400 μl of mineral oil into each well as an overlay for the culture medium.
3. The tip of a BLS aggregation needle (BLS Ltd., Budapest, Hungary; see Note 4) should be sterilized using 70 % ethanol.
4. Place the tip of the device, through the mineral oil and culture medium, against the bottom of the well, and apply rotational and slight downward pressure on the device tip, creating conical depressions in the surface of the dish. With practice, the size of the individual wells should be approximately 250 μm wide by 200 μm deep.

5. Place the lids onto each labeled culture dish and equilibrate in the incubator overnight.

For culture, pipette one embryo into each well using the smallest volume of culture medium possible. Return the dish to the incubator, taking care not to induce lateral movement of the dish, as the wells are shallow comparatively, and embryos may migrate from the small individual wells.

3.5.2. Format 2: Commercial WOW Dishes Based on the 35 mm Standard Flat Sterile Petri Dish Style

Suggested Procedure for Preparation

1. Slowly pipette 50 μl of culture medium onto the surface of the central well that contains the WOW chambers.
2. Pipette 3 ml mineral oil into the dish, dispensing the mineral oil away from the central well.
3. Inspect the WOW's for the presence of air bubbles, as these may have formed during pipetting of the culture medium.
4. Displace any air bubbles (see Note 5).
5. Place the lids onto each labeled culture dish and equilibrate in the incubator overnight.

For culture, pipette one embryo into each of the WOW chambers using the smallest volume of culture medium possible. There are ten WOW wells are arranged in a 3 × 3 grid with one additional chamber offset centrally. Replace the lid and carefully move the dish to the incubator, taking care not to induce lateral movement of the dish, as the wells are comparatively shallow, and embryos may migrate out of the small individual wells (see Note 6).

3.5.3. EmbryoSlide™ Culture Dish

The format of this dish is suited for embryo culture or for use immediately after ICSI; conventional insemination of oocytes should be done in cell wells, test tubes, or microdrops under oil.

Specifications from the Manufacturer

"12-Numbered wells for incubation of individual embryos in droplets with 25 μl media."

"Separate compartments for embryos and media, under a common oil reservoir."

"Fully compatible with standard and inverted microscopes."

"Standard slide format (25 mm × 75 mm)."

The product is described as a series of 12 wells formed at the bottom of a "polymer slide and cover", where there is a "common oil reservoir" which holds a volume of 1.2 ml oil as an overlay to cover all 12 of the wells. Each well is also designed with a microwell, a 0.2 mm depression to secure the position of each embryo for automated imaging.

4. Notes

1. Dehydration of microdrops is more of a problem with smaller volumes, e.g., for individual embryo culture (102). Smaller microdrops may be beneficial with regard to embryo density; however, there are comments in several of the publications regarding dehydration of microdrops that negatively affected embryo development. Dishes should be prepared on a cool surface, and placement of mineral oil should be done as soon as the microdrops are pipetted for each dish, although depending upon the volume dispensed, this may not protect the osmolality of the culture medium. An alternate approach for preparation of any microdrop, using any standard sterile plastic petri dish (but not the Embryo Corral® dish) is to first pipette the desired volume of culture medium and mineral oil, as described. The microdrops are then aspirated, through the oil, with a sterile pipette tip, effectively removing most of the culture medium. The desired volume of fresh culture medium is then pipetted back through the oil overlay, onto the sites of the previous microdrops. The final microdrop volume will be slightly more than the target volume.

   The volume of mineral oil recommended, 12 ml/60 mm dish (or 13 ml for the GPS® dishware) is the maximum volume of mineral oil, as larger volumes can spill over the rim with movement of the dish. Smaller volumes of mineral oil are acceptable, as long as the microdrops are completely covered, but the overlay acts as gas and heat sink, and maximizing volume of mineral oil will take advantage of this property.

   For dishes smaller than 60 mm, e.g., 35 mm, the concept is the same although fewer microdrops can be used in a dish, and the volume of mineral oil dispensed for the overlay is substantially less as well, 3 ml per dish maximum.

2. The volume of a flat-bottom well in this style of dish is approximately 2.0 ml, leaving no room for an oil overlay. Pipetting a minimum of 0.5 ml, up to 1.0 ml medium into the flat-bottom wells (or the center well), and covering the medium with 0.5 ml mineral oil is feasible. Using 1.0 ml medium and 1.0 ml mineral oil would fill the well to near capacity, and risk pushing oil and/or medium outside of the well with lateral movement of the dish. For the conical wells a volume of 200 μl medium, as suggested by the manufacturer, can be covered by 0.5 ml mineral oil though a volume of up to 1.0 ml mineral oil would provide a greater heat and atmosphere buffer.

   If your incubator does not have a source of humidity (and that would be rare) the flat-bottom wells should not be left uncovered, without oil, as suggested in Subheading 3 unless an osmolality trial determines that the lid of the dish functions to maintain humidity. The conical wells, prepared as suggested require, and are protected by the oil overlay.

3. A comment on the volume of medium for each of the four quadrants of the inner wells of the Embryo Corral® dish is warranted. The manufacturer recommends using 30 μl of culture medium in each of the four quadrants. Higher volumes of culture medium risk creation of a dome over the entire well, where an embryo would be able to migrate over the posts into an adjacent well, if, for example, the dish receives a sharp lateral movement. Do not risk deidentification of embryos, e.g., after biopsy for genetic analysis, by using incorrect volumes of culture medium per well.

   The minimum volume of medium per well should be explored by each laboratory. Note that in a well volume trial in this laboratory, 20 μl per quadrant was the absolute minimum amount of culture medium that could be used, to have any culture medium crossing over into each of the quadrants. Using 25 μl per quadrant, there was a sufficient bridging of culture medium. Do not pipette the entire volume at once, e.g., 120 μl for four quadrants, as there will be unequal distribution of medium between the wells.

4. BLS Ltd. Aggregation needles are used for preparation of WOW microwells. The manufacturer description for the product is "For preparation of aggregation plates for mouse embryo chimera production."

   This device makes small depressions in the bottom of plastic petri dishes using mechanical force. The tip of the device is sterilized with ethanol, after which the needle tip is introduced through the mineral oil overlay and culture medium onto the bottom of the dish; the needle is rotated without twisting the tip, while force is applied downwards. This results in the formation of small conical wells (WOW chambers) in the plastic surface. Also, the manufacturer states that the wells may be made inside small individual microdrops, rather than as described by Vajta et al. (60).

5. If air bubbles are found within the WOW chambers, they can be dislodged by expelling culture medium directly onto the small conical well with a calibrated volume pipette tip. Place the pipette tip directly next to the air bubble for the best results and forcefully expel the culture medium. Repeat this procedure as needed. Allow the pipette to aspirate the volume that was dispensed to help dislodge the air bubble, in order to maintain an approximate starting volume. This procedure may take some practice.

   For a complete product description see www.bls-ltd.com/needles.html.

6. The diameters of the homemade WOW's (60) are smaller (approximately 250 μm) compared to the wells of the commercially manufactured dish, where the diameter was described as rectangle of approximately 550 μm by 450 μm (87); however, the depth of the wells is more similar for both types of dishes; approximately 200 and 170 μm for the homemade and commercial dishes, respectively.

   From a practical standpoint, the rather shallow depth of the wells is more of a concern than the diameters, as embryo migration can occur in both the homemade and commercial WOW dishes, e.g., the embryos can float out of the shallow wells when the dishes are moved. Do not risk deidentification of embryos, e.g., after biopsy for genetic analysis. If WOW's are to be used for biopsied embryos, wells within individual microdrops should be manufactured by the investigator, where each WOW is physically separated and isolated, as described on the Web site for the manufacturer of the BLS Ltd. Aggregation needles.

   For the commercial version of the WOW dish, a 50 μl volume of culture medium was used to cover the centralized WOW's (87). The investigator can try removing small measured amounts of culture medium, e.g., 5 or 10 μl at a time, from the central well after dish preparation, but before placing embryos into the wells, to see if reducing the volume of medium helps to stabilize embryo migration. In a volume trial, and using mouse embryos, in this laboratory, migration of the embryos was minimal when the microdrop volume was approximately 10 μl, though it should be stated that the effort was aimed at *making* the embryos move, rather than moving the dish more cautiously, minimizing lateral movement. Effectively reducing the starting volume and shrinking the contiguous "dome" of medium that covers the wells helps to retain embryos within the wells, by the closer physical association of the culture medium–oil interface. If reducing the microdrop volume is desired, for example to 10 μl final volume, with 9–10 embryos placed individually into the WOW's, then the embryo density could effectively increase to 1:1 μl. Under these circumstances, the investigator should consider culture medium utilization per embryo over time, e.g., consider moving the embryos to a new dish after 48 h, as there are no definitive studies on the minimal volume of culture medium that is required for continuous culture of individual human embryos.

Acknowledgments

The author would like to thank Kimball Pomeroy for sharing his insights into and experiences with mammalian embryo culture, and his critical reading of this manuscript. The author would also like to acknowledge: Biogenics (Napa, CA) and John Haynes (SPI, Inc; Canton, TX) for samples of, and information pertaining to the Conical Embryo Culture Plate, and Hong Wei (Trans Ova Genetics, Sioux Center, IA) for sharing unpublished data re: the Conical Embryo Culture Plate; Hamilton Thorne Inc. (Beverly, MA) along with Gabor Vajta, Cryo Innovations Ltd. (Budapest, Hungary) for samples of, and information pertaining to the 35 mm Microwell (WOW) Embryo Culture Dish; Jacques Cohen, Tyho-Gallileo Research Laboratories (Livingston, NJ) and IVFOnline (Guelph, Ontario, Canada) for samples of, and information pertaining to the GPS® Dishware; Cryo Innovation Ltd (Budapest, Hungary), Olympus, Olympus America Inc (Center Valley, PA), and Unisense FertiliTech A/S (Aarus N, Denmark), for information pertaining to embryo video and time-lapse imaging systems.

References

1. Pool TB (2005) An update on embryo culture for human assisted reproductive technology: media, performance, and safety. Sem Reprod Med 23:309–318
2. Lane M, Gardner DK (2007) Embryo culture medium: which is the best? Clin Obstet Gynecol 21:83–100
3. Gardner DK (2008) Dissection of culture media for embryos: the most important and less important components and characteristics. Reprod Fertil Dev 20:9–18
4. Biggers JD, Summers MC (2008) Choosing a culture medium: making informed choices. Fertil Steril 90:473–483
5. Vajta G et al (2010) Embryo culture: can we perform better than nature? Reprod Biomed Online 20:453–469
6. Khosla S et al (2001) Epigenetic and experimental modifications in early mammalian development: Part II. Culture of preimplantation embryos and its long-term effects on gene expression and phenotype. Hum Reprod Update 7:419–427
7. Longergan P et al (2004) Effect of post-fertilization culture environment on the incidence of chromosome aberrations in bovine blastocysts. Biol Reprod 71:1096–1100
8. Fleming TP et al (2004) The embryo and its future. Biol Reprod 71:1046–1054
9. Fernandez-Gonzalez R et al (2007) Suboptimal in vitro culture conditions: an epigenetic origin of long-term health effects. Mol Reprod Dev 74:1149–1156
10. Giritharan G et al (2007) Effect of in vitro fertilization on gene expression and development of mouse preimplantation embryos. Reproduction 134:63–72
11. Thompson JG, Mitchell M, Kind KL (2007) Embryo culture and long-term consequences. Reprod Fertil Dev 19:43–52
12. Morgan HD et al (2008) The culture of zygotes to the blastocyst stage changes the postnatal expression of an epigenetically labile allele, Agouti Viable Yellow, in mice. Biol Reprod 79:618–623
13. Dumoulin JC et al (2010) Effect of in vitro culture of human embryos on birthweight of newborns. Hum Reprod 25:605–612
14. Orsi NM, Reischl JB (2007) Mammalian embryo co-culture: trials and tribulations of a misunderstood method. Theriogenology 67:441–458
15. Guerin P, Menezo Y (2011) Review: role of tubal environment in preimplantation embryogenesis: application to co-culture assays. Zygote 19:47–54
16. Stojkovic M et al (1995) Secretion of biologically active interferon tau by in vitro-derived bovine trophoblastic tissue. Biol Reprod 53:1500–1507
17. Stojkovic M et al (1997) Support for the development of bovine embryos in vitro by secretions of bovine trophoblastic vesicles derived in vitro. J Reprod Fertil 111:191–196
18. Hashiyada Y, Okada M, Imai K (2005) Transition of the pregnancy rate of bisected bovine embryos after co-transfer with trophoblastic vesicles prepared from in vivo-cultured in vitro-fertilized embryos. J Reprod Dev 51:749–756
19. Senatore EM et al (2010) Improved in vitro development of OPU-derived bovine (*Bos taurus*) embryos by group culture with agaraose-embedded helper embryos. Theriogenology 74:1643–1651

20. Schini SA, Bavister BD (1988) Development of golden hamster embryos through the two-cell block in chemically defined medium. J Exp Zool 245:111–115
21. Reed ML (2006) Communication skills of embryos maintained in group culture – the autocrine/paracrine debate. Clin Embryol 9:5–19
22. Van Voorhis GJ et al (2010) What do consistently high-performing in vitro fertilization programs in the U.S. do? Fertil Steril 94:1346–1349
23. Novo S et al (2010) A novel embryo identification system by direct tagging of mouse embryos using silicon-based barcodes. Hum Reprod 26:96–105
24. Georgiou AS et al (2005) Gametes alter the oviductal secretory proteome. Mol Cell Proteomics 4:1785–1796
25. Georgiou AS et al (2007) Modulation of the oviductal environment by gametes. J Proteome Res 6:4656–4666
26. Lee K-F, Yeung WSB (2006) Gamete/embryo-oviduct interactions: implications on in vitro culture. Hum Fertil 9:137–143
27. Aviles M, Gutierrez-Adan A, Coy P (2010) Oviductal secretions: will they be key factors for IVF? Mol Hum Reprod 16:896–906
28. Paria BC, Dey SK (1990) Preimplantation embryo development in vitro: cooperative interactions among embryos and role of growth factors. Proc Natl Acad Sci USA 87:4756–4760
29. Dardik A, Smith RM, Schultz RM (1992) Colocalization of transforming growth factor-α and a functional epidermal growth factor receptor (EGFR) to the inner cell mass and preferential localization of the EGFR on the basolateral surface of the trophectoderm in the mouse blastocyst. Dev Biol 154:396–409
30. Stoddart NR, Roudebush WE, Fleming SD (2001) Exogenous platelet-activating factor stimulates cell proliferation in mouse preimplantation embryos prior to the fourth cell cycle and shows isoform-specific stimulatory effects. Zygote 9:261–268
31. Terada Y et al (1996) Expression of growth hormone receptor in mouse preimplantation embryos. Mol Hum Reprod 12:879–881
32. Paria BC, Song H, Dey SK (2001) Implantation: molecular basis of embryo-uterine dialogue. Int J Dev Biol 45:597–605
33. Sher G et al (2004) Expression of sHLA-G in supernatants of individually cultured 46-h embryos: a potentially valuable indicator of 'embryo competency' and IVF outcome. Reprod Biomed Online 9:74–78
34. Purnell ET et al (2006) Influence of the preimplantation embryo development (PED) gene on embryonic platelet-activating factor (PAF) levels. J Assist Reprod Genet 23:269–273
35. O'Neill C (2008) The potential roles for embryotrophic ligands in preimplantation embryo development. Hum Reprod Update 14:275–288
36. Richter KS (2008) The importance of growth factors or preimplantation embryo development and in-vitro culture. Curr Opin Obstet Gynecol 20:292–304
37. Roberts RM et al (2008) Interferons and the maternal-conceptus dialog in mammals. Semin Cell Dev Biol 19:170–177
38. Bazer FW et al (2008) Interferons and progesterone for establishment and maintenance of pregnancy: interactions among novel cell signaling pathways. Reprod Biol 8:179–211
39. Contramaestre AP et al (2008) Secretion of stem cell factor and granulocyte–macrophage colony-stimulating factor by mouse embryos in culture: influence of group culture. Zygote 16:297–301
40. Banerjee P, Fazleabas AT (2010) Endometrial response to embryonic signals in the primate. Int J Dev Biol 54:295–302
41. Siemieniuch MJ et al (2010) Are glucocorticoids auto- and/or paracrine factors in early bovine embryo development and implantation? Reprod Biol 10:249–256
42. Cerkiene Z, Eidukaite A, Usoniene A (2010) Immune factors in human embryo culture and their significance. Medicina (Kaunas) 46:233–239
43. Neira JA et al (2010) Effect of the association of IGF-I, IGF-II, bFGF, TGF-beta1, GM-CSF, and LIF on the development of bovine embryos produced in vitro. Theriogenology 15:595–604
44. Houghton FD et al (2002) Non-invasive amino acid turnover predicts human embryo developmental capacity. Hum Reprod 17:999–1005
45. Houghton FD, Leese HJ (2004) Metabolism and developmental competence of the preimplantation embryo. Eur J Obstet Gynecol Reprod Biol 1(115 Suppl 1):S92–S96
46. Brison DR et al (2004) Identification of viable embryos in IVF by non-invasive measurement of amino acid turnover. Hum Reprod 19:2319–2324
47. Booth PJ et al (2005) Amino acid depletion and appearance during porcine preimplantation embryo development in vitro. Reproduction 130:655–668

48. Nagy ZP (2008) Non-invasive assessment of embryo viability by metabolomic profiling of culture media ('metabolomics'). Reprod Biomed Online 17:502–507
49. Seli E et al (2010) Noninvasive metabolomic profiling as an adjunct to morphology for noninvasive embryo assessment in women undergoing single embryo transfer. Fertil Steril 94:535–542
50. Seli E et al (2008) Noninvasive metabolomic profiling of embryo culture media using proton nuclear magnetic resonance correlates with reproductive potential of embryos in women undergoing in vitro fertilization. Fertil Steril 90:2183–2189
51. Marhuenda-Egea FC et al (2010) A crucial step in assisted reproduction technology: human embryo selection using metabolomic evaluation. Fertil Steril 94:772–774
52. Moessner J, Dodson WC (1995) The quality of human embryo growth is improved when embryos are cultured in groups rather than separately. Fertil Steril 64:1034–1035
53. Almagor M et al (1996) Pregnancy rates after communal growth of preimplantation human embryos in vitro. Fertil Steril 66:394–397
54. Spyropoulou I, Karamalegos C, Bolton VN (1999) A prospective randomized study comparing the outcome of in-vitro fertilization and embryo transfer following culture of human embryos individually or in groups before embryos transfer on day 2. Hum Reprod 14:76–79
55. Rijinders PM, Jansen CA (1999) Influence of group culture and culture volume on the formation of human blastocysts: a prospective randomized study. Hum Reprod 14: 2333–2337
56. Jones GM et al (1998) Evolution of a culture protocol for successful blastocyst development and pregnancy. Hum Reprod 13:169–177
57. Behr B et al (1999) Preliminary clinical experience with human blastocyst development in vitro without co-culture. Hum Reprod 14: 454–457
58. Ali J (2004) Continuous ultra micro-drop (cUMD) culture yield higher pregnancy and implantation rates than either larger-drop culture or fresh-medium replacement. Clin Embryol 7:1–23
59. Kapiteijin K et al (2006) Human embryo-conditioned medium stimulates in vitro endometrial angiogenesis. Fertil Steril 85(Suppl 1):1232–1239
60. Vajta G et al (2008) The well-of-the-well system: an efficient approach to improve embryo development. Reprod Biomed Online 17:73–81
61. Rebollar-Lazaro I, Matson P (2010) The culture of human cleavage stage embryos alone or in groups: effect upon blastocyst utilization rates and implantation. Reprod Biol 10:227–234
62. Ebner T et al (2010) Group culture in human is superior to individual culture in terms of blastulation, implantation and life birth. Reprod Biomed Online 21:762–768
63. Gardner DK, Lane M (2001) Embryo culture. In: Gardner DK, Weissman A, Howles CM, Shoham Z (eds) Textbook of assisted reproductive techniques. Laboratory and clinical perspectives. Martin Dunitz Ltd, The Livery House, London, UK
64. Gardner DK, Lane M (2004) Culture of the mammalian preimplantation embryo. In: Gardner DK, Lane M, Watson AJ (eds) A laboratory guide to the mammalian embryo. Oxford University Press, Oxford, New York, USA
65. de Oliveira ATD, Lopes RFF, Rodrigues JL (2005) Gene expression and developmental competence of bovine embryos produced in vitro under varying embryo density conditions. Theriogenology 64:1559–1572
66. Nagao Y, Iijima R, Saeki K (2008) Interaction between embryos and culture conditions during in vitro development of bovine early embryos. Zygote 16:127–133
67. Salahuddin S et al (1995) Fertilization and early embryology: effects of embryo density and co-culture of unfertilized oocytes on embryonic development of in-vitro fertilized mouse embryos. Hum Reprod 9:2382–2385
68. Larson MA, Kubisch HM (1999) The effects of group size on development and interferon-τ secretion by in-vitro fertilized and cultured bovine blastocysts. Hum Reprod 14:2075–2079
69. Alikani M, Olivennes F, Cohen J (1993) Microsurgical correction of partially degenerate mouse embryos promotes hatching and restores their viability. Hum Reprod 8:1723–1728
70. Alikani M et al (1999) Human embryos fragmentation in vitro and its implication for pregnancy and implantation. Fertil Steril 71: 836–842
71. Rienzi L et al (2005) Developmental potential of fully intact and partially damaged cryopreserved embryos after laser-assisted removal of necrotic blastomeres and post-thaw culture selection. Fertil Steril 84:888–894
72. Melin J et al (2009) In vitro embryo culture in defined, sub-microliter volumes. Dev Dyn 238:950–955
73. Vutyavanich T et al (2011) Effect of embryo density and microdrop volume on the blastocyst

development of two-cell embryos. Fertil Steril 95:1435–1439

74. Hughes PM et al (2009) Mouse embryo assay (MEA): group versus individual culture effects on the sensitivity for peroxides in mineral oil. Fertil Steril 92(Suppl):S231
75. Smith GD, Swain JE, Bormann CL (2011) Microfluidics for gametes, embryos, and embryonic stem cells. Sem Reprod Med 29:5–14
76. Matsuura K et al (2010) Improved development of mouse and human embryos using a tilting embryo culture system. Reprod Biomed Online 20:358–364
77. Isachenko E et al (2010) Mechanical agitation during the in vitro culture of human preimplantation embryos drastically increases the pregnancy rate. Clin Lab 56:569–576
78. Koike T et al (2010) In-vitro culture with a tilting device in chemically defined media during meiotic maturation and early development improves the quality of blastocysts derived from in-vitro matured and fertilized porcine oocytes. J Reprod Dev 56:552–557
79. Kim MS et al (2009) A microfluidic in vitro cultivation system for mechanical stimulation of bovine embryos. Electrophoresis 30: 3276–3282
80. Heo YS et al (2010) Dynamic microfunnel culture enhances mouse embryo development and pregnancy rates. Hum Reprod 25: 613–622
81. Trimarchi JR et al (2000) A non-invasive method for measuring preimplantation embryo physiology. Zygote 8:15–24
82. Trimarchi JR et al (2000) Noninvasive measurement of potassium efflux as an early indicator of cell death in mouse embryos. Biol Reprod 62:1866–1874
83. Lopes AS, Lane M, Thompson JG (2010) Oxygen consumption and ROS production are increased at the time of fertilization and cell cleavage in bovine zygotes. Hum Reprod 25:2762–2733
84. Pribenszky C et al (2010) Prediction of in-vitro developmental competence of early cleavage-stage mouse embryos with compact time-lapse equipment. Reprod Biomed Online 20:371–379
85. Wong CC et al (2010) Non-invasive imaging of human embryos before embryonic genome activation predicts development to the blastocyst stage. Nat Biotech 28:1115–1124
86. Sugimura S et al (2010) Time-lapse cinematography-compatible polystyrene based micro well culture system: a novel tool for tracking the development of individual bovine embryos. Biol Reprod 83:970–978
87. Pribenszky C et al (2010) Pregnancy achieved by transfer of a single blastocyst selected by time-lapse monitoring. Reprod Biomed Online 21:533–536
88. Stokes PJ, Abeydeera LR, Leese HJ (2005) Development of porcine embryos in vivo and in vitro; evidence for 'cross talk' in vitro. Dev Biol 284:62–71
89. Gopichandran N, Leese HJ (2006) The effect of paracrine/autocrine interactions on the in vitro culture of bovine preimplantation embryos. Reproduction 131:269–277
90. Spindler RI, Wildt DE (2002) Quality and age of companion felid embryos modulate enhanced development by group culture. Biol Reprod 66:167–173
91. Spindler RE et al (2006) Improved felid embryo development by group culture is maintained with heterospecific companions. Theriogenology 66:82–92
92. Neal MS, et al. (2008) Sibling zygote comparison to determine the effect of group culture on human embryos. Poster presentation for the annual Canadian fertility and andrology (CFAS) meeting. Calgary, Alberta Nov. 27 and 28
93. Reed ML et al (2009) Continuous uninterrupted single medium culture without medium renewal vs. sequential media culture: a sibling embryo study. Fertil Steril 92:1783–1786
94. O'Neill C (1997) Evidence for the requirement of autocrine growth factors for development of mouse preimplantation embryos in vitro. Biol Reprod 56:229–237
95. O'Neill C (1998) Autocrine mediators are required to act on the embryo by the 2-cell stage to promote normal development and survival of mouse preimplantation embryos in vitro. Biol Reprod 58:1303–1309
96. Vajta G et al (2000) New method for culture of zona-included or zona-free embryos: the well of the well (WOW) system. Mol Reprod Dev 55:256–264
97. Feltrin C et al (2005) In vitro bovine embryo development after nuclear transfer by handmade cloning using a modified WOW culture system. Reprod Fertil Dev 18:126
98. Hoelker M et al (2010) Effect of embryo density on in vitro developmental characteristics of bovine preimplantation embryos with respect to micro and macroenvironments. Reprod Domest Anim 45: e138–e145
99. Hoelker M et al (2009) Effect of the microenvironment and embryo density on developmental characteristics and gene expression profile of bovine preimplantative embryos cultured in vitro. Reproduction 137:415–425

100. Lane M, Gardner DK (1992) Effect of incubation volume and density on the development and viability of mouse embryos in vitro. Hum Reprod 7:558–562
101. Thouas GA, Jones GM, Trounson AO (2006) The 'GO' system – a novel method of microculture for in vitro development of mouse zygotes to the blastocyst stage. Reproduction 126:161–169
102. Swain JE et al (2010) Environmental factors and manual manipulations during preparation influence embryo culture medium osmolality. Fert Steril 94:S32

Chapter 17

Culture Systems: Air Quality

Theodore Thomas*

Abstract

Poor laboratory air quality is a known hazard to the culture of human gametes and embryos. Embryologists and chemists have employed analytical methods for identifying and measuring bulk and select air pollutants to assess the risk they pose to the embryo culture system. However, contaminant concentrations that result in gamete or embryotoxicity are poorly defined. Combating the ill effects of poor air quality requires an understanding of how toxicants can infiltrate the laboratory, the incubator, and ultimately the culture media. A further understanding of site-specific air quality can then lead to the consideration of laboratory design and management strategies that can minimize the deleterious effects that air contamination may have on early embryonic development in vitro.

Key words: Air quality, Volatile organic compound, VOC, Laboratory design, Laboratory air, In vitro fertilization, IVF, Airborne contamination, Embryology, Embryotoxicity

1. Introduction

The importance of air quality in the culture of human gametes and embryos began to emerge in the 1990s when reproductive biologists performing human in vitro fertilization (IVF) procedures began to associate episodic declines in clinical outcome with construction or renovation in or near the laboratory. Early anecdotal information led to the theory that laboratory air could be contaminated with volatile compounds or particulates derived from construction activity, outgassing of new materials, or other sources that could diffuse into the culture medium and interfere with gamete and embryo development.

In the first published investigations of air quality in IVF laboratories, clinical laboratory scientists and analytical chemists

*The author would like to disclose relationships with Alpha Environmental, Inc., as a client in 2005 and 2006, and with Origio a/s, as a scientific advisor.

Gary D. Smith et al. (eds.), *Embryo Culture: Methods and Protocols*, Methods in Molecular Biology, vol. 912,
DOI 10.1007/978-1-61779-971-6_17,

employed quantitative methods established in the air quality field to measure volatile organic compounds (VOCs) including alcohols, glycols, phenols, amines, aldehydes, ketones, and chlorinated aromatic compounds. These studies indicated that laboratory and incubator air in most IVF facilities in the late 1990s included volatile contaminates that frequently exceeded those found in many commercial indoor spaces (1, 2). Although these results were clearly a cause for concern, they were not easy to interpret. The effects of bulk or individual air contaminants on gamete and embryo development were then and continue to be poorly understood. Controlled exposure of human embryos to VOCs is unethical and limited data from animal studies (2, 3) does not readily extrapolate to a variety of human embryo culture systems that differ in media composition and volume, oil overlay, and other strategies. Nevertheless, results taken largely from case reports, retrospective, and noncontrolled studies have been used to suggest a few general concepts. The total VOC concentration, which is a predictor of morbidity in humans of all ages (sick building syndrome) (4), is also negatively correlated with clinical outcome in the IVF lab (1, 5). Certain molecular families, such as the aldehydes, are highly toxic to essentially all animal cells (6) and have solubility characteristics that should permit their penetration through an oil layer and into the culture medium. Finally, particulates composed of organic, inorganic, or biological matter remain a threat to the IVF laboratory (7, 8).

Despite the scientific uncertainty associated with the air quality risk, IVF scientists and indoor air experts began around the year 2000 to deploy design concepts aimed at reducing the uptake and distribution of airborne contamination. Many design features were borrowed from clean rooms in the other technologically demanding spaces (1, 7, 9). About the same time, engineering solutions for the removal of a wide range of VOCs from laboratory air became commercially available. Abstracts have accumulated in which improved IVF outcomes follow either the redesign of laboratory ventilation systems or the installation of VOC filtration/inactivation systems (10–14) but this has not been the universal experience (15, 16). Furthermore, many published abstracts are short on detail and difficult to evaluate since new or renovated spaces are often accompanied by other operational changes such as new incubators.

This chapter considers how laboratorians in assisted reproductive technologies can evaluate air quality in their own setting and consider the range of options for defending gametes and embryos from potential airborne toxicity.

1.1. Characterization and Evaluation of Air Pollutants

Outdoor and indoor air quality are both relevant to the function of the ART laboratory.

1.1.1. Outdoor Air

Outdoor air is a primary source of air for the laboratory. The United States Environmental Protection Agency (EPA), under the Clean Air Act updated in 1990, established National Ambient Air Quality Standards (17) for outdoor air. Seven primary or "criteria" pollutants (carbon monoxide, lead, nitrogen dioxide, sulfur dioxide, ozone, and particulate matter in 2.5 and 10 μm sizes) were marked as sentinel measures of air quality that affect human health. Approximately 189 additional hazardous air pollutants (HAPs) have been named under the Integrated Urban Air Toxics Strategy (18). Criteria pollutants and HAPs are the subject of ongoing EPA monitoring programs at stations across the USA and their appearance in outdoor air is associated with the relative presence of heavy industry (mining, manufacturing), power generating facilities, agricultural activity (livestock and pesticides), and population centers (emissions from the combustion of fossil fuels).

Data from the EPA monitoring programs can be a good starting point to gauge the quality of local outdoor air, especially when monitoring stations are geographically relevant to the site of the IVF laboratory (8). EPA data on criteria pollutants and HAPs are released daily or in summary form. State offices of the EPA can provide information on the location of monitoring stations and how local air quality data can be accessed. In most instances, however, site-specific testing is required to gain a more thorough understanding of the quality of the outside air available to the laboratory (see below).

1.1.2. Indoor Air

The EPA supports indoor air quality research broadly and conducts an internal research program at the National Risk Management Research Laboratory (NRMRL) in Research Triangle Park, NC, but has not developed extensive quality standards to date. The World Health Organization (WHO) and the US Occupational Safety and Health Administration (OSHA) have set permissible exposure limits for only a limited set of indoor air contaminants (19, 20). Indoor air quality guidelines and standards are largely based on epidemiological studies of fetal, neonatal, pediatric, and adult humans that are of uncertain relevance to cultured human gametes and preimplantation embryos which lack most of the defense mechanisms present in the advanced organism. It has therefore fallen to IVF laboratorians to acquire the knowledge, tools, and assistance necessary to evaluate and manage lab air quality in their own setting. Some laboratories have collaborated with indoor air specialists who are now active in sufficient number to support organizations such as the International Society of Indoor Air Quality and Climate (ISIAQ) which, since 1991, has organized conferences and published a peer-reviewed journal, *Indoor Air*.

1.2. The Introduction of Airborne Contamination into the IVF Laboratory

The principal vectors of airborne contamination in the IVF lab are (1) the air delivered by the heating, ventilation, and air conditioning or HVAC system; (2) diffusion of volatiles into the lab from adjacent rooms and hallways; (3) off-gassing of materials, equipment, and people in the laboratory; and (4) medical gases that enter the incubator chamber.

1.2.1. HVAC Air

The source air arriving at the HVAC unit can be derived from 100 % outside air, from recirculation of existing indoor air, or a combination of the two. The HVAC unit itself can be a source of particulates and VOCs. Fan lubricants, ductwork coatings and sealers, and other components can off gas for extended periods of time. Systems with inadequate sealing can aspirate particulate, microbial, and volatile components from unregulated spaces. HVAC systems often operate at humidity levels that, without attention to hygiene, can support the growth of microbes that release spores, particulates, and VOCs into the air.

1.2.2. VOCs Generated in the IVF Lab

VOCs generated inside the laboratory are initially high following construction or renovation from outgassing of solvents, paints, adhesives, construction materials, cabinetry, furnishings, and new equipment. The appearance of contamination in laboratory air is related to the vapor pressure of the contaminant, ambient temperature, and the surface area from which it is released. Formaldehyde found in some composite materials and adhesives has a low vapor pressure (3,890 Torr), boils at room temperature, and will be readily released according to the exposed surface area. The initial aldehyde concentration of the author's lab following construction was significant (>17 μg/m^3 by TO-11 testing) but dropped tenfold after heating the space to 32 °C (90 °F) for 15 days.

In an established, operational laboratory, the sustained release of VOCs is expected from consumables such as culture ware, packaging materials, paper, toner and ink cartridges, markers, etc. Cardboard packaging, if allowed into the laboratory, is a vector for chemical, particulate, and microbial contaminants. Personnel can introduce VOCs through personal product usage, clothing, and respirators emissions.

1.2.3. VOCs from Adjacent Spaces

VOCs may enter the laboratory from adjacent spaces through doors, wall and ceiling joints, light fixtures, sprinkler systems, drywall cutouts, etc. Recirculating air handling systems can exacerbate this form of contaminant uptake. IVF labs situated in surgical settings can be especially vulnerable due to the proximal use and storage of strong cleaning and sterilizing agents (often containing aldehydes), operating room consumables, and uncaptured anesthetic gases.

1.2.4. Incubators

Incubators and their contents can be important sources of volatiles and particulates given the proximity to the culture dish. New incubators in particular can produce >100 times the concentration of

VOCs found in mature incubators, presumably from gaskets, plastic components, fan lubricants, etc. (21). Tube racks, equipment, culture dishes, or virtually any item placed inside the incubator chamber can introduce volatile off-gassing.

1.2.5. Medical Gases

Vendors supplying medical gases used in IVF laboratories in the USA must comply with standards of purity established by the United States Pharmacopeia (USP) and National Formulary (NF). USP grades of carbon dioxide and oxygen, and nitrogen NF, must exceed minimum purity levels of 99 % by volume, but often reach levels of purity of 99.65–99.95 %. Schimmel and colleagues were concerned with the nature of medical gas contaminants and reported the presence of benzene, freons, alcohols, acetone, acetaldehyde, and chlorinated organics in their cylinders of USP grade CO_2 (21). Some medical gas vendors also perform routine testing for contaminants such as ammonia, hydrogen sulfide, nitric oxide, carbon monoxide, nitrogen dioxide, and sulfur dioxide. Further study is needed to evaluate the risks that bottled gas contaminants may pose to the embryo culture system, especially as the field transitions to low oxygen culture paradigms in which the majority of the incubator chamber environment is derived from packaged medical gases.

1.3. Airborne Toxicant Intrusion into the Culture Media

Once inside the laboratory or incubator, the passage of a volatile toxicant into the culture medium depends on its concentration and solubility characteristics. The partition coefficient of a molecule is a measure of differential solubility of the compound between two phases. In open culture systems (media with direct exposure to air), toxicants penetrate into culture media according to the air to water partition coefficient. Practically speaking, most volatile molecular species will pass into aqueous culture media to some degree. Open culture systems are therefore especially vulnerable to airborne toxicity.

The majority of assisted reproduction programs overlay the culture medium with mineral or paraffin oil which introduces an additional solubility barrier between the air and the culture medium. With oil overlay airborne toxicants must be soluble in both oil and water to enter into the culture media. The toxic risk of a given airborne molecule can be estimated from the partition coefficients for air to oil (vegetable oil) and oil (octanol) to water. This information is often obtainable from the CRC Handbook of Chemistry and Physics (22).

Although partitioning coefficients can help identify the airborne threats most likely to penetrate to the culture medium, threshold levels at which contaminants cause harm to cultured cells have not been determined in most cases. The reactivity of an individual component is a useful measure of the ability to do cellular harm and is related in part to the half-life of the molecule. The shorter the half-life, the more reactive a molecule is likely to be and the more likely to interfere with biochemical processes in the cell. Chemical reactivity information for more than 3,000 compounds is assembled from peer-reviewed literature on the Hazardous Substances Database managed by the US National Library of Medicine (23).

1.4. Quantitative Methods for Studying Airborne Contamination

The following tests are among the common ones used to evaluate air quality. Testing for volatile contaminates generally requires equipment and expertise found in accredited referral laboratories. These services can be contracted directly or can be facilitated by air quality consultants.

1.4.1. EPA TO-15

EPA TO-15 is a broad-spectrum quantitative VOC test capable of deciphering 97 of the 187 HAPs listed by the EPA (24). VOCs are defined here as organic compounds having a vapor pressure greater than 10^{-1} Torr at 25 °C and 760 mmHg. Sample collection is easily performed by any laboratorian using an evacuated, chemically clean stainless steel canister. A flow controller is used to regulate the uptake of six liters of air over 8–24 h. The canister is then sealed, labeled, and forwarded to the reference laboratory with chain of custody documentation. Most VOCs are stable at room temperature for several weeks. The air specimen is then cryoconcentrated and subjected to gas chromatographic separation. Peaks are identified and analyzed by mass spectroscopy tuned with blanks and controls (4-bromofluorobenzene). The tandem use of gas chromatographic retention time and the generally unique mass fragmentation patterns captured by the detector enhance the ability to accurately identify sample unknowns.

1.4.2. EPA TO-11

EPA TO-11 is a test specifically designed to analyze formaldehyde and 14 other members of the aldehyde and ketone (carbonyl) family (25). A flow pump and purged tubing are used to pass air through dinitrophenylhydrazine (DPNH)-coated silica tubes at a rate of 0.5–2 l/min for 8–24 h. Sample tubes are then capped, labeled, refrigerated along with batch-matched blank DPNH tubes, and sent by overnight courier to the reference lab. Separation and analysis of the now stable hydrazone derivative compounds are performed by isocratic reverse-phase high-performance liquid chromatography (HPLC) with ultraviolet (UV) detection and appropriate use of blanks and controls.

1.4.3. Particulates

Particles ranging in size from 0.5 to 25.0 μm are typically analyzed by instruments that measure dispersion of light generated by a laser beam. Instruments traceable to NIST standards may be acquired for this purpose. Particle counting services are available through indoor air consultants and often by providers of hood certification services. Counters should be mobile enough to analyze particle counts at outlets delivering air into the laboratory.

Bioactive particles can be collected and cultured on blood agar plates. Colonies of bacteria or fungi present on exposed plates can be identified by reference laboratories specializing in microbiology.

1.4.4. Handheld Devices

Handheld devices designed for the detection and quantitation of bulk VOCs or particulates are commercially available (www.research-instruments.com; http://vocmeter.com/; http://galson-labs.com).

1.5. Designs for the Control of Airborne Contamination

The design and management of the IVF laboratory and culture system can reduce the vulnerability of cultured gametes and embryos to airborne forms of toxicity.

1.5.1. Facility Isolation

A well-designed IVF laboratory includes sealed outer walls, ceiling and floor joints, doors, passages, and drywall perforations. This allows for the delivery of highly controlled air by the HVAC and exhaust of used air without appreciable uptake of contamination from adjacent spaces. Facility isolation is most easily accomplished in new construction but is also feasible in some measure as a retrofit. Tracer studies can be used to assess the degree of lab isolation or the effectiveness of implemented remedies (26).

1.5.2. HVAC

Many HVAC design elements can favorably impact air quality but can also be among the more expensive means of addressing air quality improvement. Several of the following concepts are included in the guidelines for IVF laboratories published by the American Society for Reproductive Medicine (27) and the European Society for Human Reproduction and Embryology (28).

Laboratory-Specific Air Handling

The laboratory is ideally placed on a stand-alone air handling system in order to avoid loss of design and operational control which is often the case with a shared ventilation system.

Source Air

Outside air is typically cleaner than indoor air, often by a factor of ten or more. Air quality testing can be performed at the proposed or existing laboratory site to confirm this general principle and to locate the most appropriate location for the source air intake of the HVAC system. Air intakes should be located to avoid exhaust outlets, refuse containers, storage units, and concentrations of vehicular traffic (ground floors, delivery docks, ambulance entrances). The best outdoor air in the author's setting was obtained on the highest vertical aspect of the building facing the prevailing wind. The conditioning of 100 % outside air without recycling of indoor air, although inefficient in some ways, should provide the highest laboratory air quality in most cases. The ability to switch to 100 % recirculated indoor air can be useful when outdoor air is compromised (parking lot resurfacing, fires, etc.).

Positive Pressure

Positive pressure ventilation is generated when the HVAC fan forces air into a room at a rate that creates a pressure differential with an adjacent space. The unit of measure is frequently the air pressure required to displace a column of water vertically (against gravity) in inches. Positive pressure can counteract the admission of uncontrolled air through doorways and other sources. Staged pressurization with the IVF suite can be very useful to isolate the embryology space (maximum pressurization) from other less demanding spaces (lower pressurization), thereby allowing the considered use of stronger reagents in areas that are downstream of the embryology incubators.

Fresh Air Turnover Rate

Fresh air turnover rate is the time required to replace the volume of air in a room with fresh (outdoor) air. This is not to be confused with replenishment by mixtures of outdoor and recycled indoor air. It is common to discuss turnover rate as the number of fresh air changes per hour. HVAC design with a high fresh air turnover rate can reduce the impact of volatile outgassing within the lab. A turnover rate of 10/h (one change every 6 min) is supportive of good indoor air quality. Higher rates are achievable and may lead to higher quality air. Increasing air turnover values can be accompanied by a commensurate increase in noise pollution in the conditioned space.

High-Efficiency Particulate Air Filtration

High-efficiency particulate air (HEPA) filters are composed of woven fibers that trap particles by a variety of mechanisms. HEPA filters are in common use in HVAC systems in IVF laboratories and typically remove the majority (>99.9 %) of particles down to a particular size depending on class, but usually in the range of 0.3 μm.

New Installation

At initial installation HVAC systems are frequently contaminated with volatile off-gassing from sealants and duct coatings, and particulates including biologicals. Meticulous cleaning and sanitation during assembly followed by burn-in at high temperature may reduce the problems associated with new materials. Commissioning of new or renovated HVAC systems should include testing of conditioned air for VOC, particulate, and biological contamination.

Hygiene

The hygiene of existing HVAC systems may degrade over time. A plausible fear is the collection of dirt and microbials on the interior of systems operating at moderate levels of relative humidity (30–50 %). Regular testing of conditioned air for particulate and/or biological contamination is a prudent practice. Particle counters and blood agar plates may be used to evaluate processed air at a short distance from the ceiling register.

Designers and Contractors

For new laboratory construction or renovation, consideration should be given to designers, mechanical contractors, and material suppliers that are versed in indoor air quality issues.

1.5.3. VOC Attenuation

Engineered solutions for VOC removal or inactivation have been deployed in IVF laboratories in various forms.

Molecular Sieving

A combination HEPA and molecular sieving cartridge is available for some incubators manufactured by Thermo Fisher and is placed adjacent to the air circulation fan.

Activated Charcoal and Potassium Permanganate Filtration Media

Pellets of activated charcoal impregnated with potassium permanganate are in widespread use in the indoor air service sector to remove VOCs from air. Charcoal that is "activated" has been treated to increase the porosity of the medium and thereby increase the surface area available for chemical adsorption of hydrocarbons and other VOCs. Potassium permanganate is a strong oxidizing agent which, in immobilized form, degrades many pollutants and is moderately effective in neutralizing the carbonyl family. Activated charcoal/$KMnO_4$ media products are commercially available for air purification in HVAC systems (Purafil, Doraville, GA; Camfil Farr, Stockholm, Sweden), in units designed for room air and incubator interiors (Coda Systems, IVFonline.com, LLC, Guelph, Ontario), as well as in medical gas supply lines (Origio Mid-Atlantic Devices, Mt Laurel, NJ).

A charcoal/$KMnO_4$ filtration device designed for the incubator chamber interior was evaluated by one IVF program in a prospective, randomized, crossover study (13). A significant improvement in clinical pregnancy rate was reported with use of the filtering device (52 % vs. 30 %). However, an open culture system was used by the authors at the time of this work (1998–1999). The absence of an oil overlay may have accentuated the beneficial effect of removing air contamination from the incubator chamber. A prospective randomized trial performed with the same device in another center but in the presence of oil overlay failed to show any effect on patient outcome (15).

Filter beds of activated carbon and permanganate have a useful life that is dependent on many factors including composition, residence time of the passing air, and contaminant load, to name a few. Filter bed media can be evaluated for contaminant loading of the carbon bed and consumption of $KMnO_4$ in order to establish an appropriate renewal schedule in a given laboratory setting. Carbon/$KMnO_4$ depletion testing can be performed by air quality consultants (Alpha Environmental, Emerson, NJ) or by analytical laboratories specializing in air quality assessment (Columbia Analytical Services, Simi Valley, CA).

Ultraviolet Light

UV light has been suggested as a means to attenuate VOC contamination and engineered solutions based on UV treatment of recirculated HVAC air have appeared in the IVF marketplace (LifeAire Systems, LLC., Allentown, PA; Zandair, Inc., Vero Beach, FL).

1.5.4. Culture System

Incubators

The IVF incubator, especially when new, is a potential source of VOCs. Mouse embryo testing of new incubators often reveals the need for a burn-in period prior to the ability of the incubator to support optimum culture results. Operation of new incubators at maximal temperature in a place remote from ongoing embryo culture may shorten the burn-in period. The water reservoir is a potential source of off-gassing, depending on water quality. It is common to use double-distilled or higher grades of water for incubator humidification. The water pan is also believed to be a reservoir for hydrophilic airborne molecules including the alcohol family. Microbial contamination of the water pan can result in the release of VOCs. Counter measures include frequent water changes and the addition of bacteriostatic substances directly to the water, such as copper ion (short section of copper pipe) or EDTA.

Oil Overlay

A primary defense of the developing gamete or embryo against airborne toxicants is the use of an oil overlay. The hydrophobic nature of oil impedes the diffusion of hydrophilic VOCs, biologicals, and particulates directly into the aqueous culture medium.

Oil can also serve as a sink for hydrophobic contaminants from the culture dish. Unconjugated (monomeric) styrene is an embryotoxin released from polystyrene dishes commonly used in IVF (29). The oil to water partition coefficient for styrene is 3.05 (logarithmic) which means that a molecule of styrene is 1,000-fold more likely to partition into the oil than into the medium. Mouse embryos exposed to culture media spiked with embryotoxic concentrations of styrene monomer (2) or other oil soluble toxins (30, 31) undergo developmental arrest in open culture but are rescued in the presence of oil overlay.

Some IVF labs intentionally place reservoirs of oil alone in incubators to serve as an extra sink for oil-soluble volatiles present in incubator air. Though this has intuitive appeal, the author is unaware of any data to support this practice.

It is important to recognize that mineral oil, though important in the defense against VOC exposure, can also be a source of unwanted contamination. Commercial vendors frequently extract water-soluble contaminants from mineral oil for IVF culture by washing procedures in order to minimize the release of unwanted contaminants into the media (32). Some vendors engage in proprietary purity testing of their product and others have adopted recommendations for cold storage of culture oil (4 °C) to counteract the potential buildup of by-products of oil peroxidation that can negatively impact gamete and embryo growth (33, 34).

1.5.5. General Laboratory Practices

The environmental load of airborne toxicants in the lab can be significantly reduced by the exercise of some simple workflow practices.

Staging and Storage Areas

Staging areas outside of the lab can be utilized to separate laboratory supplies from unnecessary packaging/containers. Many laboratories move supply inventories into the lab in measured quantities from outside storage areas to minimize new material outgassing. The use of alcohol, sperm fixatives and stains, and stronger cleaning reagents is best performed in areas where the release of volatiles cannot reach the incubators. This is especially true for sterilizing agents which often employ aldehyde components (Cidex) and the fixatives used in some methods for preimplantation genetic diagnosis.

Laboratory Access

Entry into the laboratory can be controlled to restrict access to necessary staff. Prior to entry staff should be required to (1) minimize personal product usage and to choose products that are low in volatile content, (2) wear professionally cleaned scrub clothing and hair covering, (3) wear dedicated lab shoes subjected to regular sanitation, and (4) perform appropriate hand and arm hygiene. Equipment should also be subjected to hygiene practices prior to entry.

Cleaning Agents

Among the many cleansing and sanitizing agents available for consideration, the most benign are water and hydrogen peroxide. A paste of sodium bicarbonate and water is an excellent abrasive cleanser. Many laboratories utilize stronger reagents for cleaning incubators and those with low volatility and high water solubility (rinse easily) are favorable. Use of volatile reagents such as alcohol should be considered in areas that are remote from the incubator environment or during periods when embryos are not in culture.

References

1. Cohen J et al (1997) Ambient air and its potential effects on conception in vitro. Hum Reprod 12:1742–1749
2. Hall J et al (1998) The origin, effects and control of air pollution in laboratories used for human embryo culture. Hum Reprod 13(Suppl 4):146–155
3. Johnson JE, Boone WR, Bernard RS (1993) The effects of volatile compounds (VC) on the outcome of in vitro mouse embryo culture. Fertil Steril 1993(Suppl 1):S98–S99
4. Anderson K et al (1997) TVOC and health in non-industrial indoor environments. Indoor Air 7:78–91
5. Colonna-Worrilow K et al (2001) A retrospective analysis: the examination of a potential relationship between particulate and volatile organic compound (VOC) levels in a class 100 IVF cleanroom laboratory (CR) and specific parameters of embryogenesis and rates of implantation (IR). Fertil Steril 76(Suppl 1):S15
6. Committee on Aldehydes, Board of Toxicology and Environmental Hazards, National Research Council (1981) Formaldehyde and other aldehydes. National Academy Press, Washington, DC
7. Boone WR et al (1999) Control of air quality in an assisted reproduction laboratory. Fertil Steril 71:150–154

8. Legro RS et al (2010) Effect of air quality on assisted reproduction. Hum Reprod 25:1317–1324
9. Gilligan AV (2010) Establishing the IVF laboratory: a systems view. In: Carrell DT, Petersen CM (eds) Reproductive endocrinology and infertility: integrating modern clinical and laboratory practice. Springer, New York, NY
10. Dickey RP et al (2010) Effect of IVF laboratory air quality on pregnancy success. Fertil Steril 94(Suppl 4):S151
11. Higdon HL, Blackhurst DW, Boone WR (2008) Incubator management in an assisted reproductive technology laboratory. Fertil Steril 89:703–710
12. Esteves SC, Gomes AP, Verza S (2004) Control of air pollution in assisted reproductive technology laboratory and adjacent areas improves embryo formation, cleavage and pregnancy rates and decreases abortion rate: comparison between a Class 100 (ISO 5) and a Class 1,000 (ISO 6) cleanroom for micromanipulation and embryo culture. Fertil Steril 82(Suppl 2):S259
13. Mayer JF et al (1999) Prospective randomized crossover analysis of the impact of and incubator air filtration system on IVF outcomes. Fertil Steril 72(Suppl 1):S42
14. Merton JS et al (2007) Carbon-activated gas filtration during in vitro culture increased pregnancy rate following transfer of in vitro-produced bovine embryos. Theriogenology 67:1233–1238
15. Battaglia DE et al (2001) Prospective randomized trial of incubator CODA filtration units revealed no effect on outcome parameters for IVF. Fertil Steril 75(Suppl 1):S6
16. Souza MCB et al (2006) Evaluation of two incubation environment class 100 versus class 100,000 on intracytoplasmic sperm injection (ICSI) cycles outcome. Fertil Steril 86(Suppl 2):S222
17. Federal Register 40 CFR part 50
18. Federal Register 37: no. 167
19. World Health Organization (2010) WHO guidelines for indoor air quality: selected pollutants (http://www.euro.who.int/__data/assets/pdf_file/0009/128169/e94535.pdf)
20. US Occupational Safety and Health Administration (2011) Indoor air quality in commercial and institutional buildings, OSHA 3430-04
21. Schimmel T et al (1997) Removal of volatile organic compounds from incubators used for gamete and embryo culture. Fertil Steril 68(Suppl 1):S165
22. Linde D (ed) (2004) Handbook of chemistry and physics, 85th edn. CRC, Boca Raton, FL
23. http://toxnet.nlm.nih.gov/cgi-bin/sis/htmlgen?HSDB
24. McClenny WA, Holdren MW (1999) Compendium method to TO-15. Compendium of methods for the determination of toxic organic compounds in ambient air, 2nd edn. (EPA/625/R-96/010b) http://www.epa.gov/ttnamti1/files/ambient/airtox/to-15r.pdf, pp 1–67
25. Winberry WT Jr et al (1999) Compendium method to TO-11A. Compendium of methods for the determination of toxic organic compounds in ambient air, 2nd edn. (EPA/625/R-96/010b) http://www.epa.gov/ttnamti1/files/ambient/airtox/to-11ar.pdf, pp 1–56
26. Grieve PW (1991) Measuring ventilation using tracer gases. Bruel & Kjaer, Denmark
27. Practice Committee of the American Society for Reproductive Medicine, Practice Committee for the Society for Assisted Reproductive Technology (2008) Revised guidelines for human embryology and andrology laboratories. Fertil Steril 90:S45
28. Committee of the Special Interest Group on Embryology for the European Society for Human Reproduction and Embryology (2008) Revised guidelines for good practice in IVF laboratories. Hum Reprod 23:1253–1262
29. Gilligan A et al (1997) Release of volatile organic compounds such as styrene by sterile Petri dishes and flasks used for in vitro fertilization. Fertil Steril 68(Suppl 1):S52
30. Miller KF, Goldberg JM, Collins RL (1994) Covering embryo cultures with mineral oil alters embryo growth by acting as a sink for an embryotoxic substance. J Assist Reprod Genet 11(7):342–345
31. Miller KF, Pursel VG (1987) Absorption of compounds in medium by the oil covering microdrop cultures. Gamete Res 17:57–61
32. Morbeck DE et al (2010) Washing mineral oil reduces contaminants and embryotoxicity. Fertil Steril 94:2747–2752
33. Otsuki J, Nagai Y, Chiba K (2007) Peroxidation of mineral oil used in droplet culture is detrimental to fertilization and embryo development. Fertil Steril 88:741–743
34. Otsuki J, Nagai Y, Chiba K (2009) Damage of embryo development caused by peroxidized mineral oil and its association with albumin in culture. Fertil Steril 91:1745–1749

Chapter 18

Culture Systems: Mineral Oil Overlay

Dean E. Morbeck and Phoebe H. Leonard

Abstract

Mineral oil overlay microdrop is commonly used during in vitro fertilization (IVF) procedures. Though mineral oil appears homogeneous, it is an undefined product that can vary in quality. Here, we describe the history, chemistry, processing, and optimal use of mineral oil for IVF and embryo culture.

Key words: Mineral oil, Peroxides, Paraffin

1. Introduction

Mineral oil has been a standard component of embryo culture since the early 1960s. Following the initial success of culturing cleavage stage embryos to blastocyst in test tubes (1, 2), two students in the laboratory of John Biggers at the University of Pennsylvania, Ralph Gwatkin (3) and Ralph Brinster (4), developed a method which allowed more frequent observation of embryos that entailed the use of mineral oil placed over the top of media in petri dishes (5). Though culturing in test tubes is still a viable option, development of this microdrop method with oil overlay, and variations of the approach in culture dishes, was a major step towards defining the culture requirements of early cleavage stage embryos and is the most common method of embryo culture in use today. In addition to facilitating embryo assessment, the oil overlay approach allows culture in small volumes of culture media while minimizing pH, temperature, and osmotic fluctuations (5, 6). Mineral oil may also absorb hydrophobic contaminants or volatile organic compounds from the culture media (7, 8). However, despite these advantages, use of mineral oil can also be problematic.

Gary D. Smith et al. (eds.), *Embryo Culture: Methods and Protocols*, Methods in Molecular Biology, vol. 912,
DOI 10.1007/978-1-61779-971-6_18,

1.1. Types of Oil

Mineral oil is derived from crude oil, the same starting material used for gasoline, diesel fuel, and chemicals such as benzene, which is another IVF-related chemical used for polystyrene to make petri dishes. Mineral oil is one fraction of crude oil distillation and is also called mineral petrolatum, paraffin oil, and light or white mineral oil. There are three types of hydrocarbons in refined mineral oil: straight chain, cyclic and aromatic hydrocarbons. Some companies assert that paraffin oil is different than mineral oil by claiming paraffin oil is composed of only saturated, straight chain hydrocarbons. In reality, because of difficulty in separating and refining these similar components, light mineral, and paraffin oil are really mixtures of mostly straight chain, with some cyclic and aromatic hydrocarbons.

Mineral oil approved for clinical purposes, e.g., as an intestinal lubricant or for embryo culture, must be United States Pharmacopeia (USP) grade. Requirements for USP grade oil include measures of viscosity, specific gravity, sulfur content, and the amount of unsaturated, polycyclic (aromatic) hydrocarbons, but do not appear to require measures of sterility or levels of other contaminants that may be embryo toxic. The specific gravity of mineral oil can vary due to different chain lengths of the hydrocarbons (10–30 carbons) and for light mineral oil ranges from 0.818 to 0.880 g/ml. Suppliers of mineral oil to IVF laboratories select oil on the lower end of this specific gravity spectrum to provide a product with the least amount of viscosity, though at least one company sells paraffin oil having two different viscosities (specific gravity 0.84 and 0.86 g/ml). Additionally, these companies may select batches of oil and perform additional testing for contaminants to ensure suitability for use in IVF.

1.2. Oil and Contaminants

Mineral oil is the least characterized component of embryo culture and brings with it the most potential for harm. This is evident by several product recalls and unconfirmed reports of embryotoxicity with several different types of contaminants reported. In the first report of oil toxicity during embryo culture (9), the toxin was not identified but variables that affected toxicity included microdrop size and lack of zona pellucidae. Additional cases of unidentified toxins in oil have been reported during manufacturer quality control (QC) testing with murine embryos (10) as well as for oil used in bovine (11), porcine, (12) and murine (13) embryo culture.

Identification of the chemicals present in mineral oil is essential to learn how to detect and prevent further contaminations. Zinc (14), Triton X-100 (15), and peroxides (15–17) have been found in oil used for embryo culture. Triton X-100 is a nonionic detergent used in research as well as a component of cleaning products used in industrial manufacturing. The observation that this chemical was present in two different bottles of the same lot of mouse embryo tested mineral oil from Sigma Chemical Company leads to some interesting questions. This oil was extremely toxic to mouse embryos: so how did it pass the manufacturer's testing process? The observation that one bottle of oil had ten-fold higher TX-100 than the other indicates bottle-to-bottle variation. Since every bottle of a lot is not tested, it is likely that a non-contaminated bottle was used for QC testing. While Sigma's mineral oil is not USP grade, it is unknown if testing required by USP would have detected this incident of contamination. The source of the TX-100 and its variable presence raise questions about the manufacturing and bottling process. Unlike other providers of USP-grade oil for IVF, which use glass or inert PET plastic bottles, Sigma supplies their oil in polypropylene bottles. It is possible that the contamination was not from the source but was present in the bottle. These examples of isolated contamination during manufacturing should be rare, but are difficult to detect during routine QC testing.

Peroxides are perhaps the most serious contaminants found in mineral oil and arise from oxidation of double bonds within the oil. This is why saturated mineral oils are preferred and likely beneficial. Because peroxides are a product of oxidation of the oil, they can develop at any time during the shelf life of the oil (15–17). Thus, a manufacturer can perform QC testing with a mouse embryo assay (MEA) when the oil is bottled but the oil can later become toxic. Otsuki et al. reported peroxide contamination of laboratory grade mineral oil and found the degree of peroxidation was dependent on exposure to heat, UV light, extended storage, manufacturer, and lot number (16, 17). In addition to peroxides, which produce reactive oxygen species (ROS), oil that is degrading also produces alkenals (saturated aldehydes) and aldehydes (15) which have been shown to be embryotoxic when present in culture media or air (18). Since quality of oil can be affected by handling and storage, end users should be wary of bottle-to-bottle variation and take care to minimize environmental effects on oil, including handling and storage conditions.

1.3. Storage of Oil

Because of the potentially reactive nature of mineral oil, many manufacturers recommend storage at 4 °C in the dark to limit heat and photo-oxidation (Table 1). The recommended shelf life of mineral oil ranges from 5 weeks to 2 years. While the shorter shelf-life seems prudent, and in fact data from Otsuki et al. (16) suggests this is important, good quality mineral oil should be stable for at least 2 years. If the oil lacks unsaturated molecules, storage at 4 °C or in the dark is likewise unnecessary. However, until manufacturers can provide and document mineral oil of higher purity, these storage precautions should be followed.

1.4. Washing of Oil

Though all mineral oils presumably pass QC testing by the manufacturer, we do not know if the testing is sensitive enough to detect early stages of peroxidation or if the storage recommendations are

Table 1
Sources of FDA approved mineral oil for embryo culture

Company	Washed?	Filtered?	Storage (°C)	Shelf-life
Medicult	Yes	No	2–8	NA
In Vitro Care	Double	No	15–25	2 years
Vitrolife Ovoil	No	Yes	2–8	5 weeks from ship
LifeGlobal—LiteOil	Yes	Yes	15–25	2 years
LifeGlobal—Paraffin	Yes	Yes	15–25	2 years
Cooper Sage	Yes	Yes	2–8	NA
Irvine Scientific	No	Yes	15–30	2 years
Cook Medical	NA	Yes	2–8	12 weeks
Nidacon—Nidoil	NA	NA	2–40	2 years

NA data not readily available

appropriate. Sifer et al. (19) performed a randomized, controlled trial comparing oil from four different companies: CryoBioSystem, Medicult, Vitrolife, and Nidacon. With nearly 120 patients in each group, they did not observe a difference in pregnancy or implantation rate, but did obtain higher quality embryos in Vitrolife's Ovoil versus Medicult or CryoBioSystem oil. Storage time and conditions of oil while at the respective distributors before shipping to the IVF lab was unknown. Given the short shelf-life of Ovoil, it is possible that this oil had less time to oxidize. Embryo development in Ovoil was not different from Nidacon's Nidoil, which has a 24-month shelf life. Of note, Nidacon uses a 90 % expanded blastocyst rate for their MEA quality control testing, whereas other companies use a 70 or 80 % threshold. It is possible that Nidacon selects better quality oil based on their more stringent QC standards.

While most manufacturers wash mineral oil prior to packaging, the practice of washing oil has been the subject of some debate. Discussions on Embryomail indicate there is a lack of consensus among embryologists on the utility of washing oil. Fleming et al. (9) were the first to demonstrate that embryotoxic mineral oil could be detoxified by washing. Similarly, Lee et al. (13) observed improved mouse embryo development using washed oil. Morbeck et al. (15) demonstrated that the amount of TX-100 could be reduced after washing and that oil containing peroxides could be made less toxic. In the case of the peroxide-laden oil, it is not known if this was due to a reduction in one or all of the toxins present (peroxide activity, aldehydes or alkenals).

While many different washing methods are used, both by manufacturers and end-users, there is little evidence to support one method of washing over another. Otsuki et al. (17) demonstrated that albumin could act as a sort of sieve to allow contaminants from oil to enter media. Thus, the protein may be useful in removing contaminants from oil when washing. However, it was also pointed out that albumin may potentially contribute to peroxidation of the oil, and it was therefore recommended that it be excluded from washing solutions (17). In their paper on washing of TX-100 and peroxide contaminated oil, Morbeck et al. (15) compared water, HTF culture medium and HTF containing HSA as washing agents at both room temperature and 37 °C for 24 h. Washing was just as effective with water as it was with media or media plus protein. Temperature did not affect washing efficacy. For practical purposes, many manufacturers choose to wash with water, as labs use many different culture media and potential carryover or contamination of another media may be a concern.

2. Materials

1. USP-Grade Mineral Oil.
2. Culture media without protein or sterile water.
3. Nalgene 250 ml sterile receiver unit (optional; see Note 1).

3. Methods

Carry out all procedures at room temperature unless otherwise indicated.

1. Mix water or culture media (see Note 2) with oil at a 5:1 ratio. Ideally, the bottle of oil supplied by the manufacturer has enough head space to allow addition of a 20 % volume of aqueous phase. For a 250 ml bottle, add 50 ml water.
2. Tightly cap bottle and invert gently 20 times.
3. Store for 24 h before use per manufacturer's instructions (see Note 3).
   (a) Mineral oil can be used to cover microdrops or larger volumes of media (0.5–1.0 ml) in organ culture dishes or 4-well dishes.
   (b) The ratio of oil to media depends on the dish type. For organ culture dishes, typically 1 ml of oil covers 1 ml of media. Four well dishes use 0.5 ml of oil and 0.5 ml of media.

Most commonly 35 or 60 mm dishes are used for microdrops. For microdrops of 25–50 μl volume, use 4.5 ml of oil in the 35 mm dish or 9.0 ml of oil in the 60 mm dish (see Note 4).

(c) Microdrops can be made before or after the addition of mineral oil to the dish depending on the type of dish used. Microdrops are prepared after the addition of oil when using plasticware for cell culture (Falcon 3000 series). This can be done with a micropipette or a Pasteur pipette. It is easier to make microdrops before the addition of oil for untreated plasticware (Nunc IVF series; Falcon 1000 series). Due to their small size, osmolarity of microdrops can change if left exposed to room air for an extended period. This step should be performed quickly and a minimal number of dishes should be made at one time.

- Place dishes in incubator the day before intended use (see Note 5).
- Quality control testing (optional; see Note 6).

4. Notes

1. Use of a container other than the bottle in which the oil is supplied can be useful if there is not enough head-space to add the water/media. Also, a higher ratio of aqueous to oil would theoretically improve the washing process. A 250 or 500 ml plastic receiver unit can be used in place of the oil bottle.
2. Studies on washing (15) demonstrated similar efficacy among water, HTF and HTF + HSA. However, care must be taken when pipeting from a bottle of oil containing an aqueous phase to avoid accidental aspiration of the liquid below the oil. If water is inadvertently added to a culture dish, it could combine with a microdrop of media and cause a significant drop in osmolarity or solute concentration that would be embryo toxic.
3. While some reports advocate equilibrating the entire bottle in the incubator during its use, a more conservative and likely safer approach is to store the bottle in the dark at 4 °C. Extended storage at higher temperature may facilitate peroxidation. An expiration date equal to that provided for the oil or media used by the manufacturer is sufficient, whichever is less.
4. Smaller volumes of oil can be used if microdrop size is less than 20 μl.
5. 50 μl microdrops require a minimum of 8 h of equilibration to reach desired pH.

6. Due to its propensity for toxicity, a new lot of mineral oil should be tested with a suitable quality control bioassay, even if this was done by the manufacturer. A direct comparison of QC methods demonstrated that the 1-cell MEA was more sensitive to peroxide in oil than either the 2-cell MEA or a sperm survival assay (20). In addition, it is recommended that a new lot of oil be introduced into clinical use cautiously by performing "split" IVF cases, where the old lot of oil is used and compared to the new lot for actual patient samples.

References

1. Hammond J Jr (1949) Recovery and culture of tubal mouse ova. Nature 163:28
2. Whitten WK (1957) The effect of progesterone on the development of mous eggs in vitro. J Endocrinol 16:80–85
3. Gwatkin RB (1963) Effect of viruses on early mammalian development. I. Action of *Mengo encephalitis* virus on mouse ova cultivated in vitro. Proc Natl Acad Sci USA 50:576–581
4. Brinster RL (1963) A method for in vitro cultivation of mouse ova from two-cell blastocyst. Exp Cell Res 32:205–208
5. Biggers JD (1987) Pioneering mammalian embryo culture. In: Bavister BD (ed) The mammalian preimplantation embryo: regulation of growth and differentiation in vitro. Plenum Press, New York/London, pp 1–22
6. Bavister BD (1995) Culture of preimplantation embryos: facts and artifacts. Hum Reprod Update 1:91–148
7. Miller KF, Goldberg JM, Collins RL (1994) Covering embryo cultures with mineral oil alters embryo growth by acting as a sink for an embryotoxic substance. J Assist Reprod Genet 11:342–345
8. Miller KF, Pursel VG (1987) Absorption of compounds in medium by the oil covering microdrop cultures. Gamete Res 17:57–61
9. Fleming TP, Pratt HP, Braude PR (1987) The use of mouse preimplantation embryos for quality control of culture reagents in human in vitro fertilization programs: a cautionary note. Fertil Steril 47:858–860
10. Gardner DK, Reed L, Linck D, Sheehan C, Lane M (2005) Quality control in human in vitro fertilization. Semin Reprod Med 23:319–324
11. Van Soom A, Mahmoudzadeh AR, Christophe A, Ysebaert MT, de Kruif A (2001) Silicone oil used in microdrop culture can affect bovine embryonic development and freezability. Reprod Dom Anim 36:169–176
12. Shimada M, Kawano N, Terada T (2002) Delay of nuclear maturation and reduction in developmental competence of pig oocytes after mineral oil overlay of in vitro maturation media. Reproduction 124:557–564
13. Lee S, Cho M, Kim E, Kim T, Lee C, Han J et al (2004) Renovation of a drop embryo cultures system by using refined mineral oil and the effect of glucose and/or hemoglobin added to a serum-free medium. J Vet Med Sci 66:63–66
14. Erbach GT, Bhatnagar P, Baltz JM, Biggers JD (1995) Zinc is a possible toxic contaminant of silicone oil in microdrop cultures of preimplantation mouse embryos. Hum Reprod 10:3248–3254
15. Morbeck DE, Khan Z, Barnidge DR, Walker DL (2010) Washing mineral oil reduces contaminants and embryotoxicity. Fertil Steril 94:2747–2752
16. Otsuki J, Nagai Y, Chiba K (2007) Peroxidation of mineral oil used in droplet culture is detrimental to fertilization and embryo development. Fertil Steril 88:741–743
17. Otsuki J, Nagai Y, Chiba K (2009) Damage of embryo development caused by peroxidized mineral oil and its association with albumin in culture. Fertil Steril 91:1745–1749
18. Hall J, Gilligan A, Schimmel T, Cecchi M, Cohen J (1998) The origin, effects and control of air pollution in laboratories used for human embryo culture. Hum Reprod 13:146–155
19. Sifer C, Pont JC, Porcher R, Martin-Pont B, Benzacken B, Wolf JP (2009) A prospective randomized study to compare four different mineral oils used to culture human embryos in IVF/ICSI treatments. Eur J Obstet Gynecol Reprod Biol 147:52–56
20. Hughes PM, Morbeck DE, Hudson S, Fredrickson J, Walker DL, Coddington CC (2010) Peroxides in mineral oil used for in vitro fertilization: defining limits of standard quality control assays. J Assist Reprod Genet 27:87–92

Chapter 19

Culture Systems: Physiological and Environmental Factors That Can Affect the Outcome of Human ART

Don Rieger

Abstract

Many aspects of the embryo culture environment have been shown to affect embryo development and the subsequent outcomes of human ART. It is now becoming increasingly evident that embryo and later development can be affected by events and conditions that occur before, perhaps long before, the oocytes and sperm are collected and brought together in the ART laboratory. These include diet and metabolic disorders, general health and disease, physical and psychological stress, exposure to environmental estrogens and other toxins, pharmaceuticals, alcohol, smoking, and drug abuse. This paper discusses the known and potential effects of season of the year (or temperature) and environmental air pollution on the outcomes of human ART. It may be useful to advise ART patients to avoid high environmental temperature and air pollution. In addition, it is important for clinical embryologists to recognize that adverse outcomes may result from such exposures, and to incorporate this into the analysis of clinic data for the purposes of quality management.

Key words: Oocyte, Sperm, ART, Environmental temperature, Air pollution

Dedication

This paper is dedicated to Keith James Betteridge, BVSc, MVSc, PhD, FRCVS, Professor Emeritus, University of Guelph. For 30 years, Keith has been a wonderful supervisor, colleague, mentor, and friend. I am indebted to him for his countless contributions to my professional and personal life. I have benefitted enormously from his extraordinary knowledge of, and passion for, embryo biology, history of the science, and the proper use of the English language. Overarching all of this is his feeling for the mystery and beauty of embryo development.

1. Introduction

Many aspects of the embryo culture environment have been shown to affect embryo development in vitro, fetal development in utero, perinatal survival, and later development (Reviewed in refs. (1–8)). However, it is now becoming increasingly evident that embryo and

Gary D. Smith et al. (eds.), *Embryo Culture: Methods and Protocols*, Methods in Molecular Biology, vol. 912, DOI 10.1007/978-1-61779-971-6_19, © Springer Science+Business Media, LLC 2012

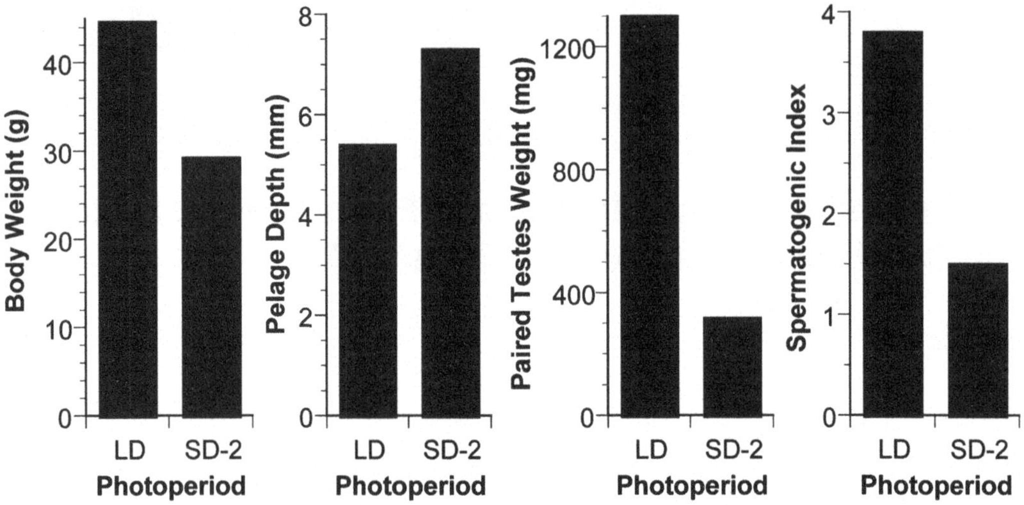

Fig. 1. The effect of maternal *preconception* photoperiod on the mean body weight, pelage depth, paired testes weight, and spermatogenic index of male pups at 49 days of age. All comparisons were significantly different ($P < 0.05$).

later development can be affected by adverse environmental events that occurred even before conception.

An example of such preconception influence is shown in Fig. 1. Lee and Zucker (9) exposed female meadow voles to different photoperiods from birth until breeding. The LD group females were exposed to daily cycles of 14 h light/10 h dark, while the SD-2 females were exposed to daily cycles of 14 h light/10 h dark until 2 weeks before breeding and then to 10 h light/14 h dark until breeding. After breeding, all the females were maintained in 10 h light/14 h dark through gestation and the resulting pups were maintained in 10 h light/14 h dark from birth onward. The only difference between the treatments was the photoperiod for 2 weeks *before conception*. The results presented in Fig. 1 are measurements made on the male pups arising from those conceptions. At 49 days of age, body weight, pelage depth, paired testes weight, and spermatogenic index were all significantly different between the male pups arising from the two preconception photoperiod treatments.

An example in humans (10) is shown in Fig. 2. Women in Cohort A were conceived and born immediately before the Dutch famine of 1944–1945. Women in Cohort D1 were exposed to the famine during the second and third trimesters in utero, women in Cohort D2 were exposed to the famine only during their first trimester in utero, and women in Cohort E were conceived and born immediately after the famine. Figure 2 shows the birth weight, gestation length and crown-rump length of children born to those women 20 or more years after the famine. Clearly, exposure to famine during the second and third trimesters in utero

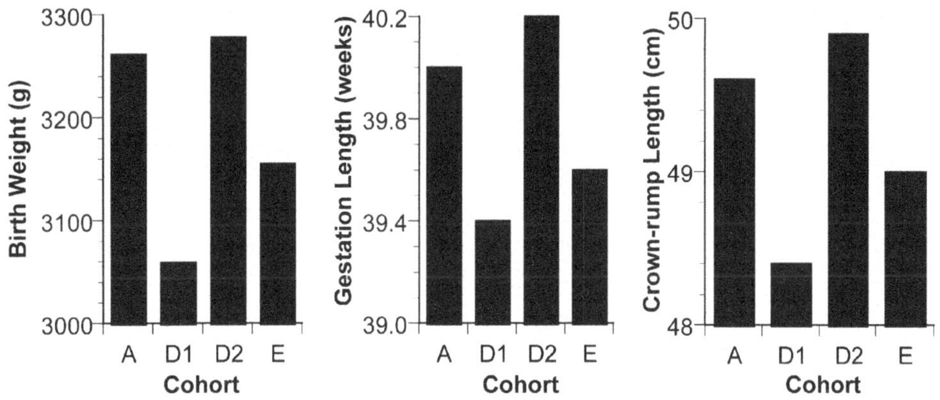

Fig. 2. The effect of maternal exposure in utero to the Dutch famine of 1944–1945 on the birth weight ($P < 0.025$), gestation length ($P < 0.10$), and crown-rump length ($P < 0.025$) of babies born to those women.

(Cohort D1) had significant detrimental effects on the development of the babies born to those women.

The exact mechanisms responsible for the effects seen in these two studies are unclear and undoubtedly very complex. However, whatever the mechanisms, these studies do show that the development of offspring can be affected by factors or events that occur before, even long before, conception. By extension, the outcome of assisted reproductive procedures (ART) including fertilization rates, embryo development in vitro, implantation, pregnancy and birth rates, and fetal and later development may also be affected by factors or events that occur before, perhaps long before, the oocytes and sperm are collected and brought together in the ART laboratory.

A great many factors have been shown to affect, or have the potential to affect, fertility and development of the offspring. These include diet and metabolic disorders, general health and disease, physical and psychological stress, exposure to environmental estrogens and other toxins, pharmaceuticals, alcohol, smoking, and drug abuse (see ref. (11)). This paper is intended as a limited overview of the literature concerning two preconception factors that have been shown to affect human fertility in general, and the outcome of ART in particular: (a) season of the year or temperature, and (b) environmental air quality.

2. Aspects of Gamete Physiology and Function Linked to Embryo Development

Sexual reproduction requires first that the sperm and oocyte come together in fertilization to produce the pronuclei, form the zygote and initiate cleavage. Development through the early cleavage stages is directed by factors in the oocyte of purely maternal origin. Once the embryonic genome is expressed, it directs all subsequent development and physiological function. This section is intended to introduce some of the aspects of gamete physiology and function that have been shown to be linked to embryonic and subsequent development.

Within the follicle, the development of the oocyte is dependent on the environment provided by the apposed and connected cumulus cells, the follicular fluid, and the follicle and ovary at large. Nutrients supplied to the oocyte pass from the perifollicular vasculature into the follicular fluid, and are then taken up by the cumulus cells. The cumulus cells form gap junctions between themselves and with the oocyte. Consequently, nutrients and signaling molecules can pass between the oocyte and the cumulus cells, and between the cumulus cells, to allow the coordination of the functions of the cumulus cells and the development of the oocyte (see ref. (12)).

The significance of those relationships is illustrated by the results of the following studies. Van Blerkhom et al. (13) measured the concentrations of vascular endothelial growth factor (VEGF) and dissolved oxygen in the follicular fluid from human oocyte collections, as well as the blood flow around the corresponding follicles. They found that the incidence of chromosomal abnormalities was inversely related to oxygen content and that embryo development was directly related to oxygen content. Oxygen content was, in turn, directly related to VEGF concentration and to perifollicular blood flow. Hamel et al. (14) related gene expression in follicular cells obtained from human oocyte collections to the ART outcome for the corresponding oocytes. They found that the expression (mRNA) of 3-beta-hydroxysteroid dehydrogenase 1, Ferredoxin 1, Serine (or cysteine) proteinase inhibitor clade E member 2, Cytochrome P450 aromatase, and Cell division cycle 42 was significantly greater in cumulus cells corresponding to oocytes that resulted in pregnancy than in cumulus cells corresponding to oocytes that did not. Adriaenssens et al. (15) similarly showed that embryo development in vitro was related to cumulus cell Activated Leukocyte Cell Adhesion Molecule and Gremlin 1 mRNA. Wang et al. (16) measured the connexin (Cx) mRNA and connexin proteins in cumulus cells obtained from human oocyte collections, as well as the gap junctional conductance between the cumulus cells. They found that Cx43 mRNA and Cx43 protein in the cells, and the gap junctional conductance between them, were directly related to rates of embryo development, implantation, and pregnancy from the corresponding oocytes. Clearly, the outcome of ART is directly related to the ability of the follicle to support the development and health of the oocyte.

Many studies have shown that experimental and clinical treatments of the donor females and/or of the cumulus–oocyte complexes in vitro can significantly affect oocyte maturation or subsequent embryo and later development (see ref. 17). However, there is relatively little information that directly relates the biological functions of the individual oocyte proper to subsequent embryo and later development, largely because there are few nondestructive measures of oocyte function.

One approach to filling that void has been to use polarized microscopy to visualize the meiotic spindle and the zona pellucida of the oocyte. Several (but not all) studies have shown that a demonstrable spindle and/or its alignment with the first polar body are associated with the potential for in vitro development and pregnancy (reviewed in ref. (18)). Shen et al. (19) found that the inner layer thickness and light retardance of the zona were significantly greater in oocytes that contributed to a pregnancy than in those that did not.

Indirect measures of the metabolic activity of the oocyte have also been shown to be related to subsequent development. Nagy et al. (20) incubated individual denuded human oocytes for 3 h before sperm injection, and used near-infrared (NIR) spectroscopy to evaluate the metabolomic profile of the spent culture medium. The metabolomic profiles were significantly different for oocytes that developed to morphologically good or poor embryos on Day 3 and 5 of embryo culture. They were also able to relate the metabolomic profile of the denuded oocyte to the success or failure of pregnancy following transfer of the resulting embryos.

The activity of glucose-6-phosphate dehydrogenase (G6PDH) has been shown to decrease with maturation of the oocyte. Brilliant cresyl blue (BCB) is a vital dye that is decolorized when reduced by G6PDH, and is used to evaluate oocyte maturation in domestic animals. Mature oocytes remain blue (BCB+) while immature oocytes do not (BCB, see ref. 21). Recent studies have shown that embryo development from BCB + oocytes is better than that from BCB-oocytes from cattle (22, 23), buffalos (24), and pigs (25). Moreover, Spikings et al. (26) found that mitochondrial replication is delayed in BCB-pig oocytes compared to BCB + oocytes, which correlated with reduced fertilization and embryo development.

Cytogenetic analysis of the polar bodies provides a measure of the global developmental potential of the oocyte. Munne et al. (27) evaluated the effect of preimplantation genetic diagnosis (PGD) on pregnancy loss in in vitro fertilization patients; the rate was significantly lower in patients when embryos diagnosed as chromosomally abnormal were not transferred, than in age-matched patients receiving embryos that were not subjected to PGD. As noted by Kuliev et al. (28), at least 95 % of embryo aneuploidy is of maternal origin, and therefore most, if not all of the pregnancy loss associated with aneuploid embryos is attributable to the oocyte. This has been confirmed by the finding that eliminating aneuploid oocytes by polar body cytogenetic analysis resulted in significantly higher implantation rates, and lower abortion rates, for IVF patients 35–39 years old (29).

Until recently, it has not been possible to evaluate the function of the specific sperm cell that fertilizes an oocyte. However, some measures of populations of sperm from which the fertilizing sperm originates have been shown to be related to subsequent embryo and later development.

Although sperm morphology is considered to be of limited of limited clinical value in predicting fertility (30), measures of morphology of samples of sperm have been linked to subsequent embryo development. Salumets et al. (31) found that cleavage rate was affected by sperm morphology, but not by sperm concentration or progressive motility. Dubey et al. (32) classified patients as either normal or teratospermic, and compared the outcomes following ICSI. There was a trend to a greater percentage of morphologically normal embryos on Day 3, and significantly greater pregnancy and implantation rates, in the normal sperm group. Check et al. (33) found that fertilization rate was directly related to sperm morphology score for conventional IVF, but not for ICSI. Overall, the fertilization rate was significantly greater for ICSI, but the implantation rate was significantly greater for conventional IVF, suggesting that some contribution of the sperm to subsequent development was affected by ICSI.

Biologically, the most important component of the sperm is the DNA that it delivers to the oocyte, and various measures of DNA damage (in populations of sperm) are being investigated for assessment of male fertility (see ref. (34)). Many have shown a negative association between sperm DNA damage and embryo quality and rates of implantation and ongoing pregnancy (see ref. (35)). In another recent review of 28 studies, Zini et al. (36) conclude that, overall, the studies do not show a consistent relationship between sperm DNA damage and embryo quality, but that it may be more significant for ICSI than for conventional IVF. Nonetheless, the fact that some studies do show a relationship between sperm DNA damage and embryo quality and subsequent development is sufficient to warrant discussion. For example, Virro et al. (37) found that the DNA fragmentation index (DFI) was not related to fertilization rate, but blastocyst, pregnancy, and ongoing pregnancy rates were greater when the DFI was less than 30 % compared with a DFI of 30 % or more. Seli et al. (38) similarly found that sperm DNA strand breakage, as measured by TUNEL labeling, was associated with a significantly lower rate of blastocyst development. Nasr-Esfahani et al. (39) assessed sperm DNA fragmentation by the Comet assay and showed that it was directly related to sperm protamine deficiency, and inversely related to the embryo cleavage score on Days 2 and 3. Meseguer et al. (40) found that sperm DNA oxidation was unrelated to fertilization rate, but was directly correlated to embryo fragmentation on Days 2 and 3, and to the proportion of arrested embryos on Day 6. Avendano et al. (41) found that sperm DNA fragmentation was negatively correlated to embryo score on Day 3, and that sperm DNA fragmentation was significantly greater in sperm samples that led to a pregnancy than in those that that did not. Most interestingly, the same group also showed that in subfertile and infertile men, morphologically normal sperm samples contained a significant percentage of sperm with DNA damage (42). Sousa et al. (43) evaluated the chromatin status by nuclear staining and DNA damage by the Tunnel assay in sperm used for ICSI or conventional IVF. Nuclear staining was directly related to DNA damage, and embryo quality and pregnancy rate were negatively related to the percentage of dark sperm nuclei. Finally, Meseguer et al. (44) found that while sperm DNA fragmentation was negatively associated with pregnancy rate after ICSI of patient oocytes, there was no such association for donor oocytes. This suggests that the negative effect of sperm DNA damage is exacerbated by poor quality oocytes or mitigated by good quality oocytes. For an extensive review of the mechanisms and implications of sperm DNA damage, (see ref. (45)).

Other aspects of sperm biology, measured in populations of sperm, have also been linked to embryo and subsequent development, including seminal reactive oxygen species (46), gene expression and activity of sperm glutathione peroxidases (47), and hyaluronic acid binding (48). Urner and Sakkas (49) showed that glucose metabolism in mouse oocytes increased markedly with entry of the sperm. They subsequently showed that the magnitude of this increase in glucose metabolism upon sperm entry into cattle oocytes was related to the known fertility of the bull. In particular, glucose metabolism through the pentose-phosphate pathway (PPP) was related to the time of onset of the first S phase in the zygote, suggesting that PPP activity is a significant function of the sperm in initiating embryo development. Most significantly, PPP activity was unrelated to rates of fertilization or embryo develop ment to 16–32 cells, but was strongly related to blastocyst development (50).

As noted above, the sperm morphology is typically evaluated for populations of sperm. Recently, however, it has become possible to evaluate the morphology of the specific sperm used for injection of an oocyte, using high magnification sperm selection (MSOME, IMSI), as originally described by Bartoov et al. (51). De Vos et al. (52) classified sperm as normal or abnormal, based on the shape and size of the head, presence of the acrosome, and lack of midpiece or tail defects. Fertilization, pregnancy, implantation and live birth rates were significantly greater for normal sperm than for abnormal sperm. However, there was no effect on embryo morphology on Day 3. Berkovitz et al. (53) injected oocytes with sperm with normal nuclei (control) or nuclei with a normal shape but large vacuoles. The presence of large vacuoles had no effect on

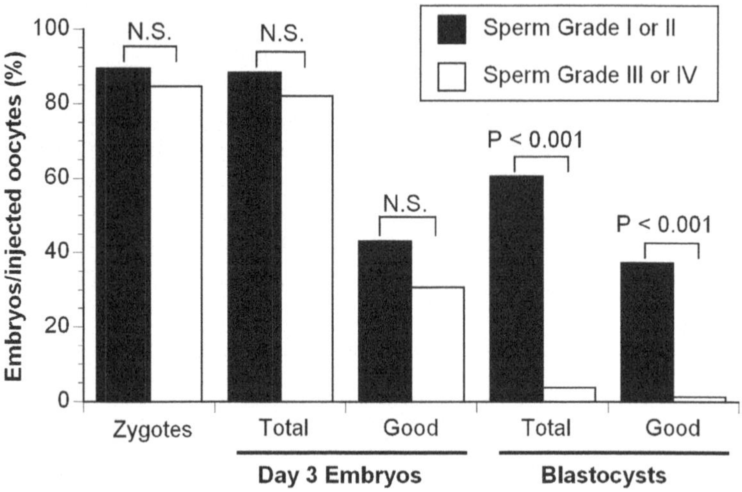

Fig. 3. The effect of the morphology of sperm injected into sibling oocytes on subsequent embryo development in vitro.

fertilization rate or embryo quality on Day 3, but resulted in significantly lower pregnancy rates and significantly higher abortion rates, compared to the controls. Antinori et al. (54) injected oocytes with either sperm selected for normal head shape and a maximum of one vacuole (IMSI), or with unselected sperm (ICSI). There was no difference in fertilization rates, but clinical pregnancy and implantation rates were significantly greater for IMSI than ICSI. The report of Vanderzwalmen et al. (55) is particularly informative, for several reasons. The injected sperm were classified into four grades (best to worst): grade I, absence of nuclear vacuoles; grade II, a maximum of two small vacuoles; grade III, more than two small vacuoles or at least one large vacuole; grade IV, large vacuoles in conjunction with abnormal head shapes or other abnormalities. Figure 3 shows the effect of the injection of Grade I/II sperm, or Grade III/IV sperm into sibling oocytes on subsequent embryo development. There was no effect of sperm grade on fertilization rate or development at Day 3, consistent with the other reports (52, 53). However, sperm grade had a highly significant effect on development to blastocyst and blastocyst quality. That the difference in embryo development was not evident until the blastocyst stage might well be related to the fact that the major onset of expression of the embryonic genome does not occur until the 4–8 cell stage in the human (56). The observation of a positive relationship between sperm DNA fragmentation and the presence of large nuclear vacuoles (57) might explain the negative relationship between sperm vacuoles and blastocyst development. It is logical to assume that sperm DNA fragmentation could result in genetic defects in the embryos and, consequently, deficiencies or errors in the expression of genes required for blastocyst and later development may occur.

A classic axiom in veterinary medicine is "no hoof, no horse"; the success of any biological process depends on its most basic underpinnings. The studies cited above demonstrate that embryonic and subsequent development is linked to the health and function of the gametes, far beyond fertilization. This is not at all surprising, but is, perhaps, not always appreciated in clinical ART. The gametes may appear normal and fertilization may occur and produce a zygote, but this does not necessarily mean that the zygote is fully equipped for development through to birth and beyond.

3. The Effects of Temperature on Gametes, Embryo Development, and Fertility

In many mammalian species, reproduction is seasonal, such that breeding occurs at the best time to ensure optimal nutritional and environmental conditions at the time of birth (58). In some species, this is controlled by the photoperiod and in others by temperature, rainfall, availability of food and other factors (59). Humans, of course, are not considered seasonal breeders, but there are distinct and consistent seasonal patterns of fertility in human populations throughout the world (60–62). In tropical and semitropical climates, much, but not all, of the effect of season on human fertility appears to be related to environmental temperature, with conception being significantly reduced during the hot summer season. In higher latitudes, there is little or no effect of temperature, and seasonal variation is likely related to photoperiod (60).

Logically, the relationship between season and fertility for natural conception could be related to frequency of intercourse and perhaps other social factors. Any such sources of variation would be unlikely to be significant for in vitro fertilization. However, the results of studies of season on the outcomes of IVF are inconsistent. Some studies have shown significant effects of season on one or more of fertilization, embryo development, and pregnancy and implantation rates (63–68), while others have not (69–72). The reasons for this inconsistency are not clear, but may reflect the large variability in sperm and oocyte quality among patients.

The question remains whether the sperm, the oocyte, or both are affected by season, and by what mechanisms. While it would be difficult to address this question in the human for ethical and practical reasons, it can be addressed by looking at other species, notably domestic cattle. As for the human, the fertility of the domestic cow, by natural breeding or artificial insemination, is significantly reduced in the hot season (73–76). Season has also been shown to have significant effects on in vitro development of cattle embryos (77, 78) and parthenotes (79).

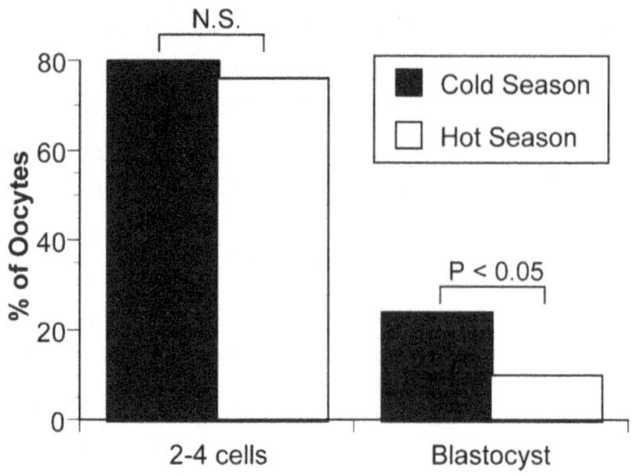

Fig. 4. The effect of season of the year on the in vitro development of cattle embryos.

3.1. Seasonal and Temperature Effects on the Oocyte

Even relatively short-term exposure of the oocyte to heat in vivo, or in vitro, can have significant effects on subsequent embryo development (80). Cows were exposed to 24 °C (thermoneutral) or 42 °C (heat stressed) for 10 h after the onset of estrus, bred by artificial insemination, and the embryos collected at Day 7. Embryo stage, quality and number of cells were all significantly reduced in the heat-stressed cows. Similar results were seen for in vitro development of embryos or parthenotes originating from oocytes collected during the cold or hot season (77–79, 81). Results from Gendelman et al. (78) are shown in Fig. 4. Note that there was no effect on early cleavage, but development to the blastocyst stage was significantly less for oocytes collected during the hot season. An effect of season early cleavage was seen in some of the other studies, but the effect on blastocyst development was more marked.

Heat treatment of cattle (82, 83) and mouse (84) oocytes for various periods during in vitro maturation also had deleterious effects on subsequent development. However, in these studies, heat treatment reduced embryo development at all stages from cleavage to blastocyst. In the case of cattle oocytes, maturation at 41 °C accelerated both nuclear maturation to MII and cytoplasmic maturation, as judged by cortical granule distribution, compared to maturation at 38.5 °C (controls) (83). Advancing the time of fertilization eliminated the effect of IVM at 41 °C on subsequent embryo development.

In the mouse study (84), the oocytes were matured in vitro at various temperatures from 37 to 41 °C and then chemically activated. Successive stages of development were increasingly sensitive to temperature: maturation was reduced at 40.7 °C, development to 2-cells at 40 °C, and development to blastocyst at 38.5 °C. Transfer of the spindles from the heat-treated oocytes to in vivo matured oocytes eliminated the effect of heat treatment on subsequent development. As for the cattle oocytes, heat treatment affected cortical granule distribution.

Other effects of heat treatment of oocytes include those on assembly and configuration of the spindle (84–86), increased DNA fragmentation and caspase activity (82), decreased glutathione (GSH/GSSG) ratio (84), decreased POU5F1 (Oct 4, transcription factor) expression (78), and changes in the lipid composition and phase transition temperature of the oocyte membrane (79).

3.2. Seasonal and Temperature Effects on Sperm

It is well established that for animals with external testes, warming the testis, even only to body core temperature, has significant deleterious effects on testicular cellular morphology and function, endocrine function, and physiology. The earlier stages of sperm development are more affected than later stages, but ultimately the concentration and morphology of sperm in the semen are adversely affected, resulting in reduced or complete loss of fertility. Matings using males that had been exposed to testicular warming resulted in increased rates of early embryo mortality, abortions, stillbirths, and fetal anomalies in laboratory rodents, rabbits, and domestic animals (reviewed by refs. (87–90)). In humans, sperm concentration, motility and morphology have been shown to be affected by season, with all three parameters at their nadir during mid-summer (91, 92).

Relatively mild and short-term exposure of males to heat can have both rapid and extended effects on fertilization and embryo survival. Burfening et al. (93) exposed male mice to 21 °C (controls) or 32 °C for 24 h and then mated them to females at from 1 to 30 days after the heating period. Fertilization rate decreased to reach a nadir with breeding at 16–20 days after heat treatment. Implantation and fetal rates dropped off sooner than fertilization rate. Similar results were obtained after exposing male mice to 32 °C for 24 h and breeding 7, 21, or 35 days later (94). There was no effect of heat treatment on fertilization, but embryo development, particularly to the blastocyst stage, was significantly reduced. Even very short-term heat exposure can have significant effects (95). The scrotums of mice were heated for 30 min in a water bath at 42 °C. Control males were left at room temperature. Sperm were collected at 16 h or 23 days after heat treatment and used for IVF. Heat treatment had no effect on development to the 2-cell or 4-cell stage, but development to blastocyst was significantly reduced (Fig. 5). It is particularly noteworthy that in the last two studies, the effect of heating was not evident until the blastocyst stage.

In vitro heat treatment of sperm also affects subsequent embryo development. Rabbit semen was incubated for 3 h at 38 or 40 °C (comparable to the rectal temperature of does maintained at 21 or 32 °C, respectively) before being used for intrauterine insemination (96). Heat treatment had no effect on fertilization rate, but

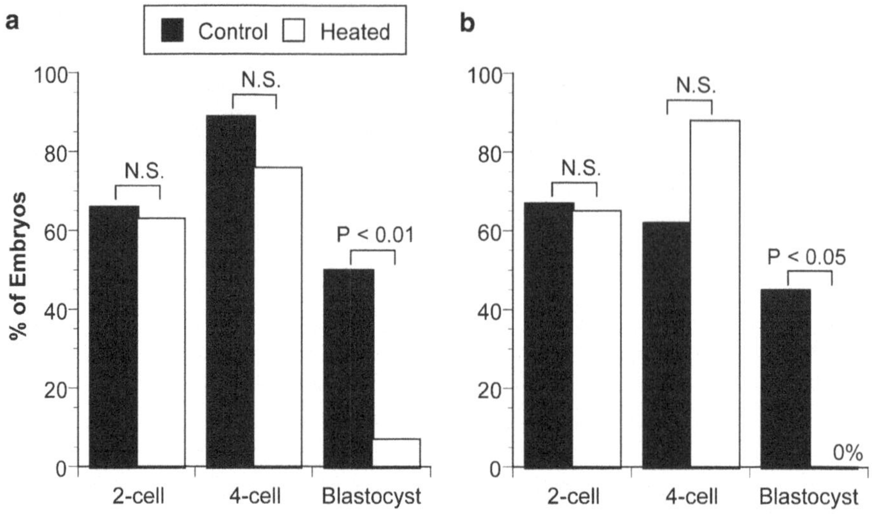

Fig. 5. The effect of scrotal warming for 30 min to 42 °C on the development of mouse embryos produced from in vitro fertilization using sperm taken (**a**) 16 h or (**b**) 23 days after heat treatment. Control males were left at room temperature.

the survival of resultant embryos was significantly less with sperm heated to 40 °C compared with 38 °C. Heat treatment of mouse sperm had similar effects (97). Sperm were either held at 37.5 °C or heated to 56 °C for 30 min before being used for ICSI. The embryos were cultured for up to 120 h to evaluate in vitro development or transferred into recipient females at the 2-cell stage to evaluate in vivo development. Again, heat treatment had no effect on fertilization rate or early cleavage, but development to blastocyst was significantly less with sperm heated to 56 °C. The implantation rate was also significantly lower with sperm heated to 56 °C, but the authors suggest that this was due to reduced preimplantation development rather than implantation failure.

As might be expected, experimental studies suggest that heating affects sperm DNA and chromatin structure (95, 98–100). In addition, heat stress of bulls resulted in a decrease in the affinity of sperm to bind heparin, which reflected the membrane integrity of the sperm, and corresponded to an increase in abnormal morphology (101). An effect of increased temperature on the sperm membrane is also suggested by the observations that it caused reduced time for capacitation in hamster sperm (102), and was associated with sperm agglutination and low percentages of intact acrosomes in boar sperm (103).

Taken together, these studies strongly support the idea that season of the year and/or temperature has very significant effects on the biological functions of the sperm and the oocyte. The result is that exposure of the gametes to increased temperatures is frankly deleterious to subsequent embryo development.

4. The Effects of Air Pollution on Gametes, Embryo Development, and Fertility

Airborne pollutants are of major concern in human health, particularly in relation to lung and heart disease (104, 105). Outdoor air pollution has been negatively linked to the pregnancy outcome of natural conception in humans (106–111). Indoor air pollution is also of concern, and, if anything, the concentration of many pollutants is greater indoors than outdoors because pollutants from building materials, furniture and other source are accumulated (112). This has received considerable attention in the design and use of air filtration and purification systems for human IVF laboratories (113). In a number of studies, the outcome of ART procedures was significantly improved when specialized filters were used to remove particulates and volatile organic contaminants from laboratory gas supplies, incubators, and the laboratory space (114–118). The results of these studies demonstrate that the developing embryo is affected by pollutants in the IVF laboratory. Increasing evidence suggests that the gametes are also affected by exposure to air pollution before they arrive at the ART laboratory.

Perin et al. (119) showed that the concentration of particulate air pollution in Sao Paulo, Brazil in the week before conception was directly related to embryo loss for both natural conception and ART. Legro et al. (120) similarly evaluated the relationships between atmospheric concentrations of particulates, SO_2, NO_2, and O_3 in the northeastern USA in the area of the patients' homes or in the area of the IVF clinic with pregnancy outcome of IVF. Increases in NO_2 at home or at the IVF clinic, were associated with a decreased likelihood of pregnancy and live birth. Fine particulate matter at the clinic was associated with decreased conception rates. A more elaborate study in Sao Paulo (121) showed that for ART, preconception particulate air pollution was unrelated to rates of oocyte maturation, fertilization, cleavage, or initial or clinical pregnancy diagnosis. However, the preconception particulate air pollution was related to the rate of pregnancy loss (Fig. 6), suggesting that environmental air pollution had effects on the gametes that were not evident during the ART procedures.

It is not clear from the results of these studies whether the effect of air pollution was on the sperm, oocytes, or both. However, in another study from Sao Paulo, female mice were exposed to filtered or unfiltered ambient air, before or during gestation, or both (122). Exposure to unfiltered (i.e., polluted) air before gestation led to changes in the placenta and decreased fetal weight. In another study, mice were exposed to filtered or unfiltered ambient Sao Paulo air during prenatal and postnatal life, postnatal life only, or neither (123). At 6 weeks of age, the end of the exposure treatments, the females were superovulated, the oocytes collected

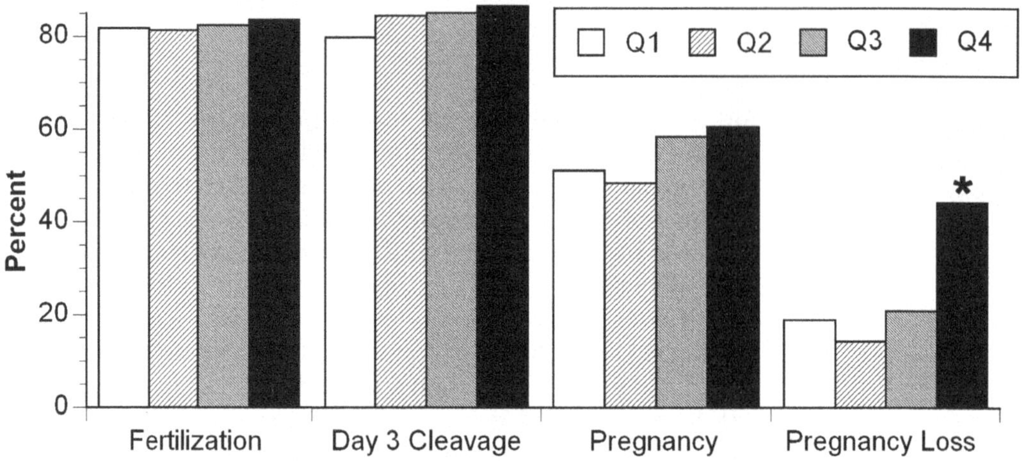

Fig. 6. The effect of preconception environmental particulate air pollution on the outcome of human ART. Particulate concentrations; Q1: ≤30.48 μg/m^3, Q2: 30.49–42.00 μg/m^3, Q3: 42.01–56.72 μg/m^3, and Q4: >56.72 μg/m^3. Particulate concentration had no effect on fertilization, Day 3 cleavage, or initial pregnancy rates, but the rate of pregnancy loss was significantly greater when the particulate concentration was >56.72 μg/m^3.

and fertilized in vitro, and the embryos cultured for 120 h. There were no effects on gestation length, litter size, birth weight or ovarian response among the exposure groups. Neither were there any effects on fertilization rate or development to blastocyst stage at 96 h of culture, or total blastocyst cell number at 120 h. However, the number of ICM cells and the ICM/TE ratios were significantly lower with prenatal or postnatal exposure to unfiltered (polluted) air. This clearly demonstrates that air pollution can have effects on the oocyte, even when the exposure occurs before birth.

Several studies have shown negative effects of environmental air pollution on human sperm morphology (124–126) and sperm DNA damage (127–129). As noted above, there is good evidence that sperm DNA damage is negatively related to embryo development and other outcomes of ART. Taken together, these two phenomena suggest that, as for its effects on the oocyte, air pollution is likely to have negative effects on the sperm that would affect the outcome of ART. However, there is limited information available that directly and specifically links the effects of environmental air pollution on the sperm alone to the outcome of ART. There is some evidence to show that cigarette smoking by the male partner (sperm source) has a negative effect on the outcome of human ART (see ref. (130)). Lichtenfels et al. (131) exposed male mice to filtered or unfiltered Sao Paulo ambient air 4 months before mating. Exposure to unfiltered (polluted) air had no effect on litter size, but the ratio of male to female pups was significantly reduced. They also found a significant decrease in the sex ratio of human babies linked to the particulate concentrations in Sao Paulo air. It therefore appears that air pollution may have effects on the sperm that ultimately affect embryo development, but the mechanisms are unclear.

5. Implications for ART Practice

Epidemiological, clinical, and experimental evidence all indicate that exposures of the biological parents (and hence their gametes) to heat and/or to air pollution can adversely affect the outcome of ART. It may, therefore, be useful to advise IVF patients to avoid such exposures. Unfortunately, this is not always possible or practical, and, moreover, they may have been exposed long before they became patients. It is nonetheless important that ART practitioners and embryologists be aware of the possible effects of temperature and air pollution as a consideration in laboratory quality management.

For example, consider the hypothetical situation shown in Fig. 7. The long-term averages for fertilization, cleavage and blastocyst development were very good. Last week however, fertilization and cleavage were still good, but blastocyst development was terrible. A superficial analysis might lead to the conclusion that there was a fault in the embryo culture system. Based on that information alone, such a conclusion would be completely unjustified. In the words of H.L. Menken, "for every problem, there is one solution which is simple, neat and wrong."

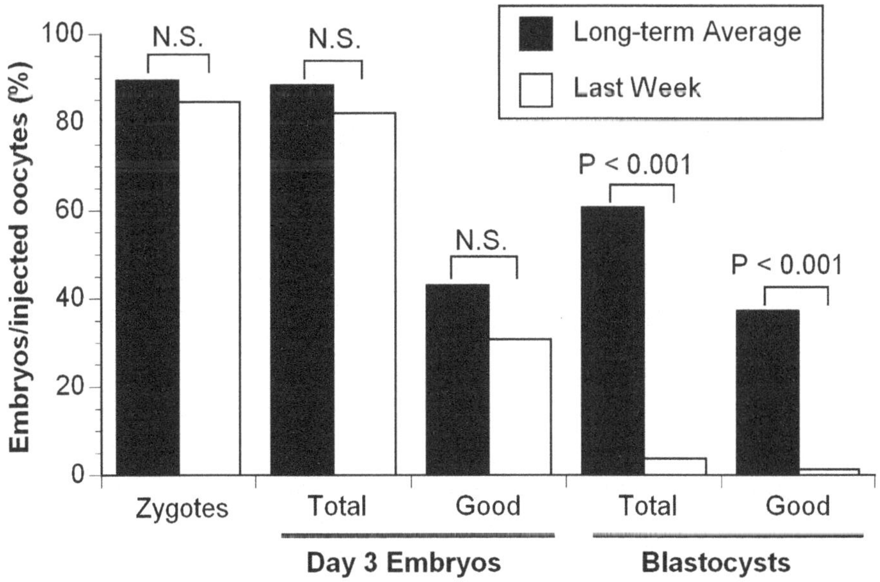

Fig. 7. A hypothetical change in the in vitro outcomes of ART in one laboratory.

In fact, Fig. 7 presents exactly the same data as were presented in Fig. 3, only the legend has been changed. The effect on in vitro development, as indicated in Fig. 3, was due to a difference in the morphological quality of the individual sperm used for injection. Consequently, if presented with the situation shown in Fig. 7, it would be at least as logical to suppose that the decrease in embryo development was due to a change in patient characteristics or sperm processing as to a fault in the embryo culture system. Such a decrease could also be the result of high preconception environmental temperature, as shown in Fig. 4 and Fig. 5. By the same token, an above average rate of spontaneous abortion for transfers done in one particular week could be the result of a spike in air pollution before fertilization, as shown in Fig. 6.

In conclusion, human ART is a very complicated process, and the outcome can be affected by a myriad of events, conditions, and factors, both in the laboratory and before the gametes are collected. A truly critical and scientific analysis of the ART outcomes for any clinic must include all of the possible factors, including preconception environmental temperature and air pollution.

Acknowledgments

My thanks to Dr. T.B. Pool (Fertility Center of San Antonio) for the many entertaining, stimulating, and enlightening discussions of human ART that led to this paper. Thanks also to Dr. Paulo Perin (CEER—Specialized Center for Human Reproduction, Sao Paulo) for bringing Lichtenfels et al. (131) to my attention.

References

1. Seamark RF, Robinson JS (1995) Potential health hazards of assisted human reproduction. Potential health problems stemming from assisted reproduction programmes. Hum Reprod 10:1321–1322
2. Rieger D (1998) Effects of the in vitro chemical environment during early embryogenesis on subsequent development. Arch Toxicol Suppl 20:121–129
3. Pool TB (2005) An update on embryo culture for human assisted reproductive technology: media, performance, and safety. Semin Reprod Med 23:309–318
4. Gardner DK, Lane M (2007) Embryo culture systems. In: Gardner DK (ed) In vitro fertilization: a practical approach. Informa Healthcare, New York, London, pp 221–282
5. Leese HJ, Sturmey RG, Baumann CG, McEvoy TG (2007) Embryo viability and metabolism: obeying the quiet rules. Hum Reprod 22:3047–3050
6. Thompson JG, Mitchell M, Kind KL (2007) Embryo culture and long-term consequences. Reprod Fertil Dev 19:43–52
7. Duranthon V, Watson AJ, Lonergan P (2008) Preimplantation embryo programming: transcription, epigenetics, and culture environment. Reproduction 135:141–150
8. Vajta G, Rienzi L, Cobo A, Yovich J (2010) Embryo culture: can we perform better than nature? Reprod Biomed Online 20:453–469
9. Lee TM, Zucker I (1988) Vole infant development is influenced perinatally by maternal photoperiodic history. Am J Physiol 255:R831–R838
10. Lumey LH (1992) Decreased birthweights in infants after maternal in utero exposure to the Dutch famine of 1944–1945. Paediatr Perinat Epidemiol 6:240–253

11. Sharpe RM, Franks S (2002) Environment, lifestyle and infertility – an inter-generational issue. Nat Cell Biol 4(Suppl):s33–s40
12. Swain JE, Pool TB (2008) ART failure: oocyte contributions to unsuccessful fertilization. Hum Reprod Update 14:431–446
13. Van Blerkom J, Antczak M, Schrader R (1997) The developmental potential of the human oocyte is related to the dissolved oxygen content of follicular fluid: association with vascular endothelial growth factor levels and perifollicular blood flow characteristics. Hum Reprod 12:1047–1055
14. Hamel M, Dufort I, Robert C, Gravel C, Leveille MC, Leader A, Sirard MA (2008) Identification of differentially expressed markers in human follicular cells associated with competent oocytes. Hum Reprod 23: 1118–1127
15. Adriaenssens T, Wathlet S, Segers I, Verheyen G, De Vos A, Van der Elst J, Coucke W, Devroey P, Smitz J (2010) Cumulus cell gene expression is associated with oocyte developmental quality and influenced by patient and treatment characteristics. Hum Reprod 25:1259–1270
16. Wang HX, Tong D, El-Gehani F, Tekpetey FR, Kidder GM (2009) Connexin expression and gap junctional coupling in human cumulus cells: contribution to embryo quality. J Cell Mol Med 13:972–984
17. Smitz JE, Thompson JG, Gilchrist RB (2011) The promise of in vitro maturation in assisted reproduction and fertility preservation. Semin Reprod Med 29:24–37
18. Heindryckx B, De Gheselle S, Lierman S, Gerris J, De Sutter P (2011) Efficiency of polarized microscopy as a predictive tool for human oocyte quality. Hum Reprod 26:535–544
19. Shen Y, Stalf T, Mehnert C, Eichenlaub-Ritter U, Tinneberg HR (2005) High magnitude of light retardation by the zona pellucida is associated with conception cycles. Hum Reprod 20:1596–1606
20. Nagy ZP, Jones-Colon S, Roos P, Botros L, Greco E, Dasig J, Behr B (2009) Metabolomic assessment of oocyte viability. Reprod Biomed Online 18:219–225
21. Goovaerts IG, Leroy JL, Jorssen EP, Bols PE (2010) Noninvasive bovine oocyte quality assessment: possibilities of a single oocyte culture. Theriogenology 74:1509–1520
22. Sugulle A, Dochi O, Koyama H (2008) Developmental competence of bovine oocytes selected by brilliant cresyl blue staining: effect of the presence of corpus luteum on embryo development. J Mamm Ova Res 25:50–55
23. Bhojwani S, Alm H, Torner H, Kanitz W, Poehland R (2007) Selection of developmentally competent oocytes through brilliant cresyl blue stain enhances blastocyst development rate after bovine nuclear transfer. Theriogenology 67:341–345
24. Manjunatha BM, Gupta PS, Devaraj M, Ravindra JP, Nandi S (2007) Selection of developmentally competent buffalo oocytes by brilliant cresyl blue staining before IVM. Theriogenology 68:1299–1304
25. Wongsrikeao P, Otoi T, Yamasaki H, Agung B, Taniguchi M, Naoi H, Shimizu R, Nagai T (2006) Effects of single and double exposure to brilliant cresyl blue on the selection of porcine oocytes for in vitro production of embryos. Theriogenology 66:366–372
26. Spikings EC, Alderson J, St John JC (2007) Regulated mitochondrial DNA replication during oocyte maturation is essential for successful porcine embryonic development. Biol Reprod 76:327–335
27. Munne S, Fischer J, Warner A, Chen S, Zouves C, Cohen J (2006) Preimplantation genetic diagnosis significantly reduces pregnancy loss in infertile couples: a multicenter study. Fertil Steril 85:326–332
28. Kuliev A, Zlatopolsky Z, Kirillova I, Spivakova J, Cieslak Janzen J (2011) Meiosis errors in over 20,000 oocytes studied in the practice of preimplantation aneuploidy testing. Reprod Biomed Online 22:2–8
29. Montag M, van der Ven K, Rosing B, van der Ven H (2009) Polar body biopsy: a viable alternative to preimplantation genetic diagnosis and screening. Reprod Biomed Online 18(Suppl 1):6–11
30. Lewis SE (2007) Is sperm evaluation useful in predicting human fertility? Reproduction 134:31–40
31. Salumets A, Suikkari AM, Mols T, Soderstrom-Anttila V, Tuuri T (2002) Influence of oocytes and spermatozoa on early embryonic development. Fertil Steril 78:1082–1087
32. Dubey A, Dayal MB, Frankfurter D, Balazy P, Peak D, Gindoff PR (2008) The influence of sperm morphology on preimplantation genetic diagnosis cycles outcome. Fertil Steril 89:1665–1669
33. Check JH, Bollendorf A, Wilson C, Summers-Chase D, Horwath D, Yuan W (2007) A retrospective comparison of pregnancy outcome following conventional oocyte insemination vs intracytoplasmic sperm injection for isolated abnormalities in sperm morphology using strict criteria. J Androl 28:607–612
34. Barratt CL, Aitken RJ, Bjorndahl L, Carrell DT, de Boer P, Kvist U, Lewis SE, Perreault

SD, Perry MJ, Ramos L, Robaire B, Ward S, Zini A (2010) Sperm DNA: organization, protection and vulnerability: from basic science to clinical applications – a position report. Hum Reprod 25:824–838
35. Lewis SE, Simon L (2010) Clinical implications of sperm DNA damage. Hum Fertil (Camb) 13:201–207
36. Zini A, Jamal W, Cowan L, Al-Hathal N (2011) Is sperm DNA damage associated with IVF embryo quality? A systematic review. J Assist Reprod Genet 28(5):391–397
37. Virro MR, Larson-Cook KL, Evenson DP (2004) Sperm chromatin structure assay (SCSA) parameters are related to fertilization, blastocyst development, and ongoing pregnancy in in vitro fertilization and intracytoplasmic sperm injection cycles. Fertil Steril 81:1289–1295
38. Seli E, Gardner DK, Schoolcraft WB, Moffatt O, Sakkas D (2004) Extent of nuclear DNA damage in ejaculated spermatozoa impacts on blastocyst development after in vitro fertilization. Fertil Steril 82:378–383
39. Nasr-Esfahani MH, Salehi M, Razavi S, Anjomshoa M, Rozbahani S, Moulavi F, Mardani M (2005) Effect of sperm DNA damage and sperm protamine deficiency on fertilization and embryo development post-ICSI. Reprod Biomed Online 11:198–205
40. Meseguer M, Martinez-Conejero JA, O'Connor JE, Pellicer A, Remohi J, Garrido N (2008) The significance of sperm DNA oxidation in embryo development and reproductive outcome in an oocyte donation program: a new model to study a male infertility prognostic factor. Fertil Steril 89:1191–1199
41. Avendano C, Franchi A, Duran H, Oehninger S (2009) DNA fragmentation of normal spermatozoa negatively impacts embryo quality and intracytoplasmic sperm injection outcome. Fertil Steril 94:549–557
42. Avendano C, Franchi A, Taylor S, Morshedi M, Bocca S, Oehninger S (2009) Fragmentation of DNA in morphologically normal human spermatozoa. Fertil Steril 91: 1077–1084
43. Sousa AP, Tavares RS, Velez de la Calle JF, Figueiredo H, Almeida V, Almeida-Santos T, Ramalho-Santos J (2009) Dual use of Diff-Quik-like stains for the simultaneous evaluation of human sperm morphology and chromatin status. Hum Reprod 24:28–36
44. Meseguer M, Santiso R, Garrido N, Garcia-Herrero S, Remohi J, Fernandez JL (2011) Effect of sperm DNA fragmentation on pregnancy outcome depends on oocyte quality. Fertil Steril 95:124–128
45. Aitken RJ (2004) Founders' Lecture. Human spermatozoa: fruits of creation, seeds of doubt. Reprod Fertil Dev 16:655–664
46. Zorn B, Vidmar G, Meden-Vrtovec H (2003) Seminal reactive oxygen species as predictors of fertilization, embryo quality and pregnancy rates after conventional in vitro fertilization and intracytoplasmic sperm injection. Int J Androl 26:279–285
47. Meseguer M, de los Santos MJ, Simon C, Pellicer A, Remohi J, Garrido N (2006) Effect of sperm glutathione peroxidases 1 and 4 on embryo asymmetry and blastocyst quality in oocyte donation cycles. Fertil Steril 86: 1376–1385
48. Parmegiani L, Cognigni GE, Bernardi S, Troilo E, Ciampaglia W, Filicori M (2010) "Physiologic ICSI": hyaluronic acid (HA) favors selection of spermatozoa without DNA fragmentation and with normal nucleus, resulting in improvement of embryo quality. Fertil Steril 93:598–604
49. Urner F, Sakkas D (1999) Characterization of glycolysis and pentose phosphate pathway activity during sperm entry into the mouse oocyte. Biol Reprod 60:973–978
50. Comizzoli P, Urner F, Sakkas D, Renard JP (2003) Up-regulation of glucose metabolism during male pronucleus formation determines the early onset of the s phase in bovine zygotes. Biol Reprod 68:1934–1940
51. Bartoov B, Berkovitz A, Eltes F (2001) Selection of spermatozoa with normal nuclei to improve the pregnancy rate with intracytoplasmic sperm injection. N Engl J Med 345:1067–1068
52. De Vos A, Van De Velde H, Joris H, Verheyen G, Devroey P, Van Steirteghem A (2003) Influence of individual sperm morphology on fertilization, embryo morphology, and pregnancy outcome of intracytoplasmic sperm injection. Fertil Steril 79:42–48
53. Berkovitz A, Eltes F, Ellenbogen A, Peer S, Feldberg D, Bartoov B (2006) Does the presence of nuclear vacuoles in human sperm selected for ICSI affect pregnancy outcome? Hum Reprod 21:1787–1790
54. Antinori M, Licata E, Dani G, Cerusico F, Versaci C, d'Angelo D, Antinori S (2008) Intracytoplasmic morphologically selected sperm injection: a prospective randomized trial. Reprod Biomed Online 16:835–841
55. Vanderzwalmen P, Hiemer A, Rubner P, Bach M, Neyer A, Stecher A, Uher P, Zintz M, Lejeune B, Vanderzwalmen S, Cassuto G, Zech NH (2008) Blastocyst development after sperm selection at high magnification is associated with size and number of

nuclear vacuoles. Reprod Biomed Online 17: 617–627
56. Telford NA, Watson AJ, Schultz GA (1990) Transition from maternal to embryonic control in early mammalian development: a comparison of several species. Mol Reprod Dev 26:90–100
57. Oliveira JB, Massaro FC, Baruffi RL, Mauri AL, Petersen CG, Silva LF, Vagnini LD, Franco JG Jr (2010) Correlation between semen analysis by motile sperm organelle morphology examination and sperm DNA damage. Fertil Steril 94:1937–1940
58. Betteridge KJ, Rieger D (1993) Embryo transfer and related techniques in domestic animals, and their implications for human medicine. Hum Reprod 8:147–167
59. Bronson FH (2009) Climate change and seasonal reproduction in mammals. Philos Trans R Soc Lond B Biol Sci 364:3331–3340
60. Lam DA, Miron JA (1994) Global patterns of seasonal variation in human fertility. Ann N Y Acad Sci 709:9–28
61. Rojansky N, Brzezinski A, Schenker JG (1992) Seasonality in human reproduction: an update. Hum Reprod 7:735–745
62. Smits LJ, Zielhuis GA, Jongbloet PH, Straatman H (1998) Seasonal variation in human fecundability. Hum Reprod 13:3520–3524
63. Stolwijk AM, Reuvers MJ, Hamilton CJ, Jongbloet PH, Hollanders JM, Zielhuis GA (1994) Seasonality in the results of in-vitro fertilization. Hum Reprod 9:2300–2305
64. Rojansky N, Benshushan A, Meirsdorf S, Lewin A, Laufer N, Safran A (2000) Seasonal variability in fertilization and embryo quality rates in women undergoing IVF. Fertil Steril 74:476–481
65. Weigert M, Feichtinger W, Kulin S, Kaali SG, Dorau P, Bauer P (2001) Seasonal influences on in vitro fertilization and embryo transfer. J Assist Reprod Genet 18:598–602
66. Vahidi A, Kalantar SM, Soleiman M, Hossein M, Arjmand A, Aflatoonian A, Karimzadeh MA, Kermaninejhad A (2004) The relationship between seasonal variability and pregnancy rates in women undergoing assisted reproductive technique. IJRM 2:82–86
67. Chang SY, Lan KC, Chen CW, Huang FJ, Tsai MY, Chang CY (2005) The influences of weather on patients with different ovarian responses in the treatment of assisted reproductive technology. J Assist Reprod Genet 22:191–198
68. Wood S, Quinn A, Troupe S, Kingsland C, Lewis-Jones I (2006) Seasonal variation in assisted conception cycles and the influence of photoperiodism on outcome in in vitro fertilization cycles. Hum Fertil (Camb) 9: 223–229
69. Fleming C, Nice L, Hughes AO, Hull MG (1994) Apparent lack of seasonal variation in implantation rates after in-vitro fertilization. Hum Reprod 9:2164–2166
70. Saltz-Greco SM, Schinfeld J, Somkuti S, Smith S, Barmat L (2004) Seasonality of spontaneous and IVF pregnancy rates, and temperature/humidity effect on IVF. Fertil Steril 82(Suppl 2):S195 (Abstract P-174)
71. Revelli A, La Sala GB, Gennarelli G, Scatigna L, Racca C, Massobrio M (2005) Seasonality and human in vitro fertilization outcome. Gynecol Endocrinol 21:12–17
72. Wunder DM, Limoni C, Birkhauser MH (2005) Lack of seasonal variations in fertilization, pregnancy and implantation rates in women undergoing IVF. Hum Reprod 20:3122–3129
73. Fuquay JW (1981) Heat stress as it affects animal production. J Anim Sci 52:164–174
74. Hansen PJ, Arechiga CF (1999) Strategies for managing reproduction in the heat-stressed dairy cow. J Anim Sci 77(Suppl 2):36–50
75. Jordan ER (2003) Effects of heat stress on reproduction. J Dairy Sci 86(E. Suppl): E104–E114
76. Hansen PJ (2009) Effects of heat stress on mammalian reproduction. Philos Trans R Soc Lond B Biol Sci 364:3341–3350
77. Al-Katanani YM, Paula-Lopes FF, Hansen PJ (2002) Effect of season and exposure to heat stress on oocyte competence in Holstein cows. J Dairy Sci 85:390–396
78. Gendelman M, Aroyo A, Yavin S, Roth Z (2010) Seasonal effects on gene expression, cleavage timing, and developmental competence of bovine preimplantation embryos. Reproduction 140:73–82
79. Zeron Y, Ocheretny A, Kedar O, Borochov A, Sklan D, Arav A (2001) Seasonal changes in bovine fertility: relation to developmental competence of oocytes, membrane properties and fatty acid composition of follicles. Reproduction 121:447–454
80. Putney DJ, Mullins S, Thatcher WW, Drost M, Gross TS (1989) Embryonic development in superovulated dairy cattle exposed to elevated ambient temperatures between the onset of estrus and insemination. Anim Reprod Sci 19:37–51
81. Rocha A, Randel RD, Broussard JR, Lim JM, Blair RM, Roussel JD, Godke RA, Hansel W (1998) High environmental temperature and

humidity decrease oocyte quality in Bos taurus but not in Bos indicus cows. Theriogenology 49:657–665
82. Roth Z, Hansen PJ (2004) Involvement of apoptosis in disruption of developmental competence of bovine oocytes by heat shock during maturation. Biol Reprod 71:1898–1906
83. Edwards JL, Saxton AM, Lawrence JL, Payton RR, Dunlap JR (2005) Exposure to a physiologically relevant elevated temperature hastens in vitro maturation in bovine oocytes. J Dairy Sci 88:4326–4333
84. Wang JZ, Sui HS, Miao DQ, Liu N, Zhou P, Ge L, Tan JH (2009) Effects of heat stress during in vitro maturation on cytoplasmic versus nuclear components of mouse oocytes. Reproduction 137:181–189
85. Ju JC, Jiang S, Tseng JK, Parks JE, Yang X (2005) Heat shock reduces developmental competence and alters spindle configuration of bovine oocytes. Theriogenology 64: 1677–1689
86. Ju JC, Tseng JK (2004) Nuclear and cytoskeletal alterations of in vitro matured porcine oocytes under hyperthermia. Mol Reprod Dev 68:125–133
87. Ulberg LC, Burfening PJ (1967) Embryo death resulting from adverse environment on spermatozoa or ova. J Anim Sci 26:571–577
88. Setchell BP (1998) The Parkes Lecture. Heat and the testis. J Reprod Fertil 114:179–194
89. Thonneau P, Bujan L, Multigner L, Mieusset R (1998) Occupational heat exposure and male fertility: a review. Hum Reprod 13:2122–2125
90. Paul C, Melton DW, Saunders PT (2008) Do heat stress and deficits in DNA repair pathways have a negative impact on male fertility? Mol Hum Reprod 14:1–8
91. Ossenbuhn S (1998) Exogenous influences on human fertility: fluctuations in sperm parameters and results of in-vitro fertilization coincide with conceptions in the normal population. Hum Reprod 13:2165–2171
92. Chen Z, Toth T, Godfrey-Bailey L, Mercedat N, Schiff I, Hauser R (2003) Seasonal variation and age-related changes in human semen parameters. J Androl 24:226–231
93. Burfening PJ, Elliott DS, Eisen EJ, Ulberg LC (1970) Survival of embryos resulting from spermatozoa produced by mice exposed to elevated ambient temperature. J Anim Sci 30:578–582
94. Zhu BK, Setchell BP (2004) Effects of paternal heat stress on the in vivo development of preimplantation embryos in the mouse. Reprod Nutr Dev 44:617–629
95. Paul C, Murray AA, Spears N, Saunders PT (2008) A single, mild, transient scrotal heat stress causes DNA damage, subfertility and impairs formation of blastocysts in mice. Reproduction 136:73–84
96. Burfening PJ, Ulberg LC (1968) Embryonic survival subsequent to culture of rabbit spermatozoa at 38° and 40°C. J Reprod Fertil 15:87–92
97. Cozzi J, Monier-Gavelle F, Lievre N, Bomsel M, Wolf JP (2001) Mouse offspring after microinjection of heated spermatozoa. Biol Reprod 65:1518–1521
98. Sailer BL, Sarkar LJ, Bjordahl JA, Jost LK, Evenson DP (1997) Effects of heat stress on mouse testicular cells and sperm chromatin structure. J Androl 18:294–301
99. Love CC, Kenney RM (1999) Scrotal heat stress induces altered sperm chromatin structure associated with a decrease in protamine disulfide bonding in the stallion. Biol Reprod 60:615–620
100. Perez-Crespo M, Pintado B, Gutierrez-Adan A (2008) Scrotal heat stress effects on sperm viability, sperm DNA integrity, and the offspring sex ratio in mice. Mol Reprod Dev 75:40–47
101. Ax RL, Gilbert GR, Shook GE (1987) Sperm in poor quality semen from bulls during heat stress have a lower affinity for binding hydrogen-3 heparin. J Dairy Sci 70:195–200
102. Bedford JM, Yanagimachi R (1991) Epididymal storage at abdominal temperature reduces the time required for capacitation of hamster spermatozoa. J Reprod Fertil 91: 403–410
103. Murase T, Imaeda N, Yamada H, Miyazawa K (2007) Seasonal changes in semen characteristics, composition of seminal plasma and frequency of acrosome reaction induced by calcium and calcium ionophore A23187 in Large White boars. J Reprod Dev 53: 853–865
104. Bernstein JA, Alexis N, Barnes C, Bernstein IL, Nel A, Peden D, Diaz-Sanchez D, Tarlo SM, Williams PB (2004) Health effects of air pollution. J Allergy Clin Immunol 114: 1116–1123
105. Kampa M, Castanas E (2008) Human health effects of air pollution. Environ Pollut 151:362–367
106. Dejmek J, Solansk I, Beneö I, Lenicek J, Sram RJ (2001) Air pollution and pregnancy outcome. In: Sram RJ (ed) Teplice program: impact of air pollution on human health. Academia, Prague, pp 127–144

107. Maroziene L, Grazuleviciene R (2002) Maternal exposure to low-level air pollution and pregnancy outcomes: a population-based study. Environ Health 1:6
108. Sram RJ, Binkova B, Dejmek J, Bobak M (2005) Ambient air pollution and pregnancy outcomes: a review of the literature. Environ Health Perspect 113:375–382
109. Ritz B, Wilhelm M, Hoggatt KJ, Ghosh JK (2007) Ambient air pollution and preterm birth in the environment and pregnancy outcomes study at the University of California, Los Angeles. Am J Epidemiol 166:1045–1052
110. Brauer M, Lencar C, Tamburic L, Koehoorn M, Demers P, Karr C (2008) A cohort study of traffic-related air pollution impacts on birth outcomes. Environ Health Perspect 116: 680–686
111. Green RS, Malig B, Windham GC, Fenster L, Ostro B, Swan S (2009) Residential exposure to traffic and spontaneous abortion. Environ Health Perspect 117:1939–1944
112. Bernstein JA, Alexis N, Bacchus H, Bernstein IL, Fritz P, Horner E, Li N, Mason S, Nel A, Oullette J, Reijula K, Reponen T, Seltzer J, Smith A, Tarlo SM (2008) The health effects of non-industrial indoor air pollution. J Allergy Clin Immunol 121:585–591
113. Hall J, Gilligan A, Schimmel T, Cecchi M, Cohen J (1998) The origin, effects and control of air pollution in laboratories used for human embryo culture. Hum Reprod 13(Suppl 4):146–155
114. Cohen J, Gilligan A, Esposito W, Schimmel T, Dale B (1997) Ambient air and its potential effects on conception in vitro. Hum Reprod 12:1742–1749
115. Forman M, Polanski V, Horvath P, Gilligan A, Rieger D (2004) Reductions in volatile organic compounds, aldehydes, and particulate air contaminants in an IVF laboratory by centralized and stand-alone air filtration systems. Fertil Steril 82(Suppl 2):S324 (Abstract P-535)
116. Merton JS, Vermeulen ZL, Otter T, Mullaart E, de Ruigh L, Hasler JF (2007) Carbon-activated gas filtration during in vitro culture increased pregnancy rate following transfer of in vitro-produced bovine embryos. Theriogenology 67:1233–1238
117. Racowsky C, Jackson KV, Nurredin A, Balint C, Shen S, de los Santos MJ, Kely JR, Pan Y (1999) Carbon-activated air filtration results in reduced spontaneous abortion rates following IVF. In: Proceedings of the 11th world congress on in-vitro fertilization and human reproductive genetics, pp O–059 (Abstract), Sydney, Australia.
118. Younis A, Carnovale D, Butler W (2009) Improvement in IVF outcome after change of CO_2 supplies to incubators. Fertil Steril 92(Suppl):S155 (Abstract P-241)
119. Perin PM, Maluf M, Czeresnia CE, Nicolosi Foltran Januario DA, Nascimento Saldiva PH (2010) Effects of exposure to high levels of particulate air pollution during the follicular phase of the conception cycle on pregnancy outcome in couples undergoing in vitro fertilization and embryo transfer. Fertil Steril 93: 301–303
120. Legro RS, Sauer MV, Mottla GL, Richter KS, Li X, Dodson WC, Liao D (2010) Effect of air quality on assisted human reproduction. Hum Reprod 25:1317–1324
121. Perin PM, Maluf M, Czeresnia CE, Januario DA, Saldiva PH (2010) Impact of short-term preconceptional exposure to particulate air pollution on treatment outcome in couples undergoing in vitro fertilization and embryo transfer (IVF/ET). J Assist Reprod Genet 27:371–382
122. Veras MM, Damaceno-Rodrigues NR, Caldini EG, Maciel Ribeiro AA, Mayhew TM, Saldiva PH, Dolhnikoff M (2008) Particulate urban air pollution affects the functional morphology of mouse placenta. Biol Reprod 79: 578–584
123. Maluf M, Perin PM, Foltran Januario DA, Nascimento Saldiva PH (2009) In vitro fertilization, embryo development, and cell lineage segregation after pre- and/or postnatal exposure of female mice to ambient fine particulate matter. Fertil Steril 92:1725–1735
124. Boggia B, Carbone U, Farinaro E, Zarrilli S, Lombardi G, Colao A, De Rosa N, De Rosa M (2009) Effects of working posture and exposure to traffic pollutants on sperm quality. J Endocrinol Invest 32:430–434
125. De Rosa M, Zarrilli S, Paesano L, Carbone U, Boggia B, Petretta M, Maisto A, Cimmino F, Puca G, Colao A, Lombardi G (2003) Traffic pollutants affect fertility in men. Hum Reprod 18:1055–1061
126. Hansen C, Luben TJ, Sacks JD, Olshan A, Jeffay S, Strader L, Perreault SD (2010) The effect of ambient air pollution on sperm quality. Environ Health Perspect 118:203–209
127. Rubes J, Rybar R, Prinosilova P, Veznik Z, Chvatalova I, Solansky I, Sram RJ (2010) Genetic polymorphisms influence the susceptibility of men to sperm DNA damage associated with exposure to air pollution. Mutat Res 683:9–15

128. Rubes J, Selevan SG, Evenson DP, Zudova D, Vozdova M, Zudova Z, Robbins WA, Perreault SD (2005) Episodic air pollution is associated with increased DNA fragmentation in human sperm without other changes in semen quality. Hum Reprod 20:2776–2783

129. Rubes J, Selevan SG, Sram RJ, Evenson DP, Perreault SD (2007) GSTM1 genotype influences the susceptibility of men to sperm DNA damage associated with exposure to air pollution. Mutat Res 625:20–28

130. Waylen AL, Metwally M, Jones GL, Wilkinson AJ, Ledger WL (2009) Effects of cigarette smoking upon clinical outcomes of assisted reproduction: a meta-analysis. Hum Reprod Update 15:31–44

131. Lichtenfels AJ, Gomes JB, Pieri PC, El Khouri Miraglia SG, Hallak J, Saldiva PH (2007) Increased levels of air pollution and a decrease in the human and mouse male-to-female ratio in Sao Paulo, Brazil. Fertil Steril 87: 230–232

Chapter 20

Culture Systems: Fluid Dynamic Embryo Culture Systems (Microfluidics)

André Monteiro da Rocha and Gary D. Smith

Abstract

The tubal/uterine lumen is a dynamic environment in which oocytes, eggs, and early embryos are submitted to different forces generated by cilia and peristaltic flow of tubal fluid. The movement of the tubal/uterine fluid, the chemical diversity, and their interaction produce a unique environment able to support embryo development and modulate gene expression.

Although culture of embryos is supported in static and low complexity chemical conditions, application of fluidic dynamics in assisted reproduction technology to improve outcomes has been in development for almost a decade. Several attempts to build devices able to facilitate fertilization and embryo culture have been made, but dynamic fluidic devices are not yet available for mass scale use in clinical embryology laboratories. Indeed, such devices for embryo culture have been constructed and they are under evaluation in IRB approved studies. Fluid dynamic devices appear to enhance embryo development and they may be innovative resources for clinical and experimental embryology laboratories. This chapter reviews the principles and results of dynamic fluid systems, and the materials and methods required to produce microfunnel dynamic culture systems for use with embryos.

Key words: Microfluidics, Embryo culture, IVF

1. Introduction

The lumen of the oviduct is a complex environment which supports fertilization of oocytes, early embryo development, and facilitates embryo transport to the uterus. This transport occurs due to the flux generated by cilia and peristaltic movements from the contraction of the smooth muscle layer. The combination of muscular and cilia actions over tubal crypts generates a turbulent flow toward the uterine horn (1). This turbulent progressive movement is responsible for homogenization of tubal secretion, dilution of

Gary D. Smith et al. (eds.), *Embryo Culture: Methods and Protocols*, Methods in Molecular Biology, vol. 912, DOI 10.1007/978-1-61779-971-6_20, © Springer Science+Business Media, LLC 2012

undesirable metabolites, and facilitation of gamete interaction (1). The velocity of tubal flow and the frequency of ciliary movement is estimated to range from 6.5 to 29 μm/s and 5 to 20 Hz, respectively (2). This dynamic microenvironment provides mechanostimulation of the embryo (3). Despite the lack of understanding on how embryo mechano-sensitivity is mediated, it is possible, that as in other cells, mechanostimulation modulates gene expression and cell fate (4, 5).

Even though in vivo embryo development occurs in a dynamic environment, embryo culture in static conditions is possible. Indeed, static systems for in vitro embryo culture are the most common methods for the growth of mammalian embryos and they are widely adopted in human infertility treatments and the production of livestock. However, studies with animal models show that, in general, in vitro-grown embryos have a lower quality and produce lower pregnancy rates than in vivo-produced embryos. It is known that exposure to the tubal/uterine environment during early development modulates metabolic pathways of embryos (6–9), which suggests that perhaps current static conditions used in vitro may not be providing an ideal culture environment.

Due to the evident discrepancy between the environment of static culture systems and of the tubal/uterine environment, researchers have theorized that embryo culture systems utilizing fluidic dynamics could recapitulate beneficial conditions and stimuli received by the embryo in in vivo conditions. This chapter reviews several methods to obtain fluidic dynamic culture and the materials and methods necessary to produce dynamic funnel devices in order to culture embryos using this innovative approach.

1.1. Generation of a Dynamic Fluid Environment

Several articles have discussed the potential applications and advantages of dynamic environments for embryo culture obtained with microfluidics (10–14). Although these reports were optimistic and drew a promising perspective, a limited number of publications actually investigated the effects of dynamic environment on embryo development (15–19). The small number of investigative reports is likely due to the relative complexity in building systems able to combine embryo safety, ease use, and well-established physical and chemical requirements of embryos with the generation of appropriate fluid dynamics.

In general, reported methods to generate fluid dynamics can be divided into three categories: (1.2) platform/dish movement, (1.3) wave fluid movement, and (1.4) designed fluid dynamics. The adoption of any one of these methods should take into account the primary goal pursued: embryo culture with an added benefit of improved development, analysis, and/or production of healthier offspring. For example, generation of fluid dynamics with platform movement or random fluid movement might be suitable for the production of embryos (16, 17), but may not allow embryo assessment beyond traditional morphological evaluation. The election of a programmable designed fluid dynamic system would allow embryo culture with fluid movement, morphological assessment of the embryos, but it also potentiates integration of other technologies to the system, such as real-time analysis of embryo metabolome and secretome (20). In this manner, the adoption of one of these methods for the production of fluidic dynamics should be based on the needs and goals of each embryology laboratory.

1.2. Platform/Dish Movement

Media in traditional embryo culture are usually held in plastic culture dishes covered with oil to avoid evaporation. This dynamic culture strategy uses the same plate system, but varies from traditional static culture in that, while in the incubator, rotation or tilting of the platform in which culture dishes are kept creates fluidic movement (17). This alters the environment in which the embryo is cultured. The utilization of dish movement to create a dynamic environment for embryo culture was reported over 35 years ago (21), although the authors used a short period culture and this early study failed to consider shear stress exerted on embryos. Embryo tolerance and resilience to shear stress were determined more recently (22).

Platform movement to generate dynamic fluid in human embryo culture was examined recently by Matsuura et al. (17). The fluid dynamic was attained by tilting the platform to produce embryo and fluid movement with a radial velocity of 1° per second up to 10°. This setting produced an embryo rolling speed of 0.3 mm/min and a shear stress of 1.5×10^{-4} dyn/cm^2. Although there was no significant increase in the rate of blastocyst formation, the number of blastomeres in embryos cultured in the tilting embryo culture system was higher than in the static culture. Whether this increased numbers of blastomeres in human embryos subjected to tilting embryo culture will convert to increased pregnancy rates still needs to be determined in clinical trials. Similar effects of tilting embryo culture system on the number of blastomeres per blastocyst were observed in porcine and murine embryos (16, 17).

Tilting platforms also were used for the movement of embryos in microchannel devices (18). Devices containing microchannels with a series of constrictions were tilted to move bovine embryos through areas of varying width. It was proposed that this resulted in simulating peristaltic forces on embryos and mimic tubal peristalsis. This system remains to be optimized and tested in order to determine the range of beneficial forces applied to the embryos (18).

1.3. Wave Fluid Movement

Mechanical stimulus through vibration seems to increase the proliferation rate and behavior of some types of somatic cells (23–25). Platform vibration can be used to modulate embryo development in vitro through fluid and embryo movement (26, 27). Pig oocytes matured in vitro under vibration (20 Hz) for 5 s every 60 min produced higher blastocyst rates than those in static culture; however, beneficial effects of embryo vibration were not observed when embryos originated from oocytes matured under static conditions (26). This might indicate that, at least for this species, vibratory movement might have a positive impact during oocyte maturation (26). More recently, wave fluid movement was applied to human embryos in a similar fashion, but with a higher frequency (44 Hz) (27). The authors state that in their experience the application of 44 Hz vibration for 10 s every 60 min resulted in increased rates of blastocyst formation and pregnancy (27). The underlying mechanisms of the effects imposed by vibratory stimulation are still unknown.

1.4. Designed Fluid Dynamic

Designed fluid dynamics are obtained by forcing fluid through a microscopic and geometrically restrained area, commonly through channels. Microfluidics devices described for embryo culture include the use of (Subheading 1.4.1) gravity gradients, (14, 28), syringe pumping (29), or (Subheading 1.4.2) peristaltic movement produced by Braille pins over a flexible surface (15).

1.4.1. Gravity Gradient and Syringe Pumping

Gravity gradient and syringe pumping of media through a microchannel are simple methods to generate microfluidic dynamics. Generally, devices designed for embryo culture in fluid dynamic systems have a single channel with barriers (30, 31) or simple arrays that can be produced through photolithography or micromolding of thick or hard elements because these methods do not require flexible membranes (28). Usually, they have relatively simple construction and operation, and can accommodate more than one function. Indeed, microchannel devices have been used for cumulus cell and zona pellucida removal and embryo culture (14).

The great advantage of these devices are their simplicity, although there are some important considerations regarding the generation of shear stress, wash out of desirable autocrine factors and also the risk of embryo entrapment into the system. Embryos confined in a microchannel might be submitted to a unidirectional flow of up to 100 nL/s at an average velocity of 2 mm/s (28) and forces up to 10^{-7} N. Further modeling and simulations with this type of channel indicate that embryos in these devices are submitted to shear stress and washout of growth factors and cytokines secreted by the embryos (15).

Detrimental and time-dependent effects of shear stress on embryos were elegantly demonstrated by Xie et al. (22). Embryos submitted to shear stress of 1.2 dyn/cm^2 had increased activity of MAPK 8/9 triggering apoptosis. These effects were noticeable after 6 h of submission to the shear force. At this stage, several embryos were able to recover from stress; however, the proportion of embryos failing to recover after the exposure to shear forces increased in a time-dependent manner and after 12 h none of the embryos were able to recover from the stress (22).

Embryos grown in microchannel devices with fluid flow are subjected to a lesser degree of shear forces than those reported in Xie et al. (22). However, it is important to recognize there period of exposure are significantly longer and therefore may impact overall shear stress and account for compromised development. (19, 28). In some instances, these reports indicate an increase of blastocyst production in microchannels culture systems (19, 32), yet data are needed in relation to pregnancy rates, number of blastomeres per embryo and subcellular changes like increased activity in MAPK 8/9.

1.4.2. Peristaltic Movement

Programming actuation of sequential Braille pins that can press against and close a flexible membrane covering a microfluidic channel can deliver different types of peristaltic flow (33). This type of flow generation permits the control of frequency of fluid displacement and speed of fluid. Many times the material utilized for construction of these microfluidic devices is the polymer polydimethylsiloxane (PDMS), because it is easy to prototype, is gas permeable, flexible, and its optical properties allow microscopic observation (15, 18, 34, 35). However, the requirement for a flexible membrane needed for Braille pin actuation can lead to considerable evaporation of media and shifts in media osmolality (36).

To circumvent evaporation and increase in osmolality, several thicknesses (0.1–10 mm) of flexible PDMS membranes and different impermeable treatments were investigated. Thicker PDMS membranes had less evaporation while PDMS membranes of 0.1 mm and 0.2 mm yielded the maximum evaporation and large osmolality increases that were detrimental to embryo development. Osmolality changes of more than 300 mosm/kg, observed in microfluidic devices with microtunnel bottoms of 0.1 mm of PDMS, impaired the development of embryos to the blastocyst stage in comparison to thicker PDMS walls (PDMS 1 mm = 72 % and PDMS 10 mm = 67 %) and control devices with glass bottoms (67 %) (36). This presented a dilemma of needing a thin layer for Braille pin actuation, yet needing to prevent the culture environment from being detrimentally affected by evaporation and osmolality shifts not supportive of embryo development. Thus, it became necessary to treat PMDS to yield a flexible yet impermeable membrane. It was determined that coating the thin walls of PMDS in fluidic devices with 2.5 μm film of parylene prevented osmolality shift and increased the rate of blastocyst formation in comparison with uncoated devices (36).

The microfunnel shape to hold embryos within a microfluidic device with peristaltic flow was tested with extensive modeling and simulation. Embryo culture in microfunnels diminishes shear stress compared to embryos within microchannels with dynamic flow. Additionally, the microfunnel increases the concentration of autocrine factors around embryos and provides mechanostimulation (15). The advantages gained by the dynamic microfunnel can be observed in mouse embryos cultured under pulsatile movement with frequency of 0.1 Hz in prototypes of microfunnel devices. The percentage of hatched blastocysts and the number of blastomeres per embryo were greater after embryo culture in microfunnel dynamic culture system in comparison to microfunnel and petri dish static culture. Furthermore, these findings were followed by implantation and pregnancy studies that found enhanced implantation and ongoing pregnancy rates in embryos culture in the microfunnel dynamic system in comparison to static culture and they closely mirrored rates obtained with in vivo produced embryos (15).

Prototypes of dynamic microfunnel culture devices are undergoing evaluation in the United States and Brazil (37) and initial reports indicate that embryos grown in dynamic microfunnel devices until day 3 of development had lower fragmentation rates in comparison to those kept under traditional static conditions. Furthermore, human embryos grown in dynamic microfunnels were more likely to develop into embryos of good quality (greater than six blastomeres with less than 20 % of fragmentation on day 3) than those cultured under traditional static conditions (37). The effect of dynamic microfunnel devices on human embryo development to the stage of blastocyst and the impact of this technology on implantation and pregnancy rates should be addressed in forthcoming research.

1.5. Final Considerations

Devices for fluid dynamic embryo culture systems can be designed for different functions in an embryology laboratory and several reports demonstrate that different types of devices produce dynamic fluid environments and enhance embryo production. Clearly, these experiments targeted the short-term goal of providing more efficient culture systems for clinical IVF laboratories and have demonstrated that advantages can be obtained through numerous technical avenues. Of great potential is the integration of technologies into a single device within the clinical-assisted reproductive technologies laboratory. For this utility, a designed and programmable dynamic culture system will be needed to facilitate numerous aspects of existing gamete/embryo clinical laboratory manipulation and cultivation, as well as allowing new abilities such as real-time bioanalysis, that will collectively improve outcomes in clinical treatment of infertility. This provides the platform for laboratory-on-a-chip devices that couple embryo culture to real-time analysis of metabolites and/or secreted factors. This type of device might have tremendous applications in the clinical embryology setting to assist embryo selection with criteria based on physiological performance, rather than morphology criteria alone. In fact, the same features that make microfluidic devices attractive to clinical embryology laboratories might make these devices interesting to experimental embryology laboratories. The ability to produce sequential and/or gradient delivery of different chemicals can be used to generate distinctive experimental designs and may lead to further improvements in fields such culture media development or cryobiology.

2. Materials

The following materials and methods relate to the utilization of a prototype system of a dynamic microfunnel device for embryo culture. Adjustments and changes should be incorporated into this protocol for the proper function of this prototype in other laboratories, as well as for the utilization of other in-house built systems for embryo culture based on the technology of microfluidic dynamics (15, 36, 38).

2.1. Materials for Construction of Dynamic Microfunnel Devices

1. PDMS prepolymer (Sylgards 184, Dow Corning).
2. Parylene C (see Note 1).
3. 200 μm Glass slide.
4. SU-8.
5. Tridecafluoro-1,1,2,2-tetrahydrooctyl-1-trichlorsilane.
6. Commercial Braille display.

2.2. Materials for Embryo Culture in Dynamic Microfunnel Devices

1. Laminar flow hood.
2. Incubator for embryos.
3. Braille pin platform (see Note 2).
4. Dynamic microfunnel device.
5. Embryo culture media.
6. Pipette for volumes up to 1 mL.
7. Pipette tips with barrier.
8. Small-bore pipette.
9. Pipetting device for small-bore pipettes.
10. Inverted microscope.
11. Stereomicroscope.

3. Methods

3.1. Construction of Dynamic Microfunnel Device

1. Model the design of your device (see Note 3).
2. Compose desired channel structures composed of SU-8 and fabricate them on a thin glass wafer using backside diffused-light photolithography.
3. Cast an 8 mm PDMS prepolymer (1:10 base-to-agent ratio) slab with microchannels and funnels against the SU-8 model.
4. Remove bubbles under vacuum for 1 h (200 mmHg) (see Note 4).
5. Cure the casted 8 mm PDMS prepolymer slabs at 60 °C for 120 min and 120 °C for 30 min.
6. Cast 0.1 mm PDMS membranes as described above.
7. Deposit parylene C (~2.5 μm) on the back side of 0.1 mm PDMS membranes using a PDS 2010 Labcoater after covering the well side with PDMS.
8. Attach PDMS slab to a thin PDMS-parylene-PDMS membrane.

3.2. Set Up of Dynamic Microfunnel Device and Embryo Culture

1. Dynamic microfunnel devices for embryo culture are gas permeable plates and they should be placed into incubators at 37 °C for at least 4 h before loading with culture media (see Note 5).
2. The dynamic microfunnel device can have two microfunnels connected by microchannels. Place media into microfunnels. The media will enter the microchannels and proceed through the channel and into the other microfunnel.
3. Search under inverted microscope or stereomicroscope for bubbles of air in the two microchannels connecting reservoir and culture microfunnels (see Note 6). Microchannels should be clear of air bubbles.
4. If the microchannels contain air bubbles, use a plastic rod to gently press the microfluidic device from the bottom and move the rod in the direction of the microfunnel to drive air bubbles out of the microchannels.
5. Once the dynamic microfunnel device is clear of air bubbles, it should be clipped on top of the Braille pin platform stored inside the incubator (see Note 7).
6. The Braille pin platform should be turned on to initiate flow of media and the whole system should run overnight before placing embryos in culture.
7. Check the dynamic microfunnel cartridge again for the presence of air bubbles before placing zygotes into the culture microfunnel (see Note 8).

8. Remove zygotes from the static culture system with the assistance of a stereomicroscope and a small-bore pipette. Place up to five embryos on a culture microfunnel. Replace the dynamic microfunnel cartridge onto the Braille pin platform and lock the safety brackets before placing back into incubator and activating the movement of Braille pins.
9. Embryo should be evaluated on day 3 or 5/6 of culture.

4. Notes

1. Parylene coating can be provided by third party companies specialized in chemical coating as Specialty Coating System.
2. Use commercially available Braille pin platform for e-mail reading (Braillex Tiny from F.H. Papenmeier GmbH & Co. KG, Germany).
3. Theoretical modeling of microfunnel devices should be performed before casting prototypes to ensure that the intended shear stress and chemical diffusions are obtained. The device utilized in Heo et al. 2010 (15) had the following dimensions:
   (a) Microchannels: length—1.5 mm, width—0.1 mm and height—0.11 mm.
   (b) Funnel: bottom radius—0.25 mm, top radius—1.77 mm, height—2.63 mm and angle in relation to the flat bottom of the funnel—60 °.
4. Avoid excessive exposure to vacuum to prevent premature curing.
5. Placing the microfluidic device for at least 4 h into an incubator at 37 °C may allow expansion prior to loading with media and is used as a strategy to avoid the formation of air bubbles in microchannels.
6. This procedure has to be carefully performed under a stereomicroscope. The presence of air bubbles should be avoided because they block the movement of media in microchannels and negate the potential benefit of dynamic culture.
7. Incubator should be set up according to standard practices of each clinical embryology laboratory, regulating temperature, and gas environment.
8. Oocyte insemination and determination of fertilization should follow the standard procedures and guidelines of each clinical embryology laboratory.

References

1. Muglia U, Motta PM (2001) A new morpho-functional classification of the Fallopian tube based on its three-dimensional myoarchitecture. Histol Histopathol 16(1):227–237
2. Paltieli Y, Weichselbaum A, Hoffman N, Eibschitz I, Kam Z (1995) Laser scattering instrument for real time in-vivo measurement of ciliary activity in human fallopian tubes. Hum Reprod 10(7):1638–1641
3. Fauci L, Dillon R (2006) Biofluidmechanics of reproduction. Annu Rev Fluid Mech 38: 371–394
4. Schumann D, Kujat R, Nerlich M, Angele P (2006) Mechanobiological conditioning of stem cells for cartilage tissue engineering. Biomed Mater Eng 16(4 Suppl):S37–S52
5. James JL, Whitley GS, Cartwright JE (2011) Shear stress and spiral artery remodelling: the effects of low shear stress on trophoblast-induced endothelial cell apoptosis. Cardiovasc Res 90(1):130–139
6. Pontes JH, Nonato-Junior I, Sanches BV, Ereno-Junior JC, Uvo S, Barreiros TR, Oliveira JA, Hasler JF, Seneda MM (2009) Comparison of embryo yield and pregnancy rate between in vivo and in vitro methods in the same Nelore (Bos indicus) donor cows. Theriogenology 71(4):690–697
7. Gil MA, Cuello C, Parrilla I, Vazquez JM, Roca J, Martinez EA (2010) Advances in swine in vitro embryo production technologies. Reprod Domest Anim 45(Suppl 2):40–48
8. Lonergan P, Pedersen HG, Rizos D, Greve T, Thomsen PD, Fair T, Evans A, Boland MP (2004) Effect of the post-fertilization culture environment on the incidence of chromosome aberrations in bovine blastocysts. Biol Reprod 71(4):1096–1100
9. Gad A, Besenfelder U, Rings F, Ghanem N, Salilew-Wondim D, Hossain MM, Tesfaye D, Lonergan P, Becker A, Cinar U, Schellander K, Havlicek V, Holker M (2011) Effect of reproductive tract environment following controlled ovarian hyperstimulation treatment on embryo development and global transcriptome profile of blastocysts: implications for animal breeding and human assisted reproduction. Hum Reprod 26(7):1693–1707
10. Smith GD, Swain JE, Bormann CL (2011) Microfluidics for gametes, embryos, and embryonic stem cells. Semin Reprod Med 29(1):5–14
11. Krisher RL, Wheeler MB (2010) Towards the use of microfluidics for individual embryo culture. Reprod Fertil Dev 22(1):32–39
12. Smith GD, Takayama S (2007) Gamete and embryo isolation and culture with microfluidics. Theriogenology 68(Suppl 1):S190–S195
13. Suh RS, Phadke N, Ohl DA, Takayama S, Smith GD (2003) Rethinking gamete/embryo isolation and culture with microfluidics. Hum Reprod Update 9(5):451–461
14. Beebe D, Wheeler M, Zeringue H, Walters E, Raty S (2002) Microfluidic technology for assisted reproduction. Theriogenology 57(1): 125–135
15. Heo YS, Cabrera LM, Bormann CL, Shah CT, Takayama S, Smith GD (2010) Dynamic microfunnel culture enhances mouse embryo development and pregnancy rates. Hum Reprod 25(3):613–622
16. Koike T, Matsuura K, Naruse K, Funahashi H (2010) In-vitro culture with a tilting device in chemically defined media during meiotic maturation and early development improves the quality of blastocysts derived from in-vitro matured and fertilized porcine oocytes. J Reprod Dev 56(5):552–557
17. Matsuura K, Hayashi N, Kuroda Y, Takiue C, Hirata R, Takenami M, Aoi Y, Yoshioka N, Habara T, Mukaida T, Naruse K (2010) Improved development of mouse and human embryos using a tilting embryo culture system. Reprod Biomed Online 20(3):358–364
18. Kim MS, Bae CY, Wee G, Han YM, Park JK (2009) A microfluidic in vitro cultivation system for mechanical stimulation of bovine embryos. Electrophoresis 30(18):3276–3282
19. Raty S, Walters EM, Davis J, Zeringue H, Beebe DJ, Rodriguez-Zas SL, Wheeler MB (2004) Embryonic development in the mouse is enhanced via microchannel culture. Lab Chip 4(3):186–190
20. Demming S, Vila-Planas J, Aliasghar Zadeh S, Edlich A, Franco-Lara E, Radespiel R, Buttgenbach S, Llobera A (2011) Poly(dimethylsiloxane) photonic microbioreactors based on segmented waveguides for local absorbance measurement. Electrophoresis 32(3–4):431–439
21. Hoppe PC, Pitts S (1973) Fertilization in vitro and development of mouse ova. Biol Reprod 8(4):420–426
22. Xie Y, Wang F, Zhong W, Puscheck E, Shen H, Rappolee DA (2006) Shear stress induces preimplantation embryo death that is delayed by the zona pellucida and associated with stress-activated protein kinase-mediated apoptosis. Biol Reprod 75(1):45–55

23. Kaupp JA, Waldman SD (2008) Mechanical vibrations increase the proliferation of articular chondrocytes in high-density culture. Proc Inst Mech Eng H 222(5):695–703
24. Ito Y, Kimura T, Nam K, Katoh A, Masuzawa T, Kishida A (2011) Effects of vibration on differentiation of cultured PC12 cells. Biotechnol Bioeng 108(3):592–599
25. Wolchok JC, Brokopp C, Underwood CJ, Tresco PA (2009) The effect of bioreactor induced vibrational stimulation on extracellular matrix production from human derived fibroblasts. Biomaterials 30(3):327–335
26. Mizobe Y, Yoshida M, Miyoshi K (2010) Enhancement of cytoplasmic maturation of in vitro-matured pig oocytes by mechanical vibration. J Reprod Dev 56(2):285–290
27. Isachenko V, Maettner R, Sterzik K, Strehler E, Kreinberg R, Hancke K, Roth S, Isachenko E (2011) In-vitro culture of human embryos with mechanical micro-vibration increases implantation rates. Reprod Biomed Online 22(6):536–544
28. Glasgow IK, Zeringue HC, Beebe DJ, Choi SJ, Lyman JT, Chan NG, Wheeler MB (2001) Handling individual mammalian embryos using microfluidics. IEEE Trans Biomed Eng 48(5): 570–578
29. Hickman DL, Beebe DJ, Rodriguez-Zas SL, Wheeler MB (2002) Comparison of static and dynamic medium environments for culturing of pre-implantation mouse embryos. Comp Med 52(2):122–126
30. Wheeler MB, Walters EM, Beebe DJ (2007) Toward culture of single gametes: the development of microfluidic platforms for assisted reproduction. Theriogenology 68(Suppl 1): S178–S189
31. Clark SG, Haubert K, Beebe DJ, Ferguson CE, Wheeler MB (2005) Reduction of polyspermic penetration using biomimetic microfluidic technology during in vitro fertilization. Lab Chip 5(11):1229–1232
32. Wheeler MB, Clark SG, Beebe DJ (2004) Developments in in vitro technologies for swine embryo production. Reprod Fertil Dev 16(1–2):15–25
33. Gu W, Zhu X, Futai N, Cho BS, Takayama S (2004) Computerized microfluidic cell culture using elastomeric channels and Braille displays. Proc Natl Acad Sci USA 101(45):15861–15866
34. Park TH, Shuler ML (2003) Integration of cell culture and microfabrication technology. Biotechnol Prog 19(2):243–253
35. Tung YC, Torisawa YS, Futai N, Takayama S (2007) Small volume low mechanical stress cytometry using computer-controlled Braille display microfluidics. Lab Chip 7(11):1497–1503
36. Heo YS, Cabrera LM, Song JW, Futai N, Tung YC, Smith GD, Takayama S (2007) Characterization and resolution of evaporation-mediated osmolality shifts that constrain microfluidic cell culture in poly(dimethylsiloxane) devices. Anal Chem 79(3):1126–1134
37. Alegretti JR, Motta ELA, Serafini P, Rocha AM, Criscuolo T, Smith GD (2011) Development of human embryos in a dynamic microfluidic culture system: results from a prospective randomized study. Hum Reprod 26(Suppl 1):i38
38. Heo YS, Jovic A, Cabrera LM, Smith GD, Takayama S (2009) Osmolality control for microfluidic embryo cell culture using hybrid polydimethylsiloxane (PDMS)-parylene membranes. In: Nahmias Y, Bhatia SN (eds) Methods in bioengineering: microdevices in biology and medicine. Artech House, Boston

Chapter 21

Human Embryo Culture Media Comparisons

Thomas B. Pool, John Schoolfield, and David Han

Abstract

Every program of assisted reproduction strives to maximize pregnancy outcomes from in vitro fertilization and selecting an embryo culture medium, or medium pair, consistent with high success rates is key to this process. The common approach is to replace an existing medium with a new one of interest in the overall culture system and then perform enough cycles of IVF to see if a difference is noted both in laboratory measures of embryo quality and in pregnancy. This approach may allow a laboratory to select one medium over another but the outcomes are only relevant to that program, given that there are well over 200 other variables that may influence the results in an IVF cycle. A study design that will allow for a more global application of IVF results, ones due to culture medium composition as the single variable, is suggested. To perform a study of this design, the center must have a patient caseload appropriate to meet study entrance criteria, success rates high enough to reveal a difference if one exists and a strong program of quality assurance and control in both the laboratory and clinic. Sibling oocytes are randomized to two study arms and embryos are evaluated on day 3 for quality grades. Inter and intra-observer variability are evaluated by kappa statistics and statistical power and study size estimates are performed to bring discriminatory capability to the study. Finally, the complications associated with extending such a study to include blastocyst production on day 5 or 6 are enumerated.

Key words: Embryo culture medium, Media comparisons, Intra-observer variability, Inter-observer variability, Kappa statistics, Power analysis, Prospective randomized trial

1. Introduction

The production of a nurturing environment in vitro is central to optimizing outcome in programs of assisted reproductive technology (ART). It is a functional system created by the interplay of essential elements, one of which is the culture medium or, in sequential strategies, culture media. There has been an increasing interest since the mid 1980s in the formulation of culture media intended specifically for the propagation of human embryos and an

Gary D. Smith et al. (eds.), *Embryo Culture: Methods and Protocols*, Methods in Molecular Biology, vol. 912, DOI 10.1007/978-1-61779-971-6_21,

attendant proliferation of commercial houses offering multiple medium formulations to the IVF community. The commercial production and sale of human embryo culture media has delivered more than just convenience to end users; it has simultaneously brought quality to even the smallest laboratory. Originally, media produced in house were made then analyzed by bioassay in order to prove quality post hoc. Often, this was as much a test of water quality as it was for component suitability and formulary accuracy. With commercial vendors now following industrial standards for good manufacturing practices, quality is built into products through the use of traceable, certified components in facilities approved by certifying bodies using strict adherence to stringent protocols. Though convenient for bringing quality to the laboratory, the evolution of so many commercially available choices of culture media has generated the need to select from a wide array of options, one or two depending upon if the laboratory employs monoculture or a sequential culture strategy. In the majority of laboratories, settling upon the ultimate medium or media for use in human ART involves a comparison, either with certain laboratory measures as endpoints or with pregnancy as the final arbiter. The literature contains many examples of such comparisons and is growing yet few have been conducted in a manor that allows for valid scientific conclusion. This is due, in large part to the complexity of variables in IVF, logistical issues in performing such studies and the inherent variability introduced with human subjects. It is the aim of this chapter not to critique the existing literature but rather to offer a methodology for conducting a valid comparison of two media, highlighting the intrinsic variables that complicate such efforts.

1.1. Why Do Commercial Formulations Need to be Compared?

The culture medium is only one component of a culture system and the performance characteristics of a medium can be influenced dramatically by other components such as gas phase, culture vessel, oil overlay, and incubator performance and recovery. Most published studies to date simply seek to identify which of two media, when dropped into the existing culture system, yields the best outcome. No effort is made in this type of study either to identify and isolate confounding variables both within and outside of the laboratory or to optimize the performance of trial media with preliminary experimentation; they are just compared head to head under existing culture conditions employed in the testing laboratory and whenever a difference emerges, the trial stops. Despite the anecdotal nature of these comparisons, there is little doubt that this approach is of value to the laboratory and more times than not, a winner will emerge in such a contest. But do such studies warrant publication? It is doubtful that all of the variables contributing to outcome can be replicated in a second setting and the results, most often, are unique to a specific clinical entity.

A much more difficult question to address in the human ART setting is the larger issue of medium formulation. Does medium "A" outperform medium "B" because the components of "A", offered at specific concentrations, foster the development of human embryos by supporting appropriate biochemical and physiological mechanisms more so than medium "B"? In reality, the bulk of the literature involves comparisons of the first type in an attempt to yield conclusions that can only be derived by designing and executing comparisons with the latter question in mind. The complexity of comparing specific formulations in human embryogenesis is illustrated clearly in the literature by the nature of the arguments made in support of one formulation over another. Are media formulated to replicate the types and concentrations of metabolites in the reproductive tract better or worse than a medium formulation that is the culmination of optimization experiments that measure direct embryonic responses to specific component manipulations? Are media developed in one animal model system more appropriate for extrapolation to human embryos than a medium developed in a different model system? There is a distinct reason why debates such as these have populated the human embryo culture literature—no culture medium has been developed from beginning to end using human embryos as the experimental system and every medium, therefore, was offered initially to human IVF laboratories based upon performance in an animal model system. It is the intent of this chapter to focus on experimental designs that address the more global question of identifying media formulations that support normal physiology leading to viability upon embryo transfer in humans.

1.2. What Are the Requisites in an ART Center for a Valid Comparison of Two Media?

Simply put, many ART centers are not in a position to conduct a meaningful comparison of two medium formulations. Success in human ART is a complicated interplay of patient selection, clinical decision making, and clinical procedural expertise in addition to laboratory experience and excellence. All must be performing in concert and at a consistently high level to have any chance of revealing subtleties in outcome due exclusively to medium composition. Programs with average or below average success rates likely lack the discriminatory power to identify such differences. So, does the center have a consistently high success rate, indicative of overall center quality? If yes, does the center have a caseload to supply an appropriate number of patients meeting study entry criteria that can be enrolled in a reasonable timeframe? Is the caseload such that study entrants can be restricted to one or similar ovarian stimulation protocols? Are stimulation protocols administered in a similar fashion by the clinical staff? Is there ongoing clinical quality assurance and quality control to document and verify the procedural stability required in the clinic to reveal differences in outcome due to laboratory components over the study period?

Another important consideration is how a given medium performs in the testing center compared to its historical performance in the literature. For example, if medium "B" only performs at a 20 % rate in your study but has historically performed at a 65 % rate in other studies, the difference in performance between "A" and "B" in your setting is doubtfully reflective of medium composition. The testing center should attempt to compare "A" and "B" only if preliminary optimization of both media indicates both are performing at their potential as documented in the literature. If a new formula, one not available commercially, is being compared to an existing, commercially available control medium, the performance of the control medium must equal its optimal performance as described by prior studies in the literature. It is obvious that a comparison of a new medium to an existing control medium, under conditions where the control medium is being used inappropriately, carries no value yet this persists as one of the recurring problems with published accounts of medium comparisons to date.

1.3. What Are the Design Elements of a Valid Study to Compare Two Culture Media?

Isolating culture medium formulation as the sole variable responsible for a significantly different outcome in human ART is extremely difficult, if it is even possible. Some of the factors that affect outcome outside of the IVF laboratory are given in Table 1 (~60 factors). Factors within the laboratory present in all IVF cases are given in Table 2 (~88 factors) and those involved with specific laboratory procedures ancillary to routine IVF are given in Table 3 (~85 factors). Out of some 233 factors that can affect outcome, the formula of the embryo culture medium is but one. The study design is crucial to stabilizing as many variables as possible and must take into account the following:

(a) Patient entrance criteria. Restrictive criteria as to age, diagnosis, BMI, ovarian reserve, and prior attempts at IVF are essential to reducing variability. In reality, these criteria should likely include not only the female, but the male as well, as sperm parameters may also significantly impact resulting embryos.

(b) Will donor oocytes be included in the study? Many centers feel that donor oocytes provide the best background upon which to conduct such studies but, in reality, are they an appropriate index population for revealing subtle differences in outcome amongst infertile couples? Embryos show remarkable developmental plasticity in that they are capable of utilizing alternative metabolic substrates and pathways under suboptimal conditions. It may well be that embryos derived from young oocyte donors carry the greatest capacity for plasticity and, thus, do not accurately reflect the compromised physiological status of embryos from patients suffering from various forms of infertility. For this reason, donor oocytes, if included, should at least be scrutinized as a subset of both

Table 1
Factors outside of the laboratory that may affect the success of IVF-ET

Domain	Factors
Patients, both	Age, general health, medical history/STDs, alcohol, smoking, environmental toxins
Patient, male	Drugs/medications, libido, abstinence interval
Patient, female	Reproductive endocrine status, cycle duration and characteristics, ovarian reserve, prior IVF attempts, surgical history/pelvic adhesions, ovarian accessibility with biopsy needle, uterine/cervical anatomy, luteal sufficiency
Ovarian stimulation	Protocol design with oral contraceptives; long, short or flare; use of antagonist; use of agonist Gonadotropins type, source, lot number, storage, conditions, reconstitution, dosage, duration Estradiol rise Compliance Ovulatory trigger hCG, rLH, agonist, dosage, interval until retrieval
Procedure room	Oocyte retrieval aspiration needles, aspiration pump/pressure, collection tubing, collection vessel, heating blocks, time of retrieval, disinfectants, anesthetic agents, skill of surgeon, air quality
Embryo transfer	Catheter, disinfectants, time, status of bladder/uterine contractions, skill of surgeon, luteal support, patient compliance

Table 2
Factors within the laboratory that may affect the success of IVF-ET for all cases

Property/procedure	Factors
For all media	Correct medium for purpose, sterility, endotoxin and bioassay tested, age, storage conditions, antibiotics and concentration, protein supplement and concentration, equilibration prior to use, pH, osmolality, temperature
Handling materials and equipment	Sterility, off-gassing, gamete/embryo safe, endotoxin and bioassay tested
Air quality	Outdoor air quality, indoor air quality, laboratory air quality, incubator air quality
Semen collection	Collection vessel, lubricants, transport conditions, time, temperature, contamination
Sperm processing	Sperm wash medium, handling pipettes, centrifuge tubes, separation media (concentration, diluent, osmolality, pH), centrifuge time/force, swim-up medium, swim-up tubes, incubator temperature, incubator gas phases, technical skill

(continued)

Table 2 (continued)

Property/procedure	Factors
Oocyte retrieval	Flush/collection media, heparin, searching dishes, warm surfaces (stages, laminar flow hoods), time, technical skill
Oocyte preparation	Oocyte washing medium, warm surfaces (stages/laminar flow hoods), exposure to light (microscopes), pipettes, culture dishes, incubator performance (temperature, gas phase, recovery), oocyte culture medium, oocyte culture oil, hyaluronidase, stripping pipettes, technical skill
Embryo culture	Cleavage-stage culture medium, embryo culture dishes, culture oil, incubator performance (temperature, gas phase, number of openings, gas recovery rate), exposure to light (microscopes), time out of incubator for evaluation, blastocyst culture medium, blastocyst culture dishes
Embryo transfer	Washing and handling media, transfer medium, warm surfaces, catheter, loading efficiency, number of embryos loaded and transferred, time

Table 3
Factors within the laboratory that may affect the success of IVF-ET for some cases; ancillary methods

Property/procedure	Factors
Assisted hatching	Handling pipettes, assisted hatching medium, assisted hatching dishes, culture oil, micromanipulator and glass tools, acid Tyrode's medium, laser calibration and performance, technical skill, procedure time
Cryopreservation-embryo	Embryo stage, washing and handling media, pipettes, warm surfaces, freezing dishes, slow freezing or vitrification protocol, cryopreservation solutions, straws or vials, calibration and performance of controlled rate freezer, cryopreservation time, technical skill during cryopreservation, thawing protocol, thawing solutions, warm surfaces, washing and handling media during thawing, thawing time, technical skill during thawing, post-thaw incubation (medium, supplements, time, conditions)
Cryopreservation-semen	Initial semen parameters (count, motility, progression, morphology), time since collection, freezing protocol, freezing solutions, calibration and performance of controlled rate freezer, storage conditions, technical skill during freezing, thawing protocol, post-thaw preparation and incubation, technical skill during thawing

(continued)

Table 3 (continued)

Property/procedure	Factors
Cryopreservation-oocytes	Handling pipettes and media, stripping dishes, stripping pipettes, hyaluronidase, time of stripping, warm surfaces, light exposure (microscope), cryopreservation protocol, calibration and performance of controlled rate freezer if via slow cooling, cryopreservation solutions, technical skill during cryopreservation, thawing protocol, thawing solutions, technical skill during thawing, incubation medium, ICSI
PGS/PGD biopsy	Embryo stage, handling pipettes, biopsy medium, biopsy dishes, culture oil, micromanipulator and tools, assisted hatching, biopsy technique and number of cells removed, procedure time, technical skill

experimental arms of a comparative study before generalizations can be made regarding medium formulation.

(c) Uniformity in clinical performance should include a restriction in the number of ovulation induction protocols employed, a unified strategy for implementing the protocols, specific criteria for triggering ovulation, use of protocol-appropriate, standardized triggers, and uniform drug source/lots. In addition, variability in embryo transfer expertise can be minimized by restricting the number of physicians involved in this stage of the study.

(d) Stability and uniformity in other components of the culture system including culture volume, medium equilibration time, pH, humidity, gas phase, incubator openings/closings, caseload per incubator, oil overlay, embryo density, medium changes, insemination interval, and insemination density.

(e) Laboratory technical skill. It should not be assumed that the technical skills of the staff are uniform unless they have been measured and documented by proficiency testing that addresses the specific steps of embryo culture reflected those in the study, including medium/culture preparation.

(f) Study period. One of the best ways to limit variables introduced both into the clinic and laboratory is to perform a prospective, randomized trial of sibling embryos over a short study period. Prolonged study periods, particularly in combination with retrospective analyses, inevitably introduce uncontrolled variability that can easily mute or eliminate small differences in outcome.

(g) Should the study include the performance of supernumerary embryos subjected to cryopreservation/thawing/transfer?

Such studies have the benefit of looking at the global effects a medium may have upon an entire IVF cycle, measuring endpoints after total embryo consumption; but this introduces new technical variables and a greatly extended time frame for complete evaluation.

Validity can only be had when all endpoint measures are properly calibrated. This is of particular importance if laboratory endpoints are selected that involve morphological assessment. It is crucial that both intra- and inter-observer variability be measured and evaluated for staff members scoring embryos during the study. Although good agreement in morphological assessment between a small number of embryologists can be had (1), it should not be assumed it is so. Seemingly straightforward morphological measures can be seen quite differently between experienced embryologists (2, 3) that can lead to significant type I and type II errors in data analysis. If claims are to be made that specific morphological features or classifications impart a clinical benefit to an embryo, statistical measures should be implemented to validate the conclusion.

Finally, statistical tests appropriate for the nature of the data generated in the study must be employed. Inherent in this consideration is a consideration of power and the enrollment of a sufficient number of patients to produce statistical power. This is true both for interim measures, such as a kappa statistic for evaluating intra- and inter-observer variability, as well as endpoint statistical scrutiny.

2. Materials

2.1. Facility Requirements, Reproductive Medicine

1. Full service center offering in vitro fertilization with sufficient caseload to enroll patients meeting entry criteria in a timely manner (see power analysis in Subheading 3).
2. Ultrasonography.
3. Access to hormone assay laboratory capable of measuring serum FSH, estradiol, progesterone, and beta-hCG.
4. Ovum retrieval facility equipped with procedure table, ultrasonography, aspiration pump, and heating blocks/warm surfaces.
5. Equipment for administering intravenous sedation including appropriate infusion media, benzodiazepine (Midazolam), Diprivan (Propofol), pulse oximeter, defibrillator.
6. Ovum retrieval needles and biopsy guide for transvaginal transducer; embryo transfer catheter.
7. Access to institutional review board for approval of research involving human subjects—needed only if culture medium is considered investigational.

2.2. Facility Requirements, IVF Laboratory

1. Full service IVF laboratory staffed 7 days per week.
2. Refrigerator, laminar flow hood/work station, trigas incubators or incubator chambers capable of receiving triple gas mixture, osmometer, pH meter, source of CO_2 gas, N_2 has or triple mix gas containing approximately 5 % CO_2, 5 % air, and 90 % N_2.
3. Lot-controlled expendable plasticware supplies for embryo culture including 100 mm dishes, 60 mm dishes, culture dishes (embryo-specific culture ware, center well, four well, etc.), 10 mL pipettes, Pasteur pipettes, transfer pipettes, and small bore pipettes with dispensing device.
4. Lot-controlled culture media delivered and stored with appropriate temperature control.
5. Lot-controlled protein supplement delivered and stored according to manufacturer's recommendations.
6. Lot-controlled sterile oil for embryo culture.
7. Stereo microscope for ovum retrieval/catheter loading.
8. Inverted microscope with Hoffman modulation contrast or equivalent optics. Microscope should have image capturing capability.
9. Access to data entry/management/storage equipment.
10. Access to statistical software or statistical analytical services.

3. Methods

A simple experimental design is presented (Fig. 1) to compare the overall distributional difference of quality grades of embryos cultured in control medium (A) compared to an experimental medium (B) on day 3 of development post insemination. Additionally, several options for calculating size effect and sample size estimation in a statistical power analysis are given. The design is then expanded to examine quality grade effects upon clinical pregnancy following embryo transfer on day 3 (Fig. 2) or upon quality grade distributions and pregnancy after extended embryo culture to day 5/6 (Fig. 3). Suggestions for measuring intra- and inter-observer variability by kappa statistics, as well as powering for such analyses are also presented.

3.1. Patients

1. Maternal age less than 35 years, undergoing initial IVF therapy for infertility due to ovulatory dysfunction, tubal factor, and male factor are included.
2. Women with elevated day 3 FSH (>10 mIU/mL), polycystic ovaries, endometriosis, and uterine factor infertility are

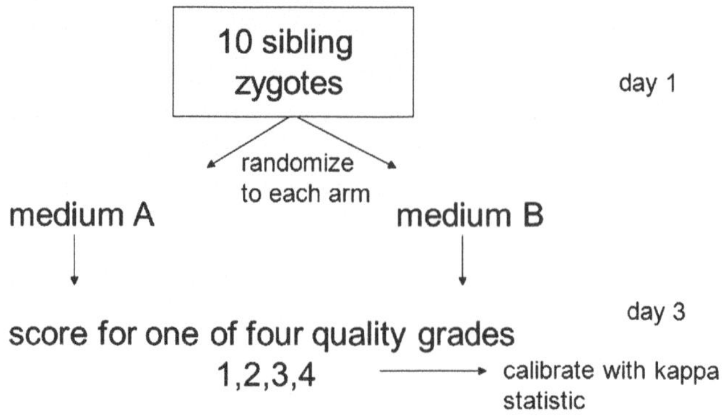

Fig. 1. Outline of study designed to measure the difference in the distributions of embryo quality grades (1–4) of day 3 embryos grown in control medium (A) or experimental medium (B). An evaluation of both intra- and inter-observer variability in the scoring of quality grades is performed with kappa statistics.

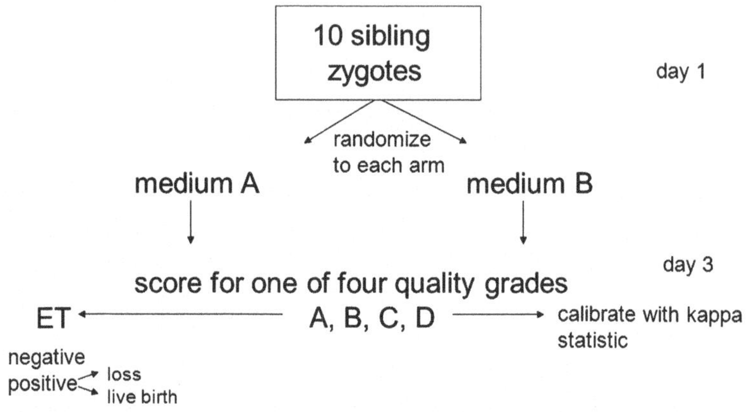

Fig. 2. The study outlined in Fig. 1 is extended to include a statistical evaluation of pregnancy, implantation, and live birth resulting from the transfer of embryo from each quality grade of embryos.

excluded as are men with a total motile count of less than 1×10^6 spermatozoa per mL.

3. All patients must sign informed consent and an authorized institutional review board must approve all protocols using investigational media. See Note 1.

3.2. Study Design

1. The study should be a prospective comparison of sibling zygotes randomized to one of two culture media (A—control; B—experimental) after the fertilization check at 18 h post-insemination or post-intracytoplasmic sperm injection (ICSI). See Note 2.
2. Oocytes are inseminated in any standardized, bicarbonate-buffered fertilization medium containing, at a minimum, glycine

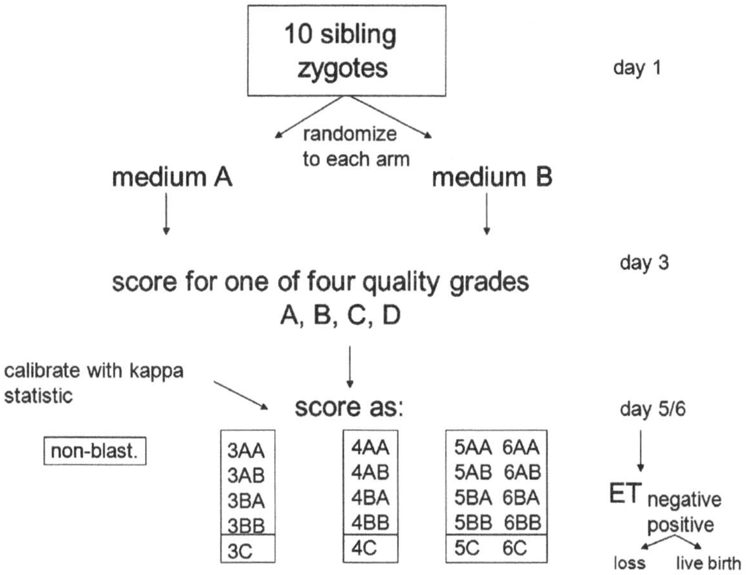

Fig. 3. Outline of the study extended to day 5 or 6. Blastocysts are scored but are placed into groupings (indicated by *boxes*) that have the same likely implantation capacity for statistical purposes. Kappa is measured on day 5/6, not 3 and transfers of the 7 implantation groupings are performed for statistical analysis as in the day 3.

and taurine and lacking EDTA. Protein supplementation should be either with human serum albumin alone or in combination with alpha and beta globulins at a final concentration of 3–5 mg/mL total protein. Insemination is performed with ~50,000 progressively motile sperm per mL. Oocytes subjected to ICSI will be placed in this same medium after injection until the fertilization check. For this example, only patients producing ten or more two pronucleate zygotes are included in the study so that a minimum of five zygotes are allocated to each medium. Centers electing to include patients producing fewer zygotes will have to make adjustments to their sample size estimation and power analysis.

3. Incubator CO_2 levels should be adjusted to yield target pH values for each medium consistent with manufacturer's recommendations or from preliminary studies yielding optimized embryonic growth. An equal number of chambers should be assigned to each medium to ensure an equal number of incubator entries during the course of the study. Oxygen tension should be set to 5 % in all chambers. Standard laboratory quality control procedures, including daily function checks on all equipment and incubators, must be performed and documented throughout the study period. Technical proficiency of the laboratory staff for embryo culture must be measured before and during the study.

4. Embryos will be scored as belonging to one of four groups based upon quality grade on day 3 of culture. Individual centers may define these categories using criteria of their choosing including, but not limited to, cell stage, symmetry, cell–cell interaction, cytoplasmic clarity, fragmentation, cytoplasmic granularity, etc. The graphical representation of the study design given in Fig. 1 reflects the identification of four distinct quality grades for day 3 embryos but does not dictate the criteria for each grade, leaving this to individual laboratories.

3.3. Statistics

1. *Power analysis and sample size estimation.* The appropriate use of statistical methods both for powering the comparison and analyzing the resulting data involve the use of a specific set of assumptions. Some may be more realistic in certain settings than others. For this reason, we present two alternative ways to estimate sample size and significance of outcomes.
2. *Calibration of observer variability with kappa statistic.* To quantify the reproducibility of the four-level quality grading system to be applied to day 3 images of treated embryos, the kappa statistic is computed (4). To control examiner bias, a regimen should be developed by the primary evaluator to train additional evaluators. Although two examiners would be sufficient to validate the primary evaluator ratings, three would be preferable because a consensus value could be obtained for each image grade. To determine if each evaluator is self-calibrated, each set of zygote image ratings should be performed twice using two different randomized orderings by each evaluator. For each pair of examiner grades, intra-examiner kappa should be computed, and for each pair of examiners, inter-examiner kappa based on the initial set of grades by each examiner should be computed.

   To compare the distributions of quality grades for mediums A and B, the Mann–Whitney *U* test is performed. To compare the proportions of zygotes resulting in live births for mediums A and B, the Z test of proportions is performed.
3. Power calculations to determine the precision of sample estimates were performed for varying numbers of subject couples using PASS 11 software (NCSS Inc., Kaysville, UT). These calculations were based on the following assumptions: (1) each couple contributes a set of 10 sibling zygotes, (2) the four quality grades are equally distributed for the entire set of day 3 zygote images, (3) for the population of examiners and day 3 zygote images, intra-examiner kappa under the null hypothesis is 0.9 and inter-examiner kappa under the null hypothesis is 0.8, (4) for the Mann–Whitney *U* test, the means of quality grades for mediums A and B is equal under the null hypothesis, (5) for the Z test of proportions, mediums A and B have an

Table 4
Lower bound for 95 % CI

# Couples	# Zygotes	Lower bound for 95 % CI (for intra-examiner kappa)	Lower bound for 95 % CI (for inter-examiner kappa)
5	50	0.70	0.55
10	100	0.77	0.63
20	200	0.81	0.68
30	300	0.83	0.71
40	400	0.84	0.72

Table 5
Detectable mean differences for quality grades

# Couples	# Zygotes	Detectable mean difference for quality grades	Detectable difference for proportion of live births
5	50	1.0	0.40
10	100	0.7	0.27
20	200	0.5	0.18
30	300	0.4	0.14
40	400	0.4	0.12

equal proportion of zygotes resulting in live births under the null hypothesis, (6) power is set at 0.9, and (7) alpha is set at 0.05. Table 4 lists the lower bound of the 95 % confidence interval of kappa based on the point estimates of 0.9 for intra-examiner ratings and 0.8 for inter-examiner ratings. Table 5 lists the mean difference between mediums A and B which is detectable by Mann–Whitney *U* test and the proportion difference between mediums A and B which is detectable by the Z test of proportions. As can be seen, a recruitment sample of five couples is not sufficient to detect an intra-examiner kappa of 0.7 as inferior to 0.9 and is not sufficient to detect a live birth proportion difference less than 0.4. Conversely, there is little to no benefit to increasing the recruitment sample from 30 couples to 40 couples.

4. *Two ordinal categorical variables and proportional odds.* When analyzing ordinal categorical data, statistical methods for two proportions should be used if the categories are dichotomous.

In the presence of more than two ordered categories, however, data can be analyzed using a *z*-test, or by logistic regression or the Mann–Whitney test assuming proportional odds. The proportional odds model assumes that the log odds ratios of cumulative classes are all equal to some common constant, say λ, across the classes. Hence, regardless of the number of categories, this single parameter λ summarizes the entire pattern of response differences between the control group (i.e., medium A) and the experimental group (i.e., medium B). For instance, a positive value of λ indicates that the experimental group exhibits a better outcome than the control group while a negative value indicates the opposite. Since λ quantifies the advantage of the experimental group over the control group, its value also represents the effect size to be detected for the purpose of power analysis. For references on the sample size calculation as well as comprehensive analysis of ordered categorical data, interested readers should refer to Whitehead (5), Agresti (6, 7), Machin et al. (8), and Julious (9).

5. *Statistical power and sample size.* For statistical formulation of this study, the null hypothesis is that the effects of two different media are equivalent (viz., H_0: $\lambda = 0$). That is, there is no differential effect of the two media on scoring embryos at day 3. The level of significance also known as alpha (α) was conventionally set to 0.05, meaning that approximately 1 test out of 20 will falsely reject the null hypothesis. This represents the probability of a type-I error, which occurs when a true null hypothesis is rejected. Alpha sets the maximum risk a researcher is willing to tolerate for testing a hypothesis in his/her experiment. For more stringent and conservative approach, alpha can be set to 0.01 which was considered in this analysis as well. On the other hand, beta (β) is the probability of a type-II error, which occurs when a false null hypothesis is not rejected. Power is the probability of rejecting a false null hypothesis, and it is equal to $1 - \beta$. Two different values of power, 0.80 and 0.90, were considered in this analysis (Table 6). Furthermore, it is known that the response proportions of the control group (i.e., medium A) have been about 30 % in grade one, 30 % in grade two, 20 % in grade three, and 20 % in grade four where the smaller the grade number is the better. Since the study focuses on identifying a difference of quality grades of embryos between two different media, the results for a two-sided test are shown in the following table with a range of values for the log odds ratio λ from 0.5 to 1.5. For convenience, it was further assumed that there would be an equal number (n) of embryos in both media.

   Table 7 describes the hypothetical response proportions of the experimental group (i.e., medium B) to result in the corresponding log odds ratio λ. The higher the effect size is,

Table 6
Power and sample size estimates to detect differential effects of two media

Effect size (λ)	Alpha (α)	Beta (β)	Power ($1-\beta$)	Sample size (n) per medium
0.5	0.01	0.1	0.9	390
		0.2	0.8	306
	0.05	0.1	0.9	276
		0.2	0.8	206
1.0	0.01	0.1	0.9	101
		0.2	0.8	79
	0.05	0.1	0.9	71
		0.2	0.8	53
1.5	0.01	0.1	0.9	47
		0.2	0.8	37
	0.05	0.1	0.9	33
		0.2	0.8	25

Table 7
Response proportion distributions for corresponding effect sizes

Effect size (λ)	Grade one	Grade two	Grade three	Grade four
0.0	0.3000	0.3000	0.2000	0.2000
0.5	0.4140	0.2980	0.1563	0.1317
1.0	0.5381	0.2649	0.1127	0.0842
1.5	0.6576	0.2129	0.0767	0.0528

the more distinctive the response pattern of embryos in medium B compared to that in medium A is with substantially higher proportions in good grades. From Table 6, one can see that to detect the effect size (or the log odds ratio) of at least 0.50 at 5 % level of significance with 80 % power, 206 embryos would be required for each medium, or 412 embryos in total for both media. This translates to more than 41 couples to be recruited since ten zygotes will be collected from each couple. If one chooses a one-sided test instead, it would require just slightly smaller sample sizes, and generally speaking, as the researcher wishes to detect even a smaller effect size, a much bigger sample size would be required to achieve the desired power at a fixed level of significance.

One should note that this is rather a simplified version of analysis upon certain assumptions to validate the statistical methodology. Nevertheless, if the experiment involves multiple hypotheses testing with more parameters to reflect the experimental setup with a finer tune, then a larger sample size than what is presented in Table 5 would be certainly necessary in order to control the overall level of significance α.

6. *Power analysis and sample size estimation for an expanded study to detect overall differences in the rate of live births among four quality grades of embryos.* It is certainly conceivable that varying culture medium composition will result in a significantly different distribution of quality grades on day of culture, but that in no way means that the transfer of embryos produced in one medium will yield a significantly different pregnancy rate from transfers using embryos from the other culture medium. One must demonstrate, either by single embryo transfers or the transfer of equal numbers of embryos with identical quality grades that pregnancy rates differ as a function of quality grade. The following is a consideration of this type of analysis. Again for mathematical derivation, it is assumed that all the zygotes regardless of the couples they are sampled independently from an identical population. On day 3, the quality of each embryo is classified into one of four quality grades and a series of single embryo transfers is performed for each grade. Then, it is observed whether or not the subject female is pregnant and if pregnant, whether the pregnancy continues to term. The implantation rate (in percent) is estimated by a ratio of the number of intrauterine sacs (an implantation) to the total number of embryos transferred for each quality grade group. The parameter of this analysis is the rate of live birth per embryo transfer for each quality grade group and it is calculated by the implantation rate times the live birth rate per implantation for each quality grade group.

7. *Chi-square test for GOF and effect size.* The Chi-square (χ^2) test for the goodness of fit is one of the most popular statistical tests for analyzing categorical data and it is commonly used to test whether sets of frequencies or proportions follow a particular pattern. To estimate the sample size required to achieve a certain statistical power, a notion of effect size, ε is introduced. Simply saying, it measures the magnitude of the test statistic to be detected. For the Chi-square test, Cohen (10) introduced the effect size formula that does not depend on the sample size and has set a small value of ε at 0.1, a medium value at 0.3, and a large value at 0.5. Since the Chi-square test is developed based on a Chi-square statistical distribution, the degrees of freedom (df) for the null distribution need to be

also specified. For testing a goodness of fit, the degrees of freedom is simply one less than the number of classes or categories. For references on the sample size calculation as well as comprehensive analysis of categorical data, interested readers should refer to Agresti (6, 7).

8. *Statistical power and sample size.* For statistical formulation of this study, the null hypothesis is that the rate of live birth per embryo transfer remains constant across the four quality grades (viz., H_0: $\pi_1 = \pi_2 = \pi_3 = \pi_4$). That is, there is no differential effect of the quality grade on successful delivery of live birth eventually. The level of significance also known as alpha (α) was conventionally set to 0.05, meaning that approximately 1 test out of 20 will falsely reject the null hypothesis. This represents the probability of a type-I error, which occurs when a true null hypothesis is rejected. Alpha sets the maximum risk a researcher is willing to tolerate for testing a hypothesis in his/her experiment. For more stringent and conservative approach, alpha can be set to 0.01, which was considered in this analysis as well. On the other hand, beta (β) is the probability of a type-II error, which occurs when a false null hypothesis is not rejected. Power is the probability of rejecting a false null hypothesis, and it is equal to $1 - \beta$. Two different values of power, 0.80 and 0.90, were considered in this analysis. The results of the analysis are shown in the following table with a range of effect size ε from 0.1 to 0.5. Here, the sample size (n) denotes the entire number of embryo transfers performed on day 3 for all four quality grades and it is calculated by the ratio of the Chi-Square (χ^2) to the effect size squared (ε^2).

   From Table 8, one can see that to detect the effect size of at least 0.10 at 5 % level of significance with 80 % power, 1,091 embryo transfers would be required in total.

   Continuing the comparison of culture media to either day 5 or day 6 in order to evaluate blastocyst production and quality complicates profoundly the power and sample size estimations, evaluation of observer variability via kappa, and the scrutiny of pregnancy outcomes. Figure 3 illustrates this point. Using the Gardner grading system (see Note 3), over 20 quality grades can be identified by scoring the cellularity of both inner cell mass and trophectoderm as well as expansion and hatching. It is likely that there is much overlap in the implantation potential of these individual categories and for simplicity and illustration we have collected blastocysts of different quality grades into groups (indicated by the boxes in Fig. 3) that share a hypothetical equal potential for implantation upon embryo transfer.

   Table 9 shows the results of an analysis assuming a more detailed classification of embryos into one of the seven implantation groups before transfer. As the classification gets finer, the sample size has to increase substantially in order to detect the same effect size.

Table 8
Power and sample size estimates to detect birth rate differences among four quality grade groups (df = 3)

Effect size (ε)	Alpha (α)	Beta (β)	Power ($1-\beta$)	Chi-square (χ^2)	Sample size (n)
0.1	0.01	0.1	0.9	19.25	1,925
		0.2	0.8	15.46	1,546
	0.05	0.1	0.9	14.18	1,418
		0.2	0.8	10.91	1,091
0.3	0.01	0.1	0.9	19.26	214
		0.2	0.8	15.48	172
	0.05	0.1	0.9	14.22	158
		0.2	0.8	10.98	122
0.5	0.01	0.1	0.9	19.25	77
		0.2	0.8	15.50	62
	0.05	0.1	0.9	14.25	57
		0.2	0.8	11.00	44

Table 9
Power and sample size estimates to detect birth rate differences among seven quality grade groups (df = 6)

Effect size (ε)	Alpha (α)	Beta (β)	Power ($1-\beta$)	Chi-square (χ^2)	Sample size (n)
0.1	0.01	0.1	0.9	23.19	2,319
		0.2	0.8	18.88	1,888
	0.05	0.1	0.9	17.42	1,742
		0.2	0.8	13.63	1,363
0.3	0.01	0.1	0.9	23.22	258
		0.2	0.8	18.90	210
	0.05	0.1	0.9	17.46	194
		0.2	0.8	13.68	152
0.5	0.01	0.1	0.9	23.25	93
		0.2	0.8	19.00	76
	0.05	0.1	0.9	17.50	70
		0.2	0.8	13.75	55

4. Conclusions

Clearly, the experimental isolation of culture medium formulation as the sole variable contributing to altered outcome following IVF is extremely difficult and, likely, impossible in the clinical setting. The intent of presenting this method is to bring recognition to many of these confounders that often go unnoticed (see Note 4). For the reasons presented here, in the consideration of extending a study to include pregnancy outcomes, it is often more convenient and expeditious to simply evaluate the potential effects of a culture medium upon laboratory measures such as cellularity and morphological grade. To date, no published study of this nature has included a calculation of kappa to monitor both intra- and inter-observer variability, a calibration and verification activity that cannot be omitted from meaningful studies. There are multiple approaches available to calculate power and estimate sample size and this, too, is an essential activity prior to initiating a media comparisons as a center can only then know if they have the caseload to conduct such a study in a reasonable time frame.

5. Notes

1. The requirement for patients to produce ten zygotes in order to enter the study will ensure that most patients will be either on a long protocol, using gonadotropin releasing hormone (GnRH) agonists in the preceding luteal phase, or one using GnRH antagonists in the follicular phase. In either case, the protocols need to be standardized as to type of gonadotropins, starting dose, strategy for adjusting gonadotropin dosage, and criteria for hCG trigger.
2. Randomizing individual patients to one of the two arms of the study would require a patient caseload that few centers have. Further, a comparison of sibling zygotes as the initial starting stage eliminates many of the patient-related variables discussed in the Introduction. Randomizing sibling zygotes after fertilization is necessary since many culture media contain components that, although beneficial to embryogenesis, are not recommended as components of insemination media.
3. Gardner and Lane (11) developed the blastocyst scoring system employed in the expanded phase of this study. It is an alphanumeric system that uses a number to characterize the formation and expansion of the blastocoel cavity, as well as stage of hatching. This is followed by two letters (A, B, or C) that describe the cellularity of the inner cell mass and the

trophectoderm, respectively. For this consideration, any blastocyst graded as "C" for either cell type is considered to be of poor quality. Other investigators may wish to consider these as having differential implantation potential so that, for example, a 4 AC embryo may not be considered as poor.

4. Enumerating all of the attendant variables, besides culture medium formulation, that can have an impact on the outcome of an IVF cycle was the initial concept of Dr. Don Rieger of Guelph, Canada and he suggested much of the content of Tables 1–3. We are indebted to him for this contribution.

References

1. Arce JC et al (2006) Interobserver agreement and intraobserver reproducibility of embryo quality assessments. Hum Reprod 21:2141–2148
2. Baxter AE et al (2006) Interobserver and intraobserver variation in day 3 embryo grading. Fertil Steril 86:1608–1615
3. de Assin R (2009) Comparison of methods to determine the assigned value in an external quality control programme for embryo evaluation. Reprod Biomed Online 19: 824–829
4. Rosner B (1995) Fundamentals of biostatistics, 4th edn. Wadsworth, Belmont, CA
5. Whitehead J (1993) Sample size calculations for ordered categorical data. Stat Med 12: 2257–2271
6. Agresti A (2002) Categorical data analysis, 2nd edn. Wiley, New York
7. Agresti A (2010) Analysis of ordinal categorical data, 2nd edn. Wiley, New York
8. Machin D, Campbell M, Fayers P, Pinol A (1997) Sample size tables for clinical studies, 2nd edn. Blackwell Science, Malden
9. Julious SA (2009) Sample sizes for clinical trials. Chapman & Hall, Boca Raton
10. Cohen J (1988) Statistical power analysis for the behavioral sciences. Lawrence Erlbaum, Hillsdale, NJ
11. Gardner DL, Lane M (2000) Embryo culture systems. In: Trounson AO, Gardner DK (eds) Handbook of in vitro fertilization, 2nd edn. CRC Press, Boca Raton

Chapter 22

Culture Systems: Embryo Culture and Monozygotic Twinning

Amy E. Sparks

Abstract

The incidence of monozygotic twinning in pregnancies achieved with assisted reproductive technologies (ART) is significantly higher than spontaneously conceived pregnancies. The factors associated with ART that predispose the embryos to splitting are not well-characterized. Assisted hatching and extended embryo culture are two ART laboratory methods that have been risk factors for monozygotic twinning. The methods and strategies that may be employed to avoid monozygotic twinning are discussed in this chapter.

Key words: Monozygotic twinning, Assisted hatching, Blastocyst culture, Single embryo transfer

1. Introduction

Multifetal gestation is one of the most common and significant complications in assisted reproductive technologies (ART). The majority of multifetal gestations that result from ART are due to the transfer and subsequent implantation of multiple embryos. Elective single embryo transfer (SET) may be employed to reduce, but not eliminate, a patient's risk of a multifetal gestation following ART. Patients undergoing SET are still at risk for a multifetal gestation due to a phenomenon called monozygotic twinning. Monozygotic twinning occurs when a single embryo divides into two or more fetuses. It is a rare event that occurs in approximately 0.4 % of all births resulting from spontaneous conceptions (1). The use of ART has been reported to increase the incidence of monozygotic twinning 2–12 times higher than natural conceptions (2–4). These findings are of great concern as monozygotic twinning is associated with a high risk of perinatal mortality, obstetrical complications, and congenital anomalies (5–8).

Gary D. Smith et al. (eds.), *Embryo Culture: Methods and Protocols*, Methods in Molecular Biology, vol. 912,
DOI 10.1007/978-1-61779-971-6_22,

The factors associated with ART that contribute to the increased incidence of monozygotic twinning are poorly understood. Identification of factors that give rise to embryos that are more susceptible to splitting is hampered by variations in ART protocols and patient populations between centers; different methods used to identify monozygotic twins; and the various times that the zygote may split, resulting in different placentation membrane classifications. The ART methods specific to embryo culture and manipulation that may be used to avoid monozygotic twinning are addressed in Subheading 3 of this chapter. Before addressing ART laboratory methods that promote embryo splitting, it is important to understand how different stages of embryo splitting lead to unique placental membrane configurations, the consequences of each membrane configuration and the challenge associated with accurately identifying monozygotic twin gestations resulting from ART.

1.1. Placental Membrane Configurations of Monozygotic Twins

An embryo may split as early as the day of fertilization (day 0) and up to 12 days post-insemination. Later stage embryos that undergo splitting will result in conjoined twins (9). The stage of development that the embryo splitting occurs affects the membrane configurations within the placenta and subsequent risk of fetal morbidity and mortality. Embryos that undergo splitting between the day of fertilization and up to 3 days post-fertilization (approximately the 8-cell stage) will lead to a twin pregnancy with two individual chorions and amnions (dichorionic–diamniotic; D–D). This type of monozygotic twin pregnancy is difficult to differentiate from a dizygotic twin pregnancy as the chorionicity and amnionicity are the same. Likewise, the risk of fetal morbidity and mortality for monozygotic D–D twin pregnancies is similar to that of dizygotic D–D twin pregnancies (10, 11).

Monozygotic twin pregnancies established when the embryo splits between the morula and blastocyst stage (4–8 days post-fertilization) will have a single chorion and two amnions (monochorionic–diamniotic; M–D). M–D monozygotic twins are at higher risk than D–D monozygotic twins for perinatal and obstetrical complications including discordant growth, twin–twin transfusion syndrome, prematurity, and perinatal death (7, 9, 12).

Embryos that split 9–12 days after fertilization are likely to give rise to monozygotic twins that share a single amnion and a single chorion. Monochorionic–monoamniotic (M–M) gestations have the highest risk of fetal mortality of the three placental membrane configurations as the shared amnion provides no barrier to prevent fetal and umbilical cord entanglement (9, 13). Less than 2 % of M–M monozygotic twin pregnancies progress to delivery (9).

1.2. Diagnosis of Monozygotic Twins

While it is clear that ART increases the risk of monozygotic twin pregnancy, the variation in the reported frequencies of monozygotic twinning following ART may be partially attributed to the method used to determine multifetal gestation zygosity and the timing of the diagnosis relative to conception (e.g., early first trimester versus birth). There is no way to account for the embryos that, after splitting, fail to implant; only reach a biochemical pregnancy; or result in an empty intrauterine gestational sac. Estimates of the incidence of monozygotic twinning following ART are typically based on ultrasound imaging of intrauterine pregnancies during the first trimester. Embryo splitting is customarily inferred when the number of fetuses detected by ultrasound exceeds the number of embryos transferred. This method may underestimate the incidence of zygotic splitting when more than one embryo has been transferred. Also ignored is the possible contribution of a spontaneous conception during the ART cycle. The likelihood of such an event is low, but has been reported following fresh and cryopreserved embryo transfer (14, 15). The risk of spontaneous conception during an ART cycle can be reduced by advising patients to abstain from intercourse during their ART cycle.

Early first trimester ultrasounds accurately identify approximately two-thirds of the monozygotic twins, primarily those with the M–D and M–M membrane configurations and D–D monozygotic twins following the transfer of a single embryo (16–18). The most accurate assessment of zygotic splitting in D–D twins occurs at birth and requires collection of DNA from like-sex twins for analysis of genetic markers to compare the DNA "fingerprint" of each twin.

1.3. ART Factors Associated with Monozygotic Twinning

There is a great deal of debate regarding which factors contribute to the increased frequency of monozygotic twinning following ART. Lacking an understanding of the mechanism of zygotic splitting, several theories have been proposed that involve ovulation induction (19–21), micromanipulation of the oocyte or embryo (16, 22–24) extended embryo culture (4, 25–28), and high order multiple gestations (6, 29, 30). Ovulation induction has been reported to lead to a greater proportion of monozygotic twin births compared to twin births resulting from spontaneous conception. Blickstein and Keith (21) hypothesized that ovarian stimulation may recruit an increased proportion of oocytes that are predisposed to splitting. The creation of an artificial gap in the zona pellucida of cleavage-stage embryos by a micromanipulation technique called assisted hatching has been associated with elevated monozygotic twinning. Alikani et al. (16) proposed that alterations to the zona pellucida prior to the formation of intercellular gap junctions between blastomeres could allow blastomeres of cleavage-stage embryos to separate and form multiple fetuses. Extended embryo culture to the blastocyst stage is the ART factor most frequently associated with monozygotic twinning. The mechanisms that predispose embryos exposed to prolonged culture to splitting include: apoptosis and subsequent cell remodeling leading to inner cell mass division at the time of hatching (25) and hardening of the zona pellucida that results in a pinching of the inner cell mass as the blastocyst escapes the zona pellucida (26). Lastly, monozygotic twinning has been frequently observed in high order multiple pregnancies resulting from implantation of two embryos with one of the two embryos splitting (6). Tummers et al. (31) reported that twin pregnancies established after the transfer of multiple embryos had a lower risk of spontaneous first trimester loss compared to pregnancies resulting from the implantation of only one of the multiple embryos transferred. They suggested that the double implantations were from better quality embryos. The continuation of dizygotic implantations through the first 12 weeks of gestation may increase the identification of monozygotic twins arising from one of the two implanted zygotes.

While the ART laboratory cannot control all factors that are likely to contribute to the increased incidence of monozygotic twinning associated with ART; there are some steps that should be taken by embryologists to assure that the embryo integrity is maintained and the stress of in vitro culture is minimized. The methods presented in this chapter are intended to aid in the avoidance of ART laboratory factors suspected to contribute to monozygotic twinning.

2. Materials

All contact materials and culture media should be tested for embryo toxicity by the bioassay method of choice and endotoxins prior to use. The items listed below are produced specifically for ART procedures by a number of manufactures. Each laboratory must determine which materials support optimal embryo development and ART outcomes in their center.

2.1. Assisted Hatching

1. Inverted microscope equipped with manipulators, micropipet holders, a stage heated to 37 °C, Hoffman Modulation Contract or DIC optics with 10×, 20×, and 40× objectives.
2. Micromanipulation medium—culture medium developed for in vitro procedures involving manipulation of gametes and embryos not requiring the use of a CO_2 incubator supplemented with 5 % albumin or 10 % serum protein supplement.
3. Washed mineral oil or paraffin oil.
4. Sterile pipets with inner diameters of ≥ 150 μM for cleavage-stage embryo handling or ≥ 300 μM for blastocyst handling.
5. A pipettor and sterile pipets for media dispensing.

6. 60 mm Petri dish, Petri dish lid or the sterile, disposable container of choice for embryo manipulation.
7. Sterile holding pipet made from borosilicate glass capillary tubes, cut and fire-polished with an inner diameter 15–20 μM, outer diameter of 80–150 μM and bent to an angle of 25–35 ° 0.5 mm from the fire-polished end of the pipet.
8. (a) *Chemical-assisted hatching*. Acidified medium (pH 2.2–2.8) and a hatching pipet made from a borosilicate glass capillary tube with blunt end, an inner diameter of 8–12 μM and bent to an angle of 25–35 ° 0.5 mm from the end of the pipet.
   (b) *Assisted hatching by laser*. 1.48 diode laser. The pulse duration must be validated for each embryo stage by the laboratory prior to use to assure optimal embryo development following assisted hatching.

3. Methods

3.1. Assisted Hatching

Assisted hatching may be performed from the 2-cell to blastocyst stage. The clinical value of assisted hatching as a method to improve embryo implantation rates continues to be debated. However, assisted hatching is required for acquisition of polar bodies from oocytes, cleavage-stage embryo blastomeres or trophectoderm cells of the blastocyst for preimplantation genetic diagnosis. The purpose of this method section is to provide assisted hatching "best practices" in order to avoid increasing the risk of monozygotic twinning.

3.1.1. Chemical-Assisted Hatching of Cleavage-Stage Embryos

1. Prepare the assisted hatching dish with 10 μl droplets of micromanipulation medium for each embryo that is to be hatched. The droplets should be elongated by dragging the pipet tip along the bottom of the dish while expelling the medium. Aliquot one 10 μl drop of acidified media away from the manipulation medium droplets (see Note 1). Quickly cover the drops with washed mineral or paraffin oil, making sure the drops are *completely* covered with oil. Warm the dish on a 37 °C surface for at least 15 min prior to micromanipulation.
2. Transfer one embryo to the end of a clean manipulation medium drop (see Note 2).
3. Lower the hatching pipet into the acidified medium and front-load the pipet.
4. Raise the acidified medium loaded pipet out of the dish.
5. Using low power magnification, focus on the embryo, and lower the holding pipet into the manipulation medium drop.

6. Position the embryo on the holding pipet so that an area of fragmentation or large perivitelline space is located opposite the holding pipet (typically at 3 o' clock). If performing blastomere biopsy, position the embryo so the desired blastomere can be accessed from the area you intend to hatch.
7. Lower the hatching pipet into the manipulation media drop. Using the 40× objective, bring the pipet tip into focus with the area of the zona pellucida to be dissolved. With the hatching pipet gently touching the zona pellucida, carefully expel the acidified medium over a 20–30 μM area (see Note 3) using a sweeping motion (see Note 4), creating a 25–30 μM hole in the zona pellucida.
8. As soon as the zona pellucida is penetrated, apply suction to the hatching pipet to remove any acidified medium from the immediate area overlying the newly formed hole. Quickly move the embryo to the opposite end of the drop.
9. Using a pipet with an inner diameter of >150 μM, remove the embryo from the micromanipulation dish. Tighter pipets may squeeze the embryo and cause the blastomeres to escape the zona pellucida. Gently wash the embryo through several drops of pre-equilibrated culture media before returning it to the culture drop for continued incubation (see Note 5).

3.1.2. Laser-Assisted Hatching of Cleavage-Stage Embryos

1. Prepare the assisted hatching dish with 10 μl droplets of micromanipulation medium for embryos that are to be hatched. Quickly cover the drops with washed mineral or paraffin oil, making sure the drops are *completely* covered with oil. Warm the dish on a 37 °C surface for at least 15 min prior to micromanipulation (see Note 6).
2. Load the embryo(s) into the micromanipulation medium drops.
3. Focus on the embryo on high power (40×) and orient the embryo to hatch at a gap between blastomeres or over an area of fragmentation. If performing blastomere biopsy, position the embryo so the desired blastomere can be accessed from the area you intend to hatch.
4. With the embryo in place, switch to the laser objective.
5. Focusing on the outer edge of the zona pellucida, fire the laser. Moving inward continue to fire until you have cut a 20–25 μM hole though the zona pellucida (see Note 7).
6. Using a pipet with an inner diameter of >150 μM, remove the embryo from the micromanipulation dish. Tighter pipets may squeeze the embryo and cause the blastomeres to escape the zona pellucida. Gently wash the embryo through several drops of pre-equilibrated culture media before returning its actual culture drop for continued incubation (see Note 5).

3.1.3. Laser-Assisted Hatching of Blastocyst Stage Embryos (see Note 8)

1. Prepare the assisted hatching dish with 10 μl droplets of micromanipulation medium. Quickly cover the drops with washed mineral or paraffin oil, making sure the drops are *completely* covered with oil. Warm the dish on a 37 °C surface for at least 15 min prior to micromanipulation (see Note 6).
2. Load the embryo(s) into the micromanipulation medium drops.
3. Focus on the embryo on high power (40×) and position the blastocyst (see Note 9). With the embryo in place, switch to the laser objective.
4. Focusing on the zona pellucida, fire the laser to create a 20 μM hole though the zona pellucida.
5. Using a pipet with an inner diameter of >250 μM, remove the embryo from the micromanipulation dish. Tighter pipets may squeeze the embryo and cause the blastomeres to escape the zona pellucida. Gently wash the embryo through several drops of pre-equilibrated culture media before returning its actual culture drop for continued incubation (see Note 5).

3.2. Embryo Culture

Embryo culture methods presented in previous chapters provide insight as to how embryologists can optimize conditions in their laboratory. Prolonged exposure to suboptimal conditions has been proposed to cause elevated rates of monozygotic twinning following extended embryo culture and blastocyst transfer (2, 4, 17, 24, 26). Despite this risk, the absence of tools to identify and select embryos with high reproductive potential at the cleavage stage, leads clinics to prefer blastocyst transfer, particularly for cases of single embryo transfer. If the theory proposed by Tummers et al. (31) that double implantation raises the risk of monozygotic twins is true, elective SET should be considered as method to reduce monozygotic twinning.

3.3. Elective Single Embryo Transfer

Programs must analyze their own data to determine which patients are good candidates for elective SET in their clinic. The methods below describe the criteria for SET at the University of Iowa Hospitals and Clinics.

1. Schedule all patients using oocytes from women under the age of 35 with six or more fertilized oocytes and women 35–37 with seven or more fertilized oocytes for a day 5 or blastocyst stage transfer.
2. Culture the embryos in groups of 3 or 4 in 50 μl drops of sequential culture media supplemented with 10 % protein under oil. All cultures are performed in 5–6 % CO_2 in air at 37°C (see Note 10).
3. Approximately 114 h post-insemination, evaluate the embryos and assign quality grades. Embryos that have a blastocoele that

comprises at least half the area of the blastocyst with the several cells in the inner cell mass, either tightly or loosely grouped, several cells forming a loose or cohesive epithelium in the trophectoderm are considered to be good quality.

4. Patients with at least one good quality blastocyst and no history of a previous failed fresh embryo transfer with the same oocyte source in the current clinic of care will undergo a single embryo transfer. If no good quality blastocysts are available, the patient may elect to transfer one or two embryos.
5. The embryo with the highest quality score is selected for transfer. Additional embryos that are graded as good quality expanded blastocysts may be cryopreserved per patient consent. Transfer of blastocysts that appear to have two inner cell masses or the inner cell mass appears to be "bridging" should be avoided. If there are no other good quality embryos available, the patient should be counseled regarding here elevated risk of monozygotic twins (see Notes 11 and 12).

4. Notes

1. As an extra precaution, draw a circle around the acidified medium drop on the underside of the dish to distinguish it from the manipulation medium drops.
2. Each manipulation medium drop is to be used once as the medium is acidified during assisted hatching (e.g., a second embryo should not be loaded into a drop that has already been exposed to the acidified medium). If assisted hatching is being performed without blastomere biopsy, more than one embryo may be loaded into the dish, but the duration of time the embryos are exposed to the acidified micromanipulation medium should be minimized.
3. The size of the hole is critical as holes ≤20 μM may prevent hatching or cause monozygotic twinning by pinching the embryo as it escapes the zona pellucida during hatching. Holes >30 μM may allow blastomere loss in pre-compaction stage embryos (32).
4. The outer shell of the zona pellucida typically dissolves quickly, but the inner layer may prove to be elastic and resistant to the acidified medium. Expanding the hole in the inner layer may require mechanical disruption with the hatching pipet as exposure to excess acidified medium may compromise embryo development.
5. The embryologists handling embryos that have undergone assisted hatching must be mindful of the artificial gap in the

zona pellucida throughout the remainder of the culture period and subsequent embryo transfer or cryopreservation. Always select a pipet with an inner diameter that provides liberal "clearance" for the embryo as it is being pipetted.

6. In contrast to using acidified medium, assisted hatching by laser does not change the chemical make-up and pH of the micromanipulation medium so the drops may be used multiple times for holding embryos. Furthermore, if embryo biopsy is *not* going to be performed, the embryologist may prefer to forego using a holding pipet.
7. Depending on the thickness of the zona pellucida, three to four pulses are typically required to cut completely through.
8. There have been no reports to date correlating an elevated risk of monozygotic twinning for blastocysts that have undergone assisted hatching. The accuracy and minimal trauma imposed by the laser make it the easiest method for assisted hatching the thin zona pellucida of the blastocyst stage embryo.
9. For trophectoderm biopsies, the blastocyst should be oriented with the ICM opposite the area to be hatched to avoid the risk that cells from the ICM are removed during biopsy. The ideal location for assisted hatching of blastocysts not undergoing biopsy is unknown. A recent study of assisted hatching in thawed blastocysts demonstrated that complete hatching was more likely if the zona was perforated near the inner cell mass (33).
10. Embryos are cultured in a cleavage medium that is glucose-, phosphate-free after the confirmation of fertilization (day 1). Unlike most sequential systems that move the embryos to a blastocyst culture medium designed for post-compaction stage embryos on day 3, we move day 3 embryos to a fresh drop of cleavage medium followed by transfer to blastocyst culture medium the morning of day 4.
11. Evidence of embryo splitting prior to embryo transfer on day 5 is rarely observed (34, 35). Meintjes et al. (35) reported establishment of a monozygotic triplet pregnancy after the transfer of a blastocyst with two inner cell masses.
12. The SET policy has been in place at our center for 6 years. During the first 3 years, the monozygotic twin pregnancy rate for patients undergoing SET was an alarming 5.8 % (9/155). During the most recent 3 years, we have been in a new laboratory and the monozygotic twin rate for SET has been 2.7 % (6/226). Only the space has changed. No other protocol or culture media changes were implemented between these two time periods. Our experience has been similar to that reported by Moeyeri et al. (36).

References

1. Bulmer MG (1970) The biology of twinning in man. Oxford University Press, Oxford
2. Jain JK et al (2004) Monozygotic twins and triplets in association with blastocyst transfer. J Assist Reprod Genet 21:103–107
3. Blickstein I, Verhoeven HC, Keith LG (1999) Zygotic splitting following assisted reproduction. N Engl J Med 340:738–739
4. Milki AA et al (2003) Incidence of monozygotic twinning with blastocyst transfer compared to cleavage-stage transfer. Fertil Steril 79:503–506
5. Harkness UF, Crombleholme TM (2005) Twin-twin transfusion syndrome: where do we go from here? Semin Perinatol 29:296–304
6. Vela G et al (2011) Monozygotic pregnancies conceived by in vitro fertilization: understanding their prognosis. Fertil Steril 95:606–610
7. Dube J, Dodds L, Armson BA (2002) Does chorionicity or zygosity predict adverse perinatal outcomes in twins? Am J Obstet Gynecol 186:579–583
8. Schinzel AGL, Smith DW, Miller JR (1979) Monozygotic twinning and structural defects. J Pediatr 95:921–930
9. Hall J (2003) Twinning. Lancet 362:735–743
10. Carroll S et al (2005) Is zygosity or chorionicity the main determinant of fetal outcome in twin pregnancies. Am J Obstet Gynecol 193:757–761
11. Lee A et al (2010) The frequency and clinical significance of twin gestations according to zygosity and chorionicity. Twin Res Hum Genet 13:609–619
12. Sebire N et al (1997) The hidden mortality of monochorionic twin pregnancies. Br J Obstet Gynaecol 104:1203–1207
13. Paek B, Shields LE (2005) Twin-to-twin transfusion syndrome: diagnosis and treatment. Curr Women Health Rev 1:43–7
14. Mains L et al (2009) Sextuplets: an unusual complication of single embryo transfer. Fertil Steril 91(3):932.e1–932.e2
15. Van der Hoorn ML et al (2011) Dizygotic twin pregnancy after transfer of one embryo. Fertil Steril 95:805.e1–805.e3
16. Alikani M et al (1994) Monozygotic twinning in the human is associated with the zona pellucida architecture. Hum Reprod 9: 1318–1321
17. Vitthala S et al (2009) The risk of monozygotic twins after assisted reproductive technology: a systematic review and meta-analysis. Hum Reprod Update 15:45–55
18. Blickstein I (2005) Estimation of iatrogenic monozygotic twinning rate following assisted reproduction: Pitfalls and caveats. Am J Obstet Gynecol 192:365–368
19. Derom C et al (1987) Increased monozygotic twinning rate after ovulation induction. Lancet 1:1236–1238
20. Derom C et al (2006) High frequency of iatrogenic monozygotic twins with administration of clomiphene citrate and a change in chorionicity. Fertil Steril 85:755–757
21. Blickstein I, Keith LG (2007) On the possible cause of monozygotic twinning: lessons from the 9-banded armadillo and from assisted reproduction. Twin Res Hum Genet 10: 394–399
22. Slotnik R, Ortega J (1996) Monoamniotic twinning and zona manipulation: a survey of U.S. IVF centers correlating zona manipulation procedures and a high-risk twinning frequency. J Assist Reprod Genet 13:381–385
23. Yakin K, Balaban B, Urman B (2003) Risks of monochorionic pregnancies after assisted hatching. Fertil Steril 79:1044–1046
24. Skiadas C et al (2008) Risk factors associated with pregnancies containing a monochorionic pair following assisted reproductive technologies. Hum Reprod 23:1366–1371
25. Menezo Y, Sakkas D (2002) Monozygotic twinning: is it related to apoptosis in the embryo? Hum Reprod 17:247–248
26. Behr B et al (2000) Blastocyst-ET and monozygotic twinning. J Assist Reprod Genet 17:349–351
27. Chang HJ et al (2009) Impact of blastocyst transfer on offspring sex ratio and the monozygotic twinning rate: a systematic review and meta-analysis. Fertil Steril 91:2381–2390
28. Kawachiya S et al (2011) Blastocyst culture is associated with an elevated incidence of monozygotic twinning after single embryo transfer. Fertil Steril 95(6):2140–2142. doi:10.1016/j.fertnstert.2010,12.018
29. Sills ES, Tucker MJ, Palermo GD (2000) Assisted reproductive technologies and monozygous twins: implications for future study and clinical practice. Twin Res 3: 217–223
30. Knopman J et al (2009) Monozygotic twinning: an eight-year experience at a large center. Fertil Steril 94:502–510
31. Tummers P, De Sutter P, Dhont M (2003) Risk of spontaneous abortion in singleton and twin pregnancies after IVF/ICS. Hum Reprod 18:1720–1723

32. Veeck LL (1999) An Atlas of human gametes and conceptuses. Parthenon Publishing, New York
33. Miyata H et al (2010) Relevance of the site of assisted hatching in thawed human blastocysts: a preliminary report. Fertil Steril 94: 2444–2445
34. Meintjes M et al (2001) Prospective identification of an in vitro-assisted monozygotic pregnancy based on a double-inner-cell-mass blastocyst. Fertil Steril 76:S172–S173
35. Payne D et al (2007) Time-lapse recording identifies human blastocysts at risk of producing monozygotic twins. Hum Reprod 22(Supp 1):i9
36. Moayeri SE et al (2007) Risk of monozygotic twinning with blastocyst transfer decreases over time: an 8-year experience. Fertil Steril 87:1028–1032

Chapter 23

Embryo Culture and Epigenetics

Brenna A. Market Velker, Michelle M. Denomme, and Mellissa R.W. Mann

Abstract

During preimplantation development, major epigenetic reprogramming occurs, erasing gametic modifications, and establishing embryonic epigenetic modifications. Given the plasticity of these modifications, they are susceptible to disruption by assisted reproductive technologies, including embryo culture. The current state of evidence is presented for the effects of embryo culture on global DNA methylation and histone modifications, retroviral silencing, X-inactivation, and genomic imprinting. Several salient points emerge from the literature; that culture in the absence of other procedures can lead to epigenetic perturbations; that all media are suboptimal; and that embryo response to in vitro culture is stochastic. We propose that embryos adapt to the suboptimal environment generated by embryo culture, including epigenetic adaptations, and that "quiet" embryos may be the least epigenetically compromised by in vitro culture.

Key words: Embryo culture, Epigenetics, Endogenous retroviral silencing, X inactivation, Genomic imprinting, DNA methylation, Histone modifications

1. Introduction

The field of epigenetics encompasses the study of heritable and reversible modifications of chromatin that influence the accessibility of genes and regulate gene transcription (1). Alterations in gene expression are not brought about by changes to the DNA sequence, yet the epigenetic information affecting gene activation or repression ultimately allows for variation in cellular phenotypes. Epigenetic mechanisms include DNA methylation and histone posttranslational modifications. These mechanisms play important roles in the establishment of cellular memory and differentiation during embryogenesis (2), and given their plasticity, are susceptible to disruption by assisted reproductive technologies (ARTs).

Gary D. Smith et al. (eds.), *Embryo Culture: Methods and Protocols*, Methods in Molecular Biology, vol. 912, DOI 10.1007/978-1-61779-971-6_23, © Springer Science+Business Media, LLC 2012

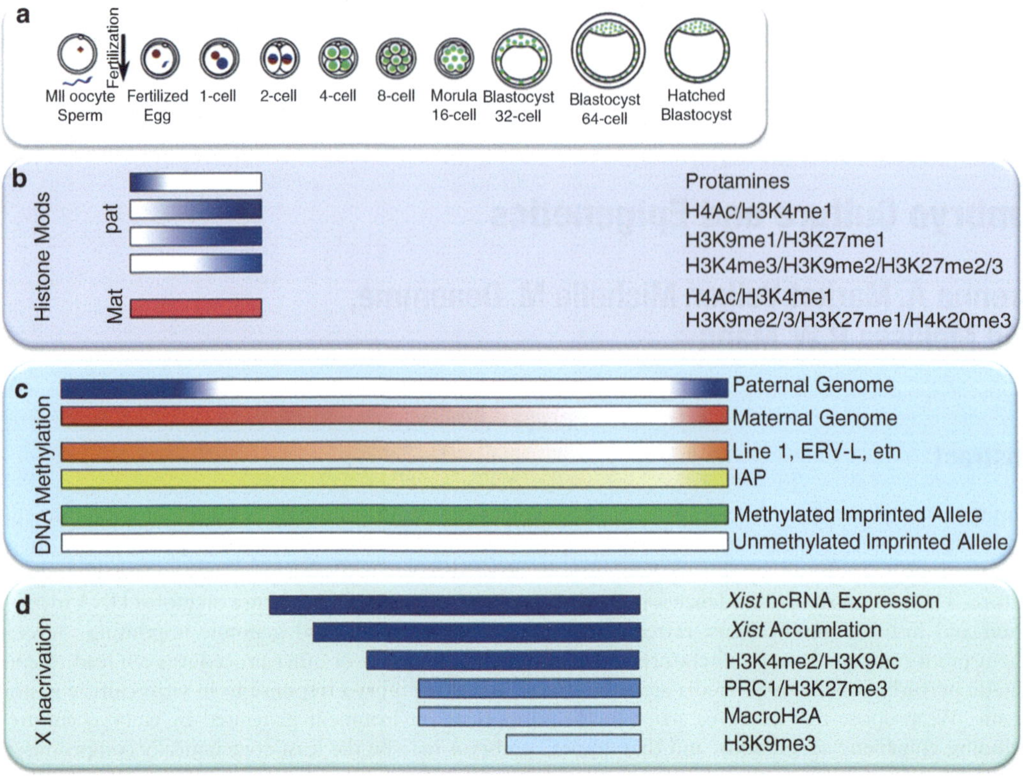

Fig. 1. (**a**) Epigenetic reprogramming during mammalian preimplantation development. Following fertilization, the maternal and paternal genomes must be epigenetically reprogrammed from a gametic to an embryonic state. Reprogramming is initiated in pronuclear stages and continues throughout preimplantation development. Maternal gametic genome (*red*), paternal gametic genome (*blue*), and embryonic genome (*green*). (**b**) Histone modifications. Protamines are actively removed from the paternal genome (*shades of blue*) immediately following fertilization. Activating histone modifications are set in the early pronuclear stages (H4Ac/H3K4me1), which are followed by acquisition of repressive histone modifications (H3K4me3/H3K9me2/HeK27me2/3). Both activating (H4Ac/H3K4me1) and repressive (H3K9me2/3/H3K27me1/H4K20me3) histone modifications are abundant throughout the maternal genome (*pink*) during pronuclear stages. (**c**) DNA Methylation. The paternal genome (*blue*) undergoes active global demethylation immediately following fertilization, while the maternal genome (*red*) is passively demethylated through subsequent cleavage divisions. Line, ERV-L, and etn (*orange*) follow a similar demethylation process, while IAPs (*yellow*) and imprinted loci (*green*) are protected from demethylation events. (**d**) X-inactivation. A cascade of repressive modifications occurs on the paternal X chromosome in female preimplantation embryos, beginning with paternal-specific *Xist* expression.

During mammalian development, there are two major phases of epigenetic reprogramming (2, 3). During gametogenesis, reprogramming consists of erasure of previous epigenetic marks that restore totipotency, followed by the establishment of sex-specific epigenetic marks. Following fertilization, major epigenetic reprogramming occurs for a second time, and includes alterations in histone post-translational modifications and DNA methylation (Fig. 1).

1.1. Histone Modifications

The nucleosome is the basic chromatin unit, containing double-stranded DNA wrapped around an octamer of core histone proteins (4). Chromatin is a dynamic structure that undergoes alterations in its packaging, affecting the accessibility of underlying genes (5). Heterochromatin is a highly condensed structure that results in gene silencing, while euchromatin is a less compacted configuration supporting gene expression. Switching between active and repressive chromatin states is facilitated by covalent posttranslational modifications of histone tails (5). Histone tail modifications, such as acetylation and methylation, confer either activating or repressive functions, depending on the type of modification and the amino acid residue that is modified. Shortly after fertilization, the paternal genome undergoes extensive remodeling that includes an exchange of protamines for histones, and acquisition of active histone modifications, including histone H4 acetylation (H4Ac) and histone H3 lysine 4 methylation (H3K4me1) (6) (Fig. 1). In early embryos, the paternal genome also acquires repressive histone modifications, including histone 3 lysine 9 and lysine 27 methylation (H3K9me2, H3K27me2, and H3K27me3). By comparison, the maternal genome possesses both active (H4Ac, H3K4me1) and repressive histone modifications (H3K9me2, H3K9me3, and H4K20me3). These covalent modifications are proposed to initiate the transcriptionally repressed state that coincides with embryonic genomic activation. This potential for chromatin bivalency, where both activating and repressive marks occupy the same stretch of chromatin, is likely a major factor in establishing the correct gene expression profile for embryonic development (7).

Switching between active and repressed chromatin states is also brought about by incorporation of specialized histone variants and chromatin remodeling proteins. One histone variant present during early mouse development is histone 3.3 (H3.3), which is first detected in the paternal pronucleus after the protamine to nucleosome exchange, and then at the 2-cell, 4-cell, and blastocyst stages in correlation with waves of embryonic gene activation (8). H3.3 is exchanged for H3 to promote transcriptionally active regions (9). In addition, chromatin remodeling proteins can change chromatin structure in a non-covalent manner. Functioning antagonistically, Polycomb group proteins act to establish a silenced chromatin state via polycomb repressive complexes (PRC1 and PRC2), while Trithorax group proteins act to promote a permissive chromatin state (10). In the mouse zygote, PRC1 and PRC2 complexes are localized to the maternal pronucleus at earlier pronuclear stages (PN0–3), and are observed at the paternal pronucleus by the latest pronuclear stage (PN5), initiating the repressive pathways for maternal and paternal constitutive heterochromatin in the embryo (11). Thus, extensive epigenetic reprogramming occurs in histone and chromatin remodeling during early development.

1.2. DNA Methylation

DNA methylation is the covalent modification of cytosines within CpG dinucleotides that is typically associated with gene silencing. DNA methyltransferases, DNMT3A, DNMT3B along with DNMT3L, act as de novo methyltransferases, establishing methylation marks. Acquisition of DNA methylation occurs in the developing gametes and is acquired differentially between the two parental genomes (12, 13). By comparison, DNMT1 is responsible for maintaining DNA methylation throughout preimplantation development through its recognition of hemi-methylated DNA. An oocyte-specific isoform, DNMT1o, is present in the mature oocyte, and localizes to the maternal pronucleus post-fertilization (14). While DNMT1o is present during all stages of preimplantation development, it is predominantly cytoplasmic, except at the 8-cell stage when it is translocated to the nucleus (15, 16). Therefore, its function in maintaining DNA methylation during preimplantation development is not completely clear. A somatic isoform, DNMT1s, which localizes to the nucleus throughout preimplantation development, is responsible for DNA methylation maintenance at repetitive elements and imprinted genes (17). Paradoxically, while the fertilized egg is highly enriched for DNMT proteins, there is a wave of demethylation during preimplantation development that erases gamete-specific methylation patterns and ensures the totipotency of the early embryo (3). The paternal genome is actively demethylated within hours after fertilization, while the maternal genome is passively demethylated during early cleavage divisions in a replication-dependent manner (Fig. 1). The end result is that the two parental genomes undergo extensive changes in global methylation during preimplantation development. Postimplantation, de novo methylation gradually increases in accordance with cellular differentiation.

In addition to these generalized epigenetic gene regulatory mechanisms, there are three specialized epigenetic phenomena undergoing dynamic regulation during early embryonic development. These phenomena are endogenous retroviral silencing, X chromosome inactivation, and genomic imprinting.

1.3. Endogenous Retroviral Silencing

Mammalian genomes are highly enriched for repetitive sequences, of which the vast majority are endogenous retroviral elements (18). To reduce the deleterious effects of retrotransposons, the host epigenetically silences these elements. As such, preimplantation development is a critical period with dynamic changes in expression and DNA methylation of retroviral elements. Mouse endogenous retroviral-like element (ERV-L) and interstitial A particles (IAP) are expressed from the 2-cell stage and are repressed by the blastocyst stage (19, 20). With regard to DNA methylation, some retro-elements are demethylated similar to the majority of the genome including human L1, mouse ERV-L, and mouse early retrotransposons (etn), while IAPs remain heavily methylated and protected from this demethylation event (14, 21, 22) (Fig. 1). DNMT1 is the methyltransferase responsible for maintenance methylation at IAPs (14, 23).

Given the dynamic nature of epigenetic retroviral regulation, a mouse model system, agouti viable yellow (A^{vy}), has been used to study environmental exposure. A^{vy} mice differ from wildtype mice, as they possess a transposable IAP element upstream of the agouti gene, which contains an alternate, constitutive promoter (24, 25). Hypomethylation of the retrotransposon allows for ectopic agouti transcription, which leads to yellow fur as well as adult-onset obesity, diabetes, and tumor development (26, 27). Hypermethylation of the transposable element silences ectopic expression, leading to wildtype agouti coat color and protection against associated pathology. These mice are ideal for environmental studies as genetically identical A^{vy} mice can exhibit an array of coat color phenotypes based on insult-induced epigenetic alterations during early development. For example, maternal exposure to deleterious environmental toxicants, such as bisphenol A, can reduce methylation levels within the fetal genome, causing hypomethylation at the A^{vy} gene (28). Conversely, maternal nutritional supplementation with methyl donors (folic acid, betaine, vitamin B12, and choline) can protect against loss of methylation, permitting hypermethylation of A^{vy} gene (29).

A second locus of interest is the *Axin1-Fused* (*Axin1*^{Fu}) gene. In this case, a transposable IAP element, which possesses a cryptic promoter, is located with intron 6 of the *Axin1* gene (30). Hypomethylation of the *Axin1*^{Fu} allele permits mutant transcripts to be expressed, generating the *Axin1*^{Fu} kinked tail phenotype, while hypermethylation of the *Axin1*^{Fu} allele silences expression from the cryptic promoter, preventing tail kinks (31). These mice are also well suited for environmental studies as genetically identical *Axin1*^{Fu} mice can exhibit an array of kinked tail phenotypes based on environmental insult during early development (31, 32). Thus, the A^{vy} and *Axin1*^{Fu} mouse models represent excellent systems to determine the epigenetic effects of environmental insult during embryo development, including embryo culture.

1.4. X-Inactivation

Dosage compensation in mammals is achieved by X chromosome inactivation, whereby XX females silence one sex chromosome to enable equal X-linked gene expression to that of XY males. The noncoding X-inactive specific transcript (*Xist*) RNA located within the X chromosome inactivation center is responsible for this process (33). During preimplantation development in the mouse, *Xist* is transcribed from the paternal X and coats the inactive X chromosome, initiating a cascade of epigenetic modifications that result in silencing of the entire chromosome (34). The propagation of X inactivation along the chromosome involves a hierarchy of epigenetic changes that are initiated with *Xist* expression (35–37)

(Fig. 1). Around the 8-cell stage, active histone modifications (H3K9Ac, H3K4me2) are lost. Chromatin remodeling complexes are then recruited to the paternal X chromosomes around the 16-cell stage, conferring the repressive histone modification histone 3 lysine 27 trimethylation (H3K27me3) (35–37). At this time, the repressive histone variant macroH2A is also incorporated into the paternal X chromosome. At the 32-cell stage, the repressive histone modification H3K9me3 is acquired on the paternal X (35–37). At the blastocyst stage, the trophectodermal lineage retains this imprinted form of X chromosome inactivation, while in the inner cell mass, the paternal X is reactivated, and random X chromosome inactivation is initiated (36, 38).

1.5. Genomic Imprinting

Genomic imprinting is an epigenetic phenomenon whereby certain genes are expressed exclusively from one parental allele (33). These imprinted genes often reside within domains, where multiple imprinted genes are coordinately regulated in *cis* by DNA elements known as imprinting centers. Genomic imprinting is the result of both acquisition of imprinting marks during gametogenesis, whereby oocytes acquire maternal-specific DNA methylation imprints and sperm obtain paternal-specific DNA methylation imprints, and of maintenance of these marks throughout development. During preimplantation development, a wave of demethylation erases gamete-specific methylation from the maternal and paternal genomes. However, DNA methylation imprinting marks are excluded from this demethylation event; the methylated allele remains methylated and protected from demethylation, while the unmethylated allele remains unmethylated (3) (Fig. 1). Maintenance of these methylation marks through preimplantation development requires expression of DNMT1 (14, 17, 39, 40), as well as changes in histone modifications, including H3K9me2 and H3K27me3 enrichment on the repressed allele (41).

Of significance is that many imprinted genes play critical roles in the development of the embryo, and their misregulation has been linked to the development of human imprinting disorders, such as Angelman syndrome (AS) and Beckwith–Wiedemann syndrome (BWS). AS is an imprinting disorder characterized by a severe mental retardation, seizures, little or no speech, an abnormal gait and frequent smiling/laughing behavior (42), while BWS is an overgrowth disorder characterized by macroglossia, abdominal wall defects, birth or postnatal weight above the 90th percentile, neonatal hypoglycemia, and an increased incidence of Wilm's tumor (43).

Given the dynamic regulation of endogenous retroviral silencing, X chromosome inactivation, and genomic imprinting during early embryonic development, these specialized epigenetic mechanisms have been the focus of studies involving assisted reproduction, with genomic imprinting the most widely investigated.

1.6. Embryo Culture

Given that epigenetic mechanisms mediate interactions between the stable intrinsic genome and variable extrinsic environment, many studies have investigated the effects of environmental insult on epigenetic gene regulation. As such, scientists and clinicians alike have questioned the environmental impact of embryo culture on epigenetic programming. While we focus on epigenetic effects of embryo culture, the following caveat should be noted. In many papers, ARTs and their observed effects are often grouped together under the umbrella of "potential consequences of ARTs", making it nearly impossible to separate effects of embryo culture from those of other techniques. Thus, one must bear in mind that the described effects of embryo culture may be attributed to more than one procedure. Given the numerous and varied forms of ARTs currently employed in research and human clinics, investigation of distinct procedures should be a clear objective of the field moving forward.

1.6.1. Epigenetic Effects of Embryo Culture

It has been proposed that embryo culture may have a global effect on DNA methylation patterns, chromatin remodeling, and gene expression. To assess the effects of in vitro culture on DNA demethylation, the dynamics of genome demethylation were assayed using immunocytochemistry with 5-methylcytosine antibodies (44, 45). Comparison between in vitro cultured and in vivo-derived mouse and rat embryos revealed a decreased ability of cultured embryos to efficiently demethylate the paternal genome, indicating that embryo culture may interfere with the earliest epigenetic reprogramming of the gamete-derived genomes.

Global changes in DNA methylation have also been analyzed. Recently, the effect of assisted reproduction on DNA methylation was assessed at more than 700 genes in placenta and cord blood from children of assisted and unassisted conceptions using a site-specific CpG methylation assay (46). A significant change in DNA methylation was observed with lower mean methylation levels in the placenta, and higher mean methylation levels in cord blood from children conceived in vitro when compared to children conceived naturally, indicating the broad effects of ART on DNA methylation.

To date, only one study has examined global changes in histone modifications. Using immunohistochemistry, H4 acetylation, H3K9 methylation, or H3S10 phosphorylation was examined in mouse embryos from the 1-cell to blastocyst stage following in vitro fertilization (47). No difference was found between in vitro cultured and in vivo-derived mouse embryos for these modifications. Given the limited nature of immunofluorescence analyses in detecting specific changes that may occur in a globally modified background, future studies should be directed towards high throughput sequence-based assays. However, the limited material available in individual preimplantation embryos makes studies utilizing these current techniques difficult at best.

In addition to global epigenetic changes, effects of embryo culture on endogenous retroviral silencing, X chromosome inactivation, and genomic imprinting have also been investigated. Two endogenous retroviral mouse models, A^{vy} and *Axin1*^{Fu}, have been employed to study the environmental impact of embryo culture on retroviral silencing. In these studies, effects of embryo culture during preimplantation development were queried by examining the phenotypic effects of postnatal A^{vy} and *Axin1*^{Fu} expression. In vitro culture was found to increase the agouti yellow and tail kink phenotypes, respectively (48, 49). This indicates that embryo culture interferes with maintenance of epigenetic silencing of IAPs, allowing an active state of the IAP allele to be propagated. In the case of the agouti yellow phenotype, the A^{vy} allele displayed a corresponding loss of DNA methylation, while for the kinked tail phenotype, the *Axin1*^{Fu} allele exhibited H3K4 dimethylation and H3K9 acetylation, two hallmarks of active chromatin (48, 49). These results again demonstrate the adverse impact of embryo culture on epigenetic gene regulation.

With respect to X chromosome inactivation and embryo culture, very few studies have been performed. In bovine, increased *XIST* transcript abundance was observed in in vitro-cultured compared to in vivo-derived embryos (50, 51). In the mouse, *Xist* was inappropriately expressed from the sole, maternally derived X chromosome in male midgestation placenta following preimplantation in vitro culture, perhaps indicating that the imprinted form of X-inactivation is disrupted by embryo culture (52).

In contrast to X chromosome inactivation, a significant body of evidence has emerged regarding the aberrant regulating of genomic imprinting following in vitro culture of mouse, human, and livestock embryos. The first study to demonstrate a link between embryo culture and genomic imprinting reported biallelic expression of the maternally expressed *H19* gene in a subset of pooled, cultured mouse embryos (53). A similar study examining *H19* imprinted expression and methylation revealed aberrant biallelic *H19* expression and loss of paternal-specific methylation in cultured mouse blastocysts (54). Since then, it has been determined that in vitro preimplantation culture of mouse embryos results in biallelic expression of the *H19* gene and loss of imprinted methylation at the *H19*, *Snrpn*, and *Peg3* genes in blastocyst stage embryo (52–58). Subsequent observations of postimplantation embryos indicate that epigenetic alterations induced by embryo culture persist. At day 9.5, following embryo culture and embryo transfer, loss of methylation and biallelic expression was observed for *H19*, *Snrpn*, *Peg3*, and *Kcnq1ot1* in extraembryonic tissues, indicating that imprinting perturbations are inherited through to midgestation, long after embryos have been removed from in vitro culture (52, 53, 59).

Genomic imprinting may also be affected by in vitro manipulation of early embryos during ARTs for the treatment of infertility/subfertility in humans. Increased incidence of the human imprinting disorders AS and BWS has been reported with the use of ARTs (60–68). More specifically, the incidence of Angelman syndrome in the general population is approximately one case per 16,000 births, with only 5% of these cases related to imprinting abnormalities (42, 60). As the prevalence of AS is low, large-scale studies containing sufficient numbers of patients have been difficult to achieve. However, seven cases of AS following the use of ARTs have been reported to date, five of which displayed imprinting abnormalities (71%) (60, 65, 68, 69). This is a significantly higher proportion in the ART population compared with the non-ART population. The specific epigenetic defect of AS has been mapped to the *Snrpn* imprinting center. Loss of methylation at the paternal *Snrpn* imprinting center results in overexpression of the paternally expressed genes of this domain and loss of expression of the maternally expressed genes.

BWS is a second imprinting disorder associated with ARTs. BWS affects about one in 13,700 children (43). As with AS, parents of children with BWS are more likely to have undergone fertility treatments than the general population, and a higher incidence of BWS is seen in ART children compared with naturally conceived children (61–63, 65, 67, 69). Two imprinted defects have been linked to BWS. With respect to the first, the maternal *H19* imprinted domain acquires a paternal epigenotype. Gain of *H19* imprinting center methylation results in silencing of *H19* expression and activation of *Igf2* expression. The second imprinting defect occurs at the *Kcnq1ot1* domain and results from loss of methylation on the maternal *Kcnq1ot1* imprinting center with ensuing biallelic expression of the long noncoding RNA *Kcnq1ot1*, and silencing of *Kcnq1* and *Cdkn1c*. The prevalence of ARTs is approximately 4 9 times greater in BWS children compared to the number of ART children in the general population. Taken together, ARTs may impose inherent risk for normal imprinted gene regulation.

Bovine and ovine animals produced through in vitro fertilization and other forms of ARTs display an increased frequency of "Large Offspring Syndrome", which is characterized by a variety of abnormalities including increased birth weight, organomegaly, polyhydraminos, hydrops fetalis, immunological defects, increased fetal/neonatal death as well as numerous skeletal and placental defects (70, 71). Interestingly, the phenotypes observed with Large Offspring Syndrome are similar to those observed in Beckwith–Wiedemann syndrome. Importantly, perturbation of the same loci involved in BWS is observed in calves with Large Offspring Syndrome. *KCNQ1OT1* is hypomethylated with a corresponding increase in *KCNQ1OT1* expression and decrease in *CDKN1C* expression (72). Large offspring sheep also exhibit loss of imprinting for the growth inhibiting *Insulin-like growth factor 2 receptor* gene (73). Furthermore, loss of maternal-specific *SNRPN* methylation and activation of normal silent maternal *SNRPN* allele was found in placenta from in vitro fertilized and cultured bovine embryos, similar to AS in humans (74). The similarities in ART-induced epigenetic errors in humans and in animal models, where subfertility is not a confounding issue, reinforces the notion that manipulation of the early embryo can lead to epigenetic perturbations which result in long-term consequences for the offspring.

One interesting observation from the above studies is that not all embryos, and not all imprinted loci in these embryos, are affected by in vitro culture. This indicates that the response of embryos to in vitro culture is stochastic. Analyses at the individual embryo level have confirmed this (52, 57, 58). Similar observations have been recently reported in ART-conceived children with BWS (66, 75). Imprinting defects at multiple imprinted loci in addition to the *Kcnq1ot1* gene were present in BWS patients whose parents had undergone some form of ART. These data suggest that developmental defects or abnormal growth in ART children might be caused by variable combinations of epigenetic perturbations at imprinted genes. The mechanism by which some embryos are able to compensate for the suboptimal preimplantation environment while others are not is unknown. However, it has been hypothesized that modulation of the expression of a group of imprinted genes (known as the "imprinted gene network") may play a role (57).

All together, these studies have confirmed maintaining preimplantation embryos outside the female reproductive tract using embryo culture affects genomic imprinting at multiple loci. In humans, infertility/subfertility may also be a contributing factor in aberrant imprinted gene regulation. Thus, it is clear that embryo culture constitutes an environmental insult to epigenetic gene regulation. However, the specific aspect of embryo culture most responsible for this epigenetic misregulation has yet to be determined.

1.6.2. Epigenetic Effects of Culture Media

Many studies have attempted to show superiority of one media or another, with respect to various measures of developmental competence (76–78). With respect to gene expression, a comparison was made between embryos cultured in KSOM augmented with amino acids (KSOMaa) with those cultured in Whitten's medium. Global expression analysis revealed that a larger number of genes were misexpressed after culture in Whitten's medium compared with culture in KSOMaa (79), suggesting aberrant epigenetic regulation of gene expression in Whitten's medium.

With regard to X chromosome inactivation, *Xist* expression has been analyzed in midgestation placental tissues from male embryos subjected to preimplantation in vitro culture. Culture in

Whitten's medium produced a greater number of embryos with ectopic *Xist* expression from the sole active X chromosome compared to culture in KSOMaa (52).

The effects of various culture media on genomic imprinting have also been evaluated. In the mouse model, while KSOMaa was better able to maintain genomic imprinting than Whitten's medium, KSOMaa has also been shown to cause disruptions in *H19*, *Snrpn*, *Peg3*, and *Ascl2* imprinting at the blastocyst stage, and *H19*, *Snrpn*, *Igf2*, *Kcnq1ot1*, *Cdkn1c*, *Mkrn3*, *Ascl2*, *Zim1*, and *Peg3* imprinting at postimplantation stage (day 9.5) (52, 54, 59). In addition, at the blastocyst stage, M16 medium was shown to cause greater perturbation of *H19* imprinting than G1.2/G2.2 (57). Human tubal fluid medium (HTF) caused loss of *H19* imprinting including aberrant histone modifications with an increase in H3K4 dimethylation on the paternal *H19* allele and an increase in H3K9 trimethylation on the maternal *H19* allele (56). In bovine, alterations in imprinting of *Snrpn* have been associated with the nonsequential SOF medium (74) as well as *Sgce* and *Ata3* in the sequential vitro cleave/vitro blast medium (80). In humans, a recent study reported loss of methylation at the *H19* locus in ~19% of a cohort of human embryos cultured in cleavage medium (81). As imprinting perturbations arise from multiple media systems, it suggests that all media are suboptimal in supporting normal genomic imprinting. Of note however, is that all the above studies employed ovarian stimulation to retrieve embryos prior to culture.

To assess the effects of various culture media on genomic imprinting in the absence of other ARTs, we performed a comparison of six embryo culture systems, Whitten's; KSOMaa; HTF; Global; Preimplantation 1/Multiblast (P1/MB); and G1v5PLUS/G2v5PLUS (G1.5/G2.5), in the mouse model system. Blastocysts cultured in each of the six culture systems experienced a loss of *H19*, *Snrpn*, and *Peg3* imprinting (58). With regard to imprinted methylation, better preservation of *H19*, *Snrpn*, and *Peg3* imprinted methylation was observed in embryos cultured in KSOMaa compared to the other media systems where a greater loss of imprinted methylation was observed. In addition, a gene-dependent response was observed with certain media supporting higher levels of imprinted methylation at some but not at all loci. More specifically, embryos cultured in KSOMaa, Global and P1/MB displayed higher levels of *H19* imprinted methylation than Whitten's, HTF and G1.5/G2.5. For *Snrpn*, KSOMaa, and Global better maintained imprinted methylation than the other media systems, and for *Peg3*, levels of imprinted methylation for KSOMaa-cultured embryos were best, while Whitten's and G1.5/G2.5 cultured embryos showed the greatest loss of imprinted methylation. Of note is that sequential media systems did not seem to confer an advantage with respect to maintenance of genomic imprinting compared to their single step counterparts, nor did medium renewal. Our results confirmed the observation that in vitro culture of preimplantation embryos alone leads to alterations in imprinted gene regulation.

1.6.3. Epigenetic Effects of Serum

The addition of serum to embryo culture media, although frequently used for somatic cell culture, has been shown to be detrimental to the development of the early embryo resulting in decreased developmental potential (82, 83). Two studies in mouse have addressed the effects of serum on genomic imprinting. The addition of serum to the culture medium was shown to have severe consequences with respect to the epigenetic integrity of the early embryo, resulting in alterations in the expression and methylation of various imprinted genes (*H19*, *Igf2*, *Grb7*, *Grb10*, and *Pegl/Mest*), a faster rate of development to the blastocyst stage, and long-term behavioral consequences (55, 84). In addition, the use of serum is highly associated with the development of Large Offspring Syndrome (71, 85, 86). As serum contains undefined growth factors and other macromolecules, the mechanism of these perturbations remains unknown. Thus, the use of serum in embryo culture is discouraged.

1.6.4. Epigenetic Effects of Oxygen Tension

In the female reproductive tract, embryos are maintained in a relatively hypoxic environment, with oxygen levels around 5% (87, 88). Culture of mammalian embryos in low oxygen improves many aspects of embryo development (89–94). These results indicate that the overall epigenetic landscape of embryos cultured in 5% oxygen is more similar to in vivo-derived embryos than those cultured in 20% oxygen (95). The effect of various oxygen concentrations on genomic imprinting in particular has not been well studied. Overall, expression levels of a subset of imprinted genes were not noted to be different in embryos cultured in high levels of oxygen post-compaction compared to those cultured in low levels. However, this investigation was not able to distinguish whether expression of these genes was from the maternal allele, paternal allele, or both (96). With the clear advantage of low oxygen tension with respect to overall embryo development, in vitro culture in low oxygen is highly recommended.

1.6.5. Multiple ARTs

As noted above, many studies examining the epigenetic effects of embryo culture have employed multiple ARTs. With this in mind, several groups have conducted research to distinguish between the effects of various ARTs. In mouse, one study tested the effects of superovulation alone, superovulation and in vitro culture following natural fertilization, and superovulation, in vitro fertilization, and in vitro culture (57). The group subjected to all three procedures showed the greatest disruption to *H19* imprinting in individual blastocysts. Similarly, we also observed that increased loss of

H19 imprinted expression as a result of embryo culture that was exacerbated by the use of superovulation (58). Finally, a study by Rivera et al. demonstrated an increase in the number of embryos with biallelic *H19*, *Cdkn1c*, *Kcnq1*, *Snrpn*, *Kcnq1ot1*, *Peg3*, and *Mkrn3* expression following superovulation with embryo transfer compared to controls, with a further increase in biallelic expression following embryo culture combined with superovulation and embryo transfer (59). Together, these studies demonstrate an additive effect with combined ART procedures producing a greater chance of epigenetic perturbations than single interventions.

In a recent study, we examined the effects solely of superovulation on genomic imprinting in a mouse model. As superovulation is thought to primarily affect oocyte development, effects were expected to be limited to maternal alleles. In addition to aberrant maternal-specific methylation, we found that paternal *H19* methylation was perturbed by superovulation (97). This suggests that superovulation affects maternal-effect gene products subsequently required for imprint maintenance during preimplantation development. Our new working model is that the effects of superovulation and embryo culture converge in the preimplantation embryo on the same epigenetic pathway that is supported by oocyte-specific product(s). In fact, it is not unreasonable that a number of environmental insults, such as oxygen tension, serum, waste by-products, and embryo transfer may perturb the same epigenetic pathway in a stochastic manner. This would be consistent with ART-associated BWS children where variable ART procedures were reported to be used, including embryo culture media (varied in type, and in glucose, amino acid and human sera albumin content), day of transfer, and specific ART method used (in vitro fertilization, intracytoplasmic sperm injection), in addition to the cause of infertility (62, 69). This is consistent with a recent study where no differences were observed in global DNA methylation and chromatin organization in human embryos generated via in vitro fertilization or intracytoplasmic sperm injection (98). Furthermore, a recent study of *H19* imprinted methylation examined intra- and interindividual variation in DNA methylation in in vitro or in vivo conceived children (99). Substantial variation was found in allele-specific *H19* methylation, with greater aberrant methylation and greater variability in abnormal methylation in the in vitro group. Furthermore, consistent with animal model studies, placenta harbored greater perturbation than cord blood or tissue. We interpret these results to mean that there is stochastic imprinting perturbation at the individual cell level following in vitro conception.

If embryo response to in vitro culture is stochastic, then the leading question is which embryo will be most compromised by suboptimal culture. In our most recent study, we investigated rates of development and loss of genomic imprinting (100). We reasoned that as culture embryos exhibit delayed development, that slower developing embryos would exhibit greater perturbation in imprinted gene regulation than their counterparts that possess faster rates of development. Contrary to our expectations, embryos with faster rates of development experienced the greatest loss of imprinting (100). Furthermore, expression of two markers of metabolic activity revealed that the fastest developing embryos were most divergent from in vivo-derived controls. Our results are consistent with the Quiet Embryo hypothesis which states that those embryos which are able to maintain a relatively quiet or low level of metabolism are most viable (101, 102). Therefore, embryos that adapt the least to their suboptimal environment, including through epigenetic alterations, will likely develop in the most normal fashion. Thus, development of culture media, which minimize the need for increased metabolic activity, will likely have a corresponding decrease in epigenetic perturbations and thus, increased overall health of the embryo.

2. Conclusions

In conclusion, all of the above studies on epigenetic gene regulation in early embryo development reinforce the point that preimplantation development is a dynamic period of epigenetic programming and suboptimal embryo culture produces epigenetic adaptations. While it is possible to generate blastocysts that appear morphologically normal, some in reality are epigenetically compromised. It is up to those pursuing research in these fields to first determine accurate, repeatable, and noninvasive techniques to detect these compromised embryos, as well as to continue to improve current techniques to minimize the amount of adaptation that embryos will require to survive and grow outside the female reproductive tract.

3. Materials

The following are solutions for the combined imprinted expression and methylation analysis for a single blastocyst stage embryo. Prepare all solutions with double-distilled autoclaved water. Use filter tips and nonstick surface microfuge tubes. Wash all pipetmen and equipment with RNAse Away (Invitrogen, Burlington, Canada). To reduce the chance of DNA and RNA contamination, change gloves often. Wipe all pipetmen, tubes, tip boxes and other reagents with RNAse Away after each change of gloves. Keep tubes angled away when open, and recap all tubes when not in use.

1. Dynabead mRNA Direct Micro kit (Dynal Biotech, Invitrogen, Burlington, Canada).

2. Ready-To-Go PCR beads (GE, Healthcare, Quebec, Canada).
3. 10% Igepal: 1 ml of Igepal (Sigma, Oakville, Canada) in 9 ml H_2O.
4. 3 M NaOH: 2.4 g NaOH in 20 ml H_2O.
5. 0.1 M NaOH: 0.5 ml 3 M NaOH in 14.5 ml H_2O.
6. 0.3 M NaOH: 1 ml of 3 M NaOH in 9 ml H_2O.
7. ITT Buffer: 50 μl of 10% Igepal and 50 μl of 10% Tween 20 in 50 ml of TE (pH 7.5).
8. 2% Low melting point (LMP) agarose: 20 mg Agarose in 1 ml H_2O. Dissolve at 70°C, then incubate at 90°C until use.
9. 2.5 M Bisulfite solution Part 1 (see Note 1): 3.8 g Sodium bisulfite, 5.5 ml H_2O, 1 ml 3 M NaOH. Dissolve at 50°C.
10. Bisulfite solution Part 2: 110 mg Hydroquinone, 1 ml H_2O. Dissolve at 50°C. When fully dissolved, mix bisulfite solution Parts 1 and 2. Store away from light at room temperature until use.

4. Methods

4.1. Prewash Dynabeads

1. Vortex Dynabeads on medium speed to resuspend. Change gloves.
2. Label one 0.2 ml thin-walled PCR tube per sample plus an additional tube for a negative control.
3. Transfer 10 μl of Dynabeads to each 0.2 ml thin-walled PCR tube.
4. Place into magnetic particle concentrator (MPC). Remove supernatant.
5. Add 100 μl Dynabead lysis buffer to each sample.
6. Vortex on low for 5–10 s, place on MPC, then remove buffer. Repeat.

4.2. Embryo Lysis

1. Retrieve individually frozen embryos from storage at -80°C (see Note 2).
2. Centrifuge briefly in microfuge to ensure embryo is at bottom of tube (~5 s at 9503–16060×*g*).
3. Add 10 μl of Dynabead lysis buffer to each sample.
4. Prior to adding embryo, place in MPC, and remove lysis buffer from prewashed Dynabeads from step 6 above.
5. Transfer entire contents of lysed embryo sample from step 3 to prewashed Dynabeads.

6. Mix gently by flicking, then centrifuge briefly at 1,520×*g* in microfuge.
7. Incubate with slow agitation on vortex for 5–10 min at room temperature to allow for hybridization of mRNA to Dynabead.
8. Centrifuge briefly at 1,520×*g*.
9. Place mRNA-Dynabead tube in MPC. Remove supernatant containing DNA to original embryo tubes, taking care not to remove any mRNA-Dynabeads.
10. Add 200 μl of Dynabead Wash Buffer A to mRNA-Dynabeads. Place in MPC and set aside until 3.3 RNA isolation.
11. Centrifuge tubes containing DNA/supernatant for 5–10 s in microfuge at 16,060×*g* to remove bubbles.
12. Add 1 μl Protease K (Sigma, Oakville, Canada) and 1 μl of 10% Igepal (see Note 3) to each DNA/supernatant tube.
13. Centrifuge briefly in microfuge at 16,060×*g* to remove any bubbles.
14. Add 300 μl of DNAse-, RNAse-, and protease-free mineral oil to each tube (see Note 4).
15. Lyse embryo by incubating for 1 h in waterbath at 50°C.

4.3. RNA Isolation

1. Retrieve mRNA-Dynabead tube containing 200 μl of Wash Buffer A.
2. Vortex on low speed for approximately 5 s. Centrifuge briefly. Place in MPC, and then remove Wash.
3. Repeat washing step with Wash Buffer A once, and Wash Buffer B three times.

4.4. Reverse Transcription and Generation of a Solid-Phase cDNA Library (see Note 5)

1. Prepare reverse transcription by mixing 2 μl 5× First Strand Buffer, 1 μl 0.1 M DTT, 0.5 μl 10 mM dNTP, 0.5 μl 40 U/μl RNaseOut (Invitrogen, Burlington, Canada), 0.25 μl Superscript II (Invitrogen, Burlington, Canada), 5.75 μl H_2O, for a 10 μl reaction.
2. Remove all of Wash Buffer B from mRNA-Dynabead tube.
3. Add 10 μl of RT mix to each sample.
4. Mix gently by flicking.
5. Centrifuge briefly. Repeat mixing and spin.
6. Incubate for 1–2 h at 42°C rotating in hybridization oven.

4.5. Agarose Bead Embedding of DNA

1. Remove Protease K-treated DNA tubes from waterbath. Place in heating block at >95°C.
2. Add 24 μl 2% LMP agarose (pre-warmed at >95°C) under mineral oil.
3. Mix gently by pipetting. Ensure the bead is well mixed.

4. Incubate for 3 min >95°C to inactivate the Protease K.
5. Incubate for 10 min on ice to allow agarose bead to harden.

4.6. Denaturation of DNA

1. Remove oil from chilled, hardened agarose bead.
2. Add 1 ml 0.1 M NaOH to each tube. Invert 5–6 times (see Note 6).
3. Incubate for 15 min at 37°C in a waterbath, inverting every 3–4 min.

4.7. Bisulfite Treatment of DNA

1. Spin gently (<1,520×*g*) (see Note 7).
2. Remove NaOH solution.
3. Add 500 μl of bisulfite solution.
4. Add 300 μl of mineral oil. Ensure that agarose bead is floating in solution (see Note 8).
5. Incubate at 50°C in a waterbath for 3.5 h (see Note 1).

4.8. Clean-Up of cDNA-Dynabead Library

1. Remove cDNA-Dynabead samples from hybridization oven.
2. Centrifuge briefly and then place in MPC.
3. Remove all RT mix.
4. Add 10 μl of ITT Buffer, flick gently to mix, centrifuge briefly, place in MPC, and remove ITT Buffer. Repeat.
5. Add 10 μl of ITT Buffer, flick gently to mix, and centrifuge briefly.
6. Incubate 1 min at 95°C in block of pre-warmed PCR machine.
7. Working with one sample at a time, centrifuge briefly, place in MPC, and remove ITT buffer.
8. Add 100 μl of ITT Buffer, flick gently to mix, centrifuge briefly, place in MPC, and remove ITT Buffer. Repeat.
9. Store cDNA-Dynabead library at 4°C (see Note 9).

4.9. Second Strand Synthesis

1. Remove cDNA-Dynabead library from storage at 4°C, add 100 μl of ITT Buffer, flick gently to mix, centrifuge briefly, place in MPC, and remove ITT Buffer. Repeat.
2. Prepare separate forward and reverse reactions according to protocols for your gene of interest.
3. Place cDNA-Dynabead library in MPC and remove ITT buffer (make sure all of liquid is removed).
4. Add forward reaction and flick gently to mix.
5. Place reaction in PCR machine, run PCR program for one cycle, according to your gene of interest.

6. Remove samples one at a time, spin down quickly in benchtop centrifuge, place in MPC, and remove all of second strand product to new tube.
7. Transfer an equivalent amount of forward second-strand mix to tubes containing pre-aliquoted reverse Mix.
8. Run PCR as per your gene of interest.
9. Rehydrate cDNA-Dynabead library by adding 100 μl of ITT Buffer, flick gently to mix, centrifuge briefly, place in MPC, and remove ITT Buffer. Repeat.
10. Store cDNA-Dynabead ibrary at 4°C (see Note 10).

4.10. Desulfonation of Bisulfite-Treated DNA

1. Remove DNA-agarose tubes from 50°C waterbath.
2. Incubate on ice for 3 min.
3. Remove bisulfite solution and mineral oil.
4. Centrifuge briefly in microfuge (<1,520×*g*).
5. Add 1 ml of TE, invert 1–2 times, centrifuge briefly, and then remove TE.
6. Add 1 ml 0.3 M NaOH, flick gently to mix, and invert 5–6 times.
7. Incubate at 37°C in a waterbath for 15 min, inverting every 3–4 min.
8. Centrifuge briefly in microfuge (<1,520×*g*).

4.11. Washing of Desulfonated DNA

1. Remove NaOH.
2. Add 1 ml of TE.
3. Incubate for 5–10 min at room temperature with shaking.
4. Spin gently, remove TE, and repeat wash once more with TE.
5. Repeat wash twice more with H_2O.
6. Check pH of washes. The last H_2O wash should have approximate pH 5. If solution is more basic, perform two addition washes with H_2O.
7. Samples are now ready for amplification of gene(s) of interest with bisulfite-specific primers (see Note 11).

5. Notes

1. BS is light sensitive. Cover bisulfite solutions, Parts I and II, with foil until ready to use. Cover all samples in foil once bisulfite mixture is added, and keep covered until after the 3.5 h incubation.

2. Embryos should be stored in a minimal amount of culture medium (1–2 μl).
3. The use of 10% Igepal is to ensure lysis of nuclear membrane as well as cell membrane.
4. Mineral oil is used to ensure that solutions do not evaporate and condense on the top of the tubes during the procedure.
5. Generation of a solid-phase library is important as it allows for reuse of Dynabeads and amplification of an essentially unlimited number of genes.
6. Invert samples gently. The agarose bead should be mixed but vigorous shaking can cause the agarose bead to break up into pieces.
7. Centrifugation of the agarose bead should not exceed 1,520×*g* to prevent breakage of agarose bead.
8. The agarose bead should be floating prior to incubation with the bisulfite solution to ensure that all sides of the agarose bead are exposed to the bisulfite solution. If the bead does not float, use a pipette tip to release it from the bottom of the tube.
9. For best results, use cDNA-Dynabead library as soon as possible. Consistent amplification has been obtained for cDNA libraries stored up to 8 months.
10. Caution: Dynabeads can be easily lost during each washing step; ensure all Dynabeads are localized to the magnetic side of the tube before removing any supernatant. Also, following repeated heating (multiple second strand syntheses), Dynabeads may clump. If this occurs, consider performing an additional washing step.
11. Set up PCR reactions. To increase PCR efficiency, add 1 μl 240 ng/ml tRNA as a carrier to PCR reaction. At 70°C, add to the 30 μl-agarose bead the required amount of water to make up 20 μl per gene(s) of interest (up to four genes). Mix agarose and water by gently pipetting. Keeping the solution at 60–70°C, mix by gently pipetting, and then split the PCR reaction into two by removing 12.5 μl into a new 0.2 ml thin-walled PCR tube. Add 12.5 μl mineral oil overlay. This allows for two independent PCR reactions. PCR amplification from the agarose bead should be performed immediately. If this is not possible, the agarose bead can be stored at 4°C up to 1 week. However, efficient amplification will decrease dramatically with each day of incubation.

References

1. Jaenisch R, Bird A (2003) Epigenetic regulation of gene expression: how the genome integrates intrinsic and environmental signals. Nat Genet 33(Suppl):245–254
2. Rivera RM (2010) Epigenetic aspects of fertilization and preimplantation development in mammals: lessons from the mouse. Syst Biol Reprod Med 56:388–404
3. Morgan HD, Santos F, Green K, Dean W, Reik W (2005) Epigenetic reprogramming in mammals. Hum Mol Genet 14(Spec No 1): R47–R58
4. Campos EI, Reinberg D (2009) Histones: annotating chromatin. Annu Rev Genet 43:559–599
5. Bannister AJ, Kouzarides T (2011) Regulation of chromatin by histone modifications. Cell Res 21:381–395
6. Shi L, Wu J (2009) Epigenetic regulation in mammalian preimplantation embryo development. Reprod Biol Endocrinol 7:59–68
7. Schultz RM (2002) The molecular foundations of the maternal to zygotic transition in the preimplantation embryo. Hum Reprod Update 8:323–331
8. Torres-Padilla ME, Bannister AJ, Hurd PJ, Kouzarides T, Zernicka-Goetz M (2006) Dynamic distribution of the replacement histone variant H3.3 in the mouse oocyte and preimplantation embryos. Int J Dev Biol 50:455–461
9. Ahmad K, Henikoff S (2002) The histone variant H3.3 marks active chromatin by replication-independent nucleosome assembly. Mol Cell 9:1191–1200
10. Schuettengruber B, Chourrout D, Vervoort M, Leblanc B, Cavalli G (2007) Genome regulation by polycomb and trithorax proteins. Cell 128:735–745
11. Puschendorf M et al (2008) PRC1 and Suv39h specify parental asymmetry at constitutive heterochromatin in early mouse embryos. Nat Genet 40:411–420
12. Hajkova P et al (2002) Epigenetic reprogramming in mouse primordial germ cells. Mech Dev 117:15–23
13. Kageyama S, Liu H, Kaneko N, Ooga M, Nagata M, Aoki F (2007) Alterations in epigenetic modifications during oocyte growth in mice. Reproduction 133:85–94
14. Kurihara Y et al (2008) Maintenance of genomic methylation patterns during preimplantation development requires the somatic form of DNA methyltransferase 1. Dev Biol 313:335–346
15. Cardoso MC, Leonhardt H (1999) DNA methyltransferase is actively retained in the cytoplasm during early development. J Cell Biol 147:25–32
16. Doherty AS, Bartolomei MS, Schultz RM (2002) Regulation of stage-specific nuclear translocation of Dnmt1o during preimplantation mouse development. Dev Biol 242:255–266
17. Cirio MC et al (2008) Preimplantation expression of the somatic form of Dnmt1 suggests a role in the inheritance of genomic imprints. BMC Dev Biol 8:9
18. Mandal PK, Kazazian HH Jr (2008) SnapShot: vertebrate transposons. Cell 135(192–192):e191
19. Kigami D, Minami N, Takayama H, Imai H (2003) MuERV-L is one of the earliest transcribed genes in mouse one-cell embryos. Biol Reprod 68:651–654
20. Svoboda P et al (2004) RNAi and expression of retrotransposons MuERV-L and IAP in preimplantation mouse embryos. Dev Biol 269:276–285
21. Kim SH et al (2004) Differential DNA methylation reprogramming of various repetitive sequences in mouse preimplantation embryos. Biochem Biophys Res Commun 324:58–63
22. Lane N et al (2003) Resistance of IAPs to methylation reprogramming may provide a mechanism for epigenetic inheritance in the mouse. Genesis 35:88–93
23. Gaudet F et al (2004) Dnmt1 expression in pre- and postimplantation embryogenesis and the maintenance of IAP silencing. Mol Cell Biol 24:1640–1648
24. Duhl DM, Vrieling H, Miller KA, Wolff GL, Barsh GS (1994) Neomorphic agouti mutations in obese yellow mice. Nat Genet 8:59–65
25. Waterland RA, Jirtle RL (2003) Transposable elements: targets for early nutritional effects on epigenetic gene regulation. Mol Cell Biol 23:5293–5300
26. Morgan HD, Sutherland HGE, Martin DIK, Whitelaw E (1999) Epigenetic inheritance at the agouti locus in the mouse. Nat Genet 23:314–318
27. Miltenberger RJ, Mynatt RL, Wilkinson JE, Woychik RP (1997) The role of the agouti gene in the yellow obese syndrome. J Nutr 127:1902S–1907S

28. Dolinoy DC, Huang D, Jirtle RL (2007) Maternal nutrient supplementation counteracts bisphenol A-induced DNA hypomethylation in early development. Proc Natl Acad Sci USA 104:13056–13061
29. Dolinoy DC, Weidman JR, Waterland RA, Jirtle RL (2006) Maternal genistein alters coat color and protects Avy mouse offspring from obesity by modifying the fetal epigenome. Environ Health Perspect 114:567–572
30. Zeng L et al (1997) The mouse fused locus encodes Axin, an inhibitor of the Wnt signaling pathway that regulates embryonic axis formation. Cell 90:181–192
31. Rakyan VK et al (2003) Transgenerational inheritance of epigenetic states at the murine Axin(Fu) allele occurs after maternal and paternal transmission. Proc Natl Acad Sci USA 100:2538–2543
32. Waterland RA et al (2006) Maternal methyl supplements increase offspring DNA methylation at Axin fused. Genesis 44:401–406
33. Verona RI, Mann MRW, Bartolomei MS (2003) Genomic imprinting: intricacies of epigenetic regulation in clusters. Annu Rev Cell Dev Biol 19:237–259
34. Thorvaldsen JL, Verona RI, Bartolomei MS (2006) X-tra! X-tra! News from the mouse X chromosome. Dev Biol 298:344–353
35. Erhardt S et al (2003) Consequences of the depletion of zygotic and embryonic enhancer of zeste 2 during preimplantation mouse development. Development 130:4235–4248
36. Okamoto I, Otte AP, Allis CD, Reinberg D, Heard E (2004) Epigenetic dynamics of imprinted X inactivation during early mouse development. Science 303:644–649
37. Mak W et al (2004) Reactivation of the paternal X chromosome in early mouse embryos. Science 303:666–669
38. Zuccotti M et al (2002) Mouse Xist expression begins at zygotic genome activation and is timed by a zygotic clock. Mol Reprod Dev 61:14–20
39. Howell CY et al (2001) Genomic imprinting disrupted by a maternal effect mutation in the Dnmt1 gene. Cell 104:829–838
40. Hirasawa R, Chiba H, Kaneda M, et al (2008) Maternal and zygotic Dnmt1 are necessary and sufficient for the maintenance of DNA methylation imprints during preimplantation development. Genes Dev 22:1607–1616
41. Kim JM, Ogura A (2009) Changes in allele-specific association of histone modifications at the imprinting control regions during mouse preimplantation development. Genesis 47: 611–616
42. Van Buggenhout G, Fryns JP (2009) Angelman syndrome (AS, MIM 105830). Eur J Hum Genet 17:1367–1373
43. Weksberg R, Shuman C, Beckwith JB (2010) Beckwith–Wiedemann syndrome. Eur J Hum Genet 18:8–14
44. Shi W, Haaf T (2002) Aberrant methylation patterns at the two-cell stage as an indicator of early developmental failure. Mol Reprod Dev 63:329–334
45. Zaitseva I, Zaitsev S, Alenina N, Bader M, Krivokharchenko A (2007) Dynamics of DNA-demethylation in early mouse and rat embryos developed in vivo and in vitro. Mol Reprod Dev 74:1255–1261
46. Katari S et al (2009) DNA methylation and gene expression differences in children conceived in vitro or in vivo. Hum Mol Genet 18:3769–3778
47. Huang JC et al (2007) Comparison of histone modifications in in vivo and in vitro fertilization mouse embryos. Biochem Biophys Res Commun 354:77–83
48. Morgan HD, Jin XL, Li A, Whitelaw E, O'Neill C (2008) The culture of zygotes to the blastocyst stage changes the postnatal expression of an epigentically labile allele, agouti viable yellow, in mice. Biol Reprod 79:618–623
49. Fernandez-Gonzalez R, Ramirez MA, Pericuesta E, Calle A, Gutierrez-Adan A (2010) Histone modifications at the blastocyst Axin1(Fu) locus mark the heritability of in vitro culture-induced epigenetic alterations in mice. Biol Reprod 83:720–727
50. Wrenzycki C et al (2002) In vitro production and nuclear transfer affect dosage compensation of the X-linked gene transcripts G6PD, PGK, and Xist in preimplantation bovine embryos. Biol Reprod 66:127–134
51. Nino-Soto MI, Basrur PK, King WA (2007) Impact of in vitro production techniques on the expression of X-linked genes in bovine (*Bos taurus*) oocytes and pre-attachment embryos. Mol Reprod Dev 74:144–153
52. Mann MR et al (2004) Selective loss of imprinting in the placenta following preimplantation development in culture. Development 131:3727–3735
53. Sasaki H, Ferguson-Smith AC, Shum ASW, Barton SC, Surani MA (1995) Temporal and spatial regulation of *H19* imprinting in normal and uniparental mouse embryos. Development 121:4195–4202
54. Doherty AS, Mann MRW, Tremblay KD, Bartolomei MS, Schultz RM (2000) Differential effects of culture on imprinted

H19 expression in the preimplantation mouse embryo. Biol Reprod 62:1526–1535
55. Khosla S, Dean W, Brown D, Reik W, Feil R (2001) Culture of preimplantation mouse embryos affects fetal development and the expression of imprinted genes. Biol Reprod 64:918–926
56. Li T et al (2005) IVF results in de novo DNA methylation and histone methylation at an Igf2-*H19* imprinting epigenetic switch. Mol Hum Reprod 11:631–640
57. Fauque P et al (2007) Assisted reproductive technology affects developmental kinetics, *H19* imprinting control region methylation and *H19* gene expression in individual mouse embryos. BMC Dev Biol 7:116
58. Market-Velker BA, Fernandes AD, Mann MR (2010) Side-by-side comparison of five commercial media systems in a mouse model: suboptimal in vitro culture interferes with imprint maintenance. Biol Reprod 83:938–950
59. Rivera RM et al (2008) Manipulations of mouse embryos prior to implantation result in aberrant expression of imprinted genes on day 9.5 of development. Hum Mol Genet 17:1–14
60. Cox GF et al (2002) Intracytoplasmic sperm injection may increase the risk of imprinting defects. Am J Hum Genet 71:162–164
61. DeBaun MR, Niemitz EL, Feinberg AP (2003) Association of in vitro fertilization with Beckwith–Wiedemann syndrome and epigenetic alterations of LIT1 and *H19*. Am J Hum Genet 72:156–160
62. Gicquel C et al (2003) In vitro fertilization may increase the risk of Beckwith–Wiedemann syndrome related to the abnormal imprinting of the KCN1OT gene. Am J Hum Genet 72:1338–1341
63. Maher ER et al (2003) Beckwith–Wiedemann syndrome and assisted reproduction technology (ART). J Med Genet 40:62–64
64. Orstavik KH et al (2003) Another case of imprinting defect in a girl with Angelman syndrome who was conceived by intracytoplasmic semen injection. Am J Hum Genet 72:218–219
65. Ludwig M et al (2005) Increased prevalence of imprinting defects in patients with Angelman syndrome born to subfertile couples. J Med Genet 42:289–291
66. Rossignol S et al (2006) The epigenetic imprinting defect of patients with Beckwith–Wiedemann syndrome born after assisted reproductive technology is not restricted to the 11p15 region. J Med Genet 43:902–907
67. Halliday J, Oke K, Breheny S, Algar E, Amor D (2004) Beckwith–Wiedemann syndrome and IVF: a case–control study. Am J Hum Genet 75:526–528
68. Sutcliffe AG et al (2006) Assisted reproductive therapies and imprinting disorders – a preliminary British survey. Hum Reprod 21: 1009–1011
69. Chang AS, Moley KH, Wangler M, Feinberg AP, Debaun MR (2005) Association between Beckwith–Wiedemann syndrome and assisted reproductive technology: a case series of 19 patients. Fertil Steril 83:349–354
70. Behboodi E et al (1995) Birth of large calves that developed from in vitro-derived bovine embryos. Theriogenology 44:227–232
71. Young LE, Sinclair KD, Wilmut I (1998) Large offspring syndrome in cattle and sheep. Rev Reprod 3:155–163
72. Hori N et al (2010) Aberrant CpG methylation of the imprinting control region KvDMR1 detected in assisted reproductive technology-produced calves and pathogenesis of large offspring syndrome. Anim Reprod Sci 122:303–312
73. Young LE et al (2001) Epigenetic change in IGF2R is associated with fetal overgrowth after sheep embryo culture. Nat Genet 27:153–154
74. Suzuki J Jr et al (2009) In vitro culture and somatic cell nuclear transfer affect imprinting of SNRPN gene in pre- and post-implantation stages of development in cattle. BMC Dev Biol 9:9
75. Lim D et al (2009) Clinical and molecular genetic features of Beckwith–Wiedemann syndrome associated with assisted reproductive technologies. Hum Reprod 24:741–747
76. Ho Y, Wigglesworth K, Eppig JJ, Schultz RM (1995) Preimplantation development of mouse embryos in KSOM: augmentation by amino acids and analysis of gene expression. Mol Reprod Dev 41:232–238
77. Gardner DK et al (1998) A prospective randomized trial of blastocyst culture and transfer in in-vitro fertilization. Hum Reprod 13:3434–3440
78. Biggers JD, Summers MC (2008) Choosing a culture medium: making informed choices. Fertil Steril 90:473–483
79. Rinaudo P, Schultz RM (2004) Effects of embryo culture on global pattern of gene expression in preimplantation mouse embryos. Reproduction 128:301–311
80. Tveden-Nyborg PY et al (2008) Analysis of the expression of putatively imprinted genes in bovine peri-implantation embryos. Theriogenology 70:1119–1128
81. Chen SL, Shi XY, Zheng HY, Wu FR, Luo C (2010) Aberrant DNA methylation of imprinted

H19 gene in human preimplantation embryos. Fertil Steril 94:2356–2358
82. Bavister BD (1995) Culture of preimplantation embryos: facts and artifacts. Hum Reprod Update 1:91–148
83. Gardner DK (1994) Mammalian embryo culture in the absence of serum or somatic cell support. Cell Biol Int 18:1163–1179
84. Fernandez-Gonzalez R et al (2004) Long-term effect of in vitro culture of mouse embryos with serum on mRNA expression of imprinting genes, development, and behavior. Proc Natl Acad Sci USA 101:5880–5885
85. Walker SK, Hartwich KM, Seamark RF (1996) The production of unusually large offspring following embryo manipulation: concepts and challenges. Theriogenology 45:111–120
86. Thompson JG, Gardner DK, Pugh PA, McMillan WH, Tervit HR (1995) Lamb birth weight is affected by culture system utilized during in vitro pre-elongation development of ovine embryos. Biol Reprod 53: 1385–1391
87. Fischer B, Bavister BD (1993) Oxygen tension in the oviduct and uterus of rhesus monkeys, hamsters and rabbits. J Reprod Fertil 99:673–679
88. Mastroianni L Jr, Jones R (1965) Oxygen tension within the rabbit fallopian tube. J Reprod Fertil 9:99–102
89. Dumoulin JC et al (1999) Effect of oxygen concentration on human in-vitro fertilization and embryo culture. Hum Reprod 14: 465–469
90. Li J, Foote RH (1993) Culture of rabbit zygotes into blastocysts in protein-free medium with one to twenty per cent oxygen. J Reprod Fertil 98:163–167
91. Orsi NM, Leese HJ (2001) Protection against reactive oxygen species during mouse preimplantation embryo development: role of EDTA, oxygen tension, catalase, superoxide dismutase and pyruvate. Mol Reprod Dev 59:44–53
92. Pabon JE Jr, Findley WE, Gibbons WE (1989) The toxic effect of short exposures to the atmospheric oxygen concentration on early mouse embryonic development. Fertil Steril 51:896–900
93. Quinn P, Harlow GM (1978) The effect of oxygen on the development of preimplantation mouse embryos in vitro. J Exp Zool 206:73–80
94. Thompson JG, Simpson AC, Pugh PA, Donnelly PE, Tervit HR (1990) Effect of oxygen concentration on in-vitro development of preimplantation sheep and cattle embryos. J Reprod Fertil 89:573–578
95. Rinaudo PF, Giritharan G, Talbi S, Dobson AT, Schultz RM (2006) Effects of oxygen tension on gene expression in preimplantation mouse embryos. Fertil Steril 86:1252–1265
96. Feil D et al (2006) Effect of culturing mouse embryos under different oxygen concentrations on subsequent fetal and placental development. J Physiol 572:87–96
97. Market-Velker BA, Zhang L, Magri LS, Bonvissuto AC, Mann MRDual (2010) Effects of superovulation: loss of maternal and paternal imprinted methylation in a dose-dependent manner. Hum Mol Genet 19:36–51
98. Santos F et al (2010) Evaluation of epigenetic marks in human embryos derived from IVF and ICSI. Hum Reprod 25:2387–2395
99. Turan N et al (2010) Inter- and intra-individual variation in allele-specific DNA methylation and gene expression in children conceived using assisted reproductive technology. PLoS Genet 6:e1001033
100. Market Velker BA, Denomme MM, Mann MR (2012) Loss of genomic imprinting in mouse embryos with fast rates of preimplantation development in culture. Biol Reprod 10;86(5):143
101. Baumann CG, Morris DG, Sreenan JM, Leese HJ (2007) The quiet embryo hypothesis: molecular characteristics favoring viability. Mol Reprod Dev 74:1345–1353
102. Leese HJ (2002) Quiet please, do not disturb: a hypothesis of embryo metabolism and viability. Bioessays 24:845–849

Index

Gary D. Smith et al. (eds.), *Embryo Culture: Methods and Protocols*, Methods in Molecular Biology, vol. 912,
DOI 10.1007/978-1-61779-971-6, © Springer Science+Business Media, LLC 2012

N

The manufacturer's authorised representative in the EU is Springer Nature Customer Service Centre GmbH, Europaplatz 3, 69115 Heidelberg, Germany. If you have any concerns regarding our products, please contact ProductSafety@springernature.com

Printed and bound by CPI Group (UK) Ltd, Croydon, CR0 4YY
30/10/2026
02242627-0013